Advanced Batteries

Materials Science Aspects

Robert A. Huggins

Advanced Batteries

Materials Science Aspects

Springer

Robert A. Huggins
Department of Materials Science and Engineering
Stanford University
Stanford, CA 94305
rhuggins@stanford.edu

ISBN: 978-0-387-76423-8 e-ISBN: 978-0-387-76424-5
DOI: 10.1007/978-0-387-76424-5

Library of Congress Control Number: 2008930484

Printed on acid-free paper

springer.com

Preface

1 Introduction

Energy is important to all of us, for a variety of reasons, but primarily because it can be useful. It can be found in a number of different forms. One readily recognizes the concepts of potential energy and kinetic energy, as well as the chemical energy in fuels, thermal energy as heat, the kinetic energy in wind and moving water, and magnetic and electrical energy in a variety of guises. But energy is often present in one form, whereas we want to use it in another form. This requires some kind of conversion mechanism or transducer device. Furthermore, energy may be available in amounts and at times and places that are different from those when and where we want to utilize it. Thus, methods to store and transport energy from place to place can be of great importance.This text has to do with the storage of energy, and there are a number of different ways in which this can be done. The title indicates that it is about electrochemical energy storage. This may appear to be misleading, for there is actually no such thing as electrochemical energy. What it actually means is that we shall deal with the general topic of the use of electrochemical means to convert between two different types of energy: electrical energy and chemical energy. This involves the use of electrochemical devices that act as *transducers*, for they convert between electrical and chemical quantities – energies, potentials, and fluxes. Such electrochemical transduction systems are often called *galvanic cells*, or in more common parlance, *batteries*. Electrical energy can be stored in electric or magnetic fields; mechanical energy can be stored in devices such as flywheels, and thermal energy can be stored in the form of heat. But the magnitudes of these forms of energy are all relatively small and the methods for their conversion into other forms are relatively unwieldy. Much larger amounts of energy can be present in the form of chemical species. This can be relatively attractive, for it can be relatively inexpensive and efficient in terms of the amount of energy stored per unit volume or weight. Thus, storage in chemical form is often a useful intermediate stage, holding energy for later use in other forms, such as electrical, heat, light, or mechanical energy. Chemical reactions can be used to convert this chemical energy into mechanical energy by the use of internal combustion engines, for example. Alternatively, electrochemical systems and devices can be used to convert this chemical energy into electrical energy. A major advantage of electrochemical transduction methods is that they can operate isothermally and thus avoid the so-called Carnot limitation. This makes it possible to achieve much greater efficiencies than are available by the use of thermal conversion processes.In many cases, electrochemical cells can also be operated in the reverse direction. Thus, it is possible to devise reversible electrochemical systems in which electrical energy is converted to chemical energy (the chemical system is charged), and the process can later be reversed to give electrical energy again (the chemical system is discharged).

2 Applications of Electrochemical Energy Storage

There is a great deal of current interest in the development of better *electrochemical systems*, or better *batteries*. There are several basic factors driving this. One of these is the concern about several important issues related to the *environment* in which we live.

It is now increasingly clear that the combustion of fossil fuels leads to the emission of species, sometimes called *greenhouse gases*, into the atmosphere, and that their accumulation leads to a significant amount of *global warming*.

In addition, the products of such combustion processes can lead to *environmental pollution* and poisonous *atmospheric smog* in urban areas. In addition to industrial sources, the use of combustion engines in an *ever-increasing number of vehicles* makes this problem worse with time. Its seriousness is not the same everywhere, being worse in areas of high population concentration and where the natural processes that cause the motion of air are most restricted. One of the locations in which this problem is particularly severe and in which there is a great deal of public and political sentiment behind efforts toward its alleviation is the Los Angeles Basin in Southern California. This led to the installation of regulations that limit the emission of specific gaseous species by vehicles in that location a number of years ago. Such regulations have become stricter with time, and similar requirements have also gradually become adopted in other locations.

Because of the poisonous nature of the gases emitted by internal combustion engines, their use inside closed buildings has long been prohibited.

There is thus a great incentive to develop *alternative methods for vehicular propulsion*. One approach to this has been the *electric automobile*, in which the motive power is supplied by the use of a *battery*, or perhaps by a nonpolluting hydrogen/oxygen *fuel cell*, or a combination of the two.

In order to provide sufficient range, as well as adequate acceleration, relatively large amounts of energy must be available in the vehicle. Thus, the amount of energy per unit weight, the *specific energy*, is an important parameter in such applications.

As a general rule of thumb, the useful range of a battery-propelled vehicle, in kilometers, is approximately two times the specific energy of its battery system, in Wh kg^{-1}.

Another important consideration is, of course, the cost. Unfortunately, the least expensive currently available batteries, based upon the Pb/PbO_2 system, have a relatively low value of specific energy, some 30–40 Wh kg^{-1}. Thus, electric vehicles using such batteries have useful ranges of only 60–80 km under normal conditions.

Although there is a great deal of activity aimed at their improvement, alternative electrochemical systems with greater specific energies are presently much more expensive.

Even though there has been a considerable amount of interest in the development of such battery-driven vehicles for some years, it has become obvious that the *current state of battery technology* is not sufficiently advanced for such *all-electric* vehicles to be price- and performance-competitive. As a result, such vehicles have found only a rather limited market to date.

More recently, a number of automobile companies have been involved in the development of vehicles propelled by *hybrid systems* in which a relatively small internal combustion engine is used to charge a modest-sized electric battery, which then acts to provide the vehicular propulsion. This approach has proven to be much more attractive than the all-electric vehicle. Although it is only a compromise in terms of the reduction in the use of fossil fuels, the reduced size of the battery in such systems greatly decreases both their weight and their cost.

Hybrid vehicles began to be introduced in the USA market in 1999, and their sales have increased rapidly. The first two manufacturers were Toyota and Honda, and data on their sales during the first 5 years are shown in Fig. 1. A number of other

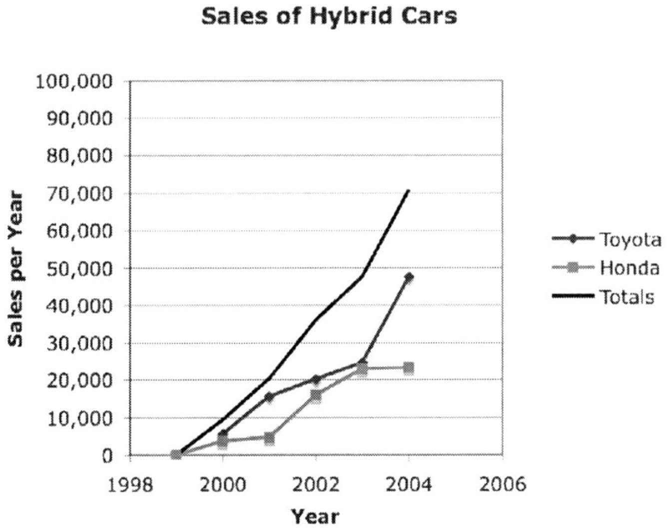

Fig. 1 Growth in the sales of hybrid automobiles in the USA in the first few years after their introduction

manufacturers have begun to introduce hybrid autos recently, and more will do this within the next few years.

In addition to the internal combustion engine–battery hybrid approach, there is currently a large development effort in several countries aimed toward use of hydrogen-driven fuel cells to provide the primary motive power in vehicles. Because of the fact that the power requirements can vary greatly with time, it may be necessary that an energy storage system also be present. Thus fuel cell–battery hybrid combinations are also being considered.

An additional consideration arises when considering hydrogen-powered vehicles. This is the question of how the hydrogen, or hydrogen-producing fuel can be carried on board the vehicle. There are several approaches to this problem being considered, but one of the most desirable would be to carry the hydrogen in the vehicle in the form of a hydrogen-containing solid that would have a high hydrogen density and also could provide the required hydrogen on demand. Such materials are often called *metal hydrides*, and are closely related to the metal hydrides that are used as the reactants in the negative electrodes of the common small *hydride/nickel batteries*. This matter will be discussed later in this text.

This trend toward hybrid vehicles is also leading to consideration of dual energy storage systems, for in addition to the obvious desire to optimize the amount of energy stored there is also the need to handle very high rates of charge and discharge. It is unrealistic to expect a single type of technology to be optimized for both of these very different types of application. This has led to research and development efforts aimed at the very high rate and high cycle-life applications, and different approaches to this problem now generally carry the labels *supercapacitors* or *ultracapacitors*.

There is an additional factor that is pushing the development of more effective batteries for use in vehicles that relates to the more traditional application of batteries for providing power for starting, lighting, and ignition – the *SLI applications*. This is the trend toward the use of more and more electrically powered functions in vehicles, such as electric windows, electrically operated seats, radios, etc. This has led the automobile manufacturers to move toward higher voltage systems. Some years ago auto batteries provided 6 V. Now they provide 12 V. Within the next few years auto electrical circuits will operate at 36 V (the batteries will be charged at 42 V). This will be more than just placing three of the current 12-V batteries in series, and changing to quite different chemistries and systems is being actively considered.

In addition to the environmentally driven issues that have been discussed here, there is growing awareness that the *availability of fossil fuels in the earth is limited*. This has led to increasing pressure toward the development of technologies that can reduce the needs for fossil energy. In addition to *energy-saving methods*, there is increasing interest in methods to enable the effective use of *alternate sources of energy*, such as *solar energy* and *wind energy* in order to reduce the dependence upon fossil fuels.

There is another feature of energy supply and consumption that also has to be taken into consideration. This is that both the utilization and the sources of energy are often not uniform with time. As an example, heating and air conditioning, as well as lighting, energy requirements generally vary with the time of day as well as with weather conditions and the time of the year. In addition, it is obvious that the energy production of some alternative energy sources, solar and wind energy devices, is intermittent. The result is that there is a need for mechanisms to coordinate the time dependence of the output and the cost of energy sources with that of their needs. The terms *load leveling* and *peak shaving* are often used in this connection, and what is needed, of course, is a mechanism for large-scale *reversible energy storage* in order to better match the time dependence of energy supplies and their use.

In addition to these environmental and economic issues, another major driving force that has greatly increased the push toward more effective energy storage devices is *market-driven*. It is the fact that the number of *portable electronic products* is increasing very rapidly; these include the *four Cs*: computers, cellular telephones, cordless power tools, and camcorders.

It is well recognized that their performance is almost always limited by the available power source, which is typically a small battery. Some such batteries are electrically rechargeable, whereas others are not. For application in such portable devices the amount of energy per unit volume, the *energy density*, and the amount of power available per unit volume, the *power density*, are often the most important criteria instead of the weight-related specific energy and power.

These power sources are typically much smaller than those necessary for vehicle propulsion, and therefore the price is not such a decisive factor. Instead, the amount of energy that can be stored in a given size and shape is often paramount. Methods of fabrication that can produce cells of small, and especially, thin shapes have become increasingly important in this type of application. The term *form factor* is often used in this connection.

Also, smaller device sizes make technological problems significantly easier to solve. These factors, as well as the ready marketability of such products, have led to a great deal of activity in recent years in a number of countries, with large-scale production of the more advanced systems occurring first in Japan, but Korea and China have now become major players. The current market is very large – billions of small batteries per year.

A further general area of application that provides an increasing market for electrochemical energy storage devices and systems relates to *security and safety*. There are a number of critical electrically powered systems in which *backup energy sources* are necessary in order to prevent major problems if the primary, or normal, source of energy is interrupted. These include large computer systems, data networks, telephone systems, and emergency lighting.

3 Changes That Have Taken Place in Recent Years

The application of batteries in these different market segments has been growing at a very high rate. Their requirements have been putting ever-increasing incentives on the development of better, and lower cost, energy storage devices and systems.

A sort of chicken-and-egg type of situation has developed, with the market demanding better properties, and when this happens it also opens up the opportunity for additional new markets.

This has led to a lot of research and development activity, and there have been a number of important technological changes in recent years. A number of these have not been just incremental improvements in already-known areas, but involve the use of new concepts, new materials, and new approaches.

An important reason for this progress has been the fact that such things as the discovery of fast ionic conduction in solids and the possibility of *solid electrolytes*, the concept of the use of materials with *insertion reactions* as high-capacity electrodes, and the discovery of materials that can produce *lithium-based batteries* with unusually high voltages have caused a number of people with backgrounds in other areas of science and technology to be drawn into this area. The result has been the infusion of new materials, concepts, and techniques into battery research and development, which used to be considered only as a part of electrochemistry.

Here is a short partial list of recent developments:

- Metal hydrides
- Lithium-carbon alloys
- Intermetallic alloys
- Lithium-transition metal oxides
- Polymeric components, in both electrolytes and electrodes
- Liquid electrodes
- Both crystalline and amorphous solid electrolytes
- Organic solvent electrolytes
- Mixed-conductor matrices
- Protective solid electrolyte interfaces in organic electrolyte systems
- Use of soft chemistry to produce nonequilibrium electrode compositions
- New fabrication methods, and new cell shapes and sizes
- Fabrication of lithium batteries in the discharged state

4 Objectives of This Book

This book deals with the *Materials Science principles* that underlie the behavior of advanced electrochemical storage systems, i.e., batteries. It focuses upon basic principles and has a tutorial flavor, and thus is quite different from the several other available books that deal with battery systems, which generally are more directed toward the detailed description of current batteries and their properties.

To be most useful to those interested in either understanding or contributing to the development of this rapidly moving area of technology, emphasis is placed *upon underlying principles* that are applicable across the spectrum of materials that are of interest as electrochemically active components in advanced batteries: the electrodes and the electrolyte.

It will be readily recognized that the content and approach in this book, *based upon materials science*, is also quite different from what is found in many books on electrochemistry. In some areas it is based upon different concepts, and utilizes different *thinking tools*. In some topic areas the terminology that will be used will also be different from that which is characteristic of much of the current electrochemical literature.

In accordance with the desire to focus on *generic principles* applicable to a variety of the more advanced materials and systems, no attempt will be made to provide a complete description of all current materials and batteries. Indeed, several major battery types will deliberately *not* be discussed. An obvious example is the common lead-acid battery, which actually has the largest market of all battery systems at the present time. In addition to the omission of this traditional battery system, little or no attention will be given to topics such as electron transfer kinetics at interfaces, and the nature of the electrical double layer at the interface between a liquid electrolyte and a solid electrode. These are both topics of central interest in conventional electrochemistry, but they have very little relevance to the key questions upon which this text will focus.

Significant changes and developments have occurred in recent years. An especially important change in thinking in this area has been the recognition of the *role played by phenomena inside solid electrode materials*, and their influence upon the interfacial conditions, and thus upon the electrochemical potential of the electrode. In many cases solid-state reactions play a critical role in determining two of the most important parameters of an electrochemical energy storage system: the potentials and the capacities of electrodes. The kinetic behavior of solid-state reactions is often very important in determining the kinetic parameters of the electrode reactions and thus of the observed overall cell behavior.

This recognition of the *importance of solid-state properties and behavior* upon the characteristics of electrochemical cells, and the major changes in battery technology in recent years have led to the entry of people into this area with backgrounds and approaches based upon materials science that are quite different from those of traditional electrochemistry.

Since this monograph is primarily intended to serve an educational role there has been no attempt to provide a thorough review of the research literature, although references to some of the early and definitive work are cited in some chapters. Reference is also made to review papers in the literature that may be helpful in some cases.

5 Thinking Tools

It is often very useful to have simple tools or other methods available that provide insight, perhaps only qualitative in some cases, into the influence of various factors in complex materials systems. In some cases, these involve graphical representations of parameters that summarize the results of either experimental or theoretical considerations and that provide insight into relationships, perhaps only qualitative in some cases, and allow one to recognize trends, etc., without the necessity of immediately looking at the details. Several such graphical *thinking tools* are used in this text. Some of them are mentioned here.

There are several types of *diagrams* that graphically represent the results of experimental observations or theoretical considerations and can act as tools that make it possible to see the important relationships that underlie what would otherwise be a bewildering and apparently very complex set of observations. These include the following:

1. Simple models of the physical configuration, e.g., what is where, related to physical processes that are taking place.
2. Equivalent circuits: These are electrical circuits whose electrical behavior is analogous to important properties of physical systems.
3. Binary phase diagrams: These show the phases and their compositions present under equilibrium conditions as a function of the overall composition and temperature in binary (two-component) systems.
4. Isothermal ternary phase diagrams: These are isothermal slices through ternary phase diagrams. They are called *Gibbs triangles*, and are often simplified by the assumption that the phases present have negligible ranges of composition, i.e., can be represented as *point phases*. In such cases these tools should more properly be called ternary *phase stability diagrams*, rather than ternary phase diagrams.
5. Defect equilibrium diagrams: These are isothermal diagrams that provide information about the concentrations of the various crystallographic and electronic defect species present in binary and ternary systems as functions of some parameters related to the overall composition, such as the activity of one of the component species.
6. Disorder domain diagrams: These diagrams indicate the predominant defect species (one negatively charged, and one positively charged) present under equilibrium conditions as a function of both the temperature and a parameter related to the overall composition.
7. Ellingham diagrams: These diagrams show the temperature dependence of the Gibbs free energy changes resulting from a number of reactions, e.g., reactions of metals with oxygen, sulfur, or chlorine.
8. Pourbaix diagrams: These diagrams indicate the ranges of stability of phases or ionic species present under equilibrium conditions as a function of electric potential and pH in aqueous systems.

6 Terminology and Conventions

It is necessary to discuss the terminology and conventions that will be used here, for some of them are different from those that have evolved over a number of years in different areas of electrochemistry.

As an example, consider the use of the terms *anode* and *cathode*. In conventional texts on electrochemistry one often finds that the anode is described as the electrode that becomes oxidized, and at which species in the electrolyte become reduced, resulting in electrons being given up to the external circuit. The cathode is then the electrode that becomes reduced, and at which species in electrolyte become oxidized and electrons are accepted from the external circuit. Looking at this matter from the standpoint of the external electronic circuit, it is clear that the anode, which supplies electrons, is the negatively charged electrode, and the cathode is the positively charged electrode.

This sounds nice and simple, and it is consistent with what happens when an electrochemical cell is being discharged. However, in an electrolytic or electrodeposition cell, in which electrical energy is used to cause chemical processes, such as electrolyte decomposition and the deposition of species on the electrodes, to occur the *terminology is reversed*. Then, the anode is where species in the electrolyte are oxidized and electrons are obtained from the external circuit. Likewise, the cathode is where species from the electrolyte are reduced and electrons are given to the electronic circuit. Nevertheless, the anode is still the negatively charged electrode and the cathode is still the positively charged electrode.

Likewise, the directions of both the ionic and electronic *reactions are reversed* when an electrochemical cell that is used as a battery is charged, rather than discharged. In this case, however, the convention is to label the electrodes in terms of their behavior during discharge, rather than during charge. Thus, the negative electrode in batteries is always called the anode and the positive electrode the cathode.

To avoid these problems, the *use of the terms anode and cathode will be avoided here*. Instead, reference will be made to the negative and positive electrodes. This has the advantage that it is easy to use a voltmeter to decide which electrode has the more positive, and which the more negative, potential, regardless of which way the current is flowing.

Here, a simple electrical cell will be represented by its three major internal components, and written as: (−) negative electrode/electrolyte/positive electrode (+), with the phase boundaries indicated by the slashes, and with the positively charged electrode always shown on the right hand side.

Unfortunately, the literature and common usage are often not consistent on this point. This can be seen by simply looking at the names of some of the common battery systems. Consider the following list (in which the electrolyte is omitted). In some cases the negative electrode is mentioned first, but in other cases the positive electrode is mentioned first. In this text the negative electrode will always be on the left side, and the positive electrode on the right side.

(−) Zn/Air (+)
(−) Pb/PbO_2 (+)
(−) $Li/LiCoO_2$ (+)
(+) Ni/Cd (−)
(+) Ni/Metal hydride (−)
(+) Ni/Zn (−)
(+) Ni/Fe (−)

As indicated earlier, charging occurs when externally provided electrical energy is converted into internal chemical energy, and the cell undergoes discharging when chemical energy is *transduced* into electrical energy. Cells are always charged by the application of an external potential difference (voltage) that is greater than the open circuit voltage of the cell. This makes the positive electrode increasingly positive relative to the negative electrode.

To be consistent, *reactions will always be written as discharge reactions*, and, although it may not always be in accordance with common usage, cells will be described with the negative electrode first.

Nonrechargeable cells (batteries) that can only convert chemical energy into electrical energy are called *primary* cells or batteries. Those that can be electrically driven in the opposite direction to form chemical products are said to be rechargeable, or *secondary* cells or batteries.

There are a number of other minor problems or inconsistencies with terminology that have grown up over a period of years. As an example, one often finds the terms *cell* and *battery* used interchangeably today, whereas *battery* traditionally has meant a group of individual cells. Incidentally, it also means a group of cannon, and a part of the island of Manhatten in New York City is called *the Battery* because it is the site where a number of cannon were placed long ago in order to protect the city against attack from the sea.

The terms *potential* and *voltage* are often confused in the literature. An *electrical potential difference* is properly called a *voltage*, yet one finds that the word *potential* is often used for both the *potential*, e.g., of an electrode, and a *potential difference*. Examples of the latter are *diffusion potential* and *cell potential*. Here, however, the term *potential* will indicate a property of a single material or electrode, and *voltage* will be used for a difference between potentials.

There are two terms that are used in electrochemistry to describe situations in which what is measured is different from what would be expected from the thermodynamics of an assumed reaction. These are *overvoltage* and *polarization*.

Overvoltage indicates a difference between an equilibrium steady-state potential of an electrode, or cell voltage, when there is no net current flowing through an electrochemical cell, and the measured potential, or voltage, when there is current flow. This difference is due to kinetic factors, and its magnitude depends upon the value of the current. The overvoltages of the two electrodes may be different, so that the observed overvoltage of a cell is a composite of the two of them. The values of overvoltages can be either positive or negative, of course.

The term *polarization*, on the other hand, is used to indicate that a measured potential is different from that which would be derived from the thermodynamic data relevant to the reaction that is assumed to be taking place. It is quite possible that the reaction that is actually happening is different from that which has been assumed, of course.

It is intended to not use these terms in this book. Instead, an effort will be made to be more explicit about the deviation of measured values from expectations.

In addition, it should be pointed out that there are several terms that were used in connection with battery technology in the past, but which have pretty much disappeared today. One of these was *depolarizer*, which was used to describe a material in one of the electrodes – typically the positive one – that acts to change, e.g., to reduce, the voltage of an electrochemical cell. This term will not be used here. Instead, emphasis will be placed upon the thermodynamic and kinetic factors that determine the measurable cell voltage.

Acknowledgments The author gladly acknowledges with gratitude the important contributions made to the development of the understanding in this area by his many students and associates in the Solid State Ionics Laboratory of the Department of Materials Science at Stanford University over many years, as well as those in the Center for Solar Energy and Hydrogen Research (ZSW) in Ulm, Germany and in the Faculty of Engineering of the Christian Albrechts University in Kiel, Germany.

Contents

Preface ... v
1 Introduction ... v
2 Applications of Electrochemical Energy Storage ... vi
3 Changes That Have Taken Place in Recent Years ... ix
4 Objectives of This Book ... x
5 Thinking Tools ... xi
6 Terminology and Conventions ... xii

1 Introductory Material ... 1
1.1 Introduction ... 1
1.2 Simple Chemical and Electrochemical Reactions ... 1
1.3 Major Types of Reaction Mechanisms ... 6
1.3.1 Reconstitution Reactions ... 6
1.3.2 Insertion Reactions ... 8
1.4 Important Practical Parameters ... 9
1.4.1 The Operating Voltage and the Concept of Energy Quality ... 10
1.4.2 The Charge Capacity ... 12
1.4.3 The Maximum Theoretical Specific Energy (MTSE) ... 13
1.4.4 Variation of the Voltage as Batteries are Discharged and Recharged ... 13
1.4.5 Cycling Behavior ... 15
1.4.6 Self-Discharge ... 16
1.5 General Equivalent Circuit of an Electrochemical Cell ... 17
1.5.1 Influence of Impedances to the Transport of Ionic and Atomic Species within the Cell ... 18
1.5.2 Influence of Electronic Leakage within the Electrolyte ... 18
1.5.3 Transference Numbers of Individual Species in an Electrochemical Cell ... 19

1.5.4 Relation between the Output Voltage and the Values of the Ionic and Electronic Transference Numbers ... 20
1.5.5 Joule Heating to Due to Self-Discharge in Electrochemical Cells ... 21
1.5.6 What If Current is Drawn from the Cell? ... 21
References ... 23

2 Principles Determining the Voltages and Capacities of Electrochemical Cells ... 25
2.1 Introduction ... 25
2.2 Thermodynamic Properties of Individual Species ... 25
2.3 A Simple Example: The Lithium/Iodine Cell ... 27
2.3.1 Calculation of the Maximum Theoretical Specific Energy ... 29
2.3.2 The Temperature Dependence of the Cell Voltage ... 29
2.4 The Shape of Discharge Curves and the Gibbs Phase Rule ... 30
2.5 The Coulometric Titration Technique ... 36
References ... 39

3 Binary Electrodes Under Equilibrium or Near-Equilibrium Conditions ... 41
3.1 Introduction ... 41
3.2 Binary Phase Diagrams ... 41
3.2.1 The Lever Rule ... 44
3.2.2 Examples of Binary Phase Diagrams ... 44
3.3 A Real Example, The Lithium: Antimony System Again ... 46
3.4 Stability Ranges of Phases ... 51
3.5 Another Example, The Lithium: Bismuth System ... 51
3.6 Coulometric Titration Measurements on Other Binary Systems ... 53
3.7 Temperature Dependence of the Potential ... 53
3.8 Application to Oxides and Similar Materials ... 55
3.9 Ellingham Diagrams ... 56
3.10 Liquid Binary Electrodes ... 57
3.11 Comments on Mechanisms and Terminology ... 60
3.12 Summary ... 62
References ... 62

4 Ternary Electrodes Under Equilibrium or Near-Equilibrium Conditions ... 65
4.1 Introduction ... 65
4.2 Ternary Phase Diagrams and Phase Stability Diagrams ... 65
4.3 Comments on the Influence of SubTriangle Configurations in Ternary Systems ... 67
4.4 An Example: The Sodium/Nickel Chloride "Zebra" System ... 70

4.5 A Second Example: The Lithium–Copper–Chlorine Ternary System ... 72
4.5.1 Calculation of the Voltages in This System ... 74
4.5.2 Experimental Arrangement for Lithium/Copper Chloride Cells ... 77
4.6 Calculation of the Maximum Theoretical Specific Energies of Li/Cucl and $Li/CuCl_2$ Cells ... 77
4.7 Specific Capacity and Capacity Density in Ternary Systems ... 78
4.8 Another Group of Examples: Metal Hydride Systems Containing Magnesium ... 78
4.9 Further Ternary Examples: Lithium–Transition Metal Oxides ... 85
4.10 Ternary Systems Composed of Two Binary Metal Alloys ... 90
4.10.1 An Example: The Li–Cd–Sn System at Ambient Temperature ... 90
4.11 What About the Presence of Additional Components? ... 91
4.12 Summary ... 91
References ... 91

5 Electrode Reactions That Deviate From Complete Equilibrium ... 93
5.1 Introduction ... 93
5.2 Stable and Metastable Equilibrium ... 93
5.3 Selective Equilibrium ... 95
5.4 Soft Chemistry (Chimie Douce) ... 96
5.5 Formation of Amorphous vs. Crystalline Structures ... 96
5.6 The Conversion of Crystalline to Amorphous Structures by Insertion Reactions ... 98
5.7 Deviations from Equilibrium for Kinetic Reasons ... 98
References ... 99

6 Insertion Reaction Electrodes ... 101
6.1 Introduction ... 101
6.2 Examples of the Insertion of Guest Species into Layer Structures ... 103
6.3 Floating and Pillared Layer Structures ... 104
6.4 More on Terminology Related to the Insertion of Species into Solids ... 104
6.5 Types of Inserted Guest Species Configurations ... 105
6.6 Sequential Insertion Reactions ... 107
6.7 Coinsertion of Solvent Species ... 110
6.8 Insertion into Materials with Parallel Linear Tunnels ... 110
6.9 Changes in the Host Structure Induced by Guest Insertion or Extraction ... 111
6.9.1 Conversion of the Host Structure from Crystalline to Amorphous ... 111

6.9.2 Dependence of the Product upon the Potential 113
6.9.3 Changes upon the Initial Extraction of the Mobile Species 114
6.10 The Variation of the Potential with Composition in Insertion Reaction Electrodes 114
6.10.1 Introduction 114
6.10.2 The Variation of the Electrical Potential with Composition in Simple Metallic Solid Solutions 116
6.10.3 Configurational Entropy of the Guest Ions 116
6.10.4 The Concentration Dependence of the Chemical Potential of the Electrons in a Metallic Solid Solution 117
6.10.5 Sum of the Effect of These Two Components upon the Electrical Potential of a Metallic Solid Solution 118
6.10.6 The Composition: Dependence of the Potential in the Case of Insertion Reactions that Involve a Two-Phase Reconstitution Reaction 119
6.11 Final Comments 122
References 122

7 Negative Electrodes in Lithium Cells 123
7.1 Introduction 123
7.2 Elemental Lithium Electrodes 123
7.3 Problems with the Rechargeability of Elemental Electrodes 124
7.3.1 Deposition at Unwanted Locations 124
7.3.2 Shape Change 124
7.3.3 Dendrites 125
7.3.4 Filamentary Growth 125
7.3.5 Thermal Runaway 126
7.4 Alternatives 127

7A Lithium–Carbon Alloys 127
7A.1 Introduction 127
7A.2 Ideal Structure of Graphite Saturated with Lithium 129
7A.3 Variations in the Structure of Graphite 130
7A.4 Structural Aspects of Lithium Insertion into Graphitic Carbons 131
7A.5 Electrochemical Behavior of Lithium in Graphite 132
7A.6 Electrochemical Behavior of Lithium in Amorphous Graphite 134
7A.7 Lithium in Hydrogen-Containing Carbons 134

7B Metallic Lithium Alloys 136
7B.1 Introduction 136
7B.2 Equilibrium Thermodynamic Properties of Binary Lithium Alloys 136
7B.3 Experiments at Ambient Temperature 137
7B.4 Liquid Binary Alloys 138
7B.5 Mixed-Conductor Matrix Electrodes 138
7B.6 Decrepitation 143
7B.7 Modification of the Micro and Nanostructure of the Electrode 145
7B.8 Formation of Amorphous Products at Ambient Temperatures .. 147
References 148

8 Convertible Reactant Electrodes 151
8.1 Introduction 151
8.2 Electrochemical Formation of Metals and Alloys from Oxides . 152
8.3 Lithium–Tin Alloys at Ambient Temperature 152
8.4 The Lithium–Tin Oxide System 153
8.5 Irreversible and Reversible Capacities 155
8.6 Other Possible Convertible Oxides 157
8.7 Final Comments 158
References 158

9 Positive Electrodes in Lithium Systems 159
9.1 Introduction 159
9.2 Insertion Reaction, Instead of Reconstitution Reaction, Electrodes 160
9.3 More than One Type of Interstitial Site or More than One Type of Redox Species 161
9.4 Cells Assembled in the Discharged State 162

9A Solid Positive Electrodes in Lithium Systems 163
9A.1 Introduction 163
9A.2 Influence of the Crystallographic Environment on the Potential. 166
9A.3 Oxides with Structures in Which the Oxygen Anions Are in a Face-Centered Cubic Array 167
9A.3.1 Materials with Layered Structures 167
9A.3.2 Materials with the Spinel Structure 169
9A.3.3 Lower Potential Spinel Materials with Reconstitution Reactions 174
9A.4 Materials in Which the Oxide Ions Are in a Close-Packed Hexagonal Array 175
9A.4.1 The Nasicon Structure 175
9A.4.2 Materials with the Olivine Structure 177
9A.5 Materials Containing Fluoride Ions 179

9A.6 Hybrid Ion Cells . . . 180
9A.7 Amorphization . . . 180
9A.8 The Oxygen Evolution Problem . . . 180
9A.9 Closing Comments . . . 186

9B Liquid Positive Electrode Reactants . . . 186
9B.1 Introduction . . . 186
9B.2 The Li/SO_2 System . . . 186
9B.3 The $Li/SOCl_2$ System . . . 188

9C Hydrogen and Water in Positive Electrode Materials . . . 188
9C.1 Introduction . . . 188
9C.2 Ion Exchange . . . 189
9C.3 Simple Addition Methods . . . 190
9C.4 Thermodynamics of the Lithium–Hydrogen–Oxygen System . . . 190
9C.5 Examples of Phases Containing Lithium That are Stable in Water . . . 191
9C.6 Materials That Have Potentials Above the Stability Window of Water . . . 192
9C.7 Absorption of Protons from Water Vapor in the Atmosphere . . . 192
9C.8 Extraction of Lithium from Aqueous Solutions . . . 193
References . . . 193

10 Negative Electrodes in Aqueous Systems . . . 197
10.1 Introduction . . . 197

10A The Zinc Electrode in Aqueous Systems . . . 197
10A.1 Introduction . . . 197
10A.2 Thermodynamic Relationships in the H–Zn–O System . . . 198
10A.3 Problems with the Zinc Electrode . . . 199

10B The "Cadmium" Electrode . . . 199
10B.1 Introduction . . . 199
10B.2 Thermodynamic Relationships in the H–Cd–O System . . . 200
10B.3 Comments on the Mechanism of Operation of the Cadmium Electrode . . . 201

10C Metal Hydride Electrodes . . . 202
10C.1 Introduction . . . 202
10C.2 Comments on the Development of Commercial Metal Hydride Electrode Batteries . . . 203
10C.3 Hydride Materials Currently Being Used . . . 203
10C.4 Disproportionation and Activation . . . 204
10C.5 Pressure–Composition Relation . . . 205
10C.6 The Influence of Temperature . . . 206

10C.7 AB_2 Alloys . . . 208
10C.8 General Comparison of These Two Structural Types . . . 209
10C.9 Other Alloys That Have Not Been Used in Commercial Batteries . . . 210
10C.10 Microencapsulation of Hydride Particles . . . 210
10C.11 Other Binders . . . 211
10C.12 Inclusion of a Solid Electrolyte in the Negative Electrode of Hydride Cells . . . 211
10C.13 Maximum Theoretical Capacities of Various Metal Hydrides . . . 211
References . . . 212

11 Positive Electrodes in Aqueous Systems . . . 213
11.1 Introduction . . . 213

11A Manganese Dioxide Electrodes in Aqueous Systems . . . 214
11A.1 Introduction . . . 214
11A.2 The Open Circuit Potential . . . 215
11A.3 Variation of the Potential During Discharge . . . 216

11B The "Nickel" Electrode . . . 216
11B.1 Introduction . . . 216
11B.2 Structural Aspects of the $Ni(OH)_2$ and NiOOH Phases . . . 217
11B.3 Mechanism of Operation . . . 218
11B.4 Relations Between Electrochemical and Structural Features . . . 220
11B.5 Self-discharge . . . 221
11B.6 Overcharge . . . 223
11B.7 Relation to Thermodynamic Information . . . 223

11C Cause of the Memory Effect in "Nickel" Electrodes . . . 226
11C.1 Introduction . . . 226
11C.2 Mechanistic Features of the Operation of the "Nickel" Electrode . . . 228
11C.3 Overcharging Phenomena . . . 230
11C.4 Conclusions . . . 232
References . . . 233

12 Other Topics Related to Electrodes . . . 235
12.1 Introduction . . . 235

12A Mixed-Conducting Host Structures into Which Either Cations or Anions Can Be Inserted . . . 235
12A.1 Introduction . . . 235
12A.2 Insertion of Species into Materials with Transition Metal Oxide Bronze Structures . . . 236

12A.3 Materials with Cubic Structures Related to Rhenium Trioxide . . 237
12A.4 Hexacyanometallates 237
12A.5 Electrochemical Behavior of Prussian Blue 241
12A.6 Various Cations Can Occupy the A Sites in the Prussian Blue Structure 244
12A.7 The Substitution of Other Species for the Fe^{3+} and the Fe^{2+} in the P and R Positions in the Prussian Blue Structure 245
12A.8 Other Materials with $x = 2$ That Have the Perovskite Structure 246
12A.9 The Electronic Properties of Members of the Prussian Blue Family 246
12A.10 Batteries with Members of the Prussian Blue Family on Both Sides 247
12A.11 Catalytic Behavior 247
12A.12 Electrochromic Behavior 248
12A.13 Insertion of Species into Graphite 249
12A.14 Insertion of Guest Species into Polymers 251
12A.15 Summary 252

12B Cells with Liquid Electrodes: Flow Batteries 252
12B.1 Introduction 252
12B.2 Redox Reactions in the Vanadium/Vanadium System 255
12B.3 Resultant Electrical Output 255
12B.4 Further Comments on the Vanadium/Vanadium Redox System . 255

12C Reactions in Fine Particle Electrodes 257
12C.1 Introduction 257
12C.2 Translation of Two-Phase Interface by Chemical Diffusion 257
12C.3 Alternative Mechanism for the Translation of Poly-Phase Interfaces 258
12C.4 Reactions in Electrodes Containing Many Small Particles 260
12C.5 Mechanism Involved in Changing the Composition of Lithium–Carbons 260
References 261

13 Potentials 263
13.1 Introduction 263

13A Potentials in and Near Solids 264
13A.1 Introduction 264
13A.2 Potential Scales 265
13A.3 Electrical, Chemical, and Electrochemical Potentials in Metals . 265
13A.4 Relation to the Band Model of Electrons in Solids 271
13A.5 Potentials in Semiconductors 272
13A.6 Interactions Between Different Materials 273

13A.7 Junctions Between Two Metals 273
13A.8 Junctions Between Metals and Semiconductors 275
13A.9 Selective Equilibrium 276

13B Reference Electrodes 276
13B.1 Introduction 276
13B.2 Reference Electrodes in Nonaqueous Lithium Systems 277
13B.2.1 Use of Elemental Lithium 277
13B.2.2 Use of Two-phase Lithium Alloys 277
13B.3 Reference Electrodes in Elevated Temperature Oxide-Based Systems 278
13B.3.1 Gas Electrodes 278
13B.3.2 Polyphase Solid Reference Electrodes 279
13B.3.3 Metal Hydride Systems 280
13B.4 Relations Between Binary Potential Scales 280
13B.5 Potentials in the Ternary Lithium–Hydrogen–Oxygen System . . 280
13B.5.1 Lithium Cells in Aqueous Electrolytes 282
13B.6 Significance of Electrically Neutral Species 282
13B.7 Reference Electrodes in Aqueous Electrochemical Systems 283
13B.8 Historical Classification of Different Types of Electrodes in Aqueous Systems 284
13B.8.1 Electrodes of the First Kind 284
13B.8.2 Electrodes of the Second Kind 285
13B.9 The Gibbs Phase Rule 287
13B.10 Application of the Gibbs Phase Rule to Reference Electrodes . . 288
13B.10.1 Nonaqueous Systems 288
13B.10.2 Aqueous Systems 288
13B.11 Systems Used to Measure the pH of Aqueous Electrolytes 291
13B.12 Electrodes with Mixed-Conducting Matrices 292
13B.13 Closing Comments 293

13C Potentials of Chemical Reactions 293
13C.1 Introduction 293
13C.2 Relation Between Chemical Redox Equilibria and the Potential and Composition of Insertion Reaction Materials 294
13C.3 Other Examples 295
13C.4 Summary 297

13D Potential and Composition Distributions Within Components of Electrochemical Cells 297
13D.1 Introduction 297
13D.2 Relevant Energy Quantities 297
13D.3 What Is Different About the Interior of Solids? 298

13D.4 Relations Between Inside and Outside Quantities 299
13D.5 Basic Flux Relations Inside Phases 299
13D.6 Two Simple Limiting Cases 300
13D.7 Three Configurations 300
13D.8 Variation of the Composition with Potential 300
13D.9 Calculation of the Concentrations of the Relevant Defects in a Binary Solid MX That Is Predominantly an Ionic Conductor 301
13D.10 Defect Equilibrium Diagrams 303
13D.11 Approximations Relevant in Specific Ranges of Composition or Activity 304
13D.12 Situation in Which an Electrical Potential Difference Is Applied Across a Solid Electrolyte Using Electrodes That Block the Entry and Exit of Ionic Species 305
13D.13 The Use of External Sensors to Evaluate Internal Quantities in Solids 307
13D.14 Another Case, a Mixed Conductor in Which the Transport of Electronic Species Is Blocked 308
13D.15 Further Comments on Composite Electrochemical Cells Containing a Mixed Conductor in Series with a Solid Electrolyte 309
13D.16 Transference Numbers of Particular Species................. 311
References 312

14 Liquid Electrolytes 315
14.1 Introduction 315
14.2 General Considerations Regarding the Stability of Electrolytes Vs. Alkali Metals 315

14A Elevated Temperature Electrolytes for Alkali Metals 317
14A.1 Introduction 317
14A.2 Lithium-Conducting Inorganic Molten Salts................. 318
14A.3 Lower Temperature Alkali Halide Molten Salts 318
14A.4 Other Modest Temperature Molten (and Solid) Salts.......... 318
14A.5 Relation Between the Potential and the Oxygen Pressure in Lithium Systems 319
14A.6 Implications for the Safety of Lithium Cells................. 320

14B Ambient Temperature Electrolytes for Lithium 321
14B.1 Introduction 321
14B.2 Organic Solvent Liquid Electrolytes 321
14B.3 Lithium Salts 322
14B.4 Ionic Liquids 323

14C Aqueous Electrolytes for Hydrogen ... 324
14C.1 Introduction ... 324
14C.2 Nafion ... 325
14C.3 Other Considerations Relating to Nafion ... 327
14C.4 Alternatives to Nafion ... 328

14D Nonaqueous Electrolytes for Hydrogen ... 328
14D.1 Introduction ... 328
14D.2 Methods Typically Used to Study Materials for Hydrogen Storage ... 329
14D.3 Potential Advantages of Electrochemical Methods ... 330
14D.4 The Amphoteric Behavior of Hydrogen ... 330
14D.5 Relationships Between the Potential and the Stability of Phases in Molten Salts ... 331
14D.6 Alkali Halide Molten Salts Containing Hydride Ions ... 332
14D.7 Solution of Hydrogen in Vanadium ... 334
14D.8 The Titanium–Hydrogen System ... 334
14D.9 Use of Low Temperature Organic-Anion Molten Salt to Study Hydrogen in Binary Magnesium Alloys ... 335
14D.10 Summary ... 336
References ... 337

15 Solid Electrolytes ... 339
15.1 Introduction ... 339

15A Solid Electrolytes: Introduction ... 340
15A.1 Introduction ... 340
15A.2 Structural Defects in Nonmetallic Solids ... 341
15A.3 Various Types of Notation That May Be Used to Describe Imperfections ... 343
15A.4 Types of Disorder ... 345

15B Mechanism and Structural Dependence of Ionic Conduction in Solid Electrolytes ... 347
15B.1 Introduction ... 347
15B.2 Characteristic Properties ... 347
15B.3 Simple Hopping Model of Defect Transport ... 350
15B.4 Interstitial Motion in Body-Centered Cubic Structures ... 352
15B.5 Rapid Ionic Motion in Other Crystal Structures ... 354
15B.6 Simple Structure-Dependent Model for the Rapid Transport of Mobile Ions ... 356
15B.7 Interstitial Motion in the Rutile Structure ... 358
15B.8 Other Materials with Unidirectional Tunnels ... 361
15B.9 Materials with the Fluorite and Antifluorite Structures ... 362
15B.10 Materials with Layer Structures ... 364

15B.11 Materials with Three-Dimensional Arrays of Tunnels 367
15B.12 Structures with Isolated Tetrahedra 367

15C Lithium Ion Conductors 368
15C.1 Introduction 368
15C.2 Materials with the Perovskite Structure 368
15C.3 Materials with the Garnet Structure 370
References 371

16 Electrolyte Stability Windows and Their Extension 375
16.1 Introduction 375
16.2 Binary Electrolyte Phases 376
16.3 Ternary Electrolyte Phases 377
16.3.1 Stability Limits Relative to Lithium 378
16.3.2 Stability Limits Relative to Oxygen 379
16.4 Summary 380

16A Composite Structures That Combine Stability Regimes 380
16A.1 Introduction 380
16A.2 Two Solid Electrolytes in Series 381
16A.3 Solid Electrolyte in Series with Aqueous Electrolyte 381
16A.4 Solid Electrolyte in Series with Molten Salt 382
16A.5 Formation of a Second Electrolyte by Topotactic Reaction Between a Liquid and a Solid Mixed Conductor Electrode 382
16A.6 Formation of a Protective Reaction Product Layer Between the Negative Electrode and the Organic Solvent Electrolyte in Lithium Cells 382

16B The SEI in Organic Solvent Systems 383
16B.1 Introduction 383
16B.2 Interaction of Organic Solvent Electrolytes with Graphite 383
16B.3 Electrolyte Additives 387

16C Combination of a Solid Electrolyte and a Molten Salt Electrolyte 388
16C.1 Introduction 388
16C.2 The Lithium–Nitrogen–Oxygen System 388
16C.3 Extension of the Effective Potential Range by the Formation of a Second Electrolyte In Situ 389
16C.4 A Primary Lithium/Carbon Cell 390
16C.5 Problems with This Concept 391
References 391

17 Experimental Methods to Evaluate the Critical Properties of Electrodes and Electrolytes . . . 393
17.1 Introduction . . . 393

17A Use of DC Methods to Determine the Electronic and Ionic Components of the Conductivity in Mixed Conductors . . . 393
17A.1 Introduction . . . 393
17A.2 Transference Numbers of Individual Species . . . 394
17A.3 The Tubandt Method . . . 395
17A.4 The DC Assymetric Polarization Method . . . 395
17A.5 Interpretation of Hebb–Wagner Asymmetric Polarization Measurements in Terms of a General Defect Equilibrium Diagram . . . 396
17A.6 DC Open Circuit Potential Method . . . 405

17B Experimental Determination of the Critical Properties of Potential Electrode Materials . . . 406
17B.1 Introduction . . . 406
17B.2 The GITT Method . . . 408
17B.3 The PITT Method . . . 408
17B.4 The FITT Method . . . 409
17B.5 The WITT Method . . . 411

17C Use of AC Methods to Determine the Electronic and Ionic Components of the Conductivity in Solid Electrolytes and Mixed Conductors . . . 414
17C.1 Introduction . . . 414
17C.2 Representation of the Properties of Simple Circuit Elements on the Complex Impedance Plane . . . 416
17C.3 The Influence of Electronic Leakage Through an Ionic Conductor . . . 422
17C.4 Case in Which Both Ionic and Electronic Transport Are Significant . . . 423
17C.5 Influence of an Additional Impedance Due to Transverse Internal Interfaces . . . 425
17C.6 Behavior When There Is Internal Transverse Interface Impedance as well as Partial Electronic Conduction . . . 426
17C.7 An Example . . . 429
17C.8 Summary . . . 430
References . . . 430

18 Use of Polymeric Materials As Battery Components . . . 433
18.1 Introduction . . . 433

18A Polymer Electrolytes 434
18A.1 Introduction 434
18A.2 High Molecular Weight Polymers Containing Salts 434
18A.3 Particle-Enhanced Conductivity 436
18A.4 Ionic Rubbers 437
18A.5 Hybrid Electrolytes Containing an Ionically Conducting Plasticizer 437
18A.6 Gel Electrolytes 437
References 439

19 Transient Behavior of Electrochemical Systems 441
19.1 Introduction 441

19A Transient Behavior Under Pulse Demand Conditions 442
19A.1 Introduction 442
19A.2 Electrochemical Charge Storage Mechanisms 445
19A.2.1 Electrostatic Energy Storage in the Electrical Double Layer in the Vicinity of the Electrolyte/Electrode Interface 445
19A.2.2 Underpotential Faradaic Two-Dimensional Adsorption on the Surface of a Solid Electrode 447
19A.2.3 Faradaic Deposition That Results in the Three-Dimensional Absorption of the Electroactive Species into the Bulk Solid Electrode Material by an Insertion Reaction 447
19A.2.4 Faradaically Driven Reconstitution Reactions 450
19A.3 Comparative Magnitudes of Energy Storage 450
19A.4 Importance of the Quality of the Stored Energy 452

19B Modeling Transient Behavior of Electrochemical Systems Using Laplace Transforms 453
19B.1 Introduction 453
19B.2 Use of Laplace Transform Techniques 453
19B.3 Simple Examples 455
References 456

20 Closing Comments 459
20.1 Introduction 459
20.2 Terminology 459
20.3 Major Attention Is Given to the Driving Forces and Mechanisms That Determine the Potentials, Kinetic Properties, and Capacities of Electrodes 460
20.4 Thinking Tools 461
20.5 Major Players in This Area 461
20.6 The Future 462

Index 463

Chapter 1
Introductory Material

1.1 Introduction

As mentioned in the Preface, electrochemical storage of energy involves the conversion, or *transduction*, of chemical energy into electrical energy, and *vice versa*. This is accomplished by the use of electrochemical cells, commonly known as *batteries*.

To understand how this works, it is first necessary to consider the *driving forces* that cause electrochemical transduction to occur in electrochemical cells and the major types of *reaction mechanisms* that can occur. These matters are discussed in this chapter.

This is followed by a brief description of the *important practical parameters* that are used to describe the behavior of electrochemical cells. How the basic properties of such electrochemical systems can be modeled through the use of simple *equivalent electrical circuits* is then shown.

The next chapter discusses the principles that determine the major properties of electrochemical cells, their voltages and capacities.

1.2 Simple Chemical and Electrochemical Reactions

First consider a simple *chemical reaction* between two metallic materials A and B, which react to form an electronically conducting product AB. This can be represented simply by the relation

$$A + B = AB \tag{1.1}$$

The driving force for this reaction is the difference in the values of the *standard Gibbs free energy* of the products - AB in this case, and the standard Gibbs free energies of the reactants, A and B.

$$\Delta G_r{}^\circ = \sum \Delta G_f{}^\circ(products) - \sum \Delta G_f{}^\circ(reactants) \tag{1.2}$$

If A and B are simple elements, this is called a *formation reaction*, and since the standard Gibbs free energy of formation of elements is zero, the value of the Gibbs free energy change that results per mol of the reaction is simply the *Gibbs free energy of formation* per mol of AB, that is:

$$\Delta G_r{}^\circ = \Delta G_f{}^\circ(AB) \tag{1.3}$$

Values of this parameter for many materials can be found in a number of sources [1–3].

While the morphology of such a reaction can take a number of forms, consider a simple 1-dimensional case in which the reactants are placed in direct contact and the product phase AB forms between them. The time sequence of the *evolution of the microstructure* during such a reaction is shown schematically in Fig. 1.1.

It is obvious that for the reaction product phase AB to grow, either A or B must move (diffuse) through it, to come into contact with the other reactant on the other side. If, for example, A moves through the AB phase to the B side, additional AB will form at the AB/B interface. Since some B is consumed, the AB/B interface will move to the right. As the amount of A on the A side has decreased, the A/AB interface will likewise move to the left. AB will grow in width in the middle. The action will be the same when species B, rather than species A, moves through the AB phase in this process. There are experimental methods to determine the identity of the moving species, but that is not relevant here.

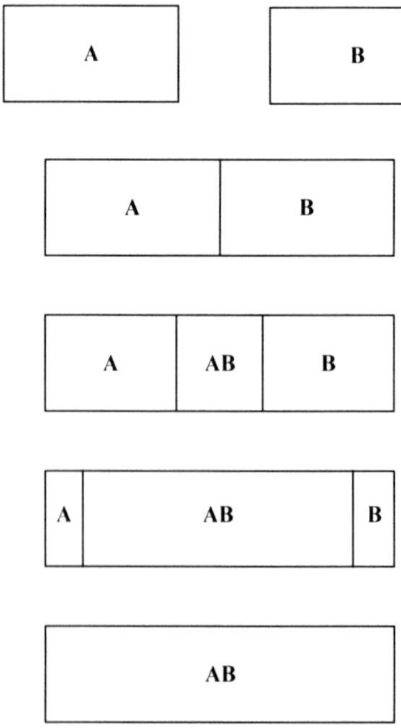

Fig. 1.1 Simple schematic model of chemical reaction of A and B to form AB, indicating how the microstructure of the system varies with time

In case this process occurs by an *electrochemical mechanism*, the time dependence of the microstructure is illustrated schematically in Fig. 1.2. Like the chemical reaction case, product *AB* must form as the result of a reaction between the reactants *A* and *B*; but an additional phase is present in the system- an electrolyte.

The *function of the electrolyte* is to *act as a filter* for the passage of ionic, but not electronic species. The electrolyte must contain ions of either *A* or *B*, or both, and be an *electronic insulator*.

The reaction between *A* and *B* involves not just ions *but electrically neutral atoms*. Hence for the reaction to proceed there must be another path whereby electrons can also move through the system. This is typically an external electrical circuit connecting *A* and *B*. If *A* is transported in the system, and the electrolyte contains A^+ ions, negatively charged electrons, e^-, must pass through the external circuit in equal numbers, or at an equal rate, to match the charge flux due to the passage of A^+ ions through the electrolyte to the other side.

For an electrochemical discharge reaction of the type illustrated in Fig. 1.2 the reaction at the interface between the phase *A* and the electrolyte can be written as

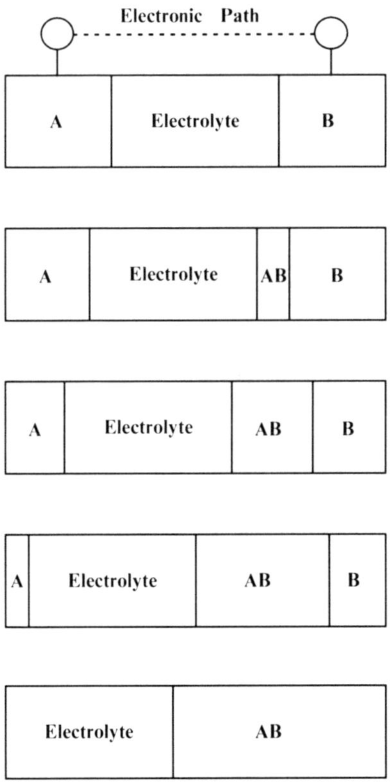

Fig. 1.2 Simple schematic model of time evolution of the microstructure during the electrochemical reaction of *A* and *B* to form *AB*, a mixed conductor. In this case it is assumed that A^+ ions are the predominant ionic species in the electrolyte. To simplify the figure, the external electronic path is shown only at the start of the reaction

$$A = A^{+} + e^{-} \tag{1.4}$$

with A^{+} ions moving into the electrolyte phase and electrons entering the external circuit through a *current collector*. There will be a corresponding reaction on the other side of the electrolyte,

$$A^{+} + e^{-} = A \tag{1.5}$$

with ions arriving at the interface from the electrolyte and electrons from the external circuit through the electronic current collector. This results in the deposition *A* atoms onto the adjacent solid phase *AB*. The A/electrolyte interface and the electrolyte/AB interface move incrementally to the left in Fig. 1.2. Interdiffusion of A and B atoms within the phase AB is necessary to ensure that its surface does not have only A atoms. In addition, this phase must be an electronic conductor.

Important consequences follow from the fact that the overall reaction is between neutral species which requires concurrent motion of either *A* or *B* ions through the electrolyte and electrons through external circuit. 1) If flow in either the electronic path or the ionic path does not occur, the entire reaction stalls. 2) If the external electrical circuit is opened and no electrons can flow through it, no ions can flow through the electrolyte, halting the reaction. 3) If flow of ions in the electrolyte is impeded - by presence of some material with very high resistance to the moving ionic species, or a loss of contact between the electrolyte and the two materials on its sides, there will be no electronic current in the external circuit.

When the electronic circuit is open, and there is no current flowing, there must be a force balance operating upon the electrically charged ions in the electrolyte. A *chemical driving force* upon the mobile ionic species within the electrolyte in one direction is simply balanced by an *electrostatic driving force* in the opposite direction.

The *chemical driving force* across the cell is due to the difference in the chemical potentials of its two electrodes. It can be expressed as the *standard Gibbs free energy change per mol of reaction*, $\Delta G_r{}^{\circ}$. This is determined by the difference between the *standard Gibbs free energies of formation* of products and reactants in the virtual chemical reaction that would occur if the *electrically neutral* materials in the two electrodes were to react chemically. It makes no difference that the reaction actually happens by the transport of ions and electrons across the electrochemical system from one electrode to the other.

The electrostatic energy per mol of an electrically charged species is $-zFE$, where E is the voltage between the electrodes, and z the *charge number* of the mobile ionic species involved in the virtual reaction. The charge number is the number of elementary charges that they transport. F is the *Faraday constant* (96,500 Coulombs per equivalent). An *equivalent* is *Avogadro's number* (one mol) of electronic charges.

The balance between the chemical and electrical forces upon the ions under open circuit conditions can hence be simply expressed as an energy balance

$$\Delta G_r{}^{\circ} = -zFE \tag{1.6}$$

The value of $\Delta G_r{}^{\circ}$ is in Joules per mol of reaction; 1 Joule is the product of one Coulomb and one Volt.

It is interesting that, a chemical reaction between neutral species in the electrodes determines the forces on charged particles in the electrolyte in the interior of an electrochemical system.

If it is assumed that the electrodes on the two sides of the electrolyte are good electronic conductors, there is an externally measurable voltage E between the points where the external electronic circuit contacts the two electrodes. This voltage allows electrical work be done by the passage of electrons in an external electric circuit, if ionic current travels through the electrolyte inside the cell.

This simple electrochemical cell acts as a transducer between chemical and electrical quantities; forces, fluxes and energy. Ideally chemical energy reduction due to the chemical reaction that takes place between A and B to form mixed-conducting AB is compensated by electrical energy transferred to the external electronic circuit.

Flow of both internal ionic species and external electrons can be reversed if a voltage is imposed in the opposite direction in the electronic path that is larger than the voltage that is the result of the driving force of the chemical reaction. Since this causes current to flow in the reverse direction, electrical energy will be consumed and the chemical energy inside the system will increase; the electrochemical system is thus being recharged.

From these considerations it is obvious that it is not important whether the ionic species are related to element A or to element B. However, the answer to this question will influence the configuration of the cell. The example illustrated schematically in Fig. 1.2 deals with a case where A^+ ions are predominant in the electrolyte. The chemical reaction proceeds with the transport of A^+ ions in the electrolyte and electrons in the external circuit from the left (A) to the right side of the cell. This involves two electrochemical reactions. On the left A atoms are converted to A^+ ions and electrons at the A/electrolyte interface. The electrons travel back through the metallic A and move into the external electronic circuit. The reverse electrochemical reaction takes place on the other side of the cell. A^+ ions from the electrolyte combine with electrons that have come through the external circuit to form neutral A at the electrolyte/AB interface.

As before, the physical locations of interfaces, the A/electrolyte interface, the electrolyte/AB interface and the AB/B interface, will move with time, as the amounts of the various species vary with the extent of the reaction.

It must be recognized that the reaction product AB will not form unless there is a mechanism which allows the newly-arrived A to react with B atoms to form AB. The transport of either A or B atoms within the AB product phase is necessary, as in the chemical reaction case illustrated in Fig. 1.1 above. If this did not happen, pure A would be deposited at the right hand electrolyte interface. The chemical composition on both sides of the electrolyte would then be the same; there would be no driving force to cause further transport of ionic species through the electrolyte, and therefore no external voltage.

If B^+ ions, rather than A^+ ions, are present in the electrolyte, so that B species flow from right to left, the direction of electron flow, and consequently the voltage polarity, in the external circuit will be opposite from that discussed above; the reaction product will form on the left side, rather than on the right side.

It is also possible, that the ions in the electrolyte are negatively charged. In that case, the electron flow in the external circuit will be in the opposite direction.

The basic driving force in an electrochemical cell is thus a *chemical reaction of neutral species* forming an *electrically neutral product*. This is why standard chemical thermodynamic data can be used to understand the equilibrium (no current, or open circuit) potentials and voltages in electrochemical cells.

For any given chemical reaction, the open circuit voltage is independent of the identity of the species in the electrolyte and the details of the reactions that take place at the electrode/electrolyte interfaces.

The situation is different when considering the kinetic behavior of electrochemical cells. For then one has to be concerned with phenomena at all of the interfaces, as well as in the electrodes, the electrolyte and the external circuit. This will be discussed in some detail later.

1.3 Major Types of Reaction Mechanisms

1.3.1 Reconstitution Reactions

A number of important chemical, and possible electrochemical, reactions exist in which some phases grow and others disappear. The result is that the microstructure of one or more of the electrode materials gets significantly changed, or *reconstituted*. *Phase diagrams* are useful thinking tools to help understand this phenomenon. They are graphical representations which indicate phases and their compositions that are present in a materials system under equilibrium conditions. These were called *constitution diagrams* in the past, and reactions in which there is a change in the identity or amounts of the phases present are designated *reconstitution reactions*. Two major types of such reactions will be briefly mentioned here, *formation reactions*, and *displacement reactions*.

1.3.1.1 Formation Reactions

The simple example that was discussed above, represented by the equation

$$A + B = AB \tag{1.7}$$

is a *formation reaction*, in which a new phase AB is formed in one of the electrodes from its atomic constituents. This can result from the transport of one of the elements, e.g. A, passing across an electrochemical cell through the electrolyte from one electrode to react with the other component in the other electrode. Since this modifies the microstructure, it represents a reconstitution reaction.

Examples of this type of formation reaction are many. Subsequent additional formation reactions can also occur whereby other phases can be formed by further reaction of an original product. As an example, electrochemical experiments were performed on the reaction of lithium with antimony at about 350°C [4]. This involved the use of an electrochemical cell in which the negative electrode was elemental lithium, and the positive electrode was initially elemental antimony. Due to the elevated temperature, the electrolyte was a lithium-conducting molten salt. On the imposition of current, lithium passed across the cell and reacted with the antimony positive electrode, changing its chemical composition. As a result, the electrical potential of the positive electrode changed. This could be observed by measurement of the voltage across the cell.

The equilibrium phase diagram for the *lithium - antimony system* is shown in Fig. 1.3. By drawing a horizontal line at 350°C, starting at the antimony side of the figure it is seen that on adding lithium, the overall composition moves into the two-phase region in which both *Sb* and the phase Li_2Sb are present under equilibrium conditions. On further addition of lithium the composition reaches 67% *Li*, or 33% *Sb*, and only the phase Li_2Sb is present. If even more lithium is added, the overall composition moves into the region in which two phases, Li_2Sb and Li_3Sb are stable. On addition of further lithium beyond the composition of the phase Li_3Sb, a region is reached in which two phases are again present, Li_3Sb and elemental lithium.

Since both the reactants and the products in the reaction that forms Li_3Sb from Li_2Sb are different from those in the first reaction, which formed Li_2Sb, the cell voltage will be different in the two composition regions. This is an important point, and the relation between the chemical composition and the electrical potential in the Li-Sb alloy system, as well as a number of others that exhibit multi-phase reactions, will be discussed in greater detail in later chapters.

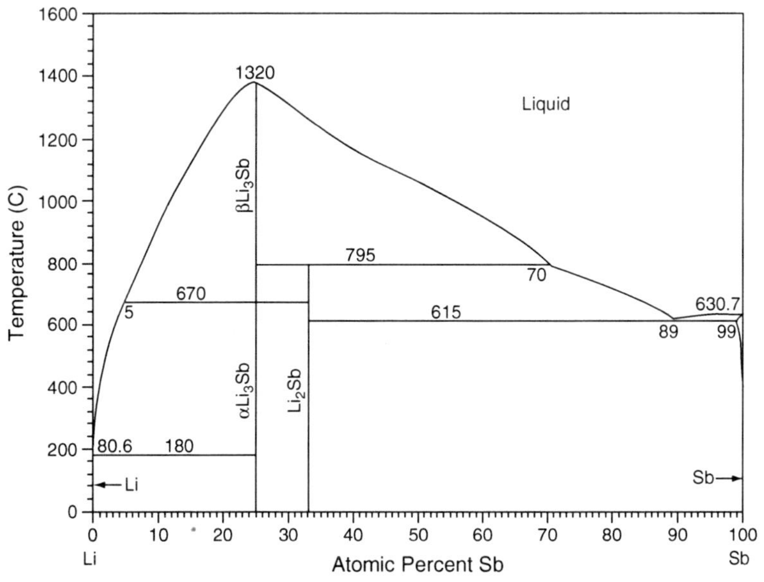

Fig. 1.3 Equilibrium phase diagram of the lithium - antimony system

It is also not necessary that both reactants in formation reactions are solids or liquids; the phase *LiCl* can result from the reaction of lithium with chlorine gas, and *ZnO* can form as a result of reaction of zinc with oxygen in air. Zn/O_2 cells, in which *ZnO* is formed, are commonly used to power hearing aids.

1.3.1.2 Displacement Reactions

Another type of reconstitution reaction involves a *displacement* process, which can be simply represented as

$$A + BX = AX + B \tag{1.8}$$

in which species *A* displaces species *B* in the simple binary phase *BX*, forming *AX*. In addition a new phase comprising elemental *B* will be formed. A driving force causing this reaction tends to occur if phase *AX* has greater stability, i.e. has a greater negative value of $\Delta G_f{}^{\circ}$, than phase *BX*. e.g.

$$Li + Cu_2O = Li_2O + Cu \tag{1.9}$$

in which the reaction of lithium with Cu_2O results in formation of two new phases, Li_2O and elemental copper.

A change in the chemical state in the electrode results in a change in its electrical potential. The relation between the chemical driving forces of these reactions, and related electrical potentials, will also be discussed in later chapters.

1.3.2 Insertion Reactions

A quite different reaction mechanism can also occur in materials in chemical and electrochemical systems. This involves the *insertion* of guest species into normally unoccupied interstitial sites in the crystal structure of an existing stable host material. Though the chemical composition of the host phase initially present can be substantially changed, this reaction does not result in a change in its identity, the basic crystal structure, or the amounts of the phases in the microstructure. In most cases, however, the addition of interstitial species to previously unoccupied locations in the structure causes a change in volume. This involves mechanical stress, and mechanical energy. The mechanical energy related to insertion and extraction of interstitial species plays a significant role in the hysteresis, and subsequent energy loss, observed in a number of reversible battery electrode reactions.

In the particular case of insertion of species into materials with layer-type crystal structures, insertion reactions are sometimes called *intercalation reactions*. Reactions in which the composition of an existing phase is changed by the incorporation of guest species can also be thought of as the solution of the guest into the host material. Such processes are also sometimes referred to as *solid solution reactions*.

Generally, incorporation of such guest species occurs *topotactically*. i.e. guest species tend to be present at specific (low energy) locations inside the crystal structure of the host species, and not randomly distributed.

A simple reaction of this type is the reaction of an amount x of species A with a phase BX to produce the product A_xBX. This can be written as

$$xA + BX = A_xBX \tag{1.10}$$

The solid solution phase can have a range of composition, i.e. a range of values of x. For example, incorporation of lithium into TiS_2 produces a product in which value of x can extend from 0 to 1. This was an important early example of an insertion reaction [5], and can be simply represented as

$$xLi + TiS_2 = Li_xTiS_2 \tag{1.11}$$

It is also possible to have a *displacement reaction* occur through the replacement of one interstitial species by another inside a stable host material. In this case, only one additional phase is sometimes formed - the material that is displaced. The term *extrusion* is used to describe this process.

In some cases, the new element or phase formed by such an *interstitial displacement process* is *crystalline*; in others, it is *amorphous*.

1.4 Important Practical Parameters

When considering the use of electrochemical energy storage systems in various applications one must be aware of the properties that might be relevant, since they are not always the same in every case.

Energy and power available per unit weight, called the *specific energy* and *specific power*, are important in some applications, such as vehicle propulsion.

On the other hand, the amount of energy that can be stored per unit volume, called the *energy density*, can be more important for other applications. This is often the case when such devices are being considered as power sources in portable electronic devices, like cellular telephones, portable computers and video camcorders.

Power per unit volume, known as *power density*, can be important for some uses, such as in cordless power tools, whereas in others the *cycle life* - the number of times that a device can be effectively recharged before its performance, e.g. its capacity has become too degraded, is critical. In addition, cost is always of concern, and sometimes is of overriding importance, even at the expense of performance.

Methods that allow the determination of the maximum theoretical values of some of these parameters, based upon the properties of the materials in the electrodes alone are described later. However, practical systems do not achieve these maximum theoretical values. One obvious reason is that a practical battery has a number of passive components that are not involved in the basic chemical reaction that acts as the energy storage mechanism. These include the electrolyte, a separator that mechanically prevents the electrodes from coming into contact, the current connectors that transport electrical current to and from the interior of the cell, and the container. In addition, effective utilization of active components in the chemical reaction is often less than optimal. Electrode reactant materials can become electronically disconnected, or shielded from the electrolyte, preventing them from participating in the electrochemical reaction, thereby rendering them passive. They add to weight and volume, but do not contribute to the transduction between electrical and chemical energy.

A rule of thumb that was used for a number of conventional aqueous electrolyte battery systems in the past was that a practical cell could only produce about 1/5–1/4 of the maximum theoretical specific energy. Optimization of a number of factors has made it possible now to exceed such values in a number of cases. In addition, the maximum theoretical values of some of the newer electrochemical systems are considerably higher than were available earlier.

Some rough values of the practical *energy density* (Wh/liter) and *specific energy* (Wh/kg) of several of the common rechargeable battery systems are listed in Table 1.1. These particular values are not definitive- they depend upon a number of operating factors and vary with the designs of different manufacturers; However, they indicate the wide range of these parameters available commercially from different technologies.

Another important parameter relating to practical use of batteries is the amount of power that they can supply, often expressed as specific power, the amount of power per unit weight. This is highly dependent on the details of design of the cell, as well as the characteristics of the reactive components. Hence, values vary over a wide range.

The characteristics of batteries are often graphically illustrated through the use of *Ragone plots*, in which the specific power is plotted against the specific energy. This type of presentation was named after D.V. Ragone, chairman of a governmental committee that wrote a report on the relative properties of different battery systems many years ago. Such a plot, showing approximate data on three current battery systems is shown in Fig. 1.4.

1.4.1 The Operating Voltage and the Concept of Energy Quality

In addition to the amount of energy stored, an important parameter of a battery system is the voltage at which it operates during discharge- when it supplies electrical energy and power- and when it is being recharged.

Table 1.1 Approximate values of the practical specific energy and energy density of some common battery systems

System	Specific energy Wh/kg	Energy density Wh/liter
Pb/PbO_2	40	90
Cd/Ni	60	130
Hydride/Ni	80	215
Li-Ion	135	320

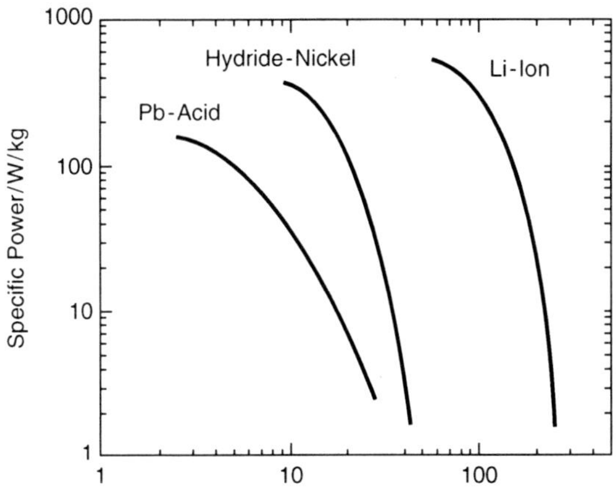

Fig. 1.4 Ragone plot showing approximate practical values of specific power and specific energy of three common battery systems

As discussed earlier in this chapter, the open circuit, or equilibrium cell voltage is primarily determined by the thermodynamics of chemical reaction between the components in the electrodes, as they supply the driving force for the transport of ions through the electrolyte, and electrons in the external circuit. During actual use, however, the operating voltage will vary from these theoretical values, depending on various kinetic factors. These will be discussed extensively later.

Another important parameter in the discussion of electrochemical energy storage, is its *quality*, and how it matches expected applications. The concept of *energy quality* is analogous to the concept of *heat quality*, well known in engineering thermodynamics.

It is well recognized that high temperature heat is more useful (e.g. has higher quality) than low temperature heat in many applications. Similarly, the usefulness of electrical energy is related to the voltage at which it is available. High voltage energy is often more useful (has higher quality) than low voltage energy. For example, in simple resistive applications the electrical power P is related to the practical (not just theoretical) voltage E and the resistance R by

$$P = E^2/R \tag{1.12}$$

The utility of an electrochemical cell in powering a light source or driving an electric motor is particularly voltage-sensitive. Because of the square relation, high voltage stored energy is superior to low voltage stored energy for such applications.

Rough energy quality rankings can be tentatively assigned to electrochemical cells on the basis of their output voltages:

3.5 – 5.5 V	High Quality Energy
1.5 – 3.5 V	Medium Quality Energy
0 – 1.5 V	Low Quality Energy

High voltage is required for a number of applications. One is the electrical system used to propel either hybrid- or all-electric vehicles. Auto manufacturers typically wish to operate such systems at over 200 V. For high voltage applications of this kind it is desirable that individual cells produce the highest possible voltage, as fewer cells are necessary. Also, 36–42 V systems are being developed for starter, lighting and ignition systems in internal combustion engine automobiles.

Despite the implications of energy quality, it is important that the voltage characteristics of electrochemical energy storage systems *match the requirements of the intended application.* It is not always necessary to have the highest possible cell voltage, as this could lead to wastage in some applications. An example is the use of batteries to power semiconductor circuits. The semiconductor industry is continually working to reduce the size of circuit components in order to get more of them per unit area of silicon wafer, i.e. to increase the packing density. The smaller the length of the gate in MOS devices, the lower the voltage required to operate the device, and the lower the Joule heat output. This latter factor is becoming particularly critical for applications in portable devices, like laptop computers.

Data on the time dependence of the typical operating voltage of low power semiconductor devices are shown in Fig. 1.5.

1.4.2 The Charge Capacity

The energy contained in an electrochemical system is the integral of the voltage multiplied by the *charge capacity*, i.e., the amount of charge available. That is,

$$Energy = \int E dq \tag{1.13}$$

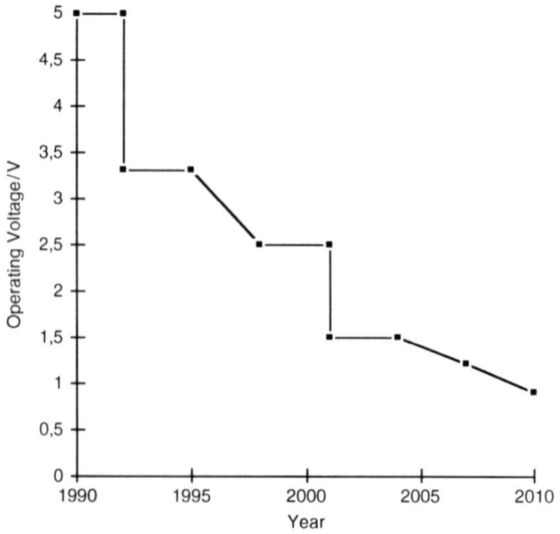

Fig. 1.5 Drop in the required voltage for semiconductor technology since 1990. The last several points indicate predictions for the future

where E is the output voltage, which can vary with the state of charge as well as kinetic parameters, and q the amount of electronic charge that can be supplied to the external circuit.

Thus it is important to know the maximum capacity, the amount of charge that can theoretically be stored in a battery. As in the case of voltage, the maximum amount of charge available under ideal conditions is a thermodynamic quantity, but of a different type. While voltage is an *intensive quantity*, independent of the amount of material present, charge capacity is an *extensive quantity*. The amount of charge that can be stored in an electrode depends on the amount of material in it. Capacity is always stated in terms of a measure such as the number of Coulombs per mol of material, per gram of electrode weight, or ml of electrode volume.

The *state of charge* is the current value of the fraction of the maximum capacity that is still available to be supplied.

1.4.3 The Maximum Theoretical Specific Energy (MTSE)

Consider a simple insertion or formation reaction that can be represented as

$$xA + R = A_xR \tag{1.14}$$

where x is the number of mols of A that reacts per mol of R. It is also the number of elementary charges per mol of R. If E is the average voltage of this reaction, the theoretical energy involved follows directly from Equation 1.13. If the energy is expressed in Joules, it is the product of the voltage in volts and the charge capacity, in Coulombs.

If W_t is the sum of the molecular weights of the reactants engaged in the reaction, the maximum theoretical specific energy (MTSE), the energy per unit weight, is simply

$$MTSE = (xE/W_t)F \tag{1.15}$$

where the $MTSE$ is in J/g, or kJ/kg, x is in equivalents per mol, E is in volts, and W_t is in g/mol. F is the Faraday constant, 96,500 Coulombs per equivalent.

Since one Watt is 1 Joule per second, one Wh is 3.6 kJ, and the value of the $MTSE$ can be expressed in Wh/kg as

$$MTSE = 26,805(xE/W_t) \tag{1.16}$$

1.4.4 Variation of the Voltage as Batteries are Discharged and Recharged

From the literature, it is seen that the voltage of most, but not all, electrochemical cells vary as their chemical energy is deleted as they are discharged. Likewise, it changes in the reverse direction when they are recharged. However, not only the voltage ranges, but also the characteristics of these state of charge - dependent changes vary widely between different electrochemical systems. It is important to understand what causes these variations.

One method of presenting this information is in terms of *discharge curves* and *charge curves*, in which the cell voltage is plotted as a function of the state of charge. These relationships can vary significantly, depending upon the rate at which the energy is extracted from, or added to the cell.

It is useful to consider the relation between the cell voltage and the state of charge under equilibrium or near-equilibrium conditions. In this case, a very important experimental technique, known as *Coulometric titration*, can provide a lot of information. This will be described in a later section.

Some examples of discharge curves under low current, or near-equilibrium conditions are shown in Fig. 1.6. They are presented here to show cell voltage as a function of the state of charge parameter. However, different battery systems have different capacities. Thus care has to be taken not to compare energies stored in different systems in this manner.

The reason for presenting near-equilibrium properties of these different cells in this way is to show that there are significant differences in the *types* of their behavior,

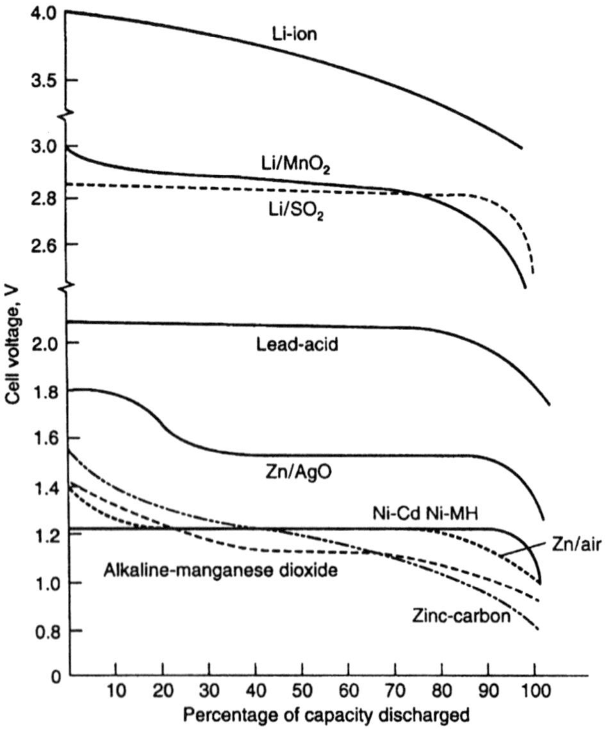

Fig. 1.6 Examples of battery discharge curves, showing variation of the voltage as a function of the fraction of their available capacity

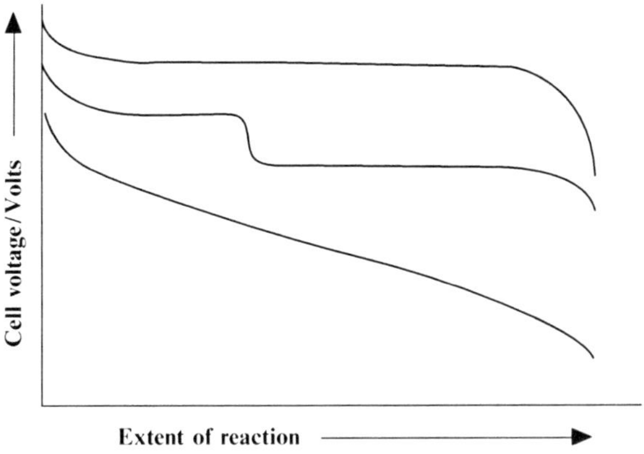

Fig. 1.7 Schematic representation of different types of discharge curves

as indicated by the shapes of their curves. It is clear that some of the discharge curves are essentially flat, others have more than one flat region, and still others have a slanted and stretched S-shape, at times with an appreciable slope. These variants can be simplified into three basic types of discharge curve shapes, as depicted in Fig. 1.7. The reasons behind their general characteristics will be discussed later.

1.4.5 Cycling Behavior

In many applications a battery is expected to maintain its major properties over many discharge - charge cycles. This can be a serious practical challenge, and is often given a lot of attention during the development and optimization of batteries. Figure 1.8 shows how the initial capacity is reduced during cycling, assuming three different values of the *Coulombic efficiency* - the fraction of the prior charge capacity that is available during the following discharge. This depends upon a number of factors, especially current and depth of discharge in each cycle.

It is seen that even minor inefficiency per cycle can have important consequences. For example, a half percent loss per cycle causes available capacity to drop to only 78% of the original value after 50 cycles. After 100 cycles, only 61% remains at that rate. The situation is worse if the Coulombic efficiency is lower.

Applications that involve many cycles of operation require cells to be designed and constructed such that the capacity loss per cycle is extremely low. This means that compromises must be made in other properties. *Supercapacitors* are expected to be used over a very large number of cycles. They typically have much lower values of specific energy than electrochemical cells which are used for applications in which the amount of energy stored is paramount.

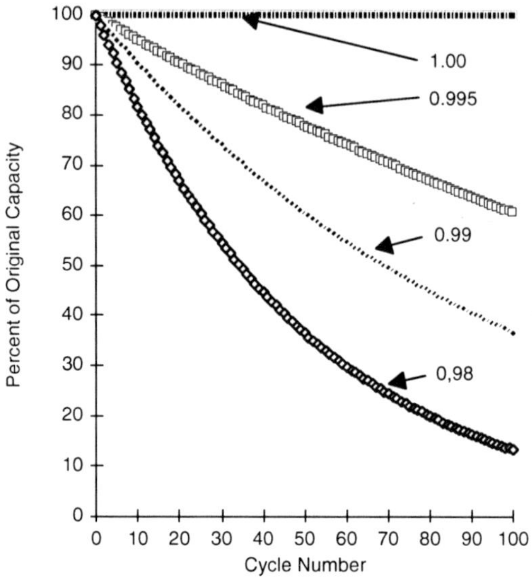

Fig. 1.8 Influence of Coulombic efficiency upon available capacity during cycling

1.4.6 Self-Discharge

Another property of importance in practical cells is *self-discharge*. This implies a decrease in available capacity with time, even without energy being taken from the cell by the passage of current through the external circuit. This is a serious practical problem in some systems, but negligible in others.

What needs to be understood at this juncture is that capacity is a property of the electrodes. Its value at any time is determined by the remaining extent of the chemical reaction between the neutral species in the electrodes. Any self-discharge mechanism that reduces the remaining capacity must involve either transport of neutral species, or concurrent transport of neutral combinations of charged species, through the cell. Since this latter process involves the transport of charged species, it is *electrochemical self-discharge*.

There are also several methods by which individual neutral species can move across a cell. These include transport through an adjacent vapor phase, cracks in a solid electrolyte, or as a dissolved gas in a liquid electrolyte. Since the transport of charged species is not involved, these processes produce *chemical self-discharge*.

It is also possible that impurities react with constituents in the electrodes or the electrolyte to reduce available capacity over time.

1.5 General Equivalent Circuit of an Electrochemical Cell

It is useful to devise electrical circuits whose electrical behavior is analogous to important phenomena in physical systems. By examination of the influence of changes in parameters in such *equivalent circuits*, they can be used as *'thinking tools'* to obtain insight into the significance of particular phenomena to the observable properties of complex physical systems. By use of this approach, the techniques of circuit analysis that have been developed for use in various branches of electrical engineering can be used in the analysis of interdependent physical phenomena.

This procedure has proved useful in some areas of electrochemistry, and will be detailed later in this text. At this point, however, it is utilized to study an ideal electrochemical cell; i.e. what happens if the electrolyte is not a perfect filter, but allows flow of some electronic current in addition to the expected ionic current. An electrochemical cell can be simply modeled as shown in Fig. 1.9, and its basic equivalent circuit is shown in Fig. 1.10.

The value of electrical equivalent of the theoretical chemical driving force is E_{th}, given by

$$E_{th} = -\Delta G_r{}^{\circ}/zF \tag{1.17}$$

as the result of the balance between the chemical and electrical forces acting upon ionic species in the electrolyte, as mentioned earlier. If there are no impedances or other loss mechanisms, the externally measurable cell voltage E_{out} is simply equal to E_{th}.

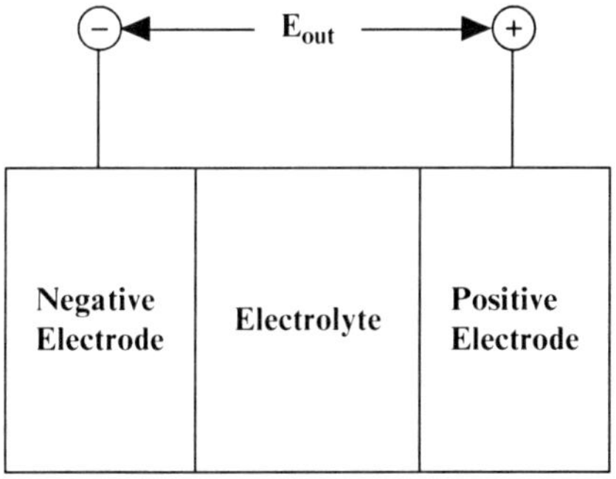

Fig. 1.9 Simplified physical model of electrochemical cell

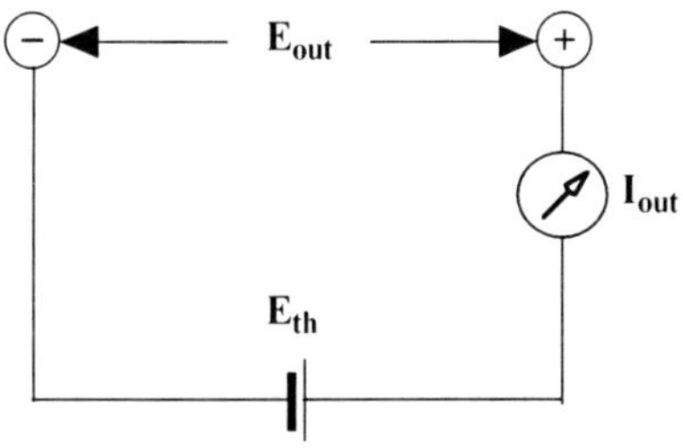

Fig. 1.10 Simple equivalent circuit model of an ideal electrochemical cell

1.5.1 *Influence of Impedances to the Transport of Ionic and Atomic Species within the Cell*

In practical electrochemical cells E_{out} is not always equal to E_{th}. There are several reasons for this disparity. Impedance always exist to the transport of electroactive ions and related atomic species across the cell e.g. resistance of electrolyte to ionic transport, or at one or both of the two electrolyte/electrode interfaces. Further, impedance to the progress of the cell reaction in some cases is related to the time-dependent solid state diffusion of the atomic species into, or out of, the electrode microstructure.

'*Impedances*' are used in this discussion instead of '*resistances*', since they can be time-dependent if time-dependent changes in structure or composition are occurring in the system. The *impedance* is the instantaneous ratio of the applied force (e.g. voltage) E_{appl} and the response (e.g. current) across any circuit element. As an example, if a voltage E_{appl} is imposed across a material that conducts electronic current I_e, the electronic impedance Z_e is given by

$$Z_e = E_{appl}/I_e \tag{1.18}$$

The inverse of the impedance is *admittance*, which is the ratio current/voltage. Under steady state (time-independent) DC conditions, the impedance and resistance of a circuit element are equivalent.

If current is flowing through the cell, there will be a voltage drop related to each impedance to the flow of ionic current within the cell. Thus, if the sum of these internal impedances is Z_i the output voltage can be written as

$$E_{out} = E_{th} - I_{out}Z_i \tag{1.19}$$

This relationship can be modeled by the simple circuit in Fig. 1.11.

1.5.2 *Influence of Electronic Leakage within the Electrolyte*

The output voltage E_{out} can also be different from the theoretical electrical equivalent of the thermodynamic driving force of the reaction between the neutral species

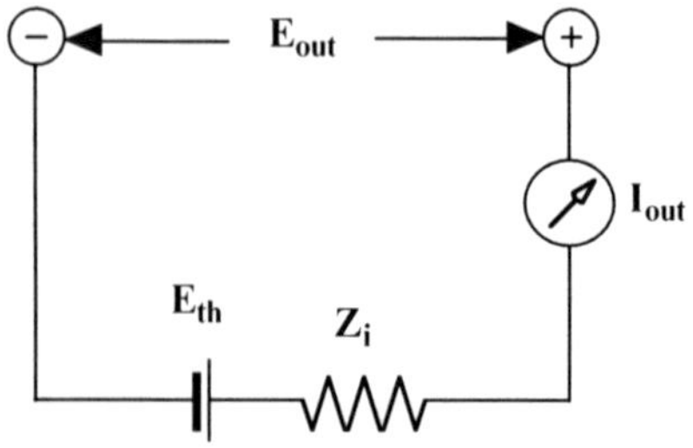

Fig. 1.11 Simple equivalent circuit for a battery or fuel cell indicating the effect of the internal ionic impedance Z_i upon the output voltage

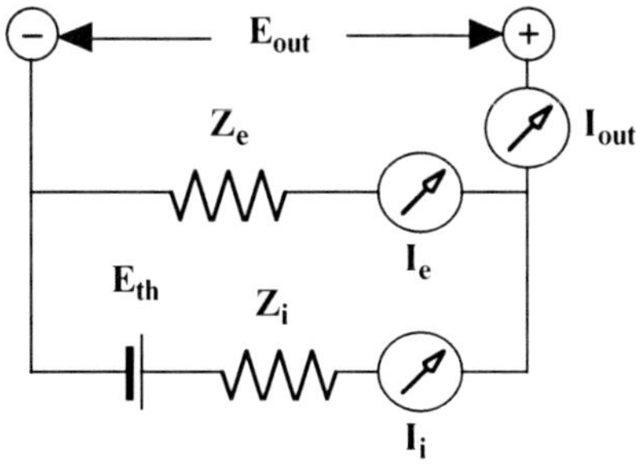

Fig. 1.12 Modified circuit including electronic leakage through the electrolyte

in the electrodes E_{th} even if there is no external current I_{out} flowing. This can be the result of electronic leakage through the electrolyte that acts to short-circuit the cell. This effect can be added to the previous equivalent circuit to give the circuit shown in Fig. 1.12.

It is evident that, even with no external current, there is an internal current related to the transport of the electronic species through the electrolyte I_e. Since the current must be the same everywhere in the lower loop, there must be a current through the electrolyte I_i with the same magnitude as the electronic current. There must be *charge flux balance* so that there is no net charge buildup at the electrodes.

The current through the internal ionic impedance Z_i generates a voltage drop, reducing the output voltage E_{out} by the product I_i Z_i, which is equal to I_eZ_e.

$$E_{out} = E_{th} - I_iZ_i \tag{1.20}$$

In addition, the fact that both ionic and electronic species flow through the cell means that this is a mechanism of *self-discharge*. This topic, discussed earlier in Sect. 4.7, results in a decrease of the available charge capacity of the cell.

1.5.3 Transference Numbers of Individual Species in an Electrochemical Cell

If more than one species can carry charge in an electrolyte, it is often of interest to know something about the relative conductivities or impedances of different species. The parameter used to describe the contributions of individual species to the transport of charge when an electrical potential difference (voltage) is applied across an electrolyte is the transference number. This is defined as the fraction of the total current that passes through the system that is carried by a particular species.

In the simple case that electrons and one type of ion can move through the electrochemical cell, we can define the transference number of ions as t_i, and electrons as t_e, where

$$t_i = I_i/(I_i + I_e) \tag{1.21}$$

and

$$t_e = I_e/(I_i + I_e) \tag{1.22}$$

and I_i and I_e are their respective partial currents upon the application of an external voltage E_{appl} across the system. The sum of the transference numbers of all mobile charge-carrying species is unity. In this case:

$$t_i + t_e = 1 \tag{1.23}$$

Instead of expressing transference numbers in terms of currents, they can also be referred to as impedances. For the case of these two species, the transport of charge by the motion of the ions under the influence of an applied voltage E_{appl},

$$t_i = (E_{appl}/Z_i)/[(E_{appl}/Z_i) + (E_{appl}/Z_e)] = Z_e/(Z_i + Z_e) \tag{1.24}$$

and likewise for electrons:

$$t_e = Z_i/(Z_i + Z_e) \tag{1.25}$$

Whereas these parameters are often thought of as properties of the electrolyte, in experiments they can also be influenced by what happens at the interfaces between the electrolyte and the electrodes, and are thus properties of the whole electrode-electrolyte system. They are solely properties of the electrolyte if there is no impedance to the transfer of either ions or electrons across the electrolyte/electrode interface or atomic and electronic species within the electrodes.

1.5.4 Relation between the Output Voltage and the Values of the Ionic and Electronic Transference Numbers

Through a simple assumption that the internal impedance is primarily due to the behavior of the ions, the general equivalent circuit of Fig. 1.12 can be rearranged to look as in Fig. 1.13.

When drawn this way, it can be readily seen that the series combination of Z_i and Z_e acts as a simple voltage divider.

If no current passes out of the system, i.e. under open circuit conditions, the output voltage is equal to the product of E_{th} and the ratio $Z_e/(Z_i + Z_e)$.

$$E_{out} = E_{th} Z_e/(Z_i + Z_e) \tag{1.26}$$

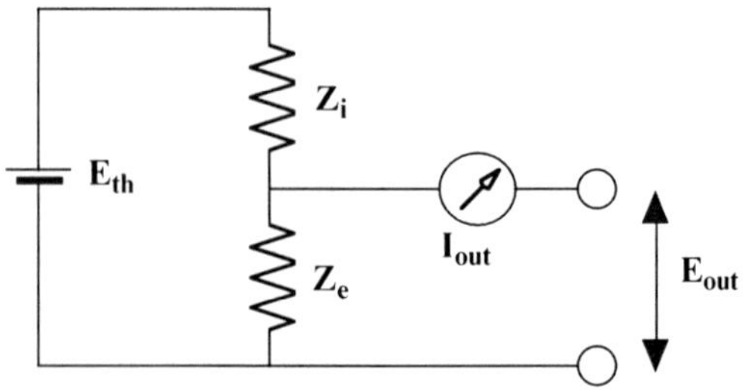

Fig. 1.13 Different representation of general equivalent circuit of Fig. 1.12

Introducing Equation 1.24, the output voltage can then be expressed as

$$E_{out} = E_{th}t_i \tag{1.27}$$

or

$$E_{out} = E_{th}(1 - t_e) \tag{1.28}$$

These are well-known and can be derived in other ways, as will be shown later. It is clear that the output voltage is optimized when t_i is as close to unity as possible.

1.5.5 Joule Heating to Due to Self-Discharge in Electrochemical Cells

Electrochemical self-discharge causes heat generation or *Joule heating*, due to the transport of charged species through the cell. The *thermal power* P_{th} caused by the passage of a current through a simple resistance R is given by

$$P_{th} = I^2R \tag{1.29}$$

However, as shown earlier, if self-discharge results from the leakage of electrons through the electrolyte there must be both electronic and ionic current with equal values. Thus the thermal power due to this type of self-discharge is:

$$P_{th} = I_i^2Z_i + I_e^2Z_e = I_e^2(Z_i + Z_e) \tag{1.30}$$

Measurements of the rate of heat generation by Joule heating under open circuit conditions can be used to evaluate the rate of self-discharge in practical cells.

1.5.6 What If Current is Drawn from the Cell?

If current is drawn from the cell into an external circuit, the normal mode of operation when chemical energy is converted into electrical energy, it flows through the ionic impedance, Z_i. This results in an additional voltage drop of I_{out} Z_i, further reducing the output voltage. If there were no electrochemical self-discharge, this can be written as

$$E_{out} = E_{th}t_i - I_{out}Z_i \tag{1.31}$$

The value of the ionic impedance of the system, Z_i, may increase with the value of the output current as the result of current-dependent impedances at the electrolyte/electrode interfaces. The difference between E_{th} and E_{out} is also known as *polarization* in electrochemical literature.

The result of the presence of current-dependent interfacial impedances to the passage of ionic species that increase Z_i is that the *effective transference number* of the ions t_i is reduced, since $t_i = Z_e/(Z_i + Z_e)$. This causes an additional reduction in the output voltage.

In addition to a *reduced output voltage*, there will also be additional *heat generation*. The total amount of Joule heating is the sum of that due to the passage of current into the external circuit I_{ext} and that due to electrochemical self-discharge.

$$P_{th} = I_{ext}^2 Z_i + I_e^2 (Z_i + Z_e) \tag{1.32}$$

In most cases, the first term is considerably larger than the second term.

Measured discharge curves vary with current density as conditions increasingly deviate from equilibrium. This is shown schematically in Fig. 1.14.

A parameter that is often used to indicate the rate at which a battery is discharged is the so-called *C-Rate*. The discharge rate of a battery is expressed as C/R, where R is the number of hours required to completely discharge its nominal capacity.

For example, if a cell has a nominal capacity of 5 Ah, discharge at the rate of C/10 would fully discharge it in 10 h. Thus the current is 0.5 A. And if the discharge rate is C/5 the discharge current is 1 A.

Although the *C-Rate* is often specified when either complete cells or individual electrodes are evaluated experimentally, and the current can be specified, this parameter is often not time-independent during real applications. If the electrical load is primarily resistive, for example, the current will decrease as the output voltage falls. This means that the *C-Rate* drops as the battery is discharged. Nevertheless, it is often important to consider the C-Rate when comparing the behavior of different materials, electrodes, and complete cells.

It is obvious that not only the average voltage, but also the charge delivered, can vary appreciably with changes in the *C-Rate*. In addition, the amount of energy that

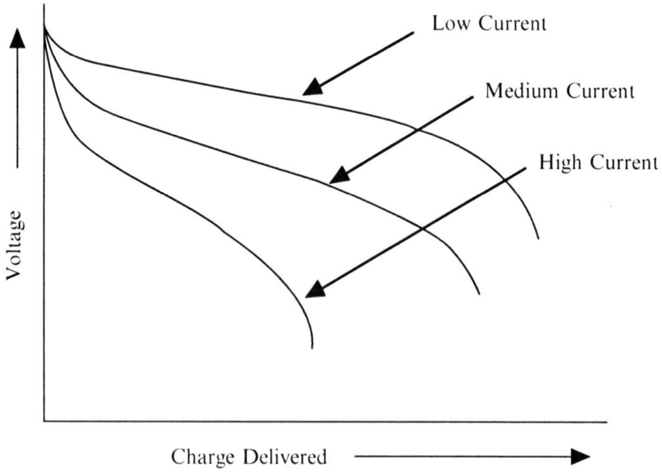

Fig. 1.14 Schematic drawing showing the influence of the current density upon the discharge curve

can be supplied, which will be seen in later chapters to be related to the area under the discharge curve, is strongly *C-Rate* dependent.

A further point that should be kept in mind is that not all of the stored energy may be useful. If the load is resistive, output power is proportional to the square of the voltage according to Equation (1.12) above; in such a case the energy that is available at lower voltages may not be beneficial.

This behavior can be understood in terms of the equivalent circuit of the battery. The internal ionic impedance Z_i - the sum of the impedances in the electrolyte and at the two electrode/electrolyte interfaces, is a function of local current density in the cell. This impedance typically varies with the state of charge. The mechanisms responsible for this behavior will be discussed later in the text.

References

1. I. Barin, Thermochemical Data of Pure Substances, (2 volumes), VCH Verlagsgesellschaft mbH, New York (1989)
2. I. Ansara and B. Sundman, in Computer Handling and Dissemination of Data, ed. by P.S. Glaeser, Elsevier Science/North-Holland, New York (1987)
3. I. Hurtado, ed. Thermodynamic Properties of Inorganic Materials by SGTE, Landolt-Boernstein Tables, Vol. 19, Group IV:Physical Chemistry, Springer, Berlin (1999)
4. W. Weppner and R.A. Huggins, J. Electrochem. Soc. 125, 7 (1978)
5. M.S. Whittingham, Science 192, 1126 (1976)

Chapter 2
Principles Determining the Voltages and Capacities of Electrochemical Cells

2.1 Introduction

In the earlier chapter we saw that the fundamental driving force across an electrochemical cell is the virtual chemical reaction that occurs if materials in the two electrodes are to react with each other. If the electrolyte is a perfect filter that allows the passage of ionic species, but not electrons, the cell voltage (when no current is passing through the system) is determined by the differences in the electrically neutral chemical compositions of the electrodes. The identity and properties of the electrolyte and the phenomena that occur at the electrode/electrolyte interfaces play no role. Likewise, it is the properties of the electrodes that determine the capacity of an electrochemical cell.

These general principles are extended further in this chapter. Emphasis is placed on the equilibrium, or near-equilibrium, state. This addresses the ideal properties of systems, which provide the upper limits for various vital parameters.

Real systems under load deviate from this behavior. As shown later, this is primarily due to kinetic factors that vary from one system to the next, and are highly dependent on the details of the materials present, the cell construction, and experimental conditions. As a result, it is difficult to obtain reproducible and quantitative experimental results, Factors that determine equilibrium, or near-equilibrium behavior are discussed here.

2.2 Thermodynamic Properties of Individual Species

Chapter 1 dealt with determining the overall driving force across a simple electrochemical cell by the change in Gibbs free energy, ΔG_r° and the virtual chemical reaction that occurs if materials in the electrodes react with one another. If there is no current flowing, this chemical driving force is balanced by an electrical driving force in the opposite direction.

R.A. Huggins, *Advanced Batteries*: Materials Science Aspects,

Individual species within the electrolyte in the cell are now considered. In open circuit conditions (and no electronic leakage) there is no net current flow. Thus there must be a *force balance* acting on all mobile species.

The thermodynamic properties of a material can be related to those of its constituents through the concept of the *chemical potential* of an individual species. The chemical potential of species i in a phase j is defined as

$$\mu_i = \partial G_j / \partial n_i \tag{2.1}$$

where G_j is the molar Gibbs free energy of phase j, and n_i the mole fraction of the i species in phase j. In integral form this is

$$\Delta\mu_i = \Delta G_j \tag{2.2}$$

Since the free energy of the phase changes with the amount of species i, the chemical potential has the same dimension as the free energy. Thus, gradients in the chemical potential of species i produce chemical forces causing i to tend to move in the direction of lower μ_i. It was shown in Chap. 1 that when there is no net flux in the electrolyte, this chemical force must be balanced by an electrostatic force, due to the voltage between the electrodes. Energy balance in the electrolyte, and therefore in the cell, can be written in terms of the single species i.

$$\Delta\mu_i = -z_i F E \tag{2.3}$$

where z_i is the number of elementary charges carried by particles (ions) of species i.

The chemical potential of a given species is related to another thermodynamic quantity, its *activity*, a_i. The defining relation is

$$\mu_i = \mu_i^\circ + RT \ln a_i \tag{2.4}$$

where μ_i° is a constant, the value of the chemical potential of species i in its standard state. R is the gas constant (8.315 J/mol deg), and T the absolute temperature.

The activity of a species is its *effective concentration*. If the activity of species i, a_i, is equal to unity, it behaves chemically like pure i. If a_i is 0.5, it implies that it is chemically composed of half species i, and is half chemically inert. In the case of a property such as vapor pressure, a material i with an activity of 0.5 has half the vapor pressure of pure i.

Consider an electrochemical cell in which the activity of species i is different in the two electrodes, $a_i(-)$ in the negative electrode, and $a_i(+)$ in the positive electrode. The difference between the chemical potential on the positive side and that on the negative side can be written as

$$\mu_i(+) - \mu_i(-) = RT[\ln a_i(+) - \ln a_i(-)] = RT \ln[a_i(+)/a_i(-)] \tag{2.5}$$

If this chemical potential difference is balanced by electrostatic energy from (2.2)

$$E = -(RT/z_i F) \ln[a_i(+)/a_i(-)] \tag{2.6}$$

This relation is called the *Nernst equation*, and relates the measurable cell voltage to the chemical difference across an electrochemical cell, i.e., it transduces between the chemical and electrical driving forces. If the activity of species *i* in one of the electrodes is a standard reference value, the *Nernst equation* provides the relative electrical potential of the other electrode.

2.3 A Simple Example: The Lithium/Iodine Cell

The thermodynamic basis for the voltage of a lithium/iodine (Li/I_2) cell is first considered. Primary (non-rechargeable) cells based on this chemical system were invented by Schneider and Moser in 1972 [1, 2], and are currently widely used to supply energy in cardiac pacemakers.

A typical configuration of this electrochemical cell employs metallic lithium as the negative electrode and a composite of iodine with about 10 wt% of poly-2-vinylpyridine (P2VP) as positive. The composite of iodine and P2VP is a charge transfer complex, with P2VP acting as an electron donor, and iodine as an acceptor. This results in the combination of high electronic conductivity; chemical properties are essentially similar to pure iodine. Reaction between Li and the (iodine, P2VP) composite produces a layer of solid LiI. This material acts as a solid electrolyte in which Li^+ ions move from the interface with a negative electrode to that with a positive electrode, where they react with iodine to form more LiI. The transport mechanism involves a flux of lithium ion vacancies in the opposite direction. Although LiI has relatively low ionic conductivity, it has negligible electronic transport, thereby meeting the requirements of an electrolyte.

This system is represented simply as

$$(-)Li/electrolyte/I_2(+) \tag{2.7}$$

The *virtual reaction* that determines the voltage is

$$Li + 1/2I_2 = LiI \tag{2.8}$$

More LiI forms between the lithium electrode and the iodine electrode as the reaction progresses. The evolution of the microstructure during discharge is shown schematically in Fig. 2.1.

The voltage across this cell under open circuit conditions is readily calculated from the balance between the chemical and electrical driving forces, as shown in Chap. 1.

$$E = -\Delta G_r / z_i F \tag{2.9}$$

where

$$\Delta G_r = \Delta G_f(LiI) \tag{2.10}$$

and z_i is +1, for the electroactive species are the Li^+ ions.

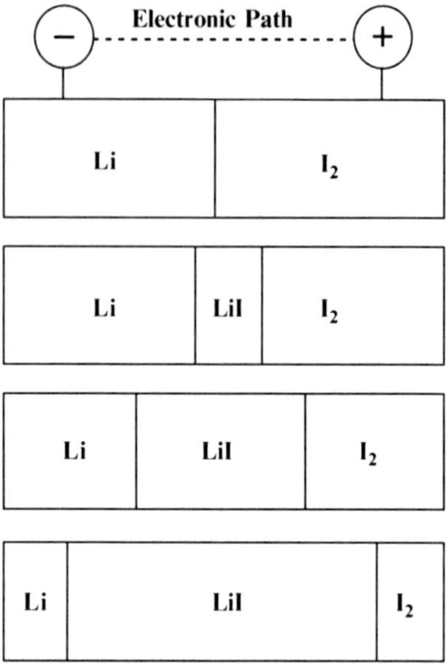

Fig. 2.1 Schematic representation of the microstructure of a Li/I_2 cell at several stages of discharge

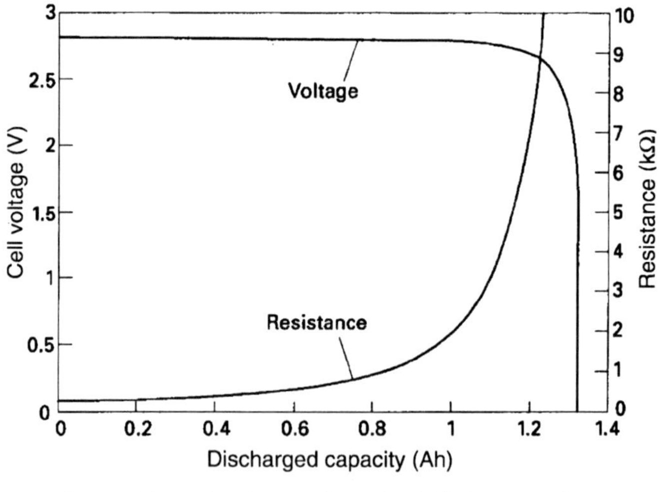

Fig. 2.2 Output voltage and internal resistance of a typical Li/I_2 battery of the type used in cardiac pacemakers

According to data in Barin [3], the Gibbs free energy of formation of LiI is −269.67 kJ/mol at 25°C. Since the value of Faraday constant is 96,500 coulombs/equiv. (mole of electronic charge), the open circuit voltage is calculated as 2.795 V at 25°C.

Data on properties of commercial Li/I_2 cells are shown in Fig. 2.2 [4]. It is seen that during its life, the voltage corresponds closely to the calculation above. It is also seen in this figure that the resistance across the cell increases with the extent of reaction, due to the increasing thickness of the solid electrolyte product that grows as the cell is discharged. Such cells are known to be *positive-electrode-limited*, i.e., positive electrode capacity is less than negative electrode capacity; this is the part of the cell that determines overall capacity.

2.3.1 Calculation of the Maximum Theoretical Specific Energy

The value of the maximum theoretical specific energy of a Li/I_2 cell is calculated from this information and the weights of the reactants. It was shown in Chap. 1 that the MTSE in W h/kg, is given by

$$MTSE = 26{,}805(xE/W_t) \tag{2.11}$$

Reactant weight W_t is the weight of a mole of Li (6.94 g) and half a mole of I_2 (126.9 g), or 133.84 g. The value of x is 1, and E is calculated as 2.795 V. Thus the value of the MTSE is 559.77 W h/kg.

This is about 15 times the value of common Pb-acid cells which are widely used as SLI batteries in automobiles, and for other purposes. Want of rechargeability, high cost of ingredients and low discharge rate unfortunately limit the range of application of Li/I_2 cells.

2.3.2 The Temperature Dependence of the Cell Voltage

As seen, the quantity that determines voltage is the Gibbs free energy change associated with the virtual cell reaction between chemical species in the electrodes. This is, however, temperature dependent and arrived at by dividing the Gibbs free energy into its enthalpy and entropy components.

$$\Delta G_r = \Delta H_r - T\Delta S_r \tag{2.12}$$

so that

$$d(\Delta G_r)/dT = -\Delta S_r \tag{2.13}$$

and

$$dE/dT = \Delta S_r/z_i F \tag{2.14}$$

The value of ΔS for the formation of LiI is given by

$$\Delta S_r(LiI) = S(LiI) - S(Li) - 1/2S(I_2) \tag{2.15}$$

Entropy data for these materials, as well as a number of others, are given in Table 2.1 These entropy values are in J/mol deg, whereas Gibbs free energy

Table 2.1 Entropy data for some species at 25 and 225°C [3]

Species	S (25°C) J/K mol
Li	29.08
Zn	41.63
H_2	130.68
O_2	205.15
Cl_2	304.32
I_2	116.14
LiF	35.66
LiCl	59.30
LiBr	74.06
LiI	85.77
H_2O (liquid)	69.95
ZnO	43.64
Species	S (225°C) J/K mol
H_2	145.74
O_2	220.69
H_2O (gas)	206.66

values are in kJ/mol. From this data, the value of ΔS_r for the formation of LiI is -1.38 J/K mol. Thus, from (2.13), the cell voltage varies only slightly, i.e. -1.43×10^{-5} V/K. As can be seen later, the temperature dependence of the voltage in many other electrochemical reactions, or other batteries, is often much greater. An example is the Zn/O_2 battery that is commonly used in hearing aids, where it is -5.2×10^{-4} V/K.

Data in Table 2.1 shows that entropy values of simple solids are considerably lower than those of liquids, and lower still than gases. This is reflected in the temperature dependence of electrochemical cells.

An example is the H_2/O_2 fuel cell where the voltage varies -1.7×10^{-3} V/K at room temperature when water, the product of the reaction, is a liquid. But at 225°C, where the product of the cell reaction is gas, steam, the variation is only -0.5×10^{-3} V/K. The resultant variation of the cell voltage from room temperature to the operating temperature of high temperature oxide-electrolyte fuel cells is shown in Fig. 2.3. Operation at high temperature results in significantly lower voltage. The theoretical open circuit voltage is 1.23 V at 25°C, but only 0.91 V at 1,025°C.

2.4 The Shape of Discharge Curves and the Gibbs Phase Rule

It was shown in Chap. 1 that the voltage of batteries often varies with the state of charge. It was pointed out that their discharge curves typically have one of three general shapes. Some are relatively flat, others have more than one relatively flat portion, while a third has a slanted or stretched-S shape, at times with a relatively large slope. The data in Fig. 2.2 show that the Li/I_2 cell falls into the first category.

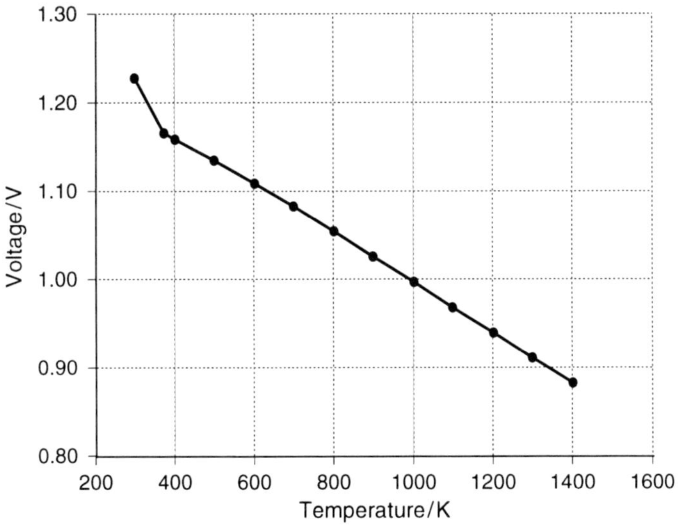

Fig. 2.3 Theoretic open circuit voltage of a H_2/O_2 fuel cell as a function of absolute temperature

To understand how the voltage across an electrochemical cell varies with the state of charge, and why it is essentially flat in the case of the Li/I_2 cell, it is useful to consider the application of the *Gibbs Phase Rule*.

The *Gibbs Phase Rule* is often written as

$$F = C - P + 2 \tag{2.16}$$

in which C is the *number of components* (e.g. elements) present, and P the *number of phases present* in this materials system in a given experiment. The quantity F is the *number of degrees of freedom*; i.e., the number of *intensive thermodynamic parameters* that must be specified to *define the system* and *all its associated properties*, one of which is the electric potential.

The application of the Phase Rule to this situation is understood by determining the thermodynamic parameters to be considered. They must be intensive variables, i.e., their values are independent of the amount of material present. For this purpose, the most useful thermodynamic parameters are temperature, overall pressure, and the chemical potential or chemical composition of each of the phases present.

This can now be applied to the Li/I_2 cell. Starting with the negative electrode there is only one phase present, Li, so P is 1. It is a single element, with only one type of atom. The number of components C is thus also equal to 1 and F is equal to 2.

This means that if values of two intensive thermodynamic parameters, like temperature and overall pressure, are specified, no degrees of freedom are left. Thus the *residual value of F* is zero, which suggests that all the intensive properties of the negative electrode system are fully defined, i.e., have fixed values.

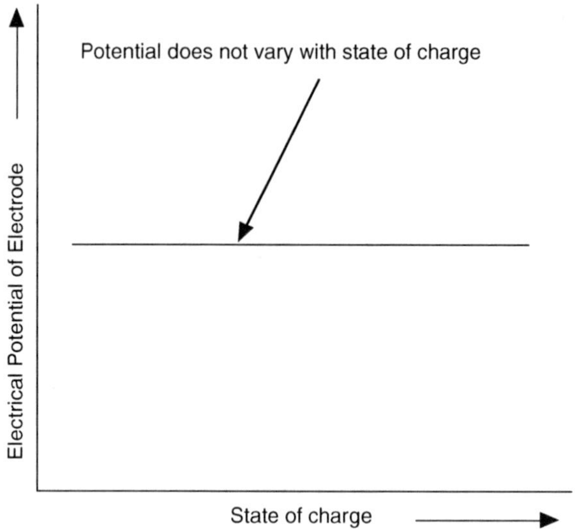

Fig. 2.4 Potential of a pure lithium electrode does not vary with the state of charge of the LiI cell

In the case of the lithium negative electrode, chemical as well as electrical potentials of all species (i.e. pure lithium), have fixed values, *regardless of the amount* of lithium present. The amount of lithium in the negative electrode decreases as the cell gets discharged and the product LiI is formed; i.e., the amount of lithium varies with the state of charge of the LiI battery. But since $F = 2$, and the residual value of F, if the temperature and total pressure are held constant, is zero, none of the intensive properties change. This means that the electrical potential of the lithium electrode is independent of the state of charge of the cell. This is shown schematically in Fig. 2.4.

If, however, some iodine *could* dissolve in the lithium, forming a solid solution, *which it does not*, the number of components in the negative electrode would be two. In a solid solution there is only one phase present. Therefore $C = 2$, $P = 1$ and $F = 3$.

In this hypothetical case, the system is not fully defined after fixing temperature and overall pressure. There is a residual value of F, i.e., one. Thus the electrical potential of the lithium–iodine alloy is not fixed, but varies, depending on other parameters, such as the amount of iodine in the Li–I solid solution. This is shown schematically in Fig. 2.5.

It is common in certain electrochemical cells, for the electrical potential of electrodes to vary with the composition, and thereby with the state of charge though this is not the case with the Li/I_2 cell. A number of examples will be discussed in subsequent chapters.

The positive electrode has only one active component (element), iodine, which is an electrochemically active phase. Thus both C and P have values of 1. Degrees of freedom is thus again 2. Therefore, the values of all intensive variables

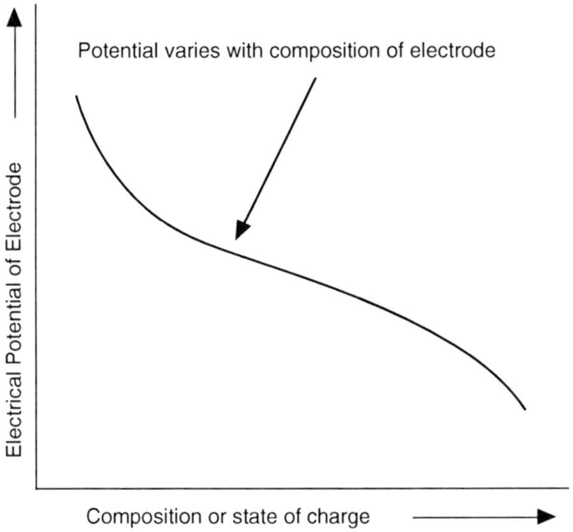

Fig. 2.5 Schematic representation of the variation of the electrical potential of an electrode as a function of its composition for the case in which the residual value of F is not zero

and associated properties, like electrical potential of the iodine electrode can be determined if values of the two independent thermodynamic parameters, temperature and total pressure are fixed.

This means that the potential of the I_2 electrode does not vary with its state of charge. As both negative and positive electrode potentials are independent of the state of charge, voltage across the cell must also be independent of the state of charge of the Li/I_2 battery. This is illustrated in Fig. 2.2.

The earlier discussion shows that the chemical potential of an element depends upon its activity, and in the case of the iodine electrode

$$\mu(I_2) = \mu^\circ(I_2) + RT \ln a(I_2) \tag{2.17}$$

where $\mu^\circ(I_2)$ is the chemical potential of iodine in its standard state, i.e. pure iodine at a pressure of one atmosphere at the temperature in question. When the activity is unity, i.e., for pure I_2,

$$\mu(I_2) = \mu^\circ(I_2) \tag{2.18}$$

Now consider the voltage of the Li/I_2 cell. This is determined by the Gibbs free energy of formation of the LiI phase, as given in (2.8) and (2.9). It is also related to the difference in the chemical potential of iodine at the two electrode/electrolyte interfaces according to the relation

$$E = -\Delta\mu(I_2)/z_i F \tag{2.19}$$

where the value of z_i is -2. Therefore the activity of iodine at the positive side of the electrolyte is unity; it is very modest at the interface on the negative electrode side. Likewise, cell voltage is related to the difference in the chemical potential of lithium at the two electrode/electrolyte interfaces:

$$E = -\Delta\mu(Li)/z_i F \qquad (2.20)$$

where the value of z_i is +1. In this case activity of lithium is unity at the negative interface, and minimal at the positive interface, where the electrolyte is in contact with I_2.

While this discussion has focused on the potential of a single electrode, the shape of the equilibrium discharge curve (voltage versus state of charge) of an electrochemical cell is the result of the change of the potentials of both electrodes as the overall reaction takes place. If the potential of one of the electrodes does not vary, the variation of the cell voltage is obviously the result of the change of the potential of the other electrode as its overall composition changes.

A number of materials are used as electrodes in electrochemical cells in which more than one reaction occurs in sequence as the overall discharge process takes place; some of these reactions are similar, while others are not.

For example, a *series of multi-phase reactions* in which the number of residual degrees of freedom is zero results in a discharge curve with a set of constant voltage plateaus. This is illustrated schematically in Fig. 2.6.

It is also possible for an electrode to undergo sequential reactions that are not similar, for example, the reaction of lithium with a spinel phase in the Li–Ti–O system. Experimental data are shown in Fig. 2.7 [5].

In this case approximately 1 Li/mol is inserted in the host spinel phase as a *solid solution reaction*. The potential varies continuously as a function of composition.

The introduction of additional lithium causes nucleation, and subsequent growth, of a second phase with rocksalt structure and a composition of approximately 2

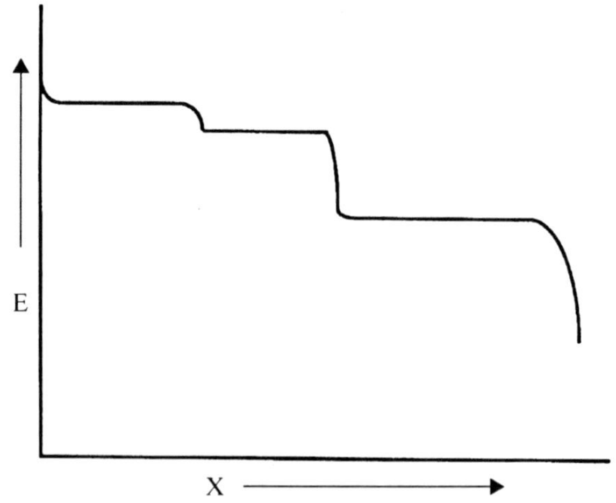

Fig. 2.6 Schematic equilibrium discharge curve of an electrode that undergoes a series of multi-phase reactions in which the residual value of F is zero

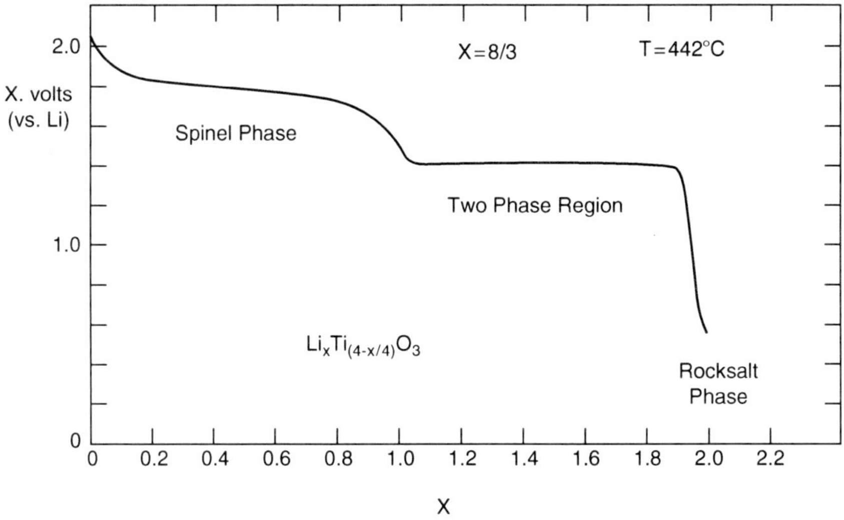

Fig. 2.7 Equilibrium discharge curve of a material in the Li–Ti–O system that initially had a composition with a spinel type of crystal structure

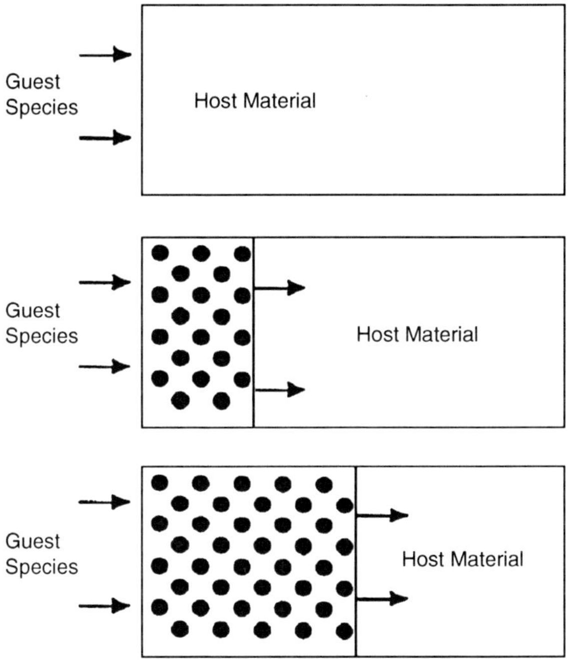

Fig. 2.8 A schematic representation of a one-dimensional moving interface reaction

Li/mol of the original host. *Reconstitution reaction* takes place when more than one lithium is added. Reconstitution reaction involves two regions within the material with different Li contents. As the reaction proceeds, the compositions of the two

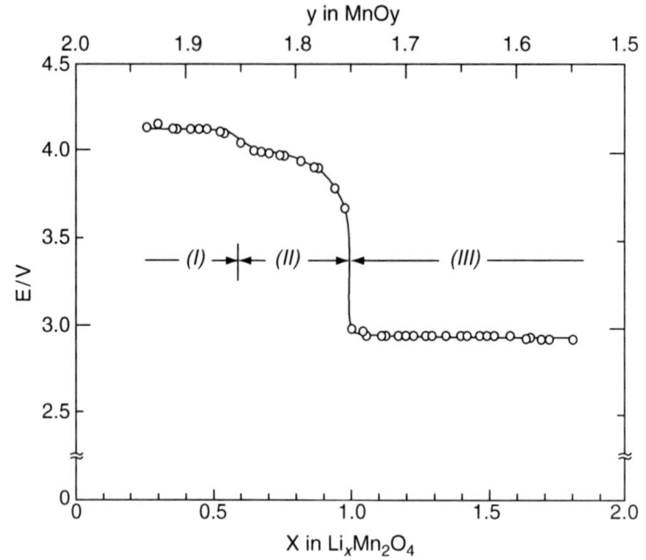

Fig. 2.9 Equilibrium discharge curve for $Li_xMn_2O_4$

phases do not change, but the relative amount of the phase with the higher Li content increases, and that of the initial solid solution phase is reduced. This occurs through the *movement of the interface* between them. This *moving interface reconstitution reaction* is schematically represented in Fig. 2.8.

An example of a series of reactions occurring when the overall composition is changed, is the Li–Mn–O system, where there is a series of three different reactions. This is seen from the shape of the equilibrium discharge curve in Fig. 2.9 [6]. There is a two-phase plateau, a single-phase solid solution region, followed by another two-phase plateau.

This interpretation is reinforced by the results of X-ray diffraction experiments shown in Fig. 2.10 [6]. It is seen that the lattice parameters remain constant within two-phase regions, and vary with the composition within the single-phase solid solution region.

2.5 The Coulometric Titration Technique

The simple examples discussed so far in this chapter assume that the requisite thermodynamic data are already known. One can calculate the open circuit voltage of an electrochemical cell from the value of Gibbs free energy of the appropriate virtual reaction, ideal capacity is determined from the *reaction's stoichiometry*.

It is also possible to do the opposite, using electrochemical measurements to obtain thermodynamic information. A useful tool for this purpose is the *Coulometric*

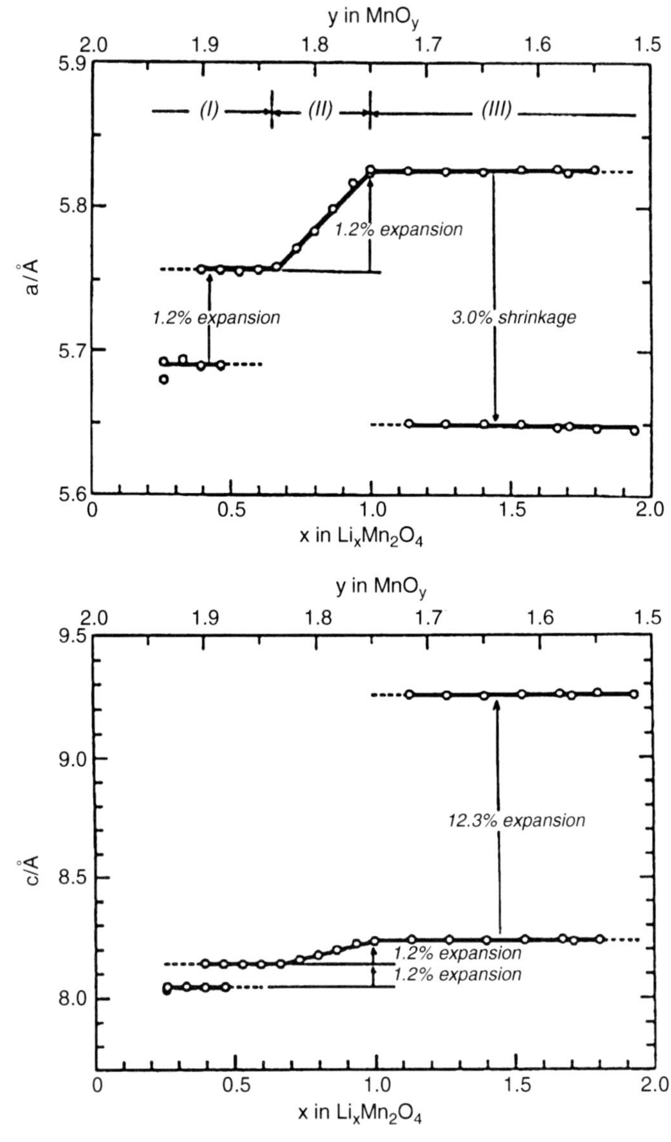

Fig. 2.10 Changes in unit cell dimensions as a function of composition in $Li_xMn_2O_4$

titration technique, which was first introduced by Wagner [7] to study the phase $Ag_{2+x}S$, which exists over a relatively narrow range of composition x. Its composition, or stoichiometry (the relative amounts of silver and sulfur) depends on the value of activity of silver within it. A simple electrochemical cell is used to change the stoichiometry and evaluate the activity of one of the species, e.g. silver in this case.

This method was further developed and applied by Weppner and Huggins [8] to the investigation of poly-phase alloy systems. It was demonstrated that the phase diagram, as well as the thermodynamic properties of the individual phases within it, can be determined by using this technique.

Consider the use of the following simple electrochemical cell to investigate properties of the *vario-stoichiometric* (the stoichiometry can have a range of values) *phase* A_yB. This can be represented schematically by

$$A/\text{Electrolyte that transports } A^+ \text{ions}/A_yB$$

Here, the element A acts as both a source and sink for electroactive species A and thermodynamic reference for component A. For simplicity, it is assumed this electrode is pure A, and therefore has activity of unity. It is also assumed that both A and A_yB are good electronic conductors, the ionic transference number in the electrolyte is unity, and the system is under isobaric and isothermal conditions.

Under these conditions, open circuit voltage E is a direct measure of chemical potential and activity of A in the phase A_yB according to (2.5) above. As the electrode of pure A has activity of unity, this relation is written as

$$E = -\Delta\mu_A/(z_{A+}F) = -(RT/z_{A+}F)\ln a(A) \tag{2.21}$$

where z_{A+} is the charge number of A^+ ions in the electrolyte, which is 1.

If a positive current is passed through the cell using an electronic source, A^+ ions are transported through the electrolyte from the left electrode to the right electrode. An equal current of electrons passes through the outer circuit because of the *requirement for charge flux balance*. The result is that the value of y in the A_yB phase is increased.

When a steady value of current I is applied for a fixed time t, the amount of charge Q that is passed across the cell is

$$Q = It \tag{2.22}$$

Number of moles of species A which is transported during this current pulse is

$$\Delta m(A) = Q/z_{A+}F \tag{2.23}$$

so that change in the value of y, the mole fraction of species A, is

$$\Delta y = \Delta m(A)/m(B) = Q/(z_{A+}Fm(B)) \tag{2.24}$$

where m(B) is the number of moles of B present in the electrode.

This method can be used to make *very minute* changes in the composition of electrode material. Inserting numbers in this equation one can see how sensitive this procedure is.

Suppose that the electrode has a weight of 5 g, and component B has a molecular weight of 100 g/mol. The value of m(B) is thus 0.05 mol. Now suppose that a current of 0.1 mA is run through the cell for 10 s. The value of Q is then 0.001 coulombs. With $z_{A+} = 1$ equiv./mol and $F = 96{,}500$ coulombs/equiv., then Δy is only 2×10^{-7}.

This is very modest. Thus it is possible to investigate the compositional dependence of the properties of phases with very narrow compositional ranges. It is very difficult to get such a high degree of compositional resolution through other techniques.

A sufficiently long time to permit the composition to become homogeneous throughout the electrode material, as evidenced by reaching a steady-state value of open circuit voltage, allows information to be obtained about the equilibrium chemical potential and activity of mobile electroactive species as a function of composition. This technique has been used to investigate a wide variety of materials of potential interest in battery systems; examples will be discussed in later chapters.

The success of this method depends on a number of assumptions (1) The electrolyte is essentially only an ionic conductor, i.e., the ionic transference number is very close to unity. (2) There is no appreciable loss of either component from the electrode material A_yB by evaporation, dissolution, or interaction with the electrical lead materials, the so-called *current collectors*. (3) The rate of *compositional equilibration via chemical diffusion* in the electrode material must be sufficiently fast. This means it may be necessary to use thin samples as electrodes to reduce the time necessary for concentration homogenization.

It should also be recognized that this Coulometric titration technique gives information about the influence of compositional changes, but not absolute composition. That will have to be determined by some other method.

References

1. Moser, J.R., US Patent 3,660,163 (1972)
2. Schneider, A.A. and Moser, J.R., US Patent 3,674,562 (1972)
3. Barin, I., *Thermochemical Data of Pure Substances* (2 volumes), VCH Verlag (1989)
4. Courtesy of Catalyst Research Corp.
5. Liebert, B.E., Weppner, W., and Huggins, R.A., in *Proceedings of the Symposium on Electrode Materials and Processes for Energy Conversion and Storage*, J.D.E. McIntyre, S. Srinivasan and F.G. Will (eds.), Electrochemical Society, Princeton, p. 821 (1977).
6. Ohzuku, T., Kitagawa, M., and Hirai, T., J. Electrochem. Soc. 137, 769 (1990)
7. Wagner, C., J. Chem. Phys. 21, 1819 (1953)
8. Weppner, W. and Huggins, R.A., J. Electrochem. Soc. 125, 7 (1978)

Chapter 3
Binary Electrodes Under Equilibrium or Near-Equilibrium Conditions

3.1 Introduction

The theoretical basis for understanding and predicting the composition-dependence of the potentials, as well as the capacities, of both binary (two element) and ternary (three element) alloys has now been established. The relevant principles of binary systems are discussed in this chapter. Ternary systems are treated in the next chapter.

Under equilibrium and near-equilibrium conditions, these important practical parameters are directly related to the thermodynamic properties and compositional ranges of the pertinent phases in the respective phase diagrams. Their behavior can be understood under dynamic conditions by simple deviations from such equilibrium conditions. In other cases, however, *metastable* phases may be present in the microstructure of an electrode whose properties are considerably different from those of *absolutely-stable* phases. The influence of *metastable* microstructures is discussed in a later chapter. In addition, it is possible that the compositional changes occurring in an electrode during the operation of an electrochemical cell can cause *amorphization* of its structure. This is also discussed later.

3.2 Binary Phase Diagrams

Phase diagrams are figures that graphically represent the equilibrium state of a chemical system. They are useful *thinking tools* to help understand the fundamental electrochemical properties of electrodes. There are various types of phase diagrams, the most common being two-dimensional plots that indicate temperature and compositional conditions for the stability of various phases and their compositions under equilibrium conditions.

A *binary phase diagram* is a two-dimensional plot of temperature vis-à-vis the overall composition of materials (*alloys*) composed of two different components (elements). It shows temperature-composition conditions for the stability and

R.A. Huggins, *Advanced Batteries*: Materials Science Aspects,

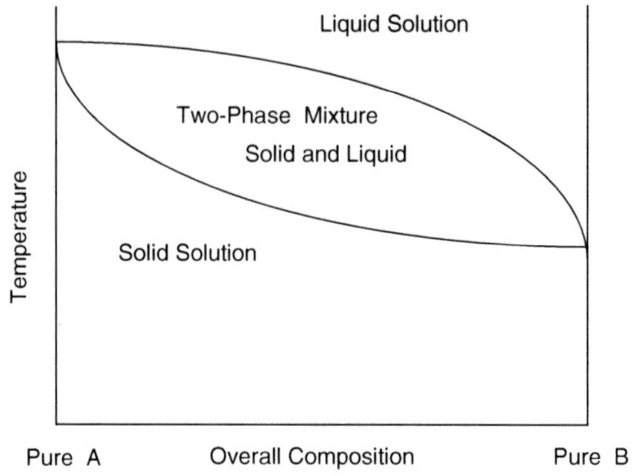

Fig. 3.1 Schematic phase diagram of binary system with complete miscibility in both the liquid and solid phases

composition ranges of the various phases that can form in a given system, and is commonly used in materials science. It is seen that there are regions in which only a single phase is stable, and others in which two phases are stable.Though not of particular importance to electrochemical experiments, there are certain special conditions under which it is possible for three phases to be present in a binary system. According to Gibbs Phase Rule, under standard pressure this can only take place at a specific temperature. Any change from that unique temperature will result in one-phase and two-phase regions. Thus at a particular temperature the three phases in equilibrium must contact a horizontal (constant temperature) line in the phase diagram. Two will touch it at the ends, and the third will only do so at a single composition between them. The significance of this statement can be seen later.

A very simple binary system is shown in Fig. 3.1. In this case it is assumed that the two components (elements) A and B are completely miscible, i.e., they can dissolve in each other over the complete range of composition, from pure element A to pure element B, both in the liquid and solid state, i.e., a liquid solution and a solid solution in different regions of temperature-composition space. Since the elements have different melting points, the temperature above which the liquid solution is stable (called the *liquidus*) will vary with composition across the diagram. The temperature below which the material is completely solid (called the *solidus*) is also composition-dependent. In any composition there is a range of temperature between the solidus and liquidus, within which two phases are present, a liquid and a solid solution.

Likewise, at any temperature between the melting points of pure A and pure B, the liquid and solid phases that are in equilibrium with each other have different compositions corresponding to two compositional limits of the two-phase region in the middle part of this type of phase diagram. This is shown schematically in Fig. 3.2. The compositions of solid and liquid phases that are in equilibrium

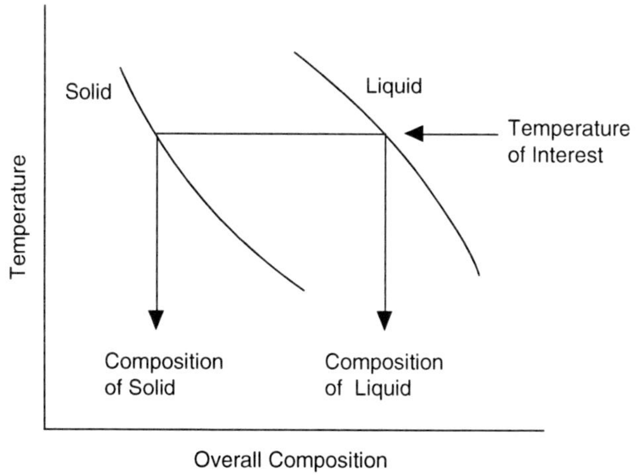

Fig. 3.2 Compositions of liquid and solid phases in equilibrium with each other at a particular temperature between the melting points of the two elements

with each other at a particular temperature of interest can be directly read off the composition scale at the bottom of the diagram. It is also obvious that these two compositions will be different for other temperatures, due to the slopes of the solidus and liquidus curves. At a fixed temperature, the relative amounts, of the two phases, though not the compositions, depend on the overall composition.

The overall composition is not limited to the range within the two-phase mixture region of the phase diagram. Consider an isothermal experiment, and suppose the overall composition were to start far to the left at the temperature of interest in the case illustrated in Fig. 3.2, perhaps as far as pure A. Continuous addition of further atoms of element B will form a solid solution of B atoms in A, whose composition will gradually rise. This continues over a relatively wide compositional range until the overall composition reaches the solidus line that signifies the one-phase/two-phase border.

Further addition of B atoms causes the overall composition to continue changing. However, the composition of the solid solution cannot become indefinitely B-rich. Instead, when the overall composition arrives at the *solidus* line, some liquid begins to form. This composition is different from that of the solid solution, as it is determined by the composition limit of the liquid solution at that temperature, the *liquidus* line. So the microstructure contains two phases with different local compositions. Further changes in the overall composition at this temperature result in the formation of more of the fixed-composition liquid solution at the expense of the fixed-composition solid solution. When the overall composition reaches that of the liquid solution (the liquidus line) there is nothing of the solid phase left. Further addition of B atoms then causes the composition of the liquid solution to gradually become more B-rich.

This behavior also occurs if all the phases in the relevant part of a phase diagram are solids. The same rules apply. In any two-phase region, at a fixed temperature, the compositions of the two end phases are constant, and variations of the overall composition are accomplished by changes in the amounts of the two fixed-composition phases. It is therefore obvious that one-phase regions are always separated by two-phase regions in such binary (two-component) phase diagrams at a constant temperature.

3.2.1 The Lever Rule

The relation between the overall composition and amounts of each of the phases present in a two-phase region of a binary phase diagram is seen by the use of a simple mechanical analog called the *lever rule*.

If two different masses M_1 and M_2 are suspended on a bar that is supported by a fulcrum, the location of the fulcrum can be adjusted so that the bar will be in balance. This is shown in Fig. 3.3. The condition for balance is the ratio of the lengths L_2 and L_1 be equal to the ratio of the masses M_1 and M_2. i.e.

$$M_1/M_2 = L_2/L_1 \tag{3.1}$$

Analogously, the amounts of the two phases in a two-phase region can be found from the lengths L_1 and L_2 on the composition scale. The ratio of the amounts of phases 1 and 2 is related to that of the deviations of their compositions L_2 and L_1 from the overall composition on the composition scale. This is illustrated in Fig. 3.4, and can be expressed as

$$Q_1/Q_2 = L_2/L_1 \tag{3.2}$$

in which Q_1 and Q_2 represent the amounts of phases 1 and 2.

3.2.2 Examples of Binary Phase Diagrams

A slightly more complicated schematic phase diagram for a hypothetical binary alloy system A–B is shown in Fig. 3.5. In this case there are four one-phase regions. The solid phases are designated as phases α, β and γ. In addition, there is a liquid

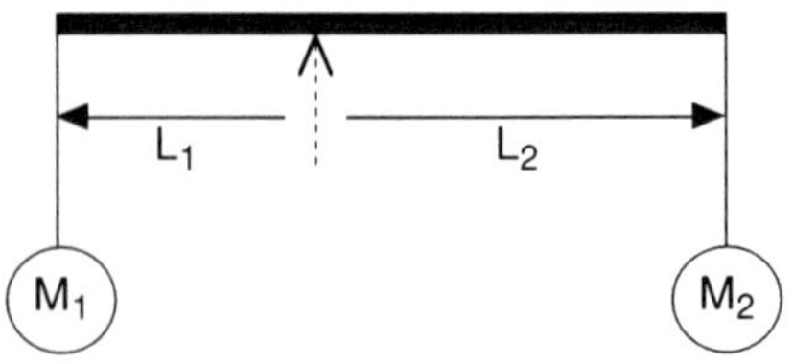

Fig. 3.3 Mechanical lever analog

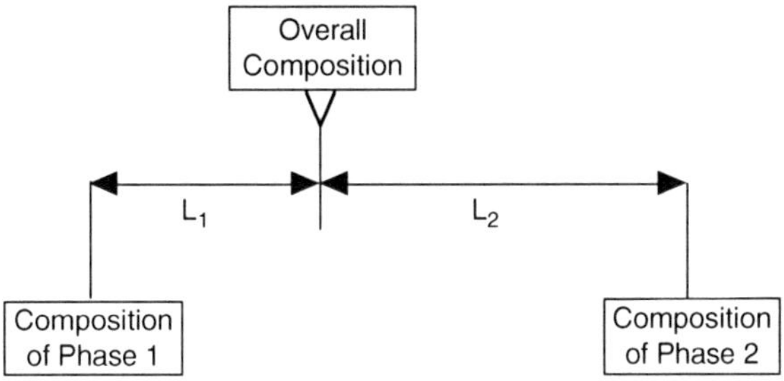

Fig. 3.4 Application of the lever rule to compositions in a two-phase region of a binary phase diagram

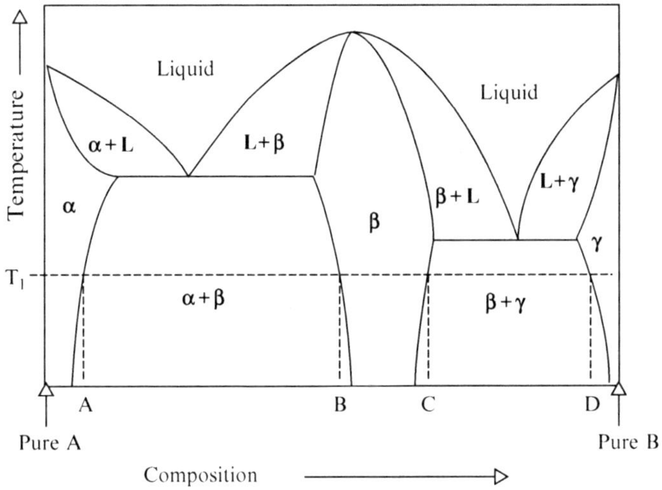

Fig. 3.5 Schematic binary phase diagram with an intermediate phase β, and solid solubility in terminal phases α and γ

phase at higher temperatures. It is seen in the figure that the single phases are all separated by two-phase regions, as the composition moves horizontally (isothermally) across the diagram.

It was shown earlier that according to the Gibbs Phase Rule, all intensive properties including electrical potential vary continually with the composition within single-phase regions in a binary system. Correspondingly, the intensive properties are composition-independent when two phases are present in a binary system. Since the equilibrium electrical potential of such an electrode, E, in an electrochemical cell is determined by chemical potential or activity of the electroactive species, it also varies with composition within single-phase regions; it is composition-independent when there are two phases present under the equilibrium conditions that are assumed.

The variation of the electrical potential with overall composition in this hypothetical system at temperature T_1 is shown in Fig. 3.6. It is seen that it alternates

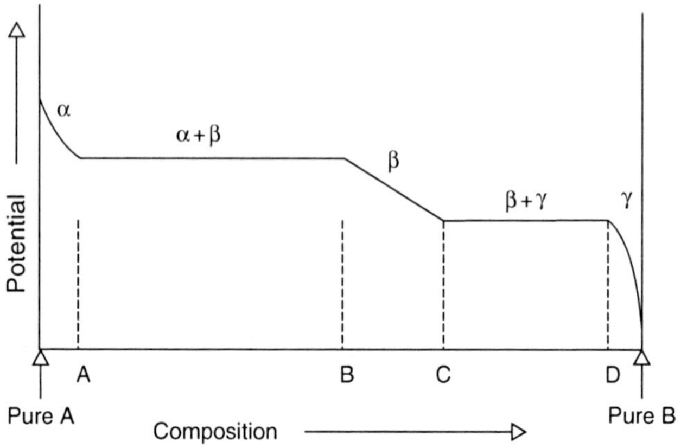

Fig. 3.6 Schematic variation of electrical potential with composition across the binary phase diagram shown in Fig. 3.5

between composition regions in which it is constant (*potential plateaus*) and those in which it varies. If B atoms are added to pure element A, the overall composition is initially in the solid solution phase α and the electrical potential varies with the composition. When α solubility limit is reached, indicated as composition A, addition of B causes the nucleation and growth of the β phase. Two phases are then present, and the potential maintains a fixed value. When the overall composition reaches B, all the α phase will have been consumed and only phase β will remain. Upon further compositional change electrical potential again becomes composition-dependent. At composition C, the upper compositional limit of the β phase at that temperature, the overall composition again enters a two-phase (β and γ) range and potential is again composition-independent. On reaching composition D the potential again varies with composition.

It is also possible for the composition ranges of phases to be quite narrow, when they are called *line phases*. As an example, a variation upon the phase diagram presented in Fig. 3.5 is shown in Fig. 3.7.

The corresponding variation of electrical potential with composition is shown schematically in Fig. 3.8. Potential drops abruptly, rather than gradually, across the narrow β phase in this case.

3.3 A Real Example, The Lithium: Antimony System Again

As a concrete example to demonstrate these principles consider the Li–Sb system, which was mentioned briefly in Chap. 1. This system has been studied both experimentally and theoretically in some detail [1–3]. Phase diagram is shown in Fig. 3.9.

Below 615°C there are two intermediate phases between Sb and Li, Li_2Sb and Li_3Sb. Both have rather narrow ranges of composition and are represented simply

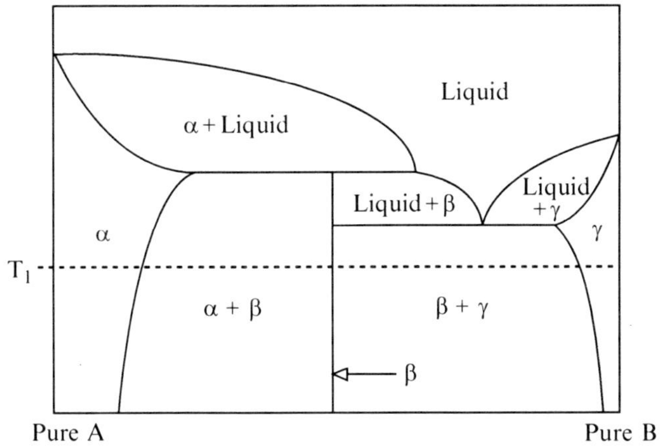

Fig. 3.7 Hhypothetical binary phase diagram in which the intermediate β phase has a small range of composition

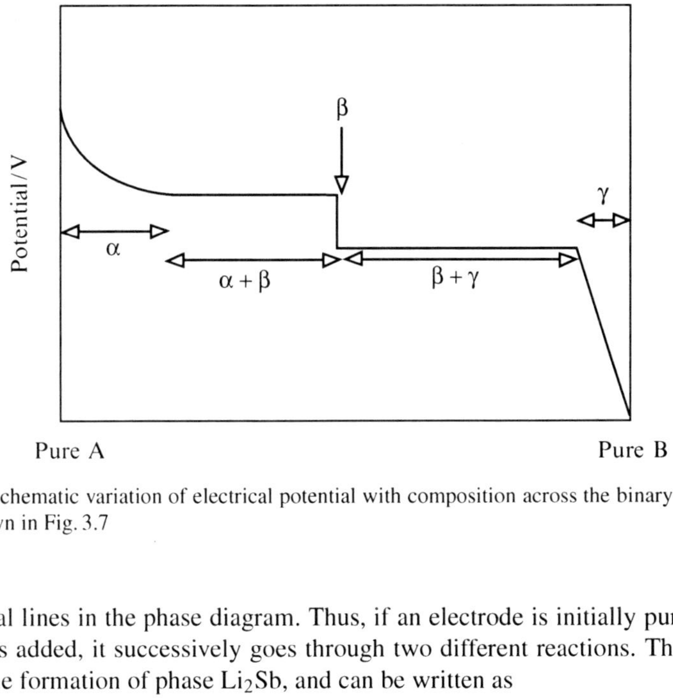

Fig. 3.8 Schematic variation of electrical potential with composition across the binary phase diagram shown in Fig. 3.7

as vertical lines in the phase diagram. Thus, if an electrode is initially pure Sb and lithium is added, it successively goes through two different reactions. The first involves the formation of phase Li_2Sb, and can be written as

$$2Li + Sb = Li_2Sb \tag{3.3}$$

On further addition of lithium, a second reaction occurs resulting in the formation of the second intermediate phase from the first. This can be written as

$$Li + Li_2Sb = Li_3Sb \tag{3.4}$$

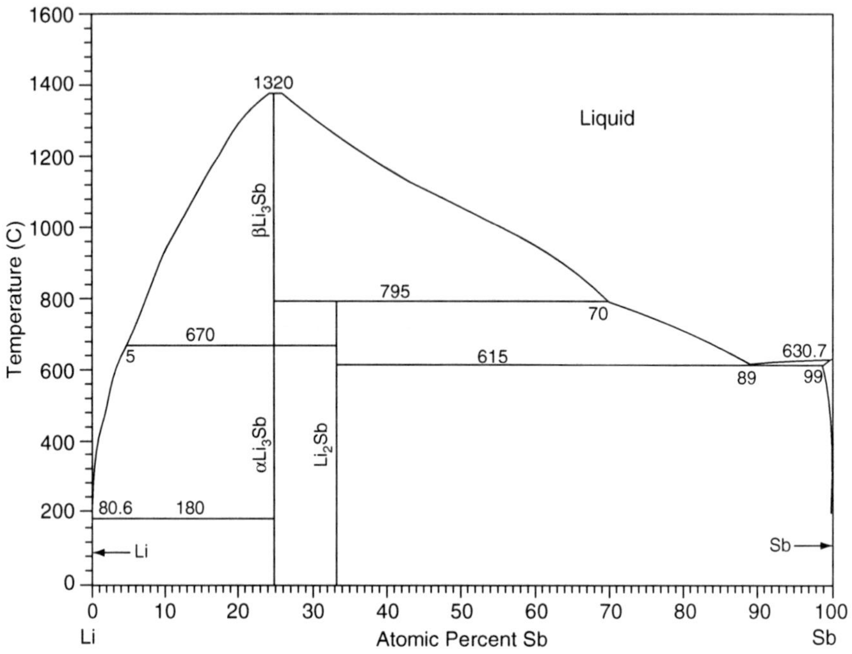

Fig. 3.9 Lithium–antimony phase diagram

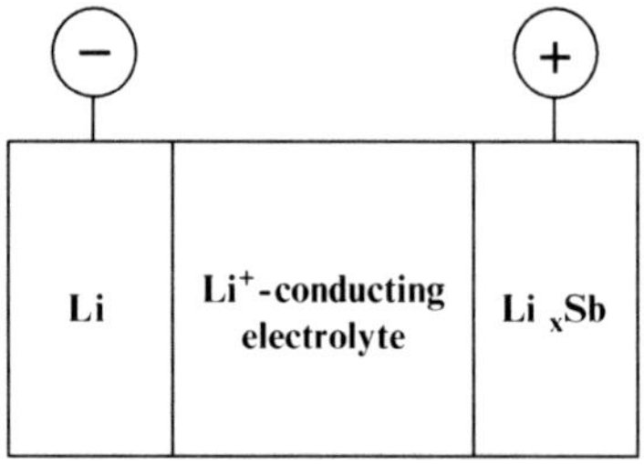

Fig. 3.10 Schematic drawing of electrochemical cell to study the Li–Sb system

This process can be studied experimentally by the use of an electrochemical cell whose initial configuration is similar to that shown schematically in Fig. 3.10.

Driving current through this cell from an external source causes the voltage between the two electrodes to be reduced from the open circuit value that it has when the positive electrode is pure antimony. Lithium will leave the negative electrode, pass through the electrolyte, and arrive at the positive electrode. If chemical diffusion rate within the Li_xSb electrode is sufficiently high relative to the rate at which lithium ions reach the positive electrode surface, this lithium is incorporated into the bulk of the electrode crystal structure, changing its composition. That is, the value of x in the positive electrode material Li_xSb increases.

If lithium is either added very slowly, or stepwise, allowing equilibrium to be attained within the positive electrode material after each step, the influence of lithium

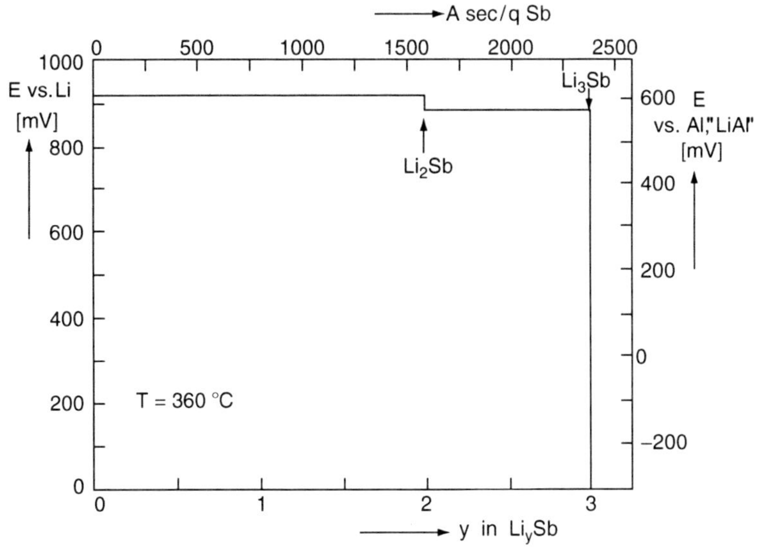

Fig. 3.11 Results from a coulometric titration experiment on the Li–Sb system at 360°C [3]

concentration in the positive electrode on its potential under equilibrium or near-equilibrium conditions can be investigated. This is the *Coulometric titration technique* discussed in Chap. 2. Data from such an experiment at 360°C is shown in Fig. 3.11.

These results are understood by considering the Gibbs Phase Rule, which was discussed in Chap. 2.

After an initial, invisibly narrow, range of solid solution, the first plateau in Fig. 3.11 corresponds to compositions in the phase diagram in which both (almost pure) Sb and the phase Li_2Sb are present. Thus it is related to the reaction in (3.3).

A very narrow composition range also exists in which only one phase, Li_2Sb, is present and the potential varies. On addition of more Li the overall composition moves into the region of the phase diagram in which two phases are again present – in this case Li_2Sb and Li_3Sb – and the potential follows along a second plateau, related to (3.4).

Potentials of the two plateaus are calculated from thermodynamic data on the standard Gibbs free energies of formation of the two phases, Li_2Sb and Li_3Sb. According to [3] these values at that temperature are −176.0 kJ/mol and −260.1 kJ/mol, respectively..

The standard Gibbs free energy change, $\Delta G_r{}^{\circ}$, related to virtual reaction (3.3) is simply that of the formation of phase Li_2Sb, $\Delta G_f{}^{\circ}(Li_2Sb)$. From this the potential of the first plateau can be calculated from

$$E - E^{\circ} = -\Delta G_r{}^{\circ}/2F \tag{3.5}$$

where E° is the potential of pure Li. This was 912 mV in the experiment.

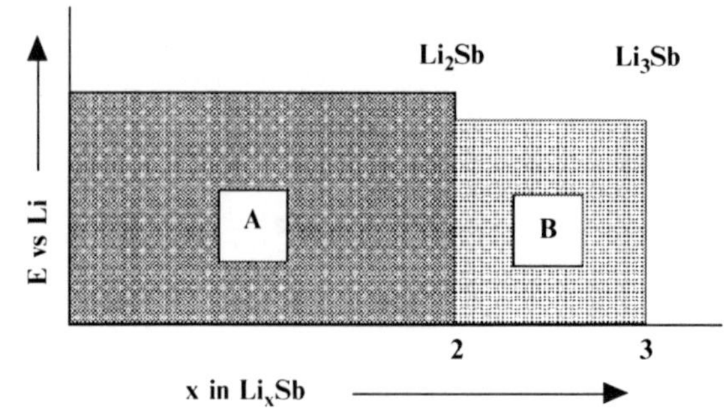

Fig. 3.12 Relation between energy stored and the titration curve in the Li–Sb system

The potential of the second plateau is related to virtual reaction (3.5), where

$$\Delta G_r{}^\circ = \Delta G_f{}^\circ(Li_3Sb) - \Delta G_f{}^\circ(Li_2Sb) \tag{3.6}$$

and in this case

$$E - E^\circ = -\Delta G_r{}^\circ / F \tag{3.7}$$

The result is that the potential of this plateau was experimentally found to be 871 mV vs pure Li.

Maximum theoretical energy that can be obtained from this alloy system is the sum of the energies involved in the two reactions. These relationships are shown schematically in Fig. 3.12.

The total energy that can be stored is proportional to the total area under the titration curve. The energy released in the first reaction is the product of the voltage of the first plateau times its capacity, i.e., the charge passed through the cell in connection with that reaction.This corresponds to the area inside rectangle A. The energy released in the second discharge reaction step is the product of its voltage and its capacity, and corresponds to the area inside rectangle B. The total energy is the sum of the two areas.

These energy values can be converted into *specific energy*, i.e., energy per unit weight. In the case of the first plateau the *maximum theoretical specific energy*, MTSE, is simply the standard Gibbs free energy of the reaction divided by the sum of the atomic weights in the product. This was found to be 1,298 kJ/kg. This can also be expressed as 360 W h/kg, since 3.6 kJ is equal to 1 W h.

The maximum theoretical specific energy can also be calculated if the composition were to only vary between compositions Li_2Sb and Li_3Sb. In this case the voltage is 871 mV and the capacity is only one mole of lithium per mole of original Li_2Sb. When calculating the MTSE, the weight of the product is that of Li_3Sb, 142.57 g/mol. The result is 589 kJ/kg and 164 W h/kg for a cell operated in this composition range.

However, if the experiment is performed starting with pure Li and pure Sb, and the energy relating to both plateaus is used, the relevant weight for both steps is the final weight of Li_3Sb. Thus the MTSE of the first reaction in the two-reaction scheme is less than it would be if it were used alone i.e., instead of 1,298 kJ/kg, it is only 1,234 kJ/kg. The total MTSE is $1,234 + 589 = 1,823$ kJ/kg.

This is similar to the result that would be obtained if it were assumed that the intermediate phase, Li_2Sb, did not form, and there is only a single voltage plateau between Li and Li_3Sb.

If the electrochemical titration curve is calculated from the experimental value of the total energy, it would have only a single plateau, and at a voltage that is the weighted average of the voltages of the two reactions that actually take place. This is a false result, due to the lack of recognition of the existence of the intermediate phase. Thus one has to be aware of all the stable phases when making voltage predictions from thermodynamic data.

3.4 Stability Ranges of Phases

While so far emphasis has been placed upon the potentials at which reactions take place, there is another important type of information to be derived from equilibrium electrochemical titration curves. The potential ranges over which the various intermediate phases are stable can be readily obtained. Since they are present at compositions between two plateaus, they are stable at all potentials between the two plateau potentials. This can be vital information if they are to be used as *mixed-conductors*. This will be described later.

3.5 Another Example, The Lithium: Bismuth System

The lithium–bismuth binary system has also been extensively explored by the use of the coulometric titration technique. The phase diagram is shown in Fig. 3.13. Note that this diagram is drawn with lithium on the right hand side, i.e., in the opposite direction from the Li–Sb diagram. This difference isnot important.

The titration curve that resulted from measurements made at 360°C is shown in Fig. 3.14.

It may be noted that there are three differences from the Li–Sb system. The phase diagram shows that there is a considerable amount of solubility of bismuth in liquid lithium at that temperature. This results in the appearance of a single-phase region in the titration curve. Also, there is a phase LiBi in the Li–Bi case, but Li_2Sb in the Li–Sb case.

In addition, the phase diagram in Fig. 3.13 indicates the solid phase "Li_3Bi" has an appreciable range of composition. This can also be seen in the titration curve. Because of the very high sensitivity in the coulometric titration technique, the electrochemical properties of this phase can be explored in greater detail. This is shown in Fig. 3.15.

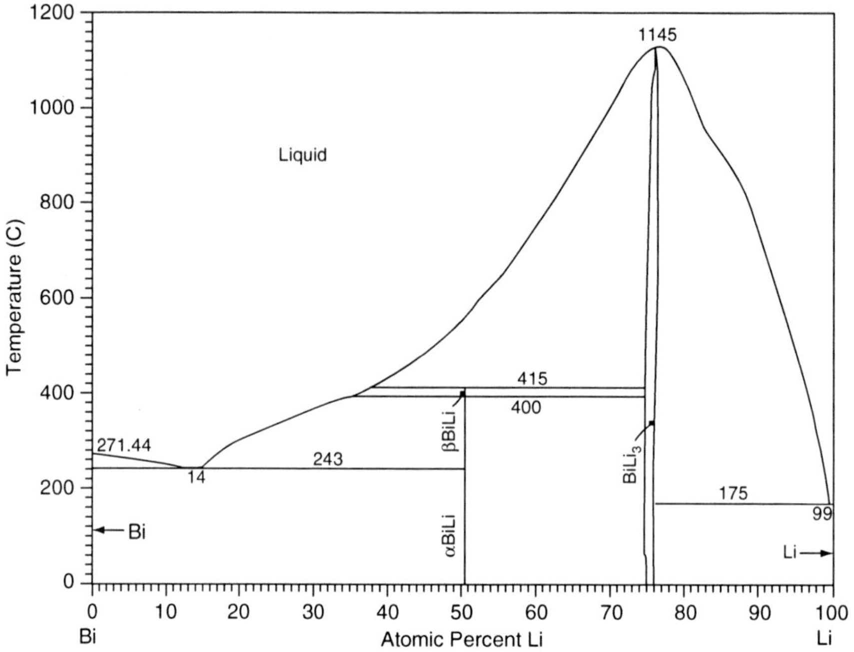

Fig. 3.13 The lithium–bismuth binary phase diagram

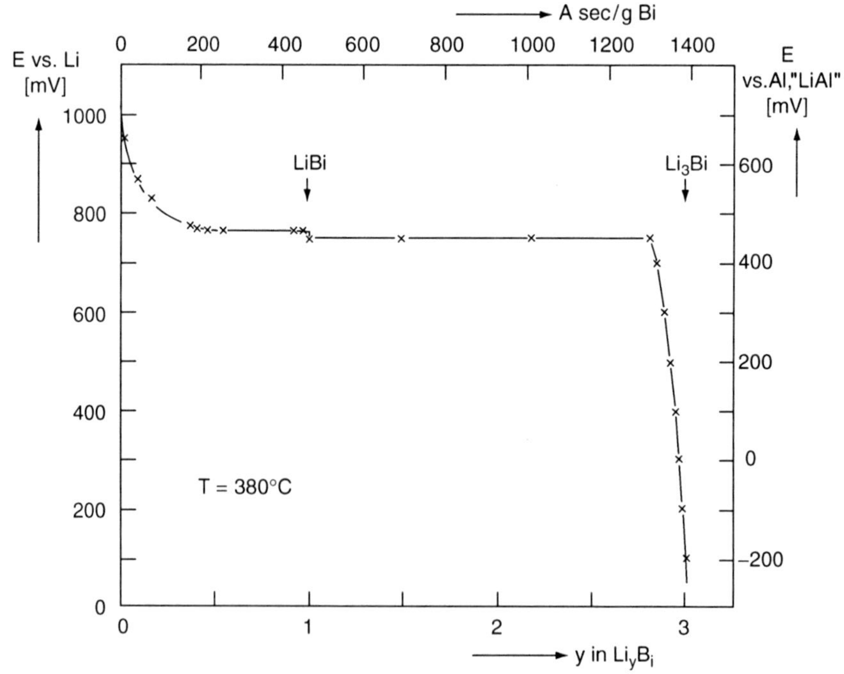

Fig. 3.14 Results from a coulometric titration experiment on the Li–Bi system at 360°C [3]

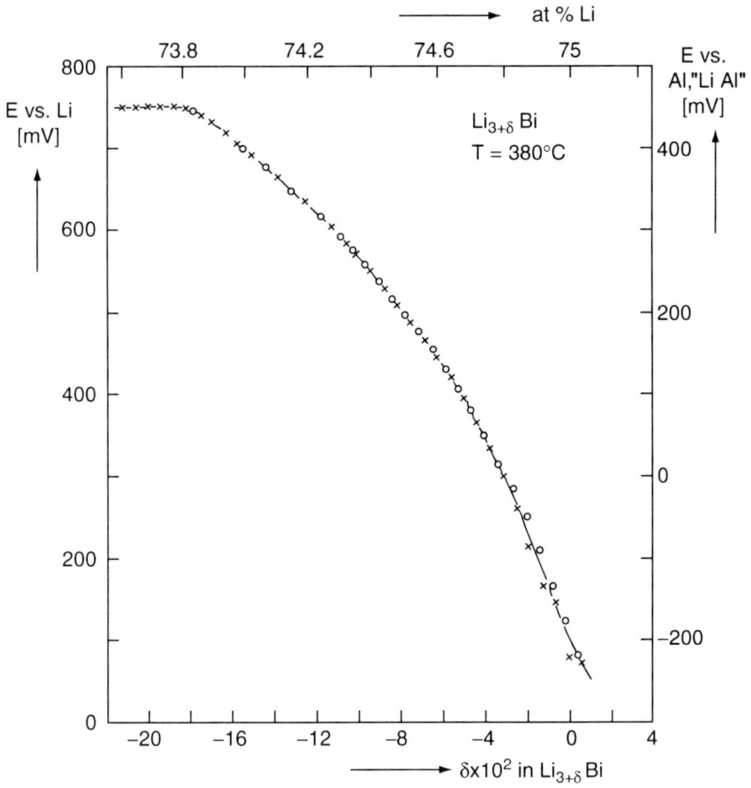

Fig. 3.15 Coulometric titration measurements within the composition range of the phase "Li_3Bi"

3.6 Coulometric Titration Measurements on Other Binary Systems

Coulometric titration experiments have been made on a number of other binary metallic systems at different temperatures. In order to obtain reliable data, experiments need to be undertaken under conditions when equilibrium can be reached within a reasonable time. This requirement is fulfilled more easily at elevated temperatures; however, in some cases, equilibrium data can be obtained at ambient temperatures. Some more examples will be discussed in later chapters.

3.7 Temperature Dependence of the Potential

The early measurements of the equilibrium electrochemical properties of binary lithium alloys and their relationship to the relevant phase diagrams were made at elevated temperatures using a LiCl–KCl molten salt electrolyte. These included experiments on the Li–Al, Li–Bi, Li–Cd, Li–Ga, Li–In, Li–Pb, Li–Sb, Li–Si, and Li–Sn systems [4–10]. This molten salt electrolyte was being used in research for the development of large scale batteries for electric vehicle propulsion and load leveling applications. Subsequently, measurements were made with lower temperature molten salts, $LiNO_3$–KNO_3 [11], and at ambient temperatures with organic solvent electrolytes. This will be discussed later.

As expected the temperature dependence of the potentials and capacities can be explained in terms of the relevant phase diagrams and thermodynamic data in all of these cases.

To demonstrate the principles involved, experimental results on materials in the Li–Sb and Li–Bi systems over a wide range of temperature are described. The results are shown in Fig. 3.16.

Each of these systems has two intermediate phases at low temperatures. The temperature dependence of the potentials of the plateaus due to the presence of two-phase equilibria in the Li–Sb system fall upon two straight lines, corresponding to the reactions

$$2Li + Sb = Li_2Sb \tag{3.8}$$

and

$$Li + Li_2Sb = Li_3Sb \tag{3.9}$$

In the Li–Bi case, however, where the comparable reactions are

$$Li + Bi = LiBi \tag{3.10}$$

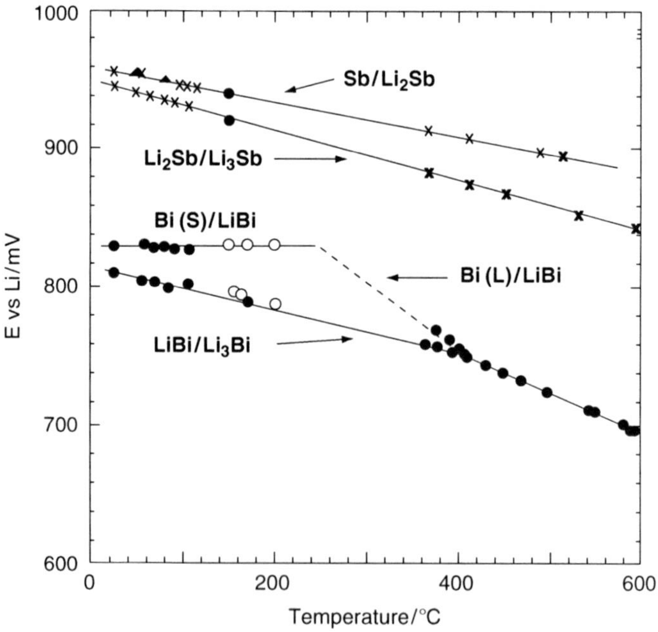

Fig. 3.16 Temperature dependence of the potentials of the two-phase plateaus in the Li–Sb and Li–Bi systems, [12]

Table 3.1 Reaction entropies in the lithium–antimony and lithium–bismuth systems

Reaction	Molar entropy of reaction (J/K mol)	Temperature range (°C)
$2Li + Sb = Li_2Sb$	−31.9	25–500
$Li + Li_2Sb = Li_3Sb$	−46.5	25–600
$Li + Bi = LiBi$	0	25–200
$2Li + LiBi = Li_3Bi$	−36.4	25–400

and

$$2Li + LiBi = Li_3Bi \tag{3.11}$$

the temperature dependence of the plateau potentials is *different*. There is a change in the slope at the *eutectic melting point* (243°C), and the data for the two plateaus converge at about 420°C, which corresponds to the fact that the LiBi phase is no longer stable above that temperature. These can be seen in the phase diagram for that system shown in Fig. 3.13. At higher temperatures there is only a single reaction;

$$3Li + Bi = Li_3Bi \tag{3.12}$$

Depending on the temperature range, the potentials of the second reaction fall along two straight line segments,. There is a significant change in slope at about 210°C, resulting in a negligible temperature dependence of the potential at low temperatures, due to the melting of bismuth.

The potentials are related to the standard Gibbs free energy change $\Delta G_r{}^\circ$ relating to the relevant reaction, and the temperature dependence of the value of $\Delta G_r{}^\circ$, is evident from the relation between the Gibbs free energy, the enthalpy, and the entropy

$$\Delta G_r{}^\circ = \Delta H_r{}^\circ - T\Delta S_r{}^\circ \tag{3.13}$$

where $\Delta H_r{}^\circ$ is the change in the standard enthalpy and $\Delta S_r{}^\circ$ is the change in the standard entropy resulting from the reaction. Thus it can be seen that

$$d\Delta G_r{}^\circ/dT = \Delta S_r{}^\circ \tag{3.14}$$

From such data, the value of the standard molar entropy changes involved in these several reactions is obtained. They are shown in Table 3.1. Thus the potentials at any temperature within this range is predictable.

3.8 Application to Oxides and Similar Materials

So far this discussion has been concerned with binary metallic alloys. However, the same principles can be applied to binary metal–oxygen systems. The Nb–O system is an example. The niobium–oxygen phase diagram is shown in Fig. 3.17.

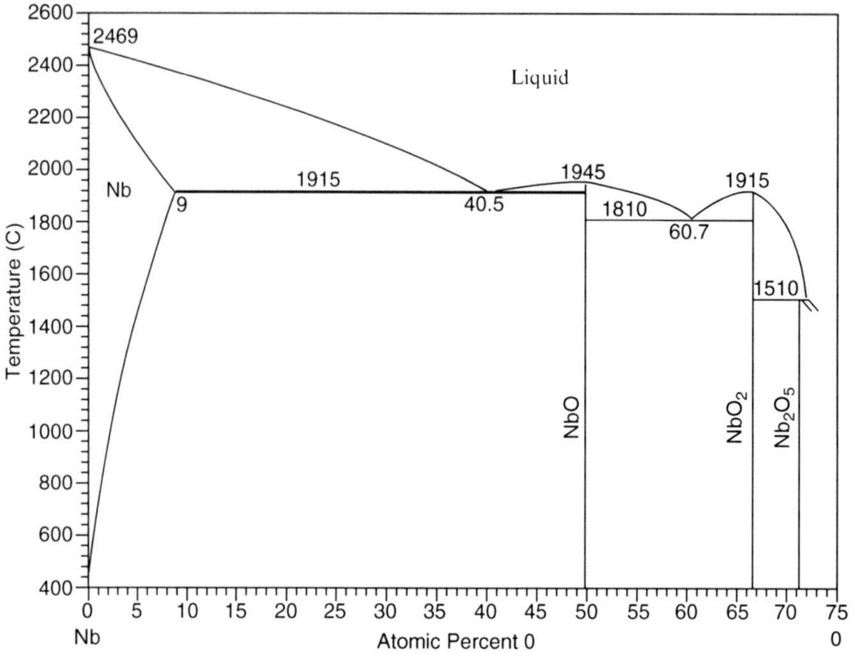

Fig. 3.17 Niobium–oxygen phase diagram

There are three intermediate phases in this system. As before, thermodynamic data can be used to calculate the potentials of the various two-phase plateaus at any temperature,. In addition, these potentials can be converted into values of the oxygen activity at the respective temperatures. Data on the plateau potentials, which define the limiting values for the stability of each of the phases, are shown for two temperatures in Fig. 3.18. In this case, the potentials are shown as voltages relative to the potential of pure oxygen at one atmosphere, at the respective temperatures.

Similar procedures can be used in the analysis of other metal–gas systems, such as iodides and chlorides.

3.9 Ellingham Diagrams

Another thinking tool sometimes used to help understand the behavior of metal–oxygen systems is an *Ellingham diagram*. These are plots of the Gibbs free energy of formation of their oxides as a function of temperature.

There are two different ways to present this information. The Ellingham diagrams often seen in textbooks are of the *integral type*, for they indicate the temperature dependence of the Gibbs free energy needed for the formation of a particular oxide from its component elements. The formation reactions are generally written on a

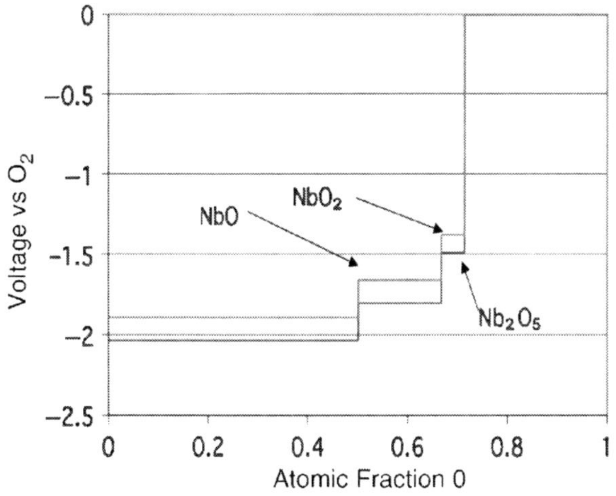

Fig. 3.18 Equilibrium potentials of the three two-phase plateaus in the Nb–O system at two temperatures

"per mole oxygen" basis, so that the lines relating to different oxides are generally parallel, as the entropy of gaseous oxygen makes the major contribution to the entropy of the formation reaction.

An equilibrium oxygen pressure scale is generally added on the right side to provide a simple graphical means for determining the oxygen partial pressure of the oxides as a function of temperature [13].

However, this information is valid only for the direct formation of an oxide from its elements; in many cases more than one oxide can be formed from a given metal, depending upon the oxygen partial pressure. For example, there are several manganese oxides: MnO, Mn_3O_4, Mn_2O_3, and MnO_2. Except for the lowest oxide, MnO, they are not formed directly by the oxidation of manganese. The higher oxides form by reaction of oxygen with lower oxides, and the relevant oxygen pressure for the formation of a given oxide is related to the reaction of oxygen with its next lower oxide.

It is much more useful to have a *difference* diagram that provides information about the oxygen pressure for the formation of given phases from their neighbors. An example that shows both types of information is given in Fig. 3.19.

3.10 Liquid Binary Electrodes

Although the discussion here involved solid binary alloys and oxides, similar principles apply to liquids in binary systems.

An important example that appeared some years ago involved the so-called *sodium–sulfur battery* that operates at about 300°C. In this case, both electrodes

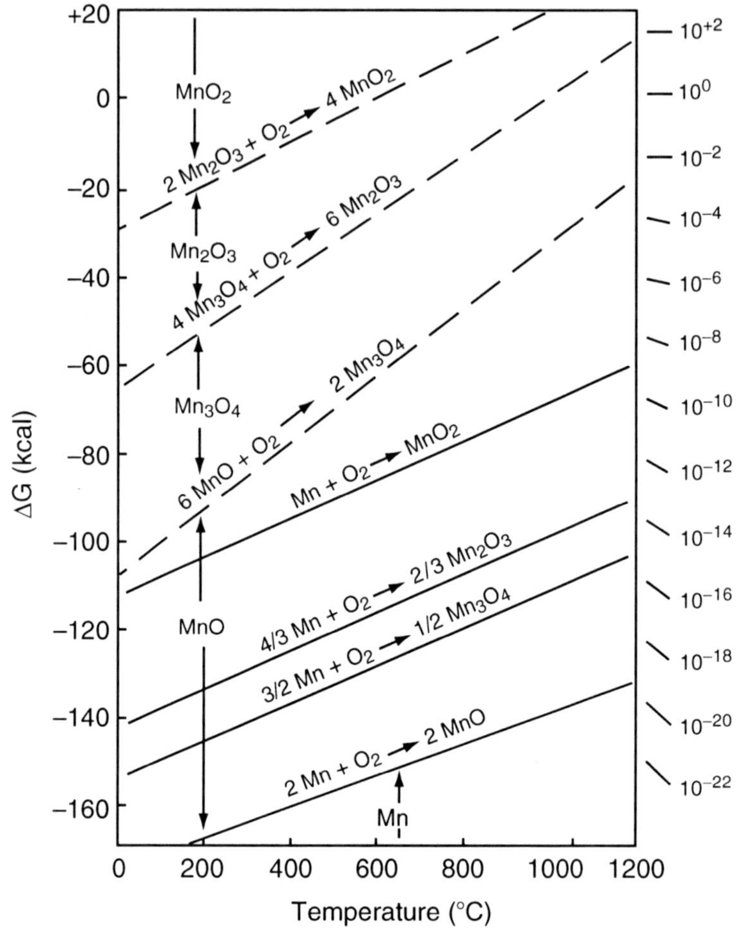

Fig. 3.19 Ellingham type diagram that shows both integral and difference data for the manganese oxide system [14]

are liquids, and the electrolyte is a solid lithium ion conductor called *sodium beta alumina*. This is described as an *L/S/L system*. It is the inverse of conventional systems, which have solid electrodes and a liquid electrolyte, i.e., *S/L/S systems*. The negative electrode is molten sodium, and the positive electrode is the product of the reaction of sodium with liquid sulfur. Thus the basic reaction can be written as

$$xNa + S = Na_xS \tag{3.15}$$

The potential of the elemental sodium is constant, independent of the amount of sodium present. The potential of the positive electrode changes as the sodium concentration varies by its transport across the cell.

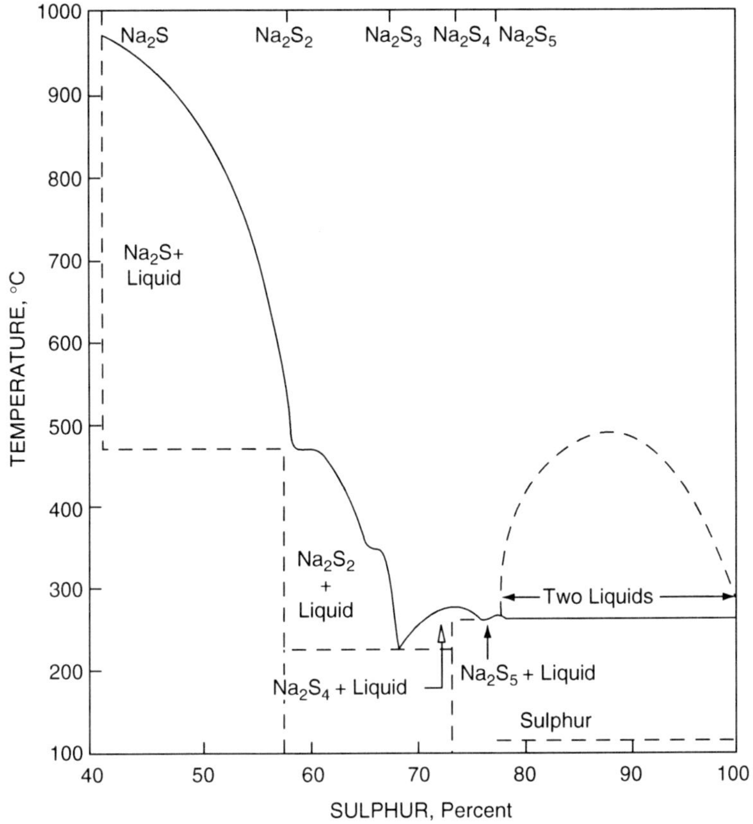

Fig. 3.20 Part of the sodium–sulfur phase diagram

The relevant portion of the Na–S phase diagram is shown in Fig. 3.20. It is seen that at about 300°C a relatively small amount of sodium can be dissolved in liquid sulfur. When this concentration is exceeded, a second liquid phase, with a composition of about 78 at% Na, is nucleated. This has a composition that is roughly $Na_{0.4}S$. As more sodium is added, the overall composition traverses the two-phase region; the amount of this liquid phase increases relative to the amount of the sulfur-rich liquid phase. Thus a potential plateau is expected over this composition range. When the sodium concentration exceeds the equivalent of $Na_{0.4}S$, the overall composition moves into a single-phase liquid range of varying potential. The maximum amount of sodium that can be used in this electrode corresponds roughly to $Na_{0.67}S$.

At higher sodium concentrations, a solid second phase begins to form from the liquid solution. This mostly forms at the interface between the solid electrolyte and the liquid electrode, preventing the ingress of more sodium, blocking further reaction. The variation of the potential with the composition of the electrode is shown in Fig. 3.21.

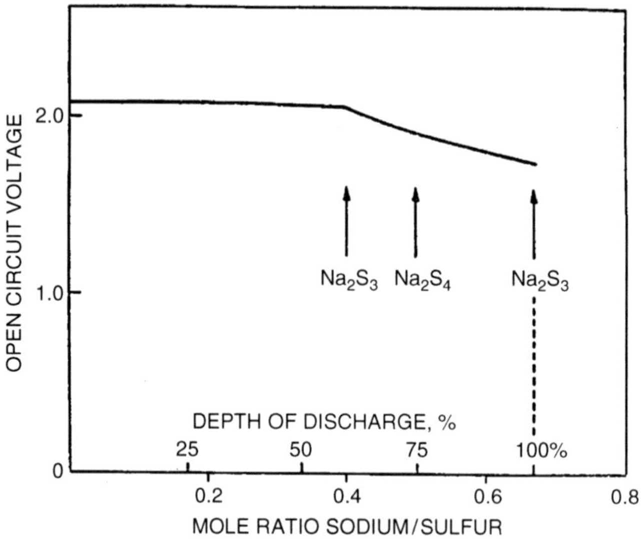

Fig. 3.21 Voltage versus pure sodium as a function of composition

3.11 Comments on Mechanisms and Terminology

It has been shown that the incorporation of species into solid electrodes can involve either a change in the composition of a single phase that is already present in the microstructure, or the nucleation and growth of an additional phase. When that species is deleted, the same two types of phenomena can occur, but in reverse. Consider how this can happen. If, for example, lithium is added to an existing phase it forms a solid solution and the composition will change, becoming more lithium-rich. If this involves more than merely the surface layer it must involve the diffusive motion of lithium atoms or ions into the crystal structure.

In many metallic alloys and ceramic materials the inserted material occupies the same type of lattice positions as the host material. This is called a *substitutional solid solution*. For the composition to change there must be a mechanism that allows atomic, or ionic, motion through the crystal structure. In substitutional solid solutions this typically involves the presence and motion of empty lattice sites, *vacancies*. Atoms can jump into these vacant lattice sites from adjacent positions in the structure. The result of a single jump is the effective motion of the vacancy in one direction, and the atom in the opposite direction. Compositional changes occur by this vacancy mechanism if one type of atom has a greater probability of jumping into adjacent vacancies than the other type and a gradient in the composition is present.

Another type of diffusion mechanism is frequently present in crystal structures in which one type of atom or ion is appreciably smaller than the other types present.

This is often the case in lithium and hydrogen systems, as these species are quite small. The smaller atoms can occupy interstitial sites between the other atoms in the crystal lattice. They can move about by hopping from one interstitial site to the next. Diffusion by this *interstitial mechanism* does not require the motion of either the other atoms or vacancies, and it is typically very much faster than the vacancy diffusion mechanism. That is, interstitial *diffusion coefficients*, or *diffusivities*, are typically greater than vacancy mechanism diffusion coefficients, or diffusivities.

Because of the large concentration of interstitial positions in most crystal structures, it is structurally possible for a large number of guest species to be inserted into the host crystal structure, if other factors, such as the electronic energy spectrum, are favorable.

If the basic structure of the host material is not appreciably altered by the *insertion* of additional guest species, ensuring there is a definite relation between the initial crystal structure and the structure that results, the reaction is called *topotactic*. This is often the case in materials of interest as electrode reactants in lithium battery systems.

Topotaxy suggests a three-dimensional relation between the structures of the *parent* and *product* structures, whereas the term *epitaxy* is used to describe a two-dimensional correspondence between two structures. Likewise, an *insertion reaction* that has a two-dimensional character is often called *intercalation*.

From this structural viewpoint it can be easily understood that there can be a limit to the concentration of interstitial guest species that can be inserted into a host crystal structure. This limit can be due to either crystallographic or electronic factors, and will not be discussed further here.

If there is a thermodynamic driving force for the incorporation of *additional* guest species than can be accommodated interstitially, this must happen by a different mechanism; a *second phase*, with a different crystal structure as well as a higher solute concentration, must be nucleated. As more and more of the guest species atoms arrive, the extent of this second phase increases, gradually replacing the interstitial phase initially present. This change in the microstructure, in which one phase is gradually replaced by another phase, is an example of a *reconstitution* reaction.

When a reconstitution reaction is taking place, the initial and product phases are both present in the microstructure. This is sometimes called a *heterophase structure*, in contrast to a *homophase* structure, in which only a single phase is present. Thus this range of compositions must be in a two-phase region of the corresponding phase diagram.

Phase diagrams are expected to provide information about the *absolutely stable* phases that tend to be present in a chemical system as a function of intensive thermodynamic variables such as temperature and composition. The term *absolutely stable* has been used to describe the *most stable equilibrium structure* possible for a given composition. On the other hand, a phase that is *stable relative to small perturbations*, and thus meets the general requirement for equilibrium, yet is not the most stable variation, is termed *metastable*.

3.12 Summary

Many batteries use binary systems as either negative or positive, or both, electrode reactants today. The theoretical limits of the potentials and capacities of such electrodes can be determined from a combination of thermodynamic data and phase diagrams. This has been demonstrated here for several examples of binary systems.

There are two general types of reactions that can take place, *homophase* reactions, in which guest atoms are inserted into an existing phase, often *topotactically*, and *reconstitution* reactions in which phases nucleate and grow in *heterophase* microstructures. The potential varies with the overall composition of an electrode in the insertion reaction homophase case, but is composition-independent when reconstitution reactions take place in heterophase microstructures.

The electrochemical titration method can be used to investigate the relevant parameters experimentally. When the composition is within a single-phase region of the relevant phase diagram, the potential varies as guest species are inserted or extracted during an electrochemical reaction. On the other hand, when the composition is within two-phase regions of the relevant phase diagram, reconstitution reactions take place, and the potential is independent of composition. Experimental results are now available for a number of systems of each type, both at elevated temperatures and at ambient temperatures.

Under equilibrium, or near-equilibrium, conditions the potentials are directly related to the values of the standard Gibbs free energies of formation of the phases involved; thus thermodynamic data can predict experimental results. Likewise, experiments can provide thermodynamic data. The temperature dependence of potential plateaus, for example, can be used to determine the standard entropy changes in the relevant reaction. Such experimental data also correlate to the stability of phases in the phase diagram. Further, the maximum theoretical specific energy of an electrochemical system can also be determined from the equilibrium electrochemical titration curve and the related thermodynamic data.

These principles are also applicable to metal oxides and liquid binary materials, as illustrated by the Nb–O and the Na–S systems. The latter is the basis for a high temperature L/S/L battery system using sodium beta alumina as a solid electrolyte.

References

1. W. Weppner and R.A. Huggins, in *Proc. of the Symposium on Electrode Materials and Processes for Energy Conversion and Storage*, ed. by J.D.E. McIntyre, S. Srinivasan and F.G. Will, Electrochemical Society, p. 833 (1977)
2. W. Weppner and R.A. Huggins, Z. Phys. Chem. N.F. 108, 105 (1977)
3. W. Weppner and R.A. Huggins, J. Electrochem. Soc. 125, 7 (1978)
4. C.J. Wen, et al., J. Electrochem. Soc. 126, 2258 (1979)
5. C.J. Wen, Ph.D. Dissertation, Stanford University (1980)
6. C.J. Wen and R.A. Huggins, J. Electrochem. Soc. 128, 1636 (1981)
7. C.J. Wen and R.A. Huggins, Mat. Res. Bull. 15, 1225 (1980)
8. M.L. Saboungi, et al., J. Electrochem. Soc. 126, 322 (1979)

9. C.J. Wen and R.A. Huggins, J. Solid State Chem. 37, 271 (1981)
10. C.J. Wen and R.A. Huggins, J. Electrochem. Soc. 128, 1181 (1981)
11. J.P. Doench and R.A. Huggins, J. Electrochem. Soc. 129, 341C (1982)
12. J. Wang, I.D. Raistrick, and R.A. Huggins, J. Electrochem. Soc. 133, 457 (1986)
13. F.D. Richardson and J.H.E. Jeffes, J. Iron Steel Inst. 160, 261 (1948)
14. N.A. Godshall, I.D. Raistrick, and R.A. Huggins, J. Electrochem. Soc. 131, 543 (1984)

Chapter 4
Ternary Electrodes Under Equilibrium or Near-Equilibrium Conditions

4.1 Introduction

The previous chapter described binary electrodes, in which the microstructure is composed of phases made up of two elements. It was pointed out that there are also cases in which three elements are present, but only partial equilibrium can be obtained in experiments, so the electrode behaves as though it were composed of two, rather than three, components.

This chapter will discuss active materials that contain three elements, but have kinetic behavior such that they behave as true ternary systems. As before, it will be seen that phase diagrams and equilibrium electrochemical titration curves are very useful thinking tools in understanding the potentials and capacities of electrodes containing such materials.

It is generally more difficult to obtain complete equilibrium in ternary systems than in binary systems, so that much of the available equilibrium, or near-equilibrium, information stems from experiments at elevated temperatures. Selective, or partial, equilibrium is much more common at ambient temperatures. This will be discussed in another chapter.

4.2 Ternary Phase Diagrams and Phase Stability Diagrams

In order to represent compositions in a three-component system one must have a figure that represents the concentrations of three components. This can be done by using a two-dimensional figure, as will be seen shortly. However, if information about the influence of temperature is also desired, a three-dimensional figure is required. This is often done in metallurgical and ceramic systems in which experiments commonly involve changes in the temperature. Most electrochemical systems operate at or near constant temperatures, so three-dimensional figures are not generally considered necessary.

R.A. Huggins, *Advanced Batteries*: Materials Science Aspects,

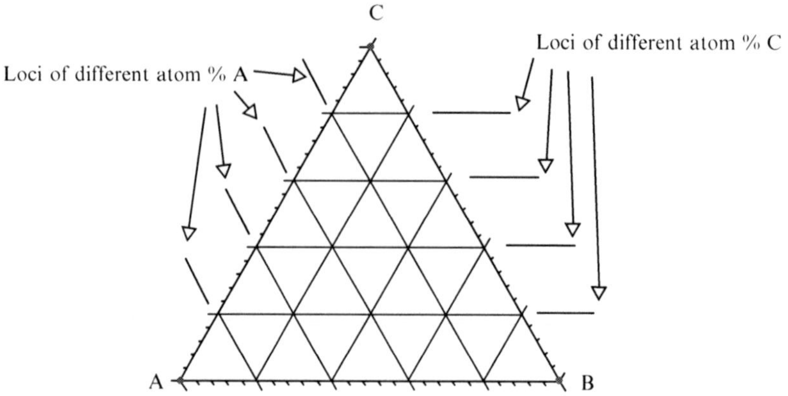

Fig. 4.1 General coordinate scheme used to depict compositions and phase equilibria in ternary systems on isothermal Gibbs triangles

Compositions in isothermal ternary systems can be represented on paper by using a triangular coordinate system. The method that is commonly employed in materials systems involves the use of *isothermal Gibbs triangles*. This scheme is illustrated in Fig. 4.1.

Compositions are expressed in terms of the atomic percentage of each of the three components, indicated as A, B, and C in this case. For the purposes of this discussion it is desirable to have elements as components, so that three elements are placed at the corners, and the atomic percentage of an element varies from zero along the opposite side to 100% at its corner. Thus the position of each point within the triangle represents the atomic fraction of each of the elements present in the system.

Although phases in ternary systems often have ranges of composition, as they do in binary systems, it is often useful to simplify the phase equilibrium situation by assuming that they act as *point phases*. That is, that they have very narrow composition ranges. The term *phase stability diagram* will be used in this discussion to describe this approximation to the actual ternary phase diagram. It will be seen that it is possible to get a large amount of useful information by the use of such an approximate isothermal Gibbs triangle.

If there are phases inside the Gibbs triangle, the influence of the Gibbs phase rule must be considered. It was shown earlier that the Gibbs phase rule can be written as

$$F = C - P + 2 \tag{4.1}$$

If the temperature and total pressure are kept constant, the number of residual degrees of freedom F will be zero when there are three phases present in a ternary system. Three phases are in equilibrium with each other within triangles inside the overall Gibbs triangle. Two phases are in equilibrium if their compositions are connected by a line, called a *two-phase tie line*. As shown in Fig. 4.2, if intermediate

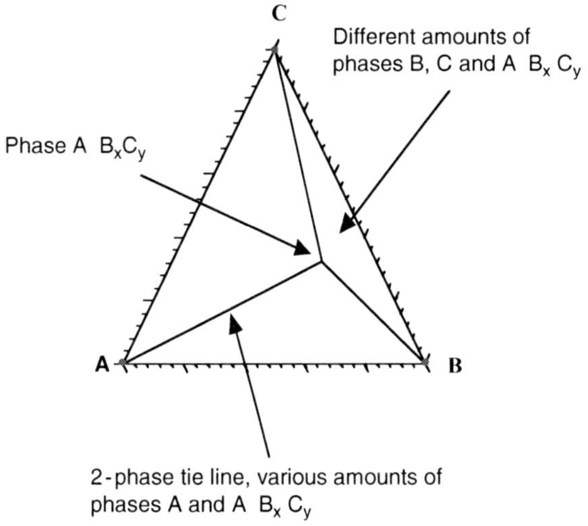

Fig. 4.2 Isothermal phase stability diagram ABC for the case in which there is a single intermediate phase whose composition is A_xB_yC

ternary phases are present, the total area within the Gibbs triangle is divided into *subtriangles* whose sides are *two-phase tie lines.*

All the compositions that lie within a given triangle have microstructures that are composed of mixtures of the three phases that are at the corners of that triangle. The overall composition determines the amounts of these different phases present, but not their compositions, for the latter are specified by the locations of the points at the corners of the triangle. Any materials having compositions that fall along one of the sides of a triangle will have microstructures composed of the two phases at the ends of that tie line. The amounts are determined by the position along the tie line. Points closer to a given end have greater amounts of the phase whose composition is at the end.

Because the compositions of the phases present within triangles are constant, determined by the locations of the corner points, all the intensive (amount-independent) thermodynamic parameters and properties are the same for all compositions inside the triangle. Important intensive properties include the chemical potentials and activities of all the components, and the electrical potential.

4.3 Comments on the Influence of SubTriangle Configurations in Ternary Systems

Binary systems can be changed to ternary systems by the addition of an additional element. As an example, consider a lithium-based binary system Li–M, in which the lithium composition can be varied. The addition of an additional element X

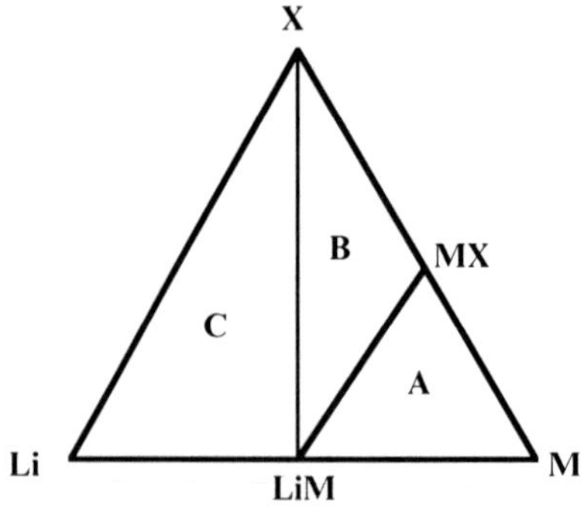

Fig. 4.3 Schematic ternary phase diagram for the Li–M–X system in which there are intermediate phases in the centers of both the Li–M and the M–X binary systems

converts this to a ternary Li–M–X system. The presence of X can result in a significant change in the potentials in the Li–M system, even if X does not react with lithium itself.

Consider a simple case. Assume that the thermodynamic properties of this system lead to a ternary phase stability diagram of the type shown in Fig. 4.3, in which it is assumed that there are two stable binary phases, LiM and MX.

If there is no X present, the composition moves along the Li–M edge of the ternary diagram, which is simply the binary Li–M system, and there will be a constant potential plateau for all compositions between pure M and LiM. The voltage vs. pure lithium in this compositional range, and therefore in triangle A of the ternary system, will be given by

$$E_A = -\Delta G_f{}^{\circ}(LiM)/F \tag{4.2}$$

What happens if lithium reacts with a material that has an original composition containing some X? The overall composition will follow a trajectory that starts at that position along the X–M side of the triangle and goes in the direction of the lithium corner of the ternary diagram. The addition of X to the M will not change the plateau potential for all compositions in triangle A. Therefore, there will be a plateau at that potential. Its length, however, will vary, depending upon the starting composition.

In addition, an additional plateau will appear at higher lithium concentrations as the overall composition enters and traverses triangle B. The potential of all compositions in that triangle will be given by

$$E_B = -(\Delta G_f{}^{\circ}(MX) - \Delta G_f{}^{\circ}(LiM))/F \tag{4.3}$$

As in the case of the binary Li–M system, when the overall composition gets into triangle C the potential will be the same as that of pure lithium.

These effects are illustrated in Figs. 4.4 and 4.5.

In Fig. 4.5 the variation of the electrode potential with overall composition is shown schematically for three different starting electrode compositions in Fig. 4.4.

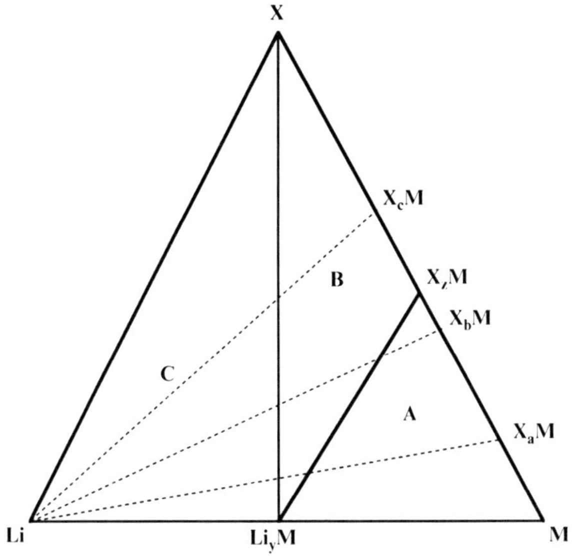

Fig. 4.4 The ternary Li–M–X system shown in Fig. 4.3, displaying the loci of the overall composition for three different initial compositions

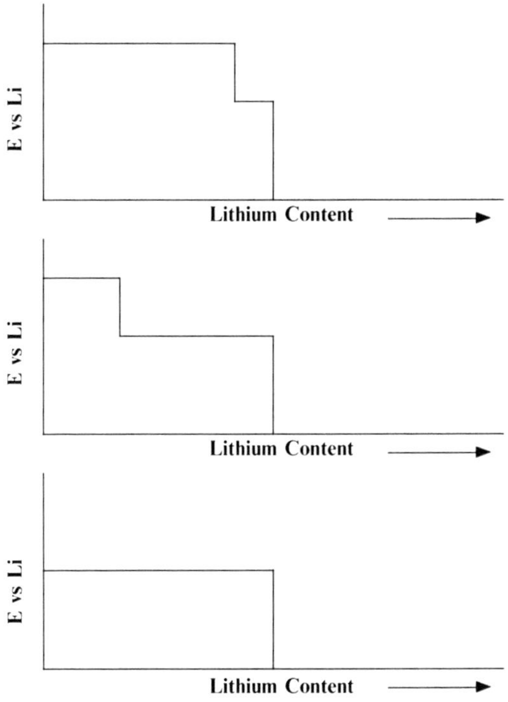

Fig. 4.5 Variation of the potential as lithium is added to electrodes with the three different starting compositions shown in Fig. 4.4. *Top* X_aM, *middle* X_bM, and *bottom* X_cM

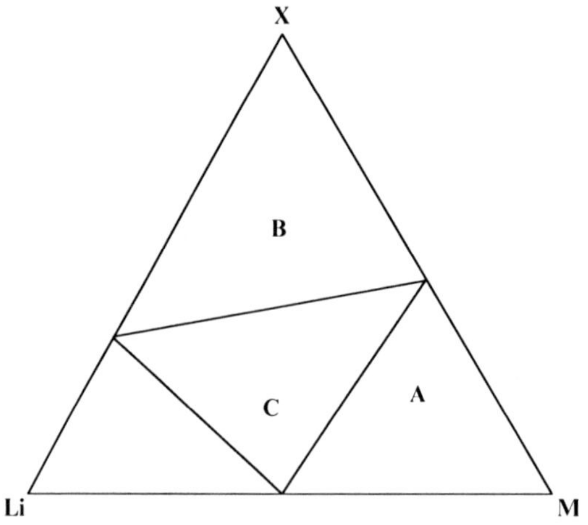

Fig. 4.6 Hypothetic ternary phase diagram in which there is one intermediate phase in each of the binary systems

In all three cases, the number of moles of lithium stored per mole of M does not change, but the weight of the electrode will change, depending upon the relative weights of M and X. In addition, the average electrode potential becomes closer to that of pure lithium. This can be either advantageous or disadvantageous, depending upon whether the material is used as a negative electrode or as a positive electrode in a lithium-based cell.

Another ternary phase configuration is shown in Fig. 4.6. In this case, it is assumed that there is also an intermediate phase in the Li–X binary system. The weight of the electrode per mole of Li will be reduced if the weight of Li_xX per mole of Li is less than that of Li per mole of Li_yM.

In practice, a binary system containing several intermediate phases may not be useful over its entire range of lithium composition, due to the change of the potential with composition. Poor diffusion kinetics in one of the intermediate phases or the terminal phase can also be deleterious.

There are many other possible configurations in ternary systems, including those containing ternary phases in the interior of the diagram. In screening possible systems for study, however, a logical starting point is to examine systems with known binary and ternary phases.

4.4 An Example: The Sodium/Nickel Chloride "Zebra" System

Some years ago an interesting battery system suddenly appeared that had been initially developed in secret in South Africa and England. It is based upon the use of the solid electrolyte sodium beta alumina, as is the Na/Na_xS system. It soon became known as the *"Zebra" cell.*

It operates at 250–300°C, and uses liquid sodium as the negative electrode, which is enclosed in a solid β-aluminum tube. At this temperature sodium is liquid, and the ionic conductivity of the β-alumina is quite high. When the cell is fully charged, the positive electrode reactant is finely powdered $NiCl_2$, which is present adjacent to the β-alumina inside a solid container. Because the contact between the solid β-alumina tube and the particles of $NiCl_2$ is only at their points of contact, a second (liquid) electrolyte, $NaAlCl_4$, is also present in the outer, positive electrode compartment, part of the cell. Thus the full surface area of the $NiCl_2$ particles acts as the electrochemical interface, which greatly increases the kinetics.

Thus this electrochemical system, when charged, has the configuration:

$$Na/Na\beta\text{-alumina}/NaAlCl_4,\ NiCl_2$$

The physical arrangement of this cell is shown schematically in Fig. 4.7.

The electrochemical behavior of a Zebra cell can be understood by consideration of the Na–Ni–Cl ternary phase diagram. Thermodynamic data indicate that there are only two binary phases in this ternary system, $NiCl_2$ and NaCl. They lie on two different sides of the ternary Na–Ni–Cl phase diagram. Since the total area must be divided into triangles, it is evident that there are two possibilities. There is either a tie line from $NiCl_2$ to the Na corner, or there is one from NaCl to the Ni corner. The decision as to which of these is stable can be determined by the direction of the virtual reaction

$$2Na + NiCl_2 = 2NaCl + Ni \tag{4.4}$$

The Gibbs free energy change in this virtual reaction is given by

$$\Delta G_r{}^\circ = 2\Delta G_f{}^\circ(NaCl) - \Delta G_f{}^\circ(NiCl_2) \tag{4.5}$$

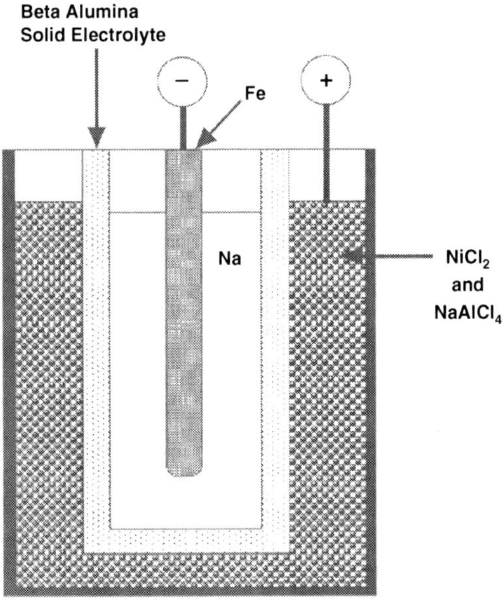

Fig. 4.7 Schematic view of the "Zebra" cell, which operates at 250–300°C

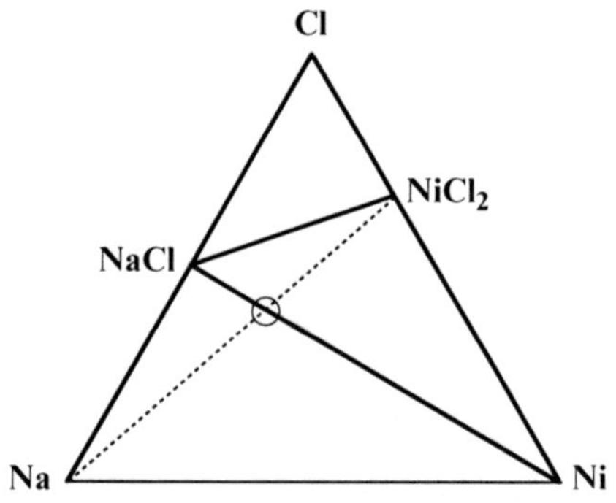

Fig. 4.8 The Na–Ni–Cl ternary phase diagram, showing the locus of the overall composition as Na reacts with $NiCl_2$

Values of the standard Gibbs free energies at 275°C of NaCl and $NiCl_2$ are -360.25 and -221.12 kJ/mol, respectively. Therefore the reaction in (4.4) will tend to go to the right, and the tie line between NaCl and Ni is more stable than the one between $NiCl_2$ and Na.

As a result, the phase stability diagram must be as shown by the solid lines in Fig. 4.8. As Na reacts with $NiCl_2$ the overall composition of the positive electrode follows the dotted line in that figure. When it reaches the composition indicated by the small circle, all the $NiCl_2$ will have been consumed, and only NaCl and Ni are present.

So long as the overall composition remains in the NaCl–$NiCl_2$–Ni triangle, the potential is constant. Its value can be calculated from the Gibbs free energy of reaction value corresponding to (4.5). The voltage of the positive electrode with respect to the pure Na negative electrode is given by

$$\Delta E = -\Delta G_r{}^{\circ}/zF \tag{4.6}$$

where $z = 2$, according to the reaction in (4.4). The result is that the potential of all compositions within that triangle in the ternary diagram, and also across the Zebra cell, is constant and equal to 2.59 V. This is also what is observed experimentally.

4.5 A Second Example: The Lithium–Copper–Chlorine Ternary System

The Li–Cu–Cl system will be used as a further example to illustrate these principles, and show how useful information can be derived from a combination of a ternary phase diagram and thermodynamic data concerning the stable phases within it.

Thermodynamic information shows that there are three stable phases within this system at 298 K, LiCl, CuCl, and $CuCl_2$. Values of their standard Gibbs free energies of formation are given in Table 4.1.

Table 4.1 Gibbs free energies of formation of phases in the Li–Cu–Cl system

Phase	$\Delta G_f{}^\circ$ at 298 K (kJ/mol)
LiCl	−384.0
CuCl	−138.7
$CuCl_2$	−173.8

All these phases fall on the edges of the isothermal Gibbs triangle. If they are assumed to be point phases, the phase stability diagram can be constructed by following a few simple rules and procedures.

1. The total area must be divided into triangles. Their edges are tie lines between pairs of phases.
2. No more than three phases can be present within a triangle. Their compositions must be at the corners.
3. Tie lines cannot cross.

The first task is to determine the stable tie lines in this system. This can be done by drawing all the possible tie lines between the stable phases on a trial basis, and then determining which of them are stable. The end result must be that the overall triangle is divided into subtriangles.

The line between LiCl and $CuCl_2$ must be stable, as there are no other possible lines that could cross it. There are four additional possibilities, lines between Li and $CuCl_2$, Li and CuCl, LiCl and CuCl, and Li and Cu. A method that can be used to determine which of these is actually stable is to write the virtual reactions between the phases at the ends of conflicting (crossing) tie lines. Which of the two pairs of phases are more stable in each case can be determined from the available thermodynamic data.

As an example, consider whether there is a tie line between LiCl and Cu or one between CuCl and Li. Both cannot be stable, for they would cross.

The virtual reaction between the pairs of possible end phases can be written as

$$LiCl + Cu = CuCl + Li \tag{4.7}$$

As before, the direction in which this virtual reaction would tend to go can be determined from the value of the standard Gibbs free energy of reaction. In this case, it is given by

$$\Delta G_r{}^\circ = \Delta G_f{}^\circ(CuCl) - \Delta G_f{}^\circ(LiCl) \tag{4.8}$$

The result is that $\Delta G_r{}^\circ$ is $(-138.7) - (-384.0) = +245.3\,\text{kJ/mol}$. Thus this reaction would tend to go to the left. This means that the combination of the phases LiCl and Cu is more stable than the combination of CuCl and Li. Thus the tie line between LiCl and Cu is stable in the phase diagram.

This implies that the tie line between LiCl and Cu is also more stable than one between $CuCl_2$ and Li, and also that a line between LiCl and CuCl exists. These conclusions can be verified by consideration of the virtual reaction between LiCl

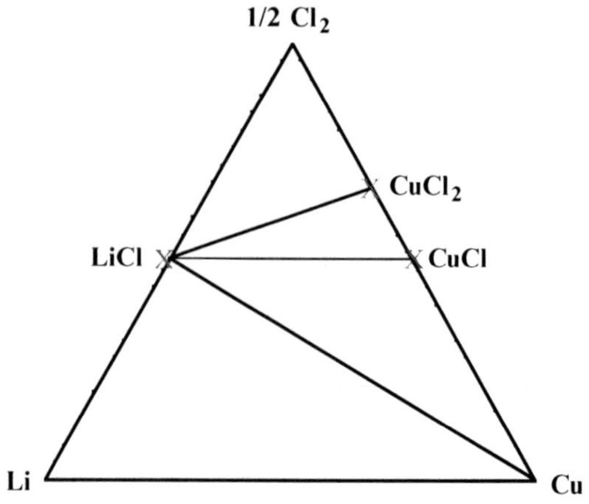

Fig. 4.9 Isothermal phase stability diagram for the Li–Cu–Cl ternary system at 25°C

and Cu, and $CuCl_2$ and Li. This reaction would be written as

$$2LiCl + Cu = CuCl_2 + 2Li \tag{4.9}$$

for which the standard Gibbs free energy of reaction is $(-173.8) - 2(-384.0) = +594.2\,\text{kJ/mol}$. Thus these conclusions were correct.

The resulting isothermal phase stability diagram for this system is shown in Fig. 4.9.

4.5.1 Calculation of the Voltages in This System

From this diagram and the thermodynamic data the voltages and capacities of electrodes in this system can also be calculated. As the first example, consider the reaction of lithium with CuCl. This reaction can be understood in terms of the ternary phase diagram as shown in Fig. 4.10.

By the addition of lithium the overall composition moves from the initial composition at the CuCl point along the dotted line toward the Li corner, as shown in Fig. 4.4. In doing so, it moves into and across the LiCl–CuCl–Cu triangle. So long as it is inside this triangle its voltage remains constant.

This voltage can be calculated from the virtual reaction that takes place by the addition of lithium as the overall composition moves into, and through, the LiCl–CuCl–Cu triangle.

$$Li + CuCl = LiCl + Cu \tag{4.10}$$

The standard Gibbs free energy change as the result of this reaction is $(-384.0) - (-138.7) = -245.3\text{kJ/mol}$. The voltage can be calculated from

$$E = -(-245.3)/[(1)(96.5)] \tag{4.11}$$

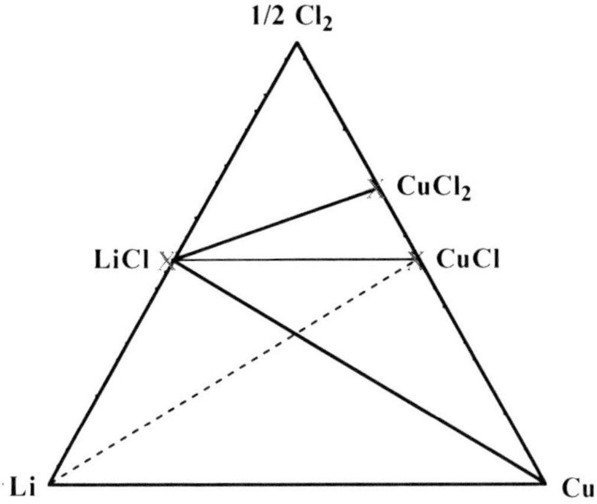

Fig. 4.10 Use of ternary phase diagram to understand the reaction of lithium with CuCl

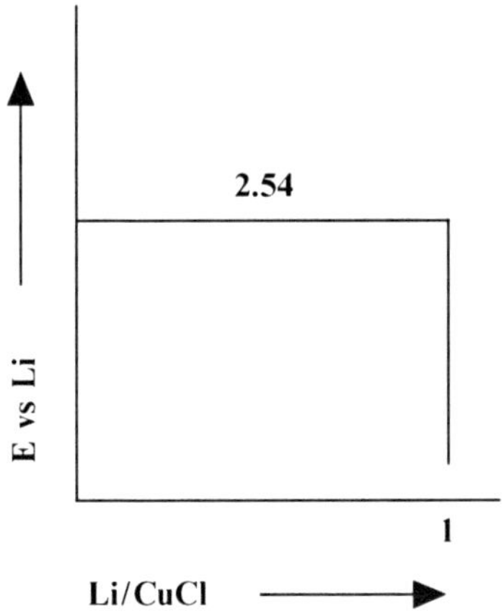

Fig. 4.11 Variation of the equilibrium voltage of Li/CuCl cell as a function of the extent of reaction

The result is 2.54 V vs. pure Li. This voltage remains constant as long as the overall composition stays in the LiCl–CuCl–Cu triangle. It is obvious from (4.10) and the phase diagram in Fig. 4.10 that up to 1 mol of Li can participate in this reaction. Thus the equilibrium titration curve, the variation of the voltage of a cell of this type as a function of composition can be drawn as in Fig. 4.11.

If, on the other hand, the positive electrode were to consist of $CuCl_2$ instead of CuCl, the overall composition would move along the dotted line shown in Fig. 4.12.

The overall composition first enters the LiCl–$CuCl_2$–CuCl triangle. The relevant virtual reaction for this triangle is

$$Li + CuCl_2 = LiCl + CuCl \tag{4.12}$$

The standard Gibbs free energy change as the result of this reaction is $(-384.0) + (-138.7) - (-173.8) = -348.9\,\mathrm{kJ/mol}$. The voltage with respect to pure Li can be calculated from

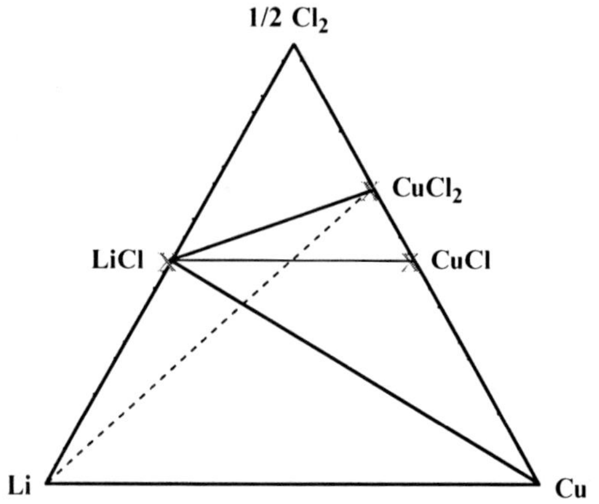

Fig. 4.12 Use of ternary phase stability diagram to understand the reaction of Li with $CuCl_2$

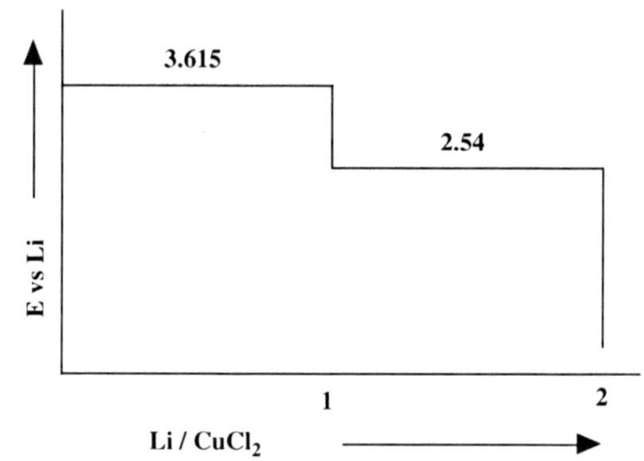

Fig. 4.13 Equilibrium titration curve for the reaction of lithium with $CuCl_2$ to form LiCl and CuCl, and then more LiCl and Cu

$$E = -\Delta G_r{}^\circ/zF = 348.9/96.5 \tag{4.13}$$

or 3.615 V vs. Li.

There will be a plateau at this voltage in the equilibrium titration curve. The LiCl cannot react further with Li. But the CuCl that is formed in this reaction can undergo a further reaction with additional lithium. When this happens the overall composition moves into and across the second triangle, whose corners are at LiCl, CuCl, and Cu. Although the reaction path is different, this is the same triangle whose voltage was calculated earlier for the reaction of lithium with CuCl. Thus the same voltage will be observed, 2.54 V vs. Li, in this second reaction, written in (4.5). The equilibrium titration curve will therefore have two plateaus, related to the two triangles that the overall composition traverses as lithium reacts with $CuCl_2$. This is shown in Fig. 4.13.

4.5.2 Experimental Arrangement for Lithium/Copper Chloride Cells

Cells based upon the reaction of lithium with either of the copper chloride phases can be constructed at ambient temperature using an electrolyte with a nonaqueous solvent, such as propylene carbonate, containing a lithium salt such as $LiClO_4$. There are a number of alternative solvents, as well as alternative salts, and this topic will be discussed in a later chapter. The important thing at the present time is that water and oxygen must be avoided, and the salt should have a relatively high solubility in the nonaqueous solvent.

4.6 Calculation of the Maximum Theoretical Specific Energies of Li/CuCl and $Li/CuCl_2$ Cells

The maximum values of specific energies that might be obtained from electrochemical cells containing either CuCl or $CuCl_2$ as positive electrode reactants can be calculated from this information.

As shown in an earlier chapter, the general relation for the maximum theoretical specific energy (MTSE) is

$$\mathrm{MTSE} = (x\mathrm{V}/\mathrm{W})(F)\,\mathrm{kJ}/\mathrm{kg} \tag{4.14}$$

where x is the number of moles of Li involved in the reaction, V the average voltage, and W the sum of the atomic weights of the reactants. F is the Faraday constant, 96,500 coulombs per mole.

In the case of a positive electrode that starts as CuCl and undergoes reaction (4.5), the sum of the atomic weights of the reactants is $(7 + 63.55 + 35.45) = 106.0\,\mathrm{g}$. The value of x is unity, and the average cell voltage is 2.54 V. Thus the MTSE is 2,312.4 kJ/kg.

This can be converted to W h/kg by dividing by 3.6, the number of kJ/W h. The result is that the MTSE can be written as 642.3 W h/kg for this reaction.

If the positive electrode starts as $CuCl_2$ and undergoes reaction (4.7) to form LiCl and CuCl the weight of the reactants is $(7 + 63.55 + (2 \times 35.45)) = 141.45\,\mathrm{g}$. The value of x is again unity, and the cell voltage was calculated to be 3.615 V. This then gives a value of MTSE of 2,466.2 kJ/kg. Alternatively, it could be expressed as 685.1 W h/kg.

If further lithium reacts with the products of this reaction, the voltage will proceed along the lower plateau, as was the case for an electrode whose composition started as CuCl. Thus additional energy is available. However, the total specific energy is not simply the sum of the specific energies that have just been calculated for the two plateau reactions independently. The reason for this is that the weight that must be considered in the calculation for the second reaction is the starting weight before the first reaction in this case.

Then, for the second plateau reaction:

$$\text{MTSE} = (1)(2.54)(96{,}500)/141.45 = 1{,}732.8\,\text{kJ/kg} \quad (4.15)$$

This is less than for the second plateau alone, starting with CuCl, which was shown earlier to be 2,312 kJ/kg. Alternatively, the specific energy content of the second plateau for an electrode that starts as $CuCl_2$ is 481.3 W h/kg instead of 642.3 W h/kg, if it were to start as CuCl.

Thus if the electrode starts out as $CuCl_2$, the total MTSE can be written as

$$\text{MTSE} = 2{,}466.2 + 1{,}732.8 = 4{,}199\,\text{kJ/kg} \quad (4.16)$$

Or alternatively, $685.1 + 481.3 = 1{,}166.4\,\text{W h/kg}$.

4.7 Specific Capacity and Capacity Density in Ternary Systems

As mentioned earlier, another parameter that is often important in battery systems is the capacity per unit weight or per unit volume. In the case of ternary systems, the capacity along a constant potential plateau is determined by the length of the path of the overall composition within the corresponding triangle. This is determined by the distance along the composition line between the binary tie lines at the boundaries of the triangles.

4.8 Another Group of Examples: Metal Hydride Systems Containing Magnesium

Binary alloys are often used as negative electrodes in hydrogen-transporting electrochemical cells. When they absorb or react with hydrogen, they are generally called *metal hydrides*. Because of the presence of hydrogen as well as the two metal components, they become ternary systems.

There is a great interest in the storage of hydrogen for a number of purposes related to the desire to reduce the dependence on petroleum. The reversible hydrogen absorption in some metal hydrides is a serious competitor for this purpose.

If the kinetics of hydrogen absorption or reaction are relatively fast, and the motion of the other constituents in the crystal structure is very sluggish, so that no structural reconstitution of the metal constituents in the microstructure takes place in the time scale of interest, such metal hydride systems can be treated as *pseudobinary systems*, i.e., hydrogen plus the metal alloy. This is the general assumption that is almost always found in the literature on the behavior of metal hydrides.

On the other hand, there are materials in which this is not the case, and the hydrogen–metal hydride combination should be treated as a ternary system.

Experiments have shown that the reaction of hydrogen with several binary magnesium alloys provides examples of such ternary systems [1, 2].

The prior examples of the reaction of lithium with the two copper chloride phases were used to illustrate how thermodynamic information can be used to determine the phase diagram and the electrochemical properties. These hydrogen/magnesium alloy systems will be discussed, however, *as reverse examples*, in which electrochemical methods can be used in order to determine the relevant phase diagrams and thermodynamic properties, as well as to determine the practical parameters of energy and capacity.

Metal hydride systems are typically studied by the use of gas absorption experiments, in which the hydrogen pressure and the temperature are the primary external variables. Electrochemical methods can generally also be employed by the use of a suitable electrolyte and cell configuration. Variation of the cell voltage can cause a change in the difference between the effective hydrogen pressure in the two electrodes. If one electrode has a fixed hydrogen activity, the hydrogen activity in the other can be varied by the use of an applied voltage. This then causes either the absorption or desorption of hydrogen. This can be expressed by the Nernst relation

$$E = (RT/zF)\Delta \ln p(H_2) \qquad (4.17)$$

where E is the cell voltage, R the gas constant, T the absolute temperature, z the charge carried by the transporting ion (hydrogen), and F the Faraday constant. The term $\Delta \ln p(\mathrm{H_2})$ is the difference in the natural logarithms of the effective partial pressures, or activities, of hydrogen at the two electrodes.

Electrochemical methods can have several advantages over the traditional pressure–temperature methods. Since no temperature change is necessary for the absorption or desorption, data can be obtained at a constant temperature. If a stable reference is used, variation of the cell voltage determines the hydrogen activity at the surface of the alloy electrode. Large changes in hydrogen activity can be obtained by the use of relatively small differences in cell voltage. Thus the effective pressure can be easily and rapidly changed over several orders of magnitude. The amount of hydrogen added to, or deleted from, an electrode can be readily determined from the amount of electrical charge that passes through the cell.

One of the important parameters in the selection of materials for hydrogen storage is the amount of hydrogen that can be stored per unit weight of host material, the specific capacity. This is often expressed as the ratio of the weight of hydrogen absorbed to the weight of the host material. Magnesium-based hydrides are considered to be potentially very favorable in this regard. The atomic weight of magnesium is quite low, 24.3 g/mol. MgH_2 contains one mole of H_2, and the ratio 2/24.3 means 8.23 w/% hydrogen. This can be readily converted to the amount of charge stored per unit weight, i.e., the number of mA h/g. One Faraday is 96,500 coulombs, or 26,800 mA h, per equivalent. The addition of two hydrogens per magnesium means that two equivalents are involved. Thus 2,204 mA h of hydrogen can be reacted per gram of magnesium.

On the other hand, one is often interested in the amount of hydrogen that can be obtained by the decomposition of a metal hydride. This means that the weight to be considered is that of the metal plus the hydrogen, rather than just the metal itself. When this is done, it is found that 7.6 w/%, or 2,038 mA h/g hydrogen can be obtained from MgH_2.

These values for magnesium are over five times those of the materials that are commonly used as metal hydride electrodes in commercial battery systems. Thus there is continued interest in the possibility of the development of useful alloys based upon magnesium. The practical problem is that magnesium forms a very stable oxide, which acts as a barrier to the passage of hydrogen. It is very difficult to prevent the formation of this oxide on the alloy surface in contact with the aqueous electrolytes commonly used in battery systems containing metal hydrides.

One of the strategies that have been explored is to put a material such as nickel, which is stable in these electrolytes, on the surface of the magnesium. It is known that nickel acts as a mixed conductor, allowing the passage of hydrogen into the interior of the alloy. However, this surface covering cannot be maintained over many charge/discharge cycles, with the accompanying volume changes.

A different approach is to use an electrolyte in which magnesium is stable, but its oxide is not. This was demonstrated by the use of a novel intermediate temperature alkali organo-aluminate molten salt electrolyte $NaAlEt_4$ [1]. The hydride salt NaH can be dissolved into this melt, providing hydride ions, H^-, that can transport hydrogen across the cell. This salt is stable in the presence of pure Na, which can then be used as a reference, as well as a counter, electrode. More information about this interesting electrolyte will be presented in Chap. 14.

This experimental method was used to study hydrogen storage in three ternary systems involving magnesium alloys, the H–Mg–Ni, H–Mg–Cu, and H–Mg–Al systems. In order to be above the melting point of this organic anion electrolyte, these experiments were performed somewhat above 140°C.

The magnesium–nickel binary phase diagram is shown in Fig. 4.14. It shows that there are two intermediate phases, Mg_2Ni and $MgNi_2$. It is also known that magnesium forms the dihydride MgH_2. These compositions are shown on the H–Mg–Ni ternary diagram shown in Fig. 4.15. Note that the ternary diagram is drawn with hydrogen at the top in this case.

In order to explore this ternary system, an electrochemical cell was used to investigate the reaction of hydrogen with three compositions in this binary alloy system, $MgNi_2$, Mg_2Ni, and $Mg_{2.35}Ni$. Thus the overall compositions of these materials moved along the dashed lines shown in Fig. 4.16 as hydrogen was added.

It was found that the voltage went to zero as soon as hydrogen was added to the phase $MgNi_2$. However, in the other cases, it changed suddenly from one plateau potential to another as certain compositions were reached. These transition compositions are indicated by the circles in Fig. 4.17. The values of the voltage vs. the hydrogen potential in the different compositions regions are also shown in that figure.

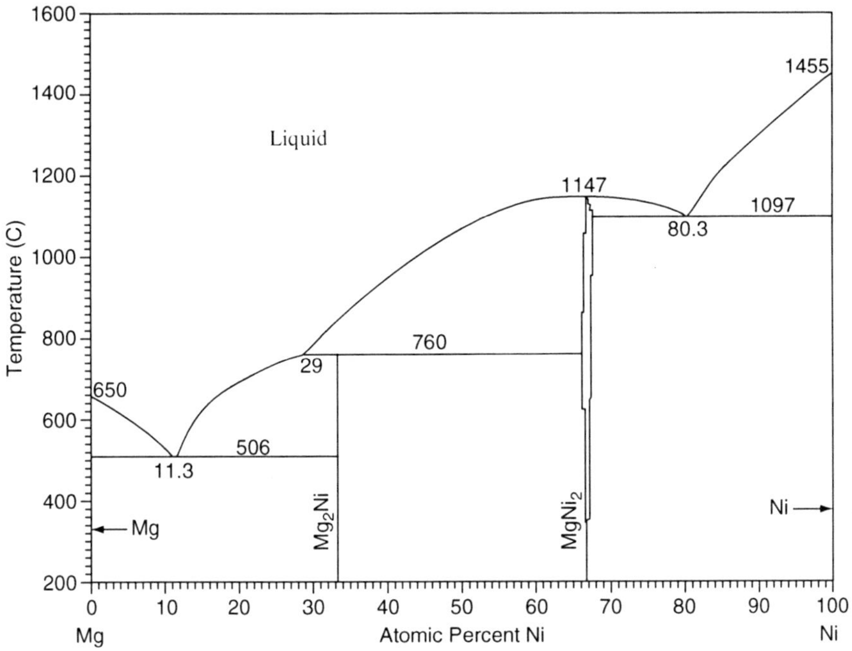

Fig. 4.14 Magnesium–nickel binary phase diagram

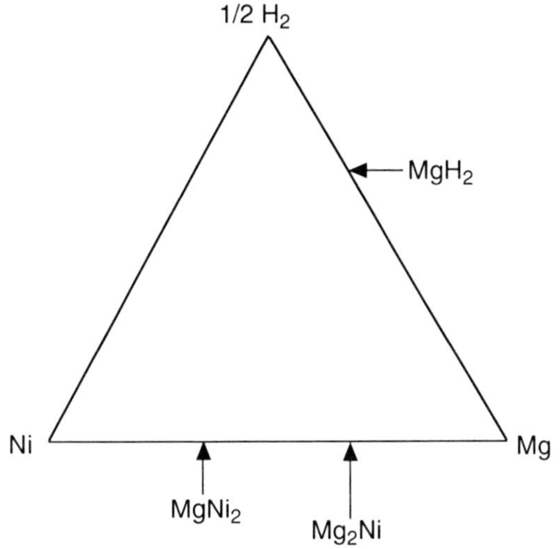

Fig. 4.15 The H–Mg–Ni ternary diagram showing only the known compositions along the binary edges

This information can be used to construct the ternary equilibrium diagram for this system. As described earlier, constant potential plateaus are found for compositions in three-phase triangles, and potential jumps occur when the composition crosses

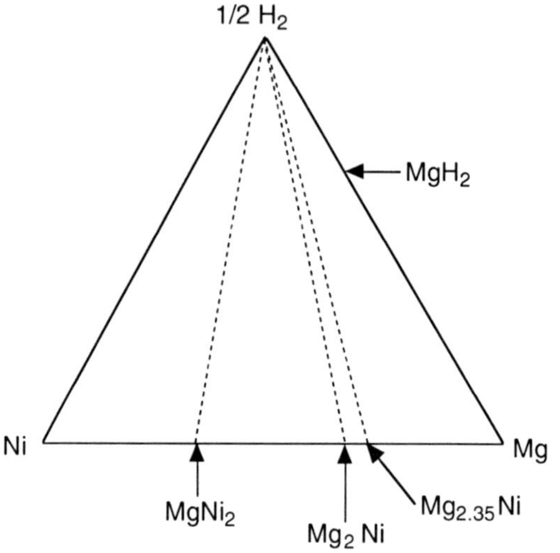

Fig. 4.16 Loci of the overall composition as hydrogen reacts with three initial Mg–Ni alloy compositions

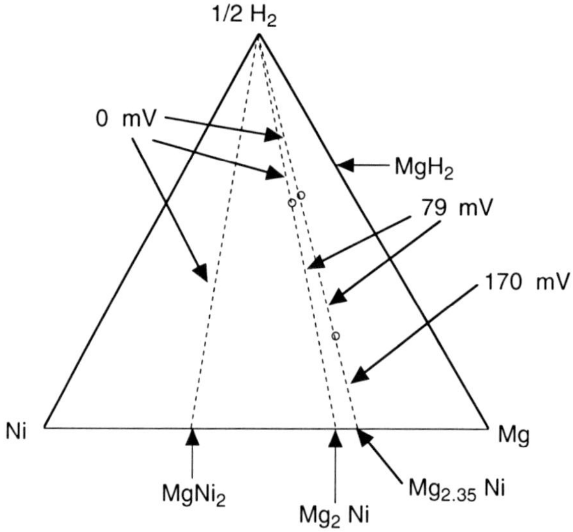

Fig. 4.17 Plateau voltages found in different composition regions

two-phase tie lines. The result is that there are no phases between $MgNi_2$ and pure hydrogen, but there must be a ternary phase with the composition Mg_2NiH_4. The resulting H–Mg–Ni ternary diagram at this temperature is shown in Fig. 4.18.

The phase Mg_2Ni reacts with four hydrogen atoms to form Mg_2NiH_4 at a constant potential of 79 mV vs. pure hydrogen. The weight of the Mg_2Ni host is 107.33 g, which is 26.83 g/mol of hydrogen atoms. This amounts to 3.73% hydrogen

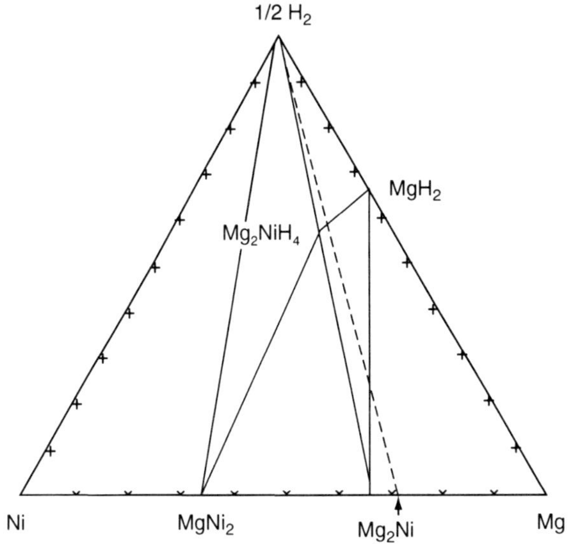

Fig. 4.18 Ternary phase stability diagram for the H–Mg–Ni system at about 140°C, derived from the compositional variation of the potential as hydrogen was reacted with three different initial binary alloy compositions

atoms stored per unit weight of the initial alloy. This is quite attractive and is considerably more than the specific capacity of the materials that are currently used in the negative electrodes of metal hydride/H_xNiO_2 batteries.

On the other hand, pure magnesium reacts to form MgH_2 at a constant potential of 107 mV vs. pure hydrogen. Because of the lighter weight of magnesium than nickel, this amounts to 8.23% hydrogen atoms per unit weight of the initial magnesium, or 7.6 w% relative to MgH_2. Thus the use of magnesium, and its conversion to MgH_2, is very attractive for hydrogen storage. There is a practical problem, however, due to the great sensitivity of magnesium to the presence of even small amounts of oxygen or water vapor in its environment.

If the initial composition is between Mg_2Ni and Mg, as is the case for the composition $Mg_{2.35}Ni$ that has been discussed earlier, there will be two potential plateaus, and their respective lengths, as well as the total amount of hydrogen stored per unit weight of the electrode, will have intermediate values, varying with the initial composition. As an example, the variation of the potential with the amount of hydrogen added to the $Mg_{2.35}Ni$ is shown in Fig. 4.19.

Similar experiments were carried out on the reaction of hydrogen with two other magnesium alloy systems, the H–Mg–Cu and H–Mg–Al systems [1]. Their ternary equilibrium diagrams were determined by using analogous methods. They are shown in Figs. 4.20 and 4.21.

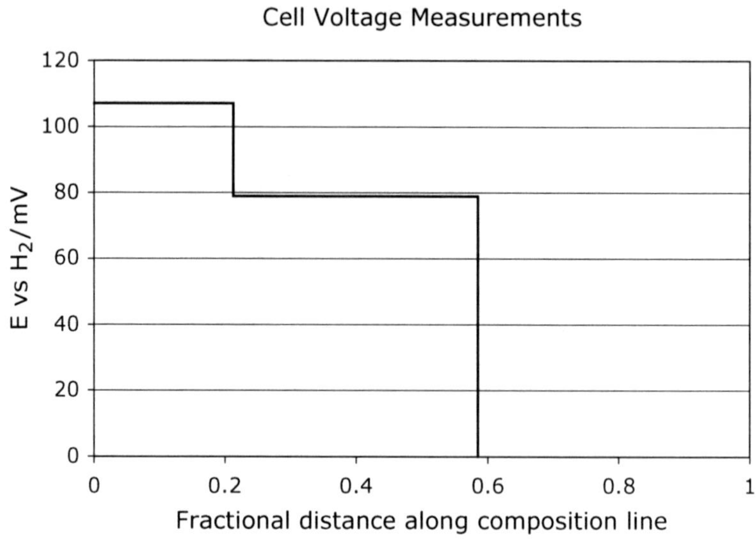

Fig. 4.19 Variation of potential as hydrogen is added to alloy with initial composition $Mg_{2.35}Ni$

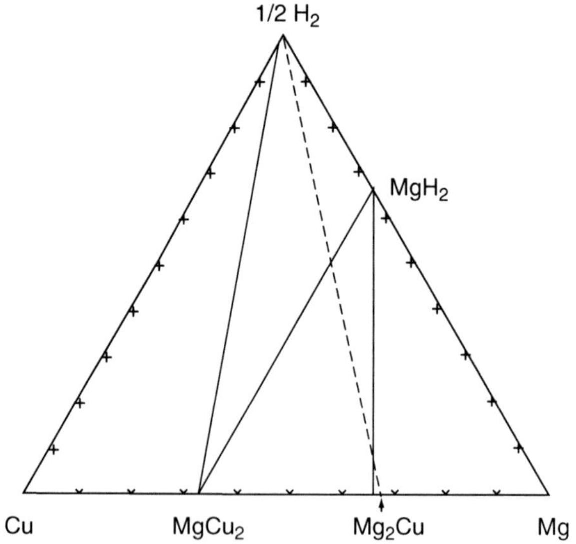

Fig. 4.20 Ternary phase stability diagram for the H–Mg–Cu system at about 140°C, derived from the compositional variation of the potential as hydrogen was reacted with different initial binary alloy compositions using organic anion molten salt electrolyte

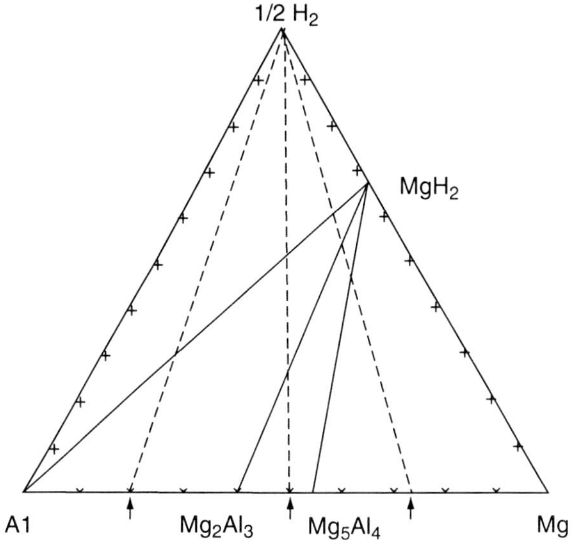

Fig. 4.21 Ternary phase stability diagram for the H–Mg–Al system at about 140°C, derived from the compositional variation of the potential as hydrogen was reacted with different initial binary alloy compositions using organic anion molten salt electrolyte

4.9 Further Ternary Examples: Lithium–Transition Metal Oxides

These same concepts and techniques have been used to investigate several lithium–transition metal oxide systems [3, 4]. They will be discussed briefly here. These examples are different from those that have been discussed thus far, for in a number of cases the initial compositions are, themselves, ternary phases, not just binary phases.

They further illustrate how electrochemical measurements on selected compositions can be used to determine the relevant phase diagrams. This makes it possible to predict the potentials and capacities of other materials within the same ternary system without having to measure them individually.

In addition, it will be seen that one can obtain a correlation between the activity of lithium, and thus the potential, and the equilibrium oxygen partial pressure, of phases and phase combinations in some cases. This provides the opportunity to predict the potentials of a number of binary and ternary materials with respect to lithium from information on the properties of relevant oxide phases alone.

Data on the ternary lithium-transition metal oxide systems that will be presented here were obtained by the use of the LiCl–KCl eutectic molten salt as electrolyte at about 400°C. They were studied at a time when there was a significant development program in the United States to develop large-scale batteries for vehicle propulsion using lithium alloys in the negative electrode and iron sulfide phases in the positive electrode. The transition metal oxides were being considered as alternatives to the sulfides.

Experiments employing this molten salt electrolyte system required the use of glove boxes that maintained both the oxygen and the nitrogen concentrations at very low levels. This salt could be used for experiments to very negative potentials, limited by the evaporation of potassium. The maximum oxygen pressure that can be tolerated is limited by the formation of Li_2O. This occurs at a partial pressure of 10^{-15} atm at 400°C. This is equivalent to 1.82 V vs. lithium at that temperature. As a result, this electrolyte cannot be used to investigate materials whose potentials are above 1.82 V relative to that of pure lithium. As will be seen in Chap. 9, many of the positive electrode materials that are of interest today operate at potentials above this limit.

The first example is the lithium–cobalt oxide ternary system. Experiments were made in which lithium was added to both the binary phase CoO and the ternary phase $LiCoO_2$. The variations of the observed equilibrium potentials as lithium was added to these phases are indicated in Fig. 4.22. It is seen that there were sudden drops from 1.807 to 1.636 V, and then to zero in the case of CoO. Starting with $LiCoO_2$, however, only one voltage jump was observed, from 1.636 to zero. Since these jumps occur when the composition crosses binary tie lines in such diagrams, it was very easy to plot the ternary figure in this case. The result is shown in Fig. 4.23, in which the values of the potential (voltage vs. pure lithium), lithium activity, cobalt activity, and oxygen partial pressure for the two relevant compositional triangles are indicated. As mentioned earlier, it was not possible to investigate the higher potential regions that are being used in positive electrodes today.

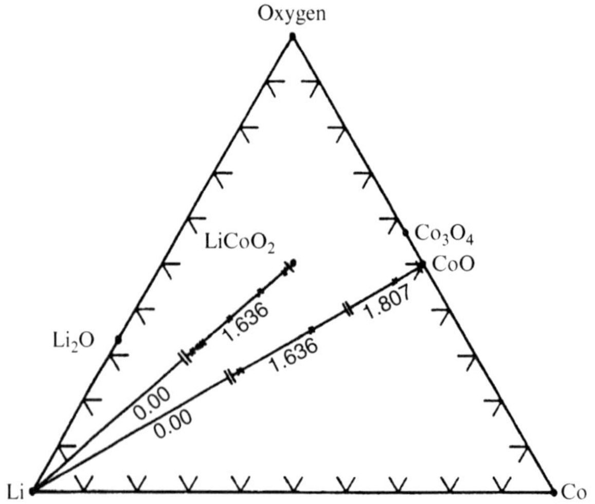

Fig. 4.22 Results of coulometric titration experiments on two compositions in the lithium–cobalt oxide system

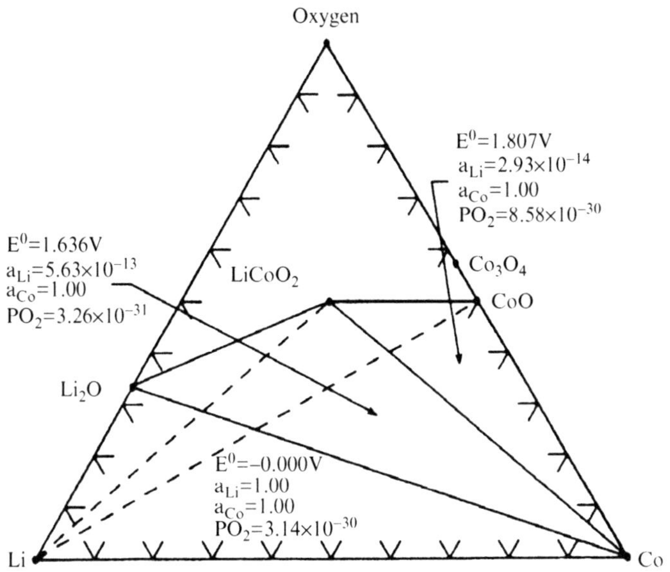

Fig. 4.23 Ternary phase stability diagram derived from the coulometric titration experiments shown in Fig. 4.22

A more complicated case is the Li–Fe–O system. Figure 4.24 shows the variation of the equilibrium potential as lithium was added to Fe_3O_4 under near-equilibrium conditions. It is seen that this is a more complex case, for after a small initial solid solution region there are three jumps in the potential.

Similar experiments were undertaken on materials with two other initial compositions, $LiFe_5O_8$ and $LiFeO_2$. From these data it was possible to plot out the whole ternary system within the accessible potential range, as shown in Fig. 4.25.

Investigation of the Li–Mn–O system produced results that are somewhat different from those in the Li–Co–O and Li–Fe–O systems. The variation of the potential as lithium was added to samples with initial compositions MnO, Mn_3O_4, $LiMnO_2$, and Li_2MnO_3 is shown in Fig. 4.26. The ternary equilibrium diagram that resulted is shown in Fig. 4.27. It is seen that all the two-phase tie lines do not go to the transition metal corner in this case. Instead, three of them lead to the composition Li_2O. Nevertheless, the principles and the experimental methods are the same.

It will be shown later, in Chap. 9, that some materials of this type behave quite differently at ambient temperature and higher potentials. In some cases lithium can be extracted from individual phases, which then act as insertion–extraction electrodes, with potentials that vary with the stoichiometry of individual phases. The principles involved in insertion–extraction reactions will be discussed later, in Chap. 6.

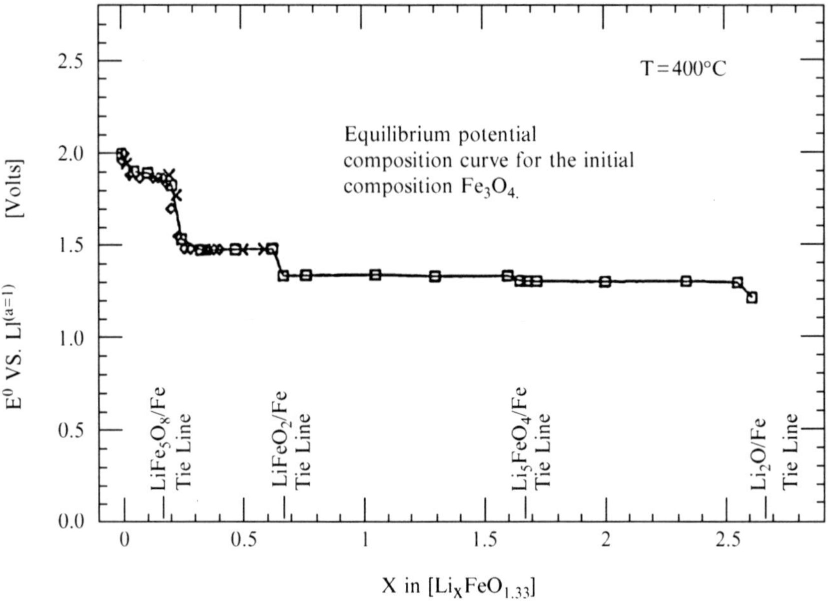

Fig. 4.24 Results of a coulometric titration experiment on a sample with an initial composition Fe_3O_4

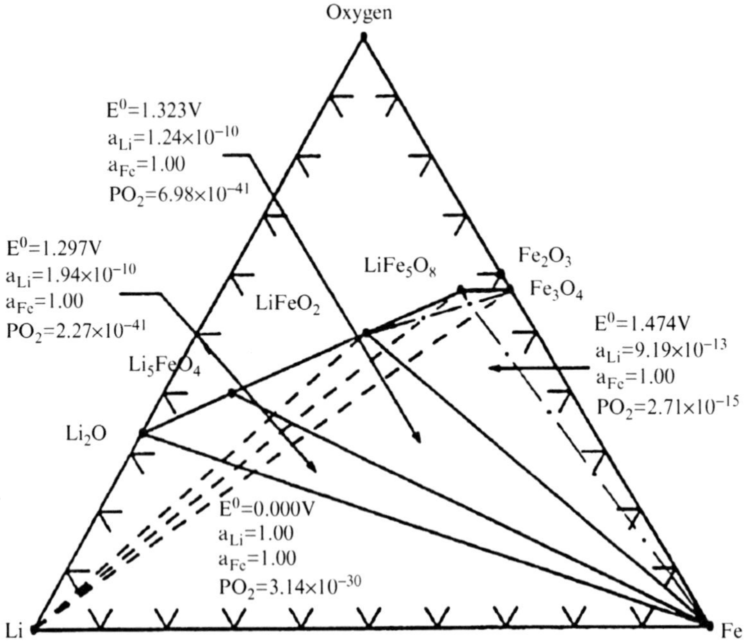

Fig. 4.25 Ternary phase stability diagram derived from coulometric titration measurements on materials in the Li–Fe–O ternary system

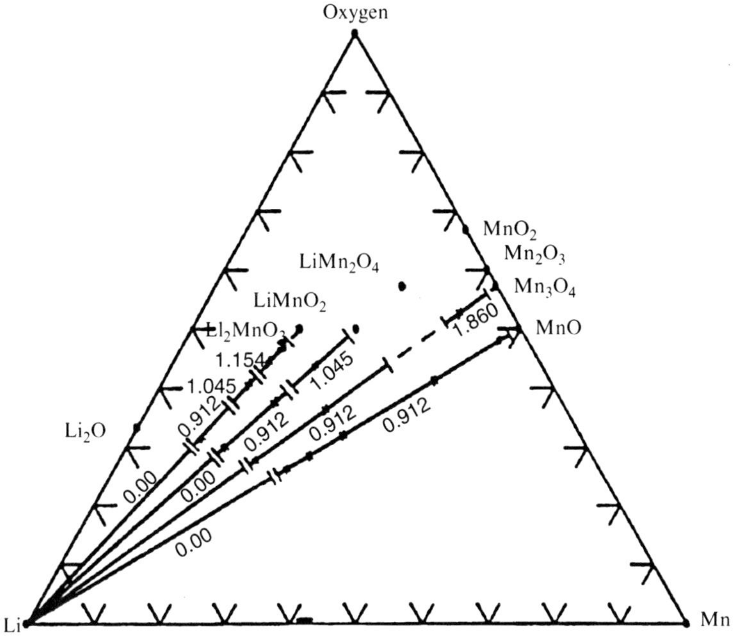

Fig. 4.26 Results of coulometric titration experiments on several phases in the Li–Mn–O ternary system

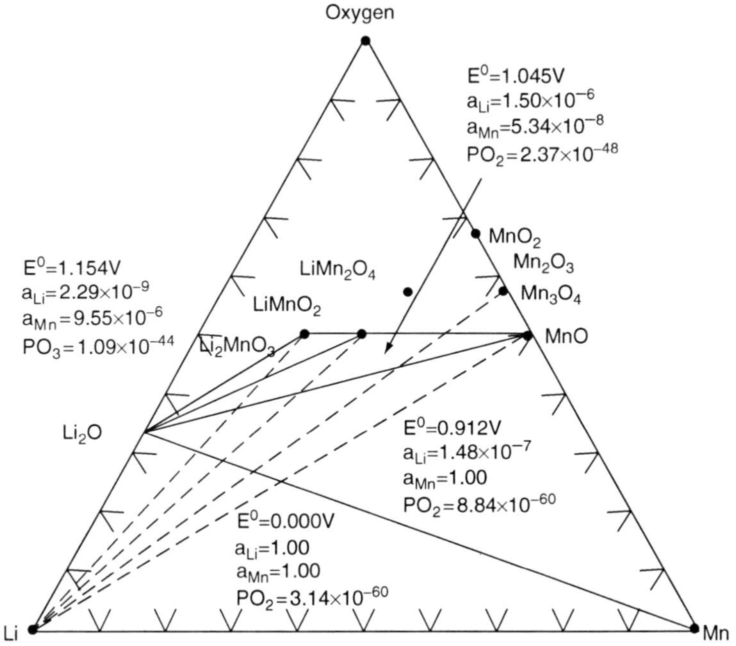

Fig. 4.27 Ternary phase stability diagram that resulted from the coulometric titration results shown in Fig. 4.26

4.10 Ternary Systems Composed of Two Binary Metal Alloys

In addition to the ternary systems that involve a nonmetal component that have been discussed thus far in this chapter, it is also possible to have ternaries in which all three components are metals. Such materials are possible candidates for use as reactants in the negative electrode of lithium battery systems.

One example will be briefly mentioned here, the Li–Cd–Sn system, which is composed of two binary lithium alloy systems, Li–Cd and Li–Sn. As will be described in Chap. 8, these, as well as a number of other binary metal alloy systems, have been investigated at ambient temperature. Their kinetic behavior is sufficiently fast that they can be used at these low temperatures. This system, as well as others, will be discussed there in connection with the important mixed-conductor matrix concept.

4.10.1 An Example: The Li–Cd–Sn System at Ambient Temperature

If the two binary phase diagrams and their related thermodynamic information are known, it is possible to predict the related ternary phase stability diagram, assuming

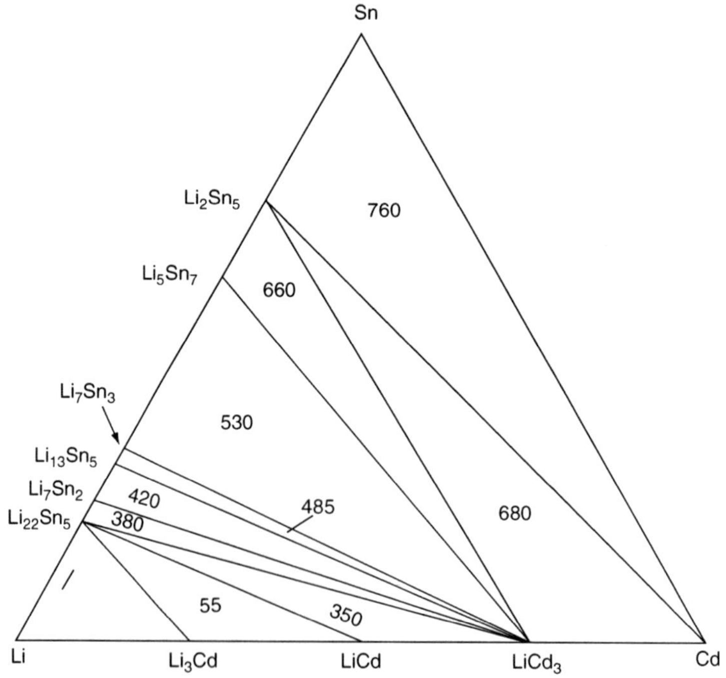

Fig. 4.28 Ternary phase stability diagram for the Li–Cd–Sn system. The numbers are the values of the voltage of all compositions in the various subtriangles relative to pure lithium

that no intermediate phases are stable. This assumption can be checked by making a relatively few experiments to measure the voltages of selected compositions. If they correspond to the predictions from the binary systems, there must be no additional internal phases. The value of this approach is that it gives a quick picture of what would happen if a third element were to be added as a dopant to a binary alloy.

As an example, the ternary phase stability diagram that shows the potentials of all possible alloys in the Li–Cd and Li–Sn system [5] is shown in Fig. 4.28.

4.11 What About the Presence of Additional Components?

Practical materials often include additional elements, either as deliberately added dopants, or as impurities. If these elements are in solid solution in the major phases present in the ternary system, they can generally be considered to cause only minor deviations from the properties of the basic ternary system. Thus it is not generally necessary to consider systems with more than three components.

4.12 Summary

This rather long chapter has shown that the ideal electrochemical behavior of ternary systems, in which the components can be solids, liquids, or gases, can be understood by the use of phase stability diagrams and theoretical electrochemical titration curves. The characteristics of phase stability diagrams can be determined from thermodynamic information, and from them the related theoretical electrochemical titration curves can be determined. Important properties, such as the maximum theoretical specific energy, can then be calculated from this information. A number of examples have been discussed that illustrate the range of application of this powerful method.

References

1. C.M. Luedecke, G. Deublein and R.A. Huggins, *Hydrogen Energy Progress V*, ed. by T.N. Veziroglu and J.B. Taylor, Pergamon Press, New York, p. 1421 (1984)
2. C.M. Luedecke, G. Deublein and R.A. Huggins, J. Electrochem. Soc. 132, 52 (1985)
3. N.A. Godshall, I.D. Raistrick and R.A. Huggins, Mat. Res. Bull. 15, 561 (1980)
4. N.A. Godshall, I.D. Raistrick and R.A. Huggins, J. Electrochem. Soc. 131, 543 (1984)
5. A.A. Anani, S. Crouch-Baker and R.A. Huggins, J. Electrochem. Soc. 135, 2103 (1988)

Chapter 5
Electrode Reactions That Deviate From Complete Equilibrium

5.1 Introduction

The example that was discussed earlier, the reaction of lithium with iodine to form LiI, dealt with elements and thermodynamically stable phases. By knowing a simple parameter, the Gibbs free energy of formation of the reaction product, the cell voltage under equilibrium and near-equilibrium conditions can be calculated for this reaction. If the cell operates under a fixed pressure of iodine at the positive electrode and at a stable temperature, the Gibbs phase rule indicates that the number of the residual degrees of freedom F in both the negative and the positive electrodes is zero. Thus the voltage is independent of the extent of the cell reaction in both cases.

This is a case in which the reaction involves species that are *absolutely stable*. The description of a phase as absolutely stable means that it is in the thermodynamic state, e.g., crystal structure, with the lowest possible value of the Gibbs free energy for its chemical composition.

5.2 Stable and Metastable Equilibrium

On the other hand, there could be several versions of a phase with different structures that might be stable in the sense that they have lower values of the Gibbs free energy than would be the case with minor changes. Such a situation, in which a phase is stable against small perturbations, is described by the term *metastable*. On the other hand, it may be less stable than the *absolutely stable* modification. This can be illustrated schematically by the use of a simple mechanical model, as is illustrated in Fig. 5.1.

This situation can also be described in terms of the changes in the potential energy of a simple block. If the block sits on its end, it is in a metastable state, and if it is tipped a small amount, its potential energy will be increased, but it will tend to revert back to its initial metastable condition. But a larger perturbation will get it

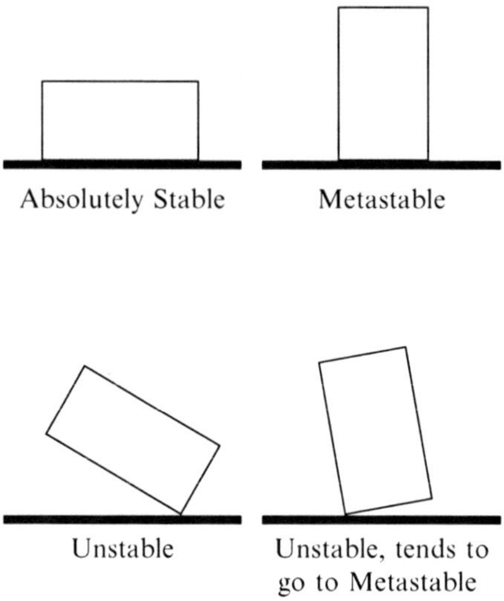

Fig. 5.1 Simple mechanical model illustrating metastable and absolutely stable states

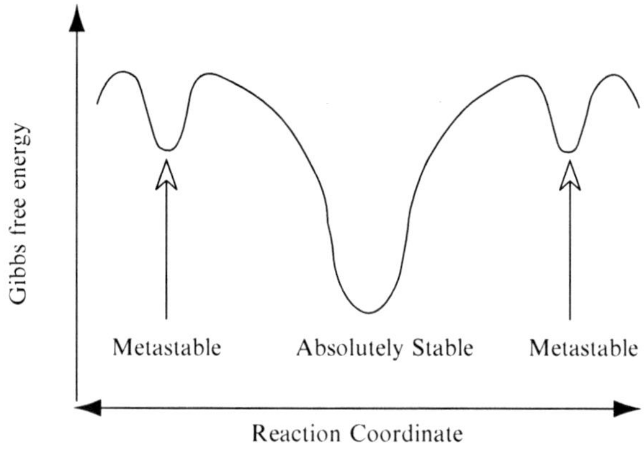

Fig. 5.2 Reaction coordinate representation of a system with metastable and absolutely stable states

over this potential energy hump so that it will tip over and land in the flat position, the absolutely stable state.

This situation can also be illustrated by the use of a reaction coordinate diagram of the type often used in discussions of chemical reaction kinetics, as shown in Fig. 5.2.

This discussion does not only apply to single phases, for it is possible to have a situation in which a material has a microstructure that consists of a metastable single phase, whereas the absolutely stable situation involves the presence of two, or perhaps more, phases. In order for the system to go from the metastable single-phase situation to the more stable *polyphase* structure it is necessary to *nucleate* the additional phase or phases as well as to change the composition of the initial metastable phase. This may be kinetically very difficult.

In the case of the Li/I system, where there is only one realistic structure for the reactant and product phases, only the absolutely stable situation has to be considered.

However, in other materials systems the situation is often different at lower temperatures from that at high temperatures, where absolutely stable phases are generally present. As will be discussed later, metastable phases and metastable crystal structures often play significant roles at ambient temperatures.

5.3 Selective Equilibrium

There is also the possibility that a material may attain equilibrium in some respects, but not in others. Some of the most interesting and important ambient temperature materials fall into this category.

A number of the reactants in ambient temperature battery systems have crystal structures that can be described as a composite consisting of a highly mobile ionic species within a relatively stable host structure.

Such structures are sometimes described as having two different *sublattices*, one of which has a high degree of mobility, and the other is highly stable, for its structural components are rigidly bound. The guest species with high mobilities are typically rather small and move about through interstitial tunnels in the surrounding rigid host structure. The species in the mobile sublattice can readily come to equilibrium with the thermodynamic forces upon them, whereas the more tightly bound parts of the host structure cannot.

The term *selective equilibrium* can be used for this situation. Under these conditions, the stable part of the crystal structure can be treated as a single component when considering the applicability of the Gibbs phase rule. An example that will be discussed later is the phase Li_xTiS_2. The structure of this material can be thought of as consisting of rigid planar slabs of covalently bonded TiS_2, with mobile lithium ions in the space between them. The lithium species readily attain equilibrium with the external environment at ambient temperatures, whereas the TiS_2 part of the structure is relatively inert so that it can be considered to be a single component. Thus at a fixed temperature and total pressure the number of residual degrees of freedom is 1. This means that the value of one additional thermodynamic parameter will determine all the intensive variables. As an example, a change in the electrical potential causes a change in the equilibrium amount of lithium in the structure, the value of x in Li_xTiS_2.

5.4 Soft Chemistry (Chimie Douce)

The term *soft chemistry*, or *chimie douce* in French, where much of the early work took place [1,2], is sometimes used to describe reactions or chemical changes that involve only the relatively mobile parts of the crystal structure, whereas the balance of the structure remains relatively unchanged. Such reactions are often highly reversible, and are discussed further in Chap. 6, that discusses *insertion reaction electrodes.*

5.5 Formation of Amorphous vs. Crystalline Structures

An amorphous structure can result when a phase is formed under conditions in which complete equilibrium and the expected crystalline structure cannot be attained. Although they may have some localized ordered arrangements, amorphous structures do not have regular long-range arrangements of their constituent atoms or ions. Amorphous structures are always less stable than the crystalline structure with the same composition. Thus they have less negative values of the Gibbs free energy of formation than their crystalline cousins.

If the phase LiM can be electrochemically synthesized by the reaction of lithium with species M, a type of reconstitution reaction, there will be a corresponding constant voltage two-phase plateau in the titration curve related to that reaction. The magnitude of the plateau voltage is determined by the Gibbs free energy of the product phase, as described earlier. Because of its less negative Gibbs free energy of formation, the potential of the plateau related to the formation of an amorphous LiM phase must always be lower than that of the corresponding crystalline version of LiM. This is illustrated schematically in Fig. 5.3.

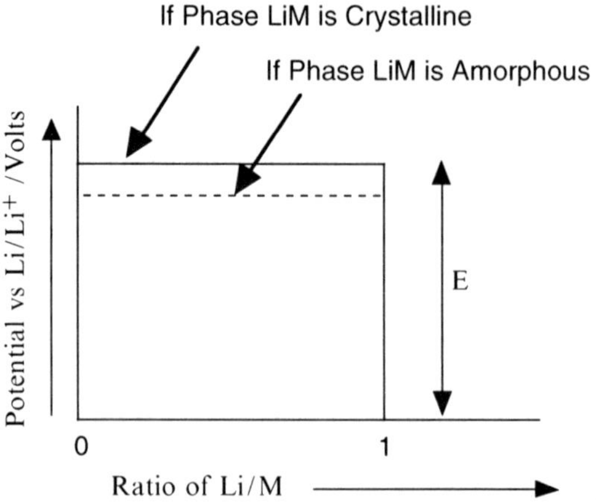

Fig. 5.3 Schematic drawing of the voltage of galvanic cell as a function of overall composition for a simple formation reaction Li + M = LiM for two cases, one in which the LiM product is crystalline, and the other in which it is amorphous

This has interesting consequences for the case in which two intermediate phases can be formed. As an example, assume that lithium can react with material M to form two phases in sequence, LiM and Li_2M. The reaction for the formation of the first phase, LiM, is

$$Li + M = LiM \tag{5.1}$$

and if the phase LiM has a very narrow range of composition, the equilibrium titration curve, a plot of potential E vs. composition, will look like that shown in Fig. 5.3.

The plateau voltage is given by

$$E = -\Delta G_f{}^{\circ}(LiM)/F \tag{5.2}$$

If additional lithium can react with LiM to form the phase Li_2M there will be an additional voltage plateau, whose potential is determined by the reaction

$$Li + LiM = Li_2M \tag{5.3}$$

This is shown schematically in Fig. 5.4.

The voltage of the second plateau is lower than that of the first and is given by

$$E = -\left[\Delta G_f{}^{\circ}(Li_2M) - \Delta G_f{}^{\circ}(LiM)\right]/F \tag{5.4}$$

But what if the first phase, LiM, is amorphous, rather than crystalline? As mentioned earlier, this means that Gibbs free energy of formation of that phase is smaller and the voltage of the first plateau is reduced.

The total Gibbs free energy of the two reactions is determined, however, by the Gibbs free energy of formation of the final phase, Li_2M. This is not changed by the formation of the intermediate phase LiM. The total area under the curve is thus a constant. The interesting result is that if the voltage of the first plateau is reduced, the voltage of the second one must be correspondingly increased. This can be depicted as in Fig. 5.5.

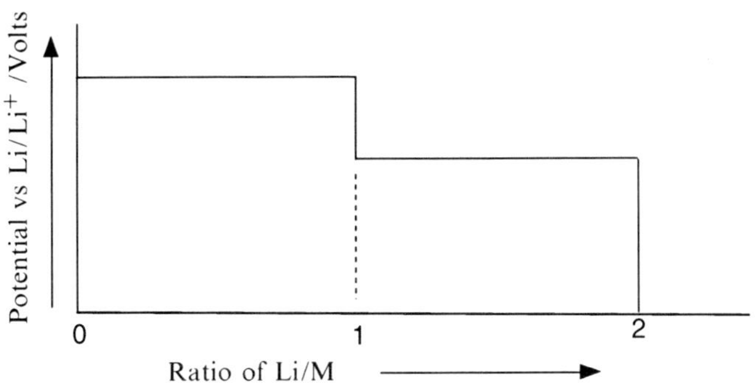

Fig. 5.4 Schematic titration curve for a sequence of two reactions of Li with M, first forming LiM, and then forming Li_2M

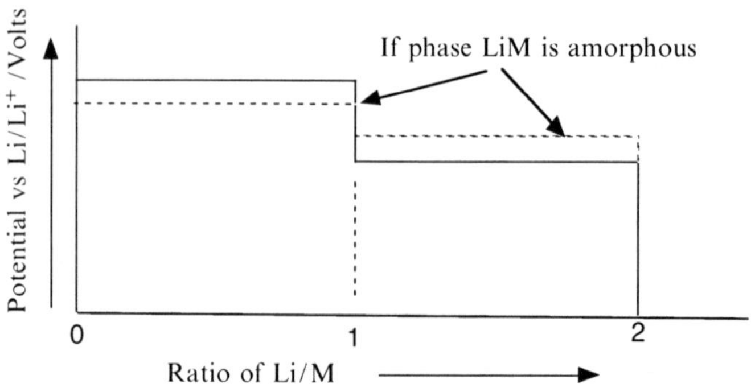

Fig. 5.5 Change in the schematic titration curve if the first product, LiM, is amorphous. The voltage of the second plateau must be higher to compensate for the reduced voltage of the first plateau

Thus the reduced stability of the intermediate phase decreases the magnitude of the step in the titration curve. Therefore the overall behavior approaches what it would be if the intermediate phase did not form at all, and there would only be one reaction, the direct formation of phase Li_2M.

5.6 The Conversion of Crystalline to Amorphous Structures by Insertion Reactions

There are a number of cases in which it has been found that the insertion of guest species into host crystal structures can cause them to become amorphous. This will be discussed in Chap. 6.

5.7 Deviations from Equilibrium for Kinetic Reasons

The observed potentials and capacities of electrodes are often displaced from those that would be expected from equilibrium thermodynamic considerations because of kinetic limitations. There may not be sufficient time to attain compositional and/or structural changes that should, in principle, occur. This is more likely to occur at lower temperatures and under higher current conditions.

The influence of increasing deviations from equilibrium conditions upon the behavior of a simple reconstitution reaction is shown schematically in Fig. 5.6. It is seen that both the potential and the apparent capacity can deviate significantly from equilibrium values.

The kinetics of electrode reactions, and methods that can be used to evaluate them, will be discussed in subsequent chapters.

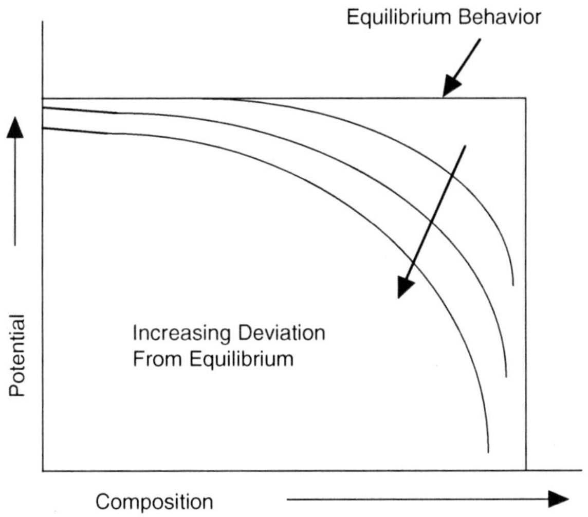

Fig. 5.6 Schematic representation of the influence of kinetic limitations upon both the potential and the capacity of an electrode reaction

References

1. J. Livage. *Le Monde*, October 26, 1977.
2. J. Rouxel, *Soft Chemistry Routes to New Materials – Chimie Douce*, Materials Science Forum, 1994, pp. 152–153.

Chapter 6
Insertion Reaction Electrodes

6.1 Introduction

The topic of *insertion reaction electrodes* did not even appear in discussions of batteries and related phenomena just a few years ago, but is a major feature of some of the most important modern battery systems today. Instead of reactions occurring on the surface of solid electrodes, as in traditional electrochemical systems, what happens *inside* the electrodes is now recognized to be of critical importance.

A few years after the surprise discovery that ions can move surprisingly fast inside certain solids, enabling their use as solid electrolytes, it was recognized that some ions can move rapidly into and out of some other (electrically conducting) materials. The first use of insertion reaction materials was for nonblocking electrodes to assist the investigation of the ionic conductivity of the (then) newly discovered ambient temperature solid electrolyte, sodium beta alumina [1–3]. Their very important use as charge-storing electrodes began to appear shortly thereafter.

This phenomenon is a key feature of the electrodes in many of the most important battery systems today, such as the lithium-ion cells. Specific examples will be discussed in later chapters.

Many examples are now known in which a mobile guest species can be *inserted into*, or *removed (extracted) from*, a *stable host crystal structure*. This phenomenon is an example of both *soft chemistry* and *selective equilibrium*, in which the mobile species can readily come to equilibrium, but this may not be true of the host, or of the overall composition. The mobile species can be atoms, ions, or molecules, and their concentration is typically determined by equilibrium with the thermodynamic conditions imposed on the surface of the solid phase.

In the simplest cases, there is little, if any, change in the structure of the host. There may be modest changes in the volume, related to bond distances, and possibly directions, but the general character of the host is preserved. In many cases the *insertion* of guest species is reversible, and they can also be *extracted* or *deleted*, returning the host material to its prior structure.

R.A. Huggins, *Advanced Batteries*: Materials Science Aspects,

The terms "*intercalation*" and "*de-intercalation*" are often used for reactions involving the insertion and extraction of guest species for the specific case of host materials that have *layer-type crystal structures*. On the other hand, "*insertion*" and "*extraction*" are more general terms. Reactions of this type are most likely to occur when the host has an open framework or a layered type of crystal structure, so that there is space available for the presence of additional small ionic species. Since such reactions involve a change in the chemical composition of the host material, they can also be called *solid solution reactions*.

Insertion reactions are generally *topotactic*, with the guest species moving into, and residing in, specific sites within the host lattice structure. These sites can often be thought of as *interstitial sites* in the host crystal lattice that are otherwise empty. The occurrence of a *topotactic* reaction implies some three-dimensional correspondence between the crystal structures of the parent and the product. On the other hand, the term *epitaxy* relates to a correspondence that is only two dimensional, such as on a surface.

It has been known for a long time that large quantities of hydrogen can be inserted into, and extracted from, palladium and some of its alloys. Palladium–silver alloys are commonly used as hydrogen-pass filters. Several types of materials with layer structures, including graphite and some clays, are also often used to remove contaminants from water by absorbing them into their crystal structures.

The most common examples of interest in connection with electrochemical phenomena involve the insertion or extraction of relatively small guest cationic species, such as H^+, Li^+, and Na^+. However, it will be shown later that there are materials in which anionic species can also be inserted into a host structure.

It should be remembered that electrostatic energy considerations dictate that only neutral species, or neutral combinations of species, can be added to, or deleted from, solids. Thus the addition of cations requires the concurrent addition of electrons, and the extraction of cations is accompanied by either the insertion of holes or the deletion of electrons. Thus this phenomenon almost always involves materials that have at least some modicum of electronic conductivity.

In some cases, however, the insertion or extraction of mobile atomic or ionic species causes irreversible changes in the structure of the host material, and the reversal of this process does not return the host to its prior structure. In extreme cases, the structure may be so distorted that it becomes *amorphous*. These matters will be discussed later.

As mentioned earlier, the term "*soft chemistry*" is now often employed to describe compositional and structural changes that result from the insertion or extraction of mobile species from relatively stable host structures.

Insertion reactions are much more prevalent at lower temperatures than at high temperatures. The mobility of the component species in the host structure generally increases rapidly with temperature. This allows much more significant changes in the overall structure to occur, leading to *reconstitution reactions*, with substantial structural changes, rather than only the motion of the more mobile species, at elevated temperatures. Reconstitution reactions typically can be thought of as involving bond breakage, atomic reorganization, and the formation of new bonds.

6.2 Examples of the Insertion of Guest Species into Layer Structures

Many materials have crystal structures that can be characterized as being composed of rather stiff *covalently bonded slabs* containing several layers of atoms. These slabs are held together by relatively strong, e.g., covalent, bonds. But adjacent slabs are bound to each other by relatively weak *van der Waals* forces. The space between the tightly bound slabs is called the *gallery space*, and additional species can reside there. Depending upon the identity, size, and charge of any inserted species present, the interslab dimensions can be varied.

Materials with the CdI_2 structure represent a simple example. They have a basic stoichiometry MX_2, and can be viewed as consisting of close-packed layers of negatively charged X ions held together by strong covalent bonding to positive M cations. In this case, the cations are octahedrally coordinated by six X neighbors, and the stacking of the X layers is hexagonal, with alternate layers directly above and below each other. This is generally described as ABABAB stacking.

This structure can be depicted as shown schematically in Fig. 6.1. Examples of materials with this type of crystal structure are CdI_2, $Mg(OH)_2$, $Fe(OH)_2$, $Ni(OH)_2$, and TiS_2. Another, simpler, way to depict these structures is illustrated in Fig. 6.2 for the case of TiS_2.

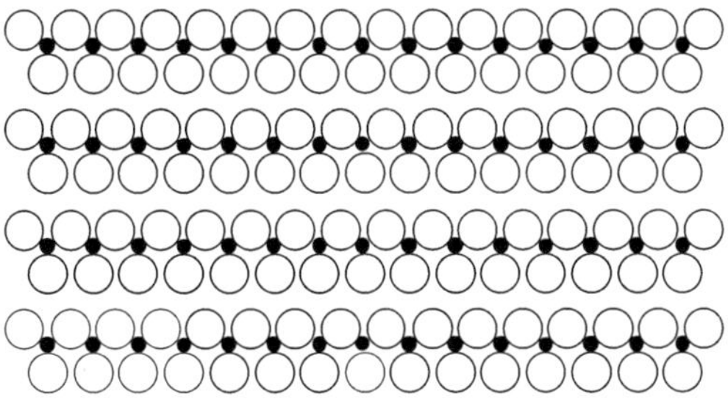

Fig. 6.1 Simple schematic model of a layer-type crystal structure with hexagonal ABABAB stacking. The empty areas between the covalently bonded slabs are called galleries

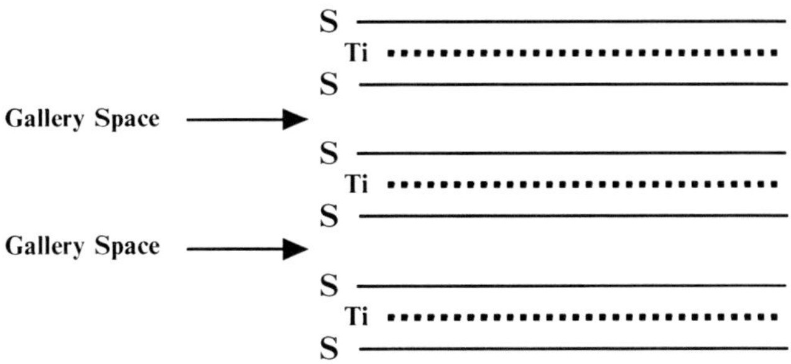

Fig. 6.2 Another type of model of a layer-type crystal structure. The example is TiS_2

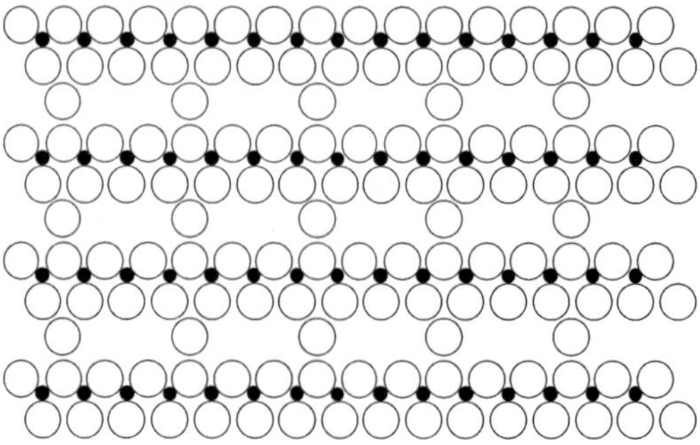

Fig. 6.3 Schematic model of pillared layer structure

6.3 Floating and Pillared Layer Structures

In many cases, the mobile species move into and through sites in the previously empty gallery space between slabs of host material that are held together only by relatively weak van der Waals forces. The slabs can then be described as floating, and the presence of guest species often results in a significant change in the interslab spacing.

However, in other cases the slabs are already rigidly connected by *pillars*, which partially fill the galleries through which the mobile species move. The pillar species are typically immobile and thus are different from the mobile guest species. Because of the presence of the static pillars, the mobile species move through a two-dimensional network of interconnected tunnels, instead of through a sheet of available sites.

The presence of pillars acts to determine the spacing between the slabs of the host material and thus the dimensions of the space through which mobile guest species can move. Examples of this kind will be discussed later. A simple schematic model of a pillared layer structure is shown in Fig. 6.3.

6.4 More on Terminology Related to the Insertion of Species into Solids

Sheets:

Single layers of atoms or ions. In the case of graphite, individual sheets are called *graphene* layers

Stacks:

Parallel sheets of chemically identical species

Slabs or Blocks:

Multilayer structures tightly bound together, but separated from other structural features

Example: covalently bonded MX_2 slabs such those shown in the CdI_2 structure

Galleries:

The spaces between slabs in which the bonding is relatively weak, and in which guest species typically reside

Pillars:

Immobile species within the galleries that serve to support the adjacent slabs and to hold them together

Tunnels:

Connected interstitial space within the host structure in which the guest species can move and reside. Tunnels can be empty, partly occupied, or fully occupied by guest species

Cavities:

Empty space larger than the size of a single atom vacancy

Windows:

Locations within the host structure through which the guest species have to move in order to go from one site to another. Windows are typically defined by structural units of the host structure

6.5 Types of Inserted Guest Species Configurations

There are several types of insertion reactions. In one case the mobile guest species randomly occupy sites within all the galleries, gradually filling them all up as the guest population increases. When this is the case the variation of the electric potential with composition indicates a single-phase solid solution reaction, and there can be transient composition gradients within the gallery space.

If, however, the presence of the guest species causes a modification of the host structure, the insertion process can occur by the motion of an interface that separates the region into which the guest species have moved from the area in which there are no, or fewer, guest species. Thermodynamically, this has the characteristics of a polyphase reconstitution reaction and occurs at a constant potential.

Alternatively, there can be two or more types of sites in the gallery space, with different energies, and the guest species can occupy an ordered array of sites, rather than all of them. When this is the case, changes in the overall concentration of mobile species require the translation of the interface separating the occupied regions from those that are not occupied, again characteristic of a constant-potential reconstitution reaction. These moving interfaces can remain planar, or they can develop geometrical roughness. Several possibilities are illustrated schematically in Figs. 6.4–6.6.

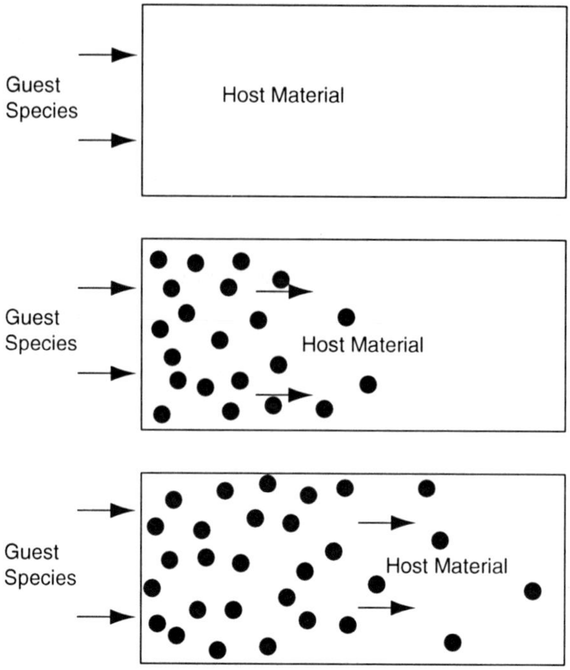

Fig. 6.4 Random diffusion of guest species into gallery space

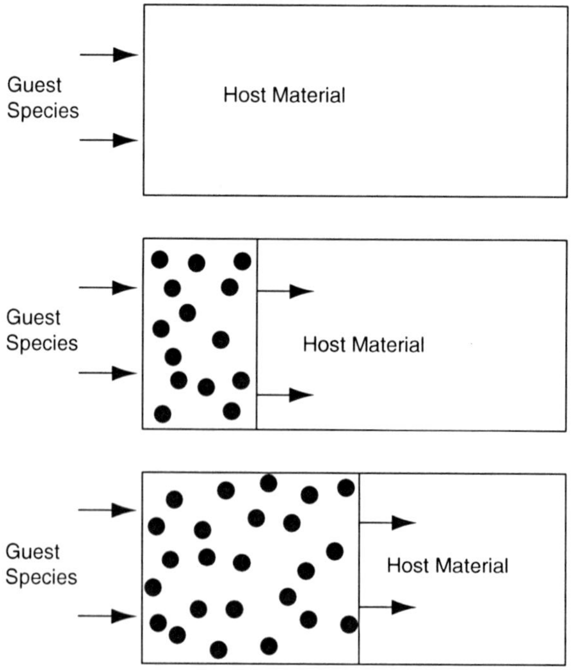

Fig. 6.5 Motion of two-phase interface when guest species is not ordered upon possible sites

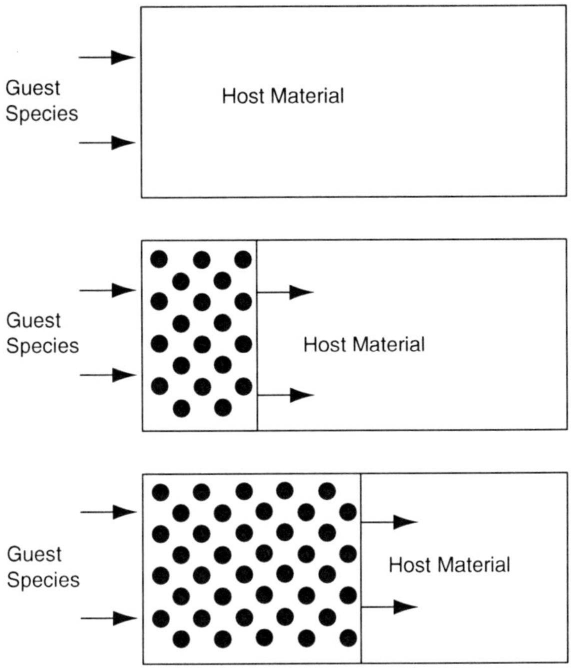

Fig. 6.6 Motion of two-phase interface when guest species is ordered upon possible sites in the gallery space

6.6 Sequential Insertion Reactions

If there are several different types of sites with different energies, insertion generally occurs on one type of site first, followed by the occupation of the other type of site. Figure 6.7 shows the potential as a function of composition during the insertion of lithium into $NiPS_3$, in which there are two types of sites available. They are occupied in sequence, with random occupation in both cases.

Another example in which there are also different types of sites available for the insertion of Li ions involves the host $K_xV_2O_5$ structure. The host crystal structure illustrating the several different types of sites for guest ions is shown schematically in Fig. 6.8 [4].

The experimentally measured coulometric titration curve for the insertion of Li ions into a member of this group of materials is shown in Fig. 6.9 [5]. It shows that the reaction involves three sequential steps. Up to about 0.4 Li can be incorporated into the first set of sites randomly. This is followed by the insertion of another 0.4 Li into another set of sites in an ordered arrangement. This means that there are two different lithium arrangements, with a moving interface between them. Thus there are two phases present, so this corresponds to a reconstitution reaction. This is then followed by another reconstitution reaction, the insertion of about one additional Li into another ordered structure.

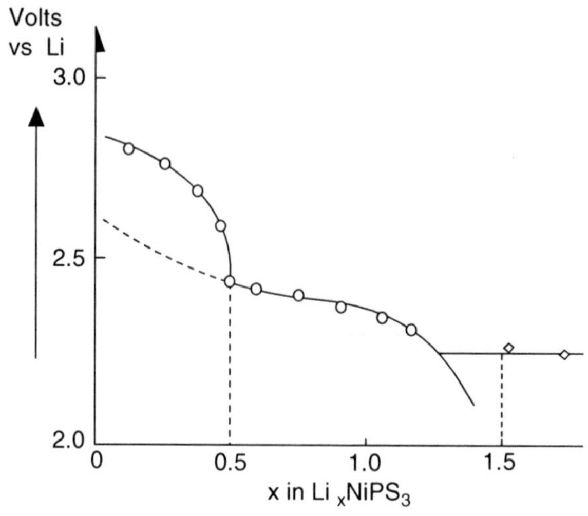

Fig. 6.7 Coulometric titration curve related to the insertion of lithium into $NiPS_3$. There is random filling of the first two types of sites. A reconstitution reaction occurs above about 1.4 Li

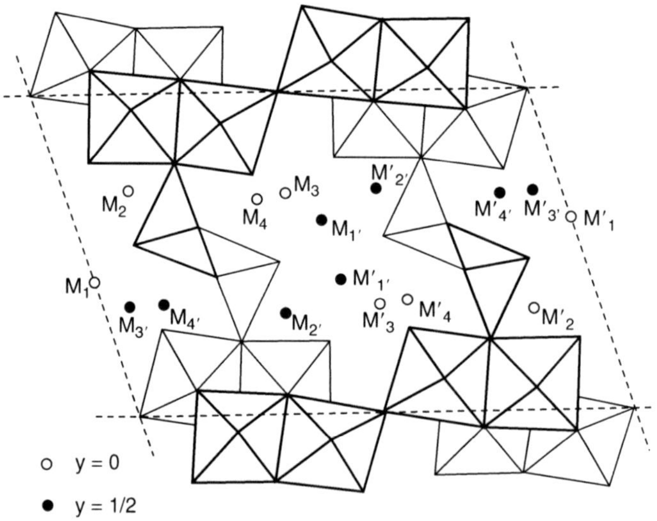

Fig. 6.8 (010) projection of the $K_xV_2O_5$ structure, showing the different types of sites for the guest species

A different type of ordered reaction involves selective occupation of particular galleries, and not others, in a material with a layered crystal structure. This phenomenon is described as "staging." If alternate galleries are occupied and intervening ones are not, the material is described as having a "second-stage" structure. If every third gallery is occupied, the structure is "third stage" and so forth. A simple model depicting staging is shown in Fig. 6.10.

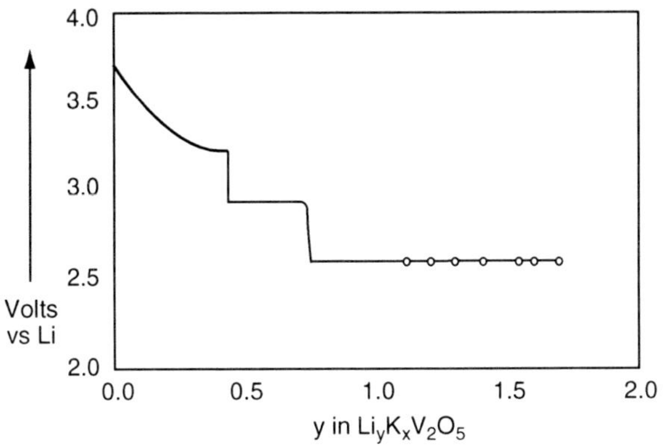

Fig. 6.9 Coulometric titration curve for the insertion of Li into $K_{0.27}V_2O_5$

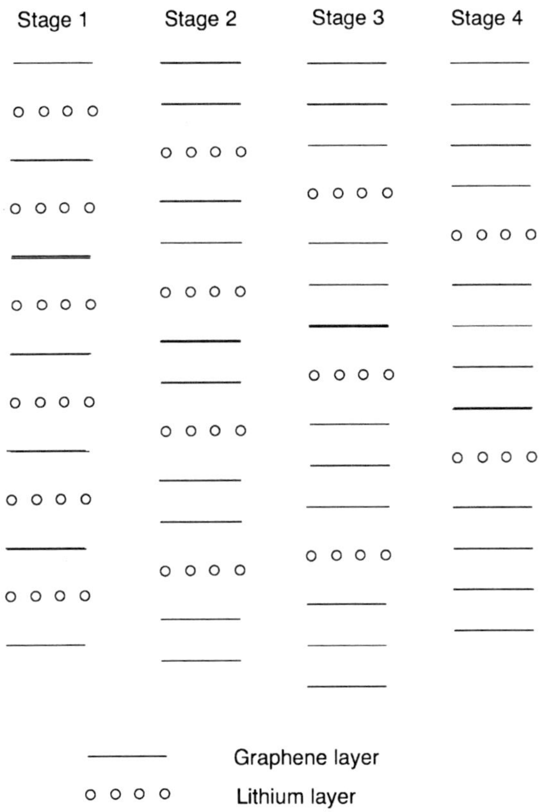

Fig. 6.10 Simple model depicting staging

6.7 Coinsertion of Solvent Species

In some cases it is found that species from the electrolyte can also move into the gallery space. This tends to be the case when the electrolyte solvent molecules are relatively small, so that they can enter without causing major disruption of the host structure. This is found to occur in some organic solvent systems, and also some aqueous electrolyte systems where the electroactive ion is surrounded by a hydration sheath. This is a matter of major concern in the case of negative electrodes in lithium systems and will be discussed at much greater length in Chap. 7.

6.8 Insertion into Materials with Parallel Linear Tunnels

The existence of staging indicates that, at least in some materials, the presence of inserted species in one part of the structure is "seen" in other parts of the structure. An interesting example of this involves the presence of mobile guest species in the material Hollandite, that has a crystal structure with parallel linear tunnels, rather than slabs.

A drawing of this structure is shown in Fig. 6.11. At low temperatures the interstitial ions within the tunnels are in an ordered arrangement upon the available sites.

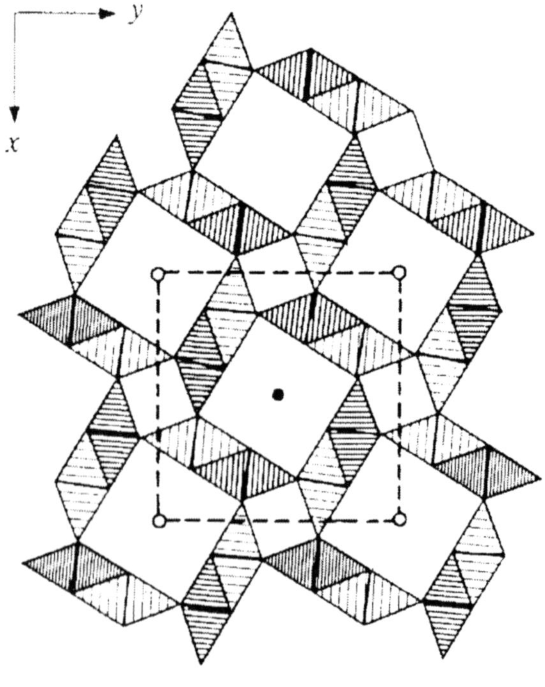

Fig. 6.11 Hollandite structure. Viewed along the *c*-axis

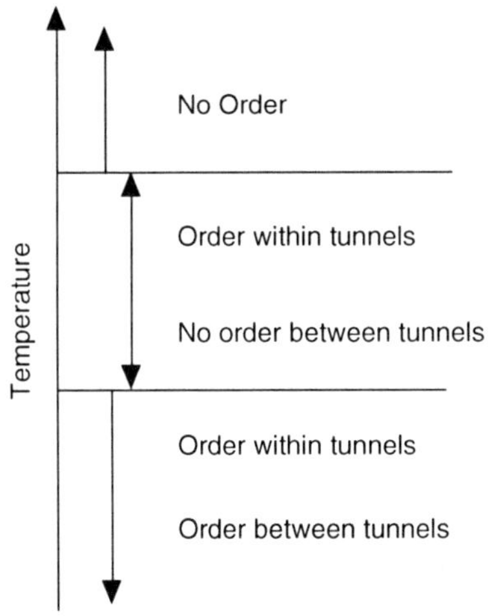

Fig. 6.12 Influence of temperature upon various types of order in structure with parallel tunnels

In addition, there is coordination between the arrangement in one tunnel with that of other nearby tunnels. Thus there is three-dimensional ordering of the guest species.

As the temperature is raised somewhat, increased thermal energy causes the ordered interaction between the mobile ion distributions in nearby tunnels to relax, although the ordering within tunnels is maintained.

At even higher temperatures the in-tunnel ordering breaks down, so that the species are distributed randomly inside the tunnels, as well. The influence of temperature is illustrated schematically in Fig. 6.12.

6.9 Changes in the Host Structure Induced by Guest Insertion or Extraction

It was mentioned earlier that the insertion or extraction of mobile guest species can cause changes in the host structure. There are several types of such structural changes that can occur. They will be briefly discussed in the following sections.

6.9.1 Conversion of the Host Structure from Crystalline to Amorphous

There are a number of examples in which an initially crystalline material becomes amorphous as a result of the insertion of guest species and the corresponding mechanical strains in the lattice. This often occurs gradually as the insertion/extraction reaction is repeated, e.g., upon electrochemical cycling. One example of this, the

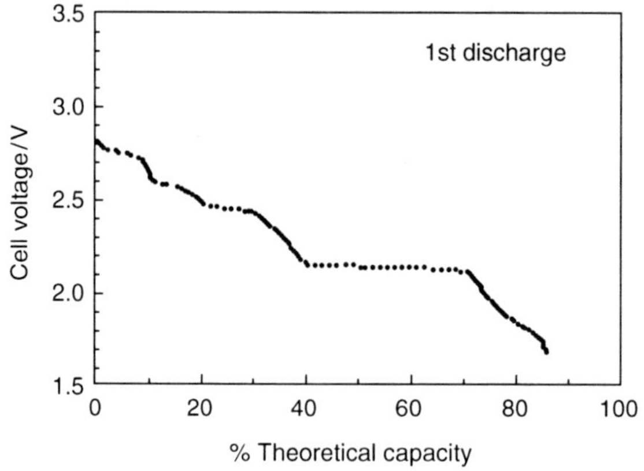

Fig. 6.13 Discharge curve observed during the initial insertion of lithium into a material that was initially V_6O_{13}

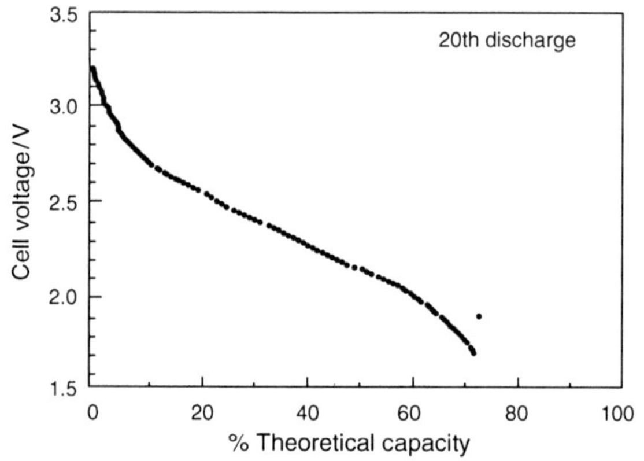

Fig. 6.14 Discharge curve observed during the 20th insertion of lithium into a material that was initially V_6O_{13}

$Li_xV_6O_{13}$ binary system, is shown in Figs. 6.13 and 6.14 [6]. In this case, the shape of the potential curve during the first insertion of lithium into crystalline V_6O_{13} shows that a sequence of reconstitution reactions take place that give rise to a series of different phases, and a discharge curve with well-defined features.

After a number of cycles, however, the discharge curve changes, with a simple monotonous decrease in potential, indicative of a single-phase insertion reaction. X-ray diffraction experiments confirmed that the structure of the material had become amorphous.

Another example of changes resulting from an insertion reaction is shown in Fig. 6.15. In this case, lithium was inserted into a material that was initially V_2O_5 [7]. The result is similar to the V_6O_{13} case, with clear evidence of the formation of a series of different phases as lithium was added. It was found that the insertion

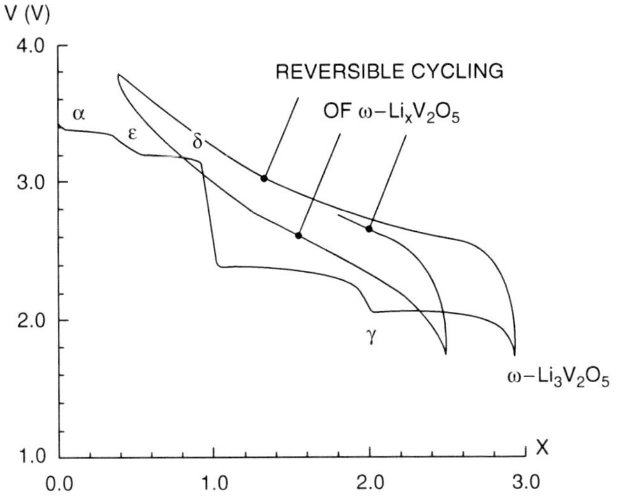

Fig. 6.15 The variation of the potential as lithium is added to V_2O_5. When the composition reached $Li_3V_2O_5$ an amorphous phase was formed

reaction was reversible, forming the ε and δ structures, if only up to about 1 Li was inserted into α V_2O_5. The addition of more lithium resulted in the formation of different structural modifications, called the γ and ω structures, which have nominal compositions of $Li_2V_2O_5$ and $Li_3V_2O_5$, respectively. These two reactions are not reversible, however.

When lithium was extracted from the ω phase, its charge-discharge curve became very different, exhibiting the characteristics of a single phase with a wide range of solid solution. The amount of lithium that could be extracted from this phase was quite large, down to a composition of about $Li_{0.4}V_2O_5$. Upon the reinsertion of lithium, the discharge curve maintained the same general form, indicating that a reversible amorphous structure had been formed during the first insertion process.

6.9.2 Dependence of the Product upon the Potential

It has been found that displacement reactions can occur in a number of materials containing silicon when they are reacted with lithium to a low potential (high lithium activity). An irreversible reaction occurs that results in the formation of fine particles of amorphous silicon in an inert matrix of a residual phase that is related to the precursor material [8, 9]. Upon cycling, the amorphous Li–Si structure shows both good capacity and reversibility.

However, it has also been shown [10] that if further lithium is inserted, going to a potential below 50 mV, a crystalline Li–Si phase forms instead of the amorphous one.

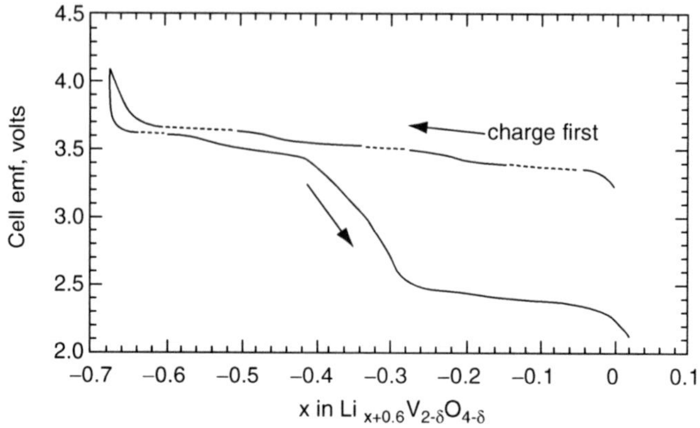

Fig. 6.16 Initial charging and discharge curves of a material with a composition about $Li_{0.6}V_2O_4$

Because of the light weight of silicon, the large amount of lithium that can react with it, and the attractive potential range, it appears as though silicon or its alloys may play an important role as a negative electrode reactant in lithium batteries in the future.

6.9.3 Changes upon the Initial Extraction of the Mobile Species

Similar phenomena can also occur during the initial extraction of a mobile species that is already present in a solid. This is shown in Fig. 6.16 for the case of a material with an initial composition of about $Li_{0.6}V_2O_4$ [11]. It can be seen that the potential starts between 3.0 and 3.5 V vs. pure Li, as is generally found for materials that have come into equilibrium with air. The reason for this will be discussed later, in Chap. 14.

The initial lithium could be essentially completely deleted from the structure, causing the potential to rise to over 4 V vs. pure lithium. When lithium was subsequently reinserted, the discharge curve had a quite different shape, indicating the presence of a reconstitution reaction resulting in the formation of an intermediate phase.

6.10 The Variation of the Potential with Composition in Insertion Reaction Electrodes

6.10.1 Introduction

The externally measured electrical potential of an electrode is determined by the electrochemical potential of the electrons within it, η_{e^-}. This is often called the

Fermi level, E_F. Since potentials do not have absolute values, they are always measured as differences. The voltage of an electrochemical cell is the electrically measured difference between the Fermi levels of the two electrodes:

$$\Delta E = \Delta \eta_{e^-}. \tag{6.1}$$

As has been demonstrated many times in this text already, the measured potential of an electrode often varies with its composition, e.g., as guests species are added to, or deleted from, a host material. The relevant compositional parameter is the chemical potential of the electrically neutral electroactive species. If this species exists as a cation M^+ within the electrode, the important parameter is the chemical potential of neutral M, μ_M. This is related to the electrochemical potentials of the ions and the electrons by

$$\mu_M = \eta_{M^+} + \eta_{e^-}. \tag{6.2}$$

Under open circuit conditions there is no flux of ions through the cell. Since the driving force for the ionic flux through the electrolyte is the gradient in the electrochemical potential of the ions, for open circuit

$$\frac{d\eta_{M^+}}{dx} = \Delta \eta_{M^+} = 0. \tag{6.3}$$

Therefore, the measured voltage across the cell is simply related to the difference in the chemical potential of the neutral electroactive species in the two electrodes by

$$\Delta E = \Delta \eta_{e^-} = \frac{-1}{z_{M^+} q} \Delta \mu_M. \tag{6.4}$$

The common convention is to express both the difference in the electrical potential (the voltage) and the difference in chemical potential as the values in the right-hand (positive) electrode less those in the left-hand electrode.

A general approach that is often used is to understand the potentials of electrons is based upon the *electron energy band model*. The critical features are the variation of the density of available states with the energy of the electrons and the filling of those states up to a maximum value that is determined by the chemical composition. The energy at this maximum value is the *Fermi level*.

In the case of metals the variation of the potential of the available states is a continuous function of the composition, and the *free-electron theory* can be used to express this relationship.

In nonmetals, semiconductors, and insulators, the density of states is not a continuous function of the chemical composition. Instead, there are potential ranges in which there are no available states that can be occupied by electrons. In the case of the simple semiconductors such as silicon or gallium arsenide, one speaks of a *valence band*, in which the states are generally fully occupied, an *energy gap* within which there are no available states, and a *conduction band* with normally empty states. The concentrations of electrons in these bands vary with the temperature due to *thermal excitation*, and can also be modified by the presence of aliovalent species, generally called *dopants*. *Optical excitation* has an effect similar to that of *thermal excitation*.

In a number of materials, particularly those in which the electronic conductivity is relatively low, it is convenient to think of the relation between the occupation of energy states and a change in the formal valence, or charge, upon particular species within the host structure. For example, the addition of an extra electron could result in a change of the formal charge of W^{6+} to W^{5+}, Ti^{4+} to Ti^{3+}, Mn^{4+} to Mn^{3+}, or Fe^{3+} to Fe^{2+} in a transition metal oxide. Such phenomena are called *redox* reactions.

These different cases will be discussed later, and it will be seen that there is a clear relationship between them.

6.10.2 The Variation of the Electrical Potential with Composition in Simple Metallic Solid Solutions

There are a number of metals in which insertion of mobile guest species can occur. As mentioned already, this can be described as a solid solution of the guest species in the host crystal structure. The important quantity controlling the potential is the variation of the chemical potential of the neutral guest species as a function of its concentration. This can be formally divided into two components, the influence of the change in the electron concentration in the host material and the effect due to a change in the concentration of the ionic guest species, M^+.

In the case of a random solid solution in a material with a high electronic conductivity the two major contributions are the contribution from the composition dependence of the Fermi level of the degenerate electron gas that is characteristic of such mixed conductors, and that due to the composition dependence of the enthalpy and configurational entropy of the guest ions in the host crystal lattice [12].

6.10.3 Configurational Entropy of the Guest Ions

If the guest ions are highly mobile and can move rapidly through the host crystal structure we may assume that all the identical crystallographic sites are equally accessible. There will be a contribution to the total free energy due to the *configurational entropy* S_c related to the random distribution of the guest atoms over the available sites. This is given by

$$S_c = -k\left(\ln\frac{x}{x_0 - x}\right), \tag{6.5}$$

where x is the concentration of guest ions and x_0 is the concentration of identical available sites. k is Boltzmann's constant. This configurational entropy contribution to the potential is the product of the absolute temperature and the entropy. This is plotted in Fig. 6.17.

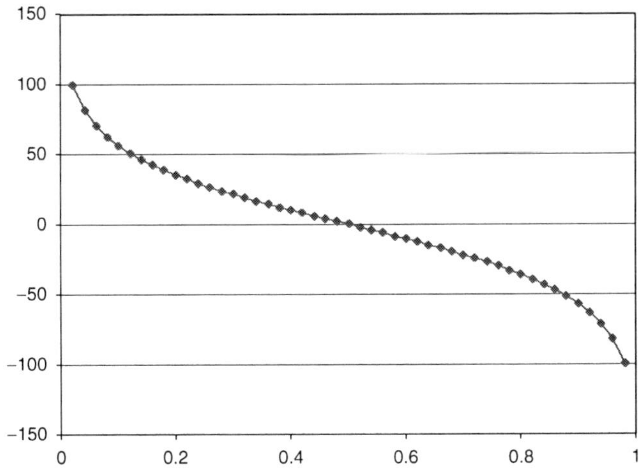

Fig. 6.17 Contribution to the potential due to the configurational entropy of a random distribution of the guest ions upon the available identical positions in a host crystal structure. The values on the abscissa are the fractional site occupation, and those on the ordinate are mV

This model assumes that there is no appreciable interaction between nearby guest species, and that there is only one type of site available for occupation.

6.10.4 The Concentration Dependence of the Chemical Potential of the Electrons in a Metallic Solid Solution

In a simple metal the electron concentration is typically sufficiently high that at normal temperatures the electron gas can be considered to be completely degenerate. Under those conditions the electrochemical potential of the electrons can be approximated by the energy of the Fermi level E_F.

In the free-electron model this can be expressed as

$$E_F = \frac{h^2}{8m\pi^2}\left(\frac{3\pi^2 N_A}{V_m}\right)^{2/3} N^{2/3}, \tag{6.6}$$

where m is the electron mass, N_A is Avogadro's number, V_m is the molar volume, and E_F is calculated from the bottom of the conduction band.

Thus the electronic contribution to the total chemical potential is proportional to the two-third the power of the guest species concentration if the simple free-electron model is valid. More generally, however, the electron mass is replaced by an effective mass m^*. This takes into account other effects, such as the nonparabolicity of the conduction band.

If one can assume that the electrons can be treated as fully degenerate, the chemical potential of the electrons is directly related to the Fermi level, E_F, which can be written as

$$E_F = (\text{Constant})\left(\frac{x^{2/3}}{m^*}\right), \tag{6.7}$$

where x is the guest ion concentration and m^* is the effective mass of the electrons.

6.10.5 Sum of the Effect of These Two Components upon the Electrical Potential of a Metallic Solid Solution

Thus the composition dependence of the electrode potential in a metallic solid solution can be written as the sum of the influence of composition upon the configurational entropy of the guest ions, and the composition dependence of the Fermi level of the electrons. This can be simply expressed as

$$E = (\text{Constant}) - \left(\frac{x^{2/3}}{m^*}\right) - \left(\frac{RT}{zF}\right)\ln\left(\frac{x}{1-x}\right). \tag{6.8}$$

This relationship is illustrated in Fig. 6.18 for several values of the electron effective mass. It also shows the influence of the value of the electron effective mass upon the

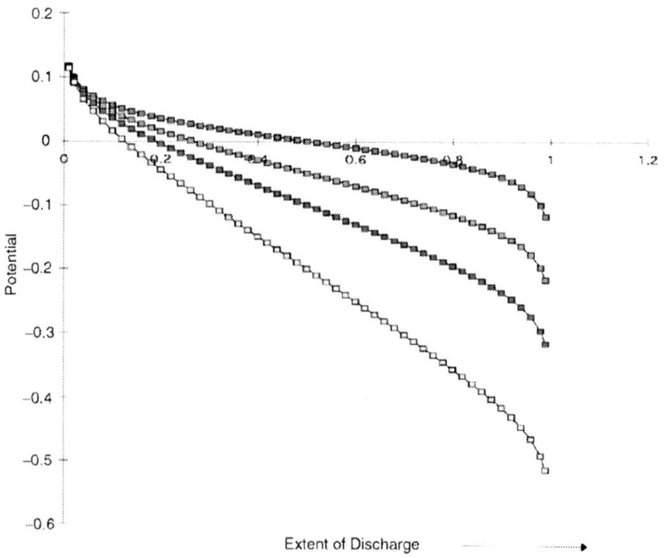

Fig. 6.18 Calculated influence of the value of the electronic effective mass upon the composition dependence of the potential in an insertion reaction in a simple metal

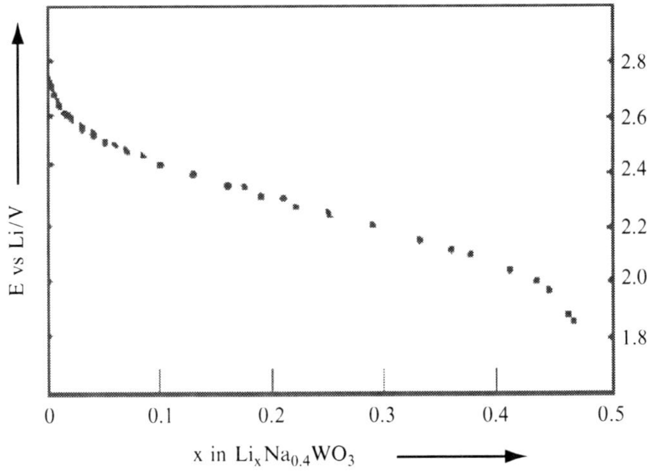

Fig. 6.19 Variation of the potential as a function of lithium concentration in $Li_xNa_{0.4}WO_3$

general slope of the curve. From (6.8) it can be seen that smaller effective masses make the first term larger, and thus results in the potential being more composition – dependent.

An example showing experimental data [12] that illustrate the general features of this model is shown in Fig. 6.19. Although the host material in this case was an oxide, it is an example of a "tungsten bronze." In this family of oxides the electron energy spectrum approximates that of a free-electron metal.

It should be remembered that although the band diagrams commonly used in discussing semiconductors are plotted with greater energy values higher, the scale of the electrical potential is in the opposite direction. This is because the energy of a charged species is the product of its charge and the electrical potential, and the charge on electrons is negative.

6.10.6 The Composition: Dependence of the Potential in the Case of Insertion Reactions that Involve a Two-Phase Reconstitution Reaction

The earlier discussion of the influence of the Gibbs Phase Rule upon the compositional variation of the potentials in electrodes pointed out that when there are two phases present in a two-component system, the potential will have a fixed, or constant, value, independent of the composition. This will also be the case for materials that act as pseudobinaries, regardless of how many different species are actually present. A number of insertion reaction materials are of this type, with one relatively mobile species inside a relatively stable host structure. If, in the time span of interest, the host structure does not undergo any changes it can be considered to be a single component thermodynamically. This is what is found in a number of materials in which the host structure is a transition metal oxide.

One example in which the potential is composition independent involves the insertion and extraction of lithium in materials with the composition $Li_4Ti_5O_{12}$, which has a defective spinel structure [13]. The general composition of spinel structure materials can be described as AB_2O_4, where the A species resides on tetrahedral sites and the B species on octahedral sites within the close-packed face-centered cubic oxygen lattice. If one assumes this general stoichiometry, one of the four lithium ions would share the octahedral sites with the titanium ions, and the other ones would reside on tetrahedral sites. Thus the composition can be written as $Li_3[LiTi_5]O_{12}$, or alternatively, $Li[Li_{1/3}Ti_{5/3}]O_4$.

It has been found that an additional lithium ion can react with this material, and this can be written as

$$Li + Li[Li_{1/3}Ti_{5/3}]O_4 = Li_2[Li_{1/3}Ti_{5/3}]O_4. \tag{6.9}$$

X-Ray diffraction data have indicated that the lithium ions now occupy octahedral sites, instead of tetrahedral sites. Since there are only as many octahedral sites available as oxide ions in this structure, they must now be all filled. This is likely why the capacity of this electrode material is limited at this composition.

These materials were prepared in air and were white in color. As with all materials prepared in air, their potential was initially near 3 V vs. lithium. When lithium was added by transfer from the negative electrode, lithium in carbon, the potential went rapidly down to 1.55 V, and remained there until the reaction was complete. Thus this insertion reaction has the characteristics of a moving-interface reconstitution reaction.

Upon deletion of the inserted lithium the potential retraced the discharge curve closely, with very little hysteresis. This is illustrated in Fig. 6.20 [13]. Because of the small volume change, negligible hysteresis, and rapid kinetics this material acts as a very attractive electrode in lithium cells. The one disadvantage is that its potential is unfortunately about half way between the negative and positive electrode potentials in most lithium batteries.

As will be discussed later, hysteresis, which leads to a difference in the composition dependence of the potential when charging and discharging, is often related to mechanical strain energy, i.e., dislocation generation and motion, as a consequence of volume changes that occur due to the insertion and extraction of the guest ions.

Another example of an insertion-driven reconstitution reaction is the reaction of lithium with $FePO_4$, which also happens readily at ambient temperature. This also has a very flat reaction potential, as shown in Fig. 6.21 [14]. In this case the material is prepared (in air) as $LiFePO_4$, and the initial reaction within the cell involves charging, i.e., deleting lithium from it. This lithium goes across the electrochemical cell and into the carbon material in the negative electrode. The reaction that occurs at the operating potential during the initial charge can be simply written as

$$LiFePO_4 = Li + FePO_4.$$

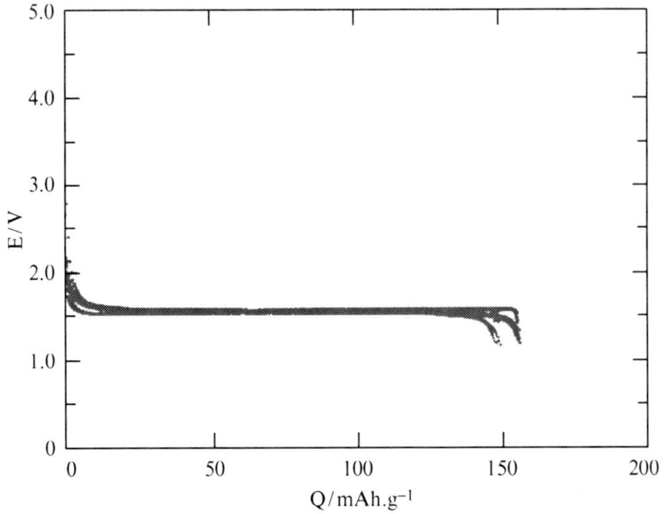

Fig. 6.20 Charge and discharge curves for two different $Li_4Ti_5O_{12}$ cells

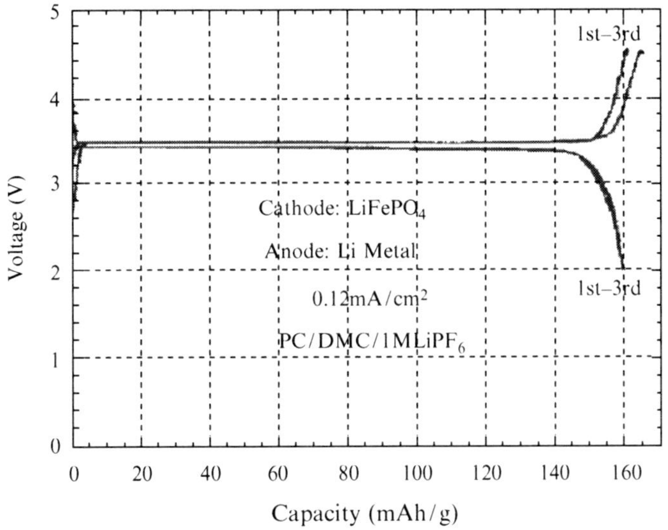

Fig. 6.21 Charge and discharge curves for the reaction of lithium with $FePO_4$

Upon discharge of the cell, the reaction goes, of course, in the opposite direction.

This material is now one of the most important positive electrode reactants in lithium batteries and will be discussed further in Chap. 10-A.

6.11 Final Comments

This chapter has been intended to be only a general introduction to the scope of insertion reactions in electrode materials. This is a very important topic and will be addressed further in the discussions of specific materials in later chapters.

References

1. M.S. Whittingham and R.A. Huggins, J. Electrochem. Soc. 118, 1 (1971)
2. M.S. Whittingham and R.A. Huggins, J. Chem. Phys. 54, 414 (1971)
3. M.S. Whittingham and R.A. Huggins, in *Solid State Chemistry,* ed. by R.S. Roth and S.J. Schneider, Nat. Bur. of Stand. Special Publication 364 (1972), p. 139
4. J. Goodenough, in *Annual Review of Matls Sci.*, Vol. 1, ed. by R.A. Huggins (1970), p. 101
5. I.D. Raistrick and R.A. Huggins, Mat. Res. Bull. 18, 337 (1983)
6. W.J. Macklin, R.J. Neat and S.S. Sandhu, Electrochim. Acta 37, 1715 (1992)
7. C. Delmas, H. Cognac-Auradou, J.M. Cocciantelli, M. Menetrier and J.P. Doumerc, Solid State Ionics 69, 257 (1994)
8. A. Netz, R.A. Huggins and W. Weppner, J. Power Sources 119–121, 95 (2003)
9. A. Netz and R.A. Huggins, Solid State Ionics 175, 215 (2004)
10. M.N. Obrovac and L. Christensen, Electrochem. Solid State Lett. 7, A93 (2004)
11. T.A. Chirayil, P.Y. Zavalij and M.S. Whittingham, J. Electrochem. Soc. 143, L193 (1996)
12. I.D. Raistrick, A.J. Mark and R.A. Huggins, Solid State Ionics 5, 351 (1981)
13. T. Ohzuku, A. Ueda and N. Yamamoto, J. Electrochem. Soc. 142, 1431 (1995)
14. A. Yamada, S.C. Chung and K. Hinokuma, J. Electrochem. Soc. 148, A224 (2001)

Chapter 7
Negative Electrodes in Lithium Cells

7.1 Introduction

Early work on the commercial development of rechargeable lithium batteries to operate at or near ambient temperatures involved the use of elemental lithium as the negative electrode reactant. As discussed later, this leads to significant problems. Negative electrodes currently employed on the negative side of lithium cells involve a solid solution of lithium in one of the forms of carbon.

Lithium cells that operate at temperatures above the melting point of lithium must necessarily use alloys instead of elemental lithium. These are generally binary or ternary metallic phases.

There is also increasing current interest in the possibility of the use of metallic alloys instead of carbons at ambient temperatures, with the goal of reducing the electrode volume, as well as achieving significantly increased capacity.

There are differences in principle between the behavior of elemental and binary phase materials as electrodes. It is the purpose of this chapter to elucidate these principles as well as to present some examples. Ternary systems will be discussed elsewhere.

7.2 Elemental Lithium Electrodes

It is obvious that elemental lithium has the lowest potential, as well as the lowest weight per unit charge, of any possible lithium reservoir material in an electrochemical cell. Materials with lower lithium activities have higher potentials, leading to lower cell voltages, and they also carry along extra elements as dead weight.

There are problems with the use of elemental lithium, however. These are due to phenomena that occur during the recharging of all electrodes composed of simple metallic elements. In the particular case of lithium, however, this is not just a matter of increasing electrode impedance and reduced capacity, as are typically found with

R.A. Huggins, *Advanced Batteries*: Materials Science Aspects,

other electrode materials. In addition, severe safety problems can ensue. Some of these phenomena will be discussed in the following sections.

7.3 Problems with the Rechargeability of Elemental Electrodes

In the case of an electrochemical cell in which an elemental metal serves as the negative electrode the process of recharging may seem to be very simple, for it merely involves the electrodeposition of the metal from the electrolyte onto the surface of the electrode. This is not the case, however.

In order to achieve good rechargeability, a consistent geometry must be maintained on both the macroscopic and the microscopic scales. Both electrical disconnection of the electroactive species and electronic short circuits must also be avoided. In addition, thermal runaway must not occur.

Phenomena related to the inherent microstructural and macrostructural instability of a growth interface and related thermal problems will now be briefly reviewed.

7.3.1 Deposition at Unwanted Locations

In the absence of a significant nucleation barrier, deposition will tend to occur anywhere at which the electric potential is such that the element's chemical potential is at, or above, that corresponding to unit activity. This means that electrodeposition may take place upon current collectors and other parts of an electrochemical cell that are at the same electrical potential as the negative electrode, as well as upon the electrode structure where it is actually desired. This was a significant problem during the period in which attempts were being made to use pure (molten) lithium as the negative electrode in high-temperature molten halide salt electrolyte cells. Another problem with these high-temperature cells was the fact that alkali metals dissolve in their halides at elevated temperatures. This leads to electronic conduction and self discharge.

7.3.2 Shape Change

Another difficulty is the *shape change* phenomenon, in which the location of the electrodeposit is not the same as that where the discharge (deplating) process took place. Thus, upon cycling the electrode metal gets preferentially transferred to new locations. For the most part, this is a problem of current distribution and hydrodynamics, rather than being a materials issue. Therefore, it will not be discussed further here.

7.3.3 Dendrites

An additional type of problem relates to the inherent instability of a flat interface on a microscopic scale during electrodeposition, even in the case of a chemically clean surface. It has been shown that there can be an electrochemical analog of the constitutional supercooling that occurs ahead of a growth interface during thermally driven solidification [1].

This will be the case if the current density is such that solute depletion in the electrolyte near the electrode surface causes the local gradient of the element's chemical potential in the electrolyte immediately adjacent to the solid surface to be positive. Under such a condition there will be a tendency for any protuberance upon the surface to grow at a faster rate than the rest of the interface. This leads to exaggerated surface roughness, and eventually to the formation of either dendrites or filaments. In more extreme cases, it leads to the nucleation of solid particles in the liquid electrolyte ahead of the growing solid interface.

This is also related to the inverse phenomenon, the formation of a flat interface during electropolishing, as well as the problem of morphology development during the growth of an oxide layer upon a solid solution alloy [2, 3]. Another analogous situation is present during the crystallization of the solute phase from liquid metal solutions.

The protuberances upon a clean growing interface can grow far ahead of the general interface, often developing into dendrites. A general characteristic of dendrites is a tree-and-branches type of morphology, which has very distinct geometric and crystallographic characteristics, due to the orientation dependence of either the surface energy or the growth velocity.

7.3.4 Filamentary Growth

A different phenomenon that is often mistakenly confused with dendrite formation is the result of the presence of a reaction product layer upon the growth interface if the electrode and electrolyte are not stable in the presence of each other. The properties of these layers can have an important effect upon the behavior of the electrode. In some cases they may be useful solid electrolytes and allow electrodeposition by ionic transport through them. Such layers upon negative electrodes in lithium systems have been given the name *SEI* and will be discussed in a later chapter. But in other cases reaction product layers may be ionically blocking and thus significantly increase the interfacial impedance.

Interfacial layers often have defects in their structure that can lead to local variations in their properties. Regions of reduced impedance can cause the formation of deleterious filamentary growths upon recharge of the electrode. This is an endemic problem with the use of organic solvent electrolytes in contact with lithium electrodes at ambient temperatures.

When a protrusion grows ahead of the main interface the protective reaction product layer will typically be locally less thick. This means that the local impedance to the passage of ionic current is reduced, resulting in a higher current density and more rapid growth in that location. This behavior can be exaggerated if the blocking layer is somewhat soluble in the electrolyte, with a greater solubility at elevated temperatures. When this is the case, the higher local current leads to a higher local temperature and a greater solubility. The result is then a locally thinner blocking layer and an even higher local current.

Furthermore, the current distribution near the tip of a protrusion that is well ahead of the main interface develops a 3-dimensional character, leading to even faster growth than the main electrode surface, where the mass transport is essentially 1 dimensional. Especially in relatively low concentration solutions, this leads to a runaway type of process, so that the protrusions consume most of the solute, and grow farther and farther ahead of the main, or bulk, interface.

This phenomenon can result in the metal deposit having a hairy or spongy character. During a subsequent discharge step, the protrusions often get disconnected from the underlying metal, so that they cannot participate in the electrochemical reaction, and the rechargeable capacity of the electrode is reduced.

This unstable growth is a major problem with the rechargeability of elementary negative electrodes in a number of electrochemical systems and constitutes an important limitation upon the development of rechargeable lithium batteries using elemental lithium as the negative electrode reactant.

7.3.5 Thermal Runaway

The organic solvent electrolytes that are typically used in lithium batteries are not stable in the presence of high lithium activities. This is a common problem when using elemental lithium negative electrodes in contact with electrolytes containing organic cationic groups, regardless of whether the electrolyte is an organic liquid or a polymer [4].

They react with lithium and form either crystalline or amorphous product layers upon the surface of the electrode structure. These reactions are exothermic and cause local heating. Experiments using an *accelerating rate calorimeter* have shown that this problem increases dramatically as cells are cycled, presumably due to an increase in the surface area of the lithium due to morphological instability during repetitive recharging [5]. This is a fundamental difficulty with elemental lithium electrodes and has led to serious safety problems.

The exothermic formation of reaction product films also occurs when carbon or alloy electrodes are used that operate at potentials at which the electrolyte reacts with lithium. However, if their morphology is constant the surface area does not change substantially, so that it can lead to heating, but typically does not lead to thermal runaway at the negative electrode.

7.4 Alternatives

Because of the safety and cycle life problems with the use of elemental lithium, essentially all rechargeable lithium batteries use lithium–carbon alloys as negative electrode reactants today. This topic is discussed in Sect. 7.4.

A considerable amount of research attention is now being given to the possibility of the use of metallic lithium alloys instead of the graphites, because of the expectation that this may lead to significant increases in capacity. The large volume changes that accompany increased capacity present a significant problem, however. These matters as well as the possibility of the use of novel micro- or nanostructures to alleviate this difficulty will be discussed in Sect. 7.4.

7A Lithium–Carbon Alloys

7A.1 Introduction

Lithium–carbons are currently used as the negative electrode reactant in the very common small rechargeable lithium batteries used in consumer electronic devices. As will be seen in this chapter, a wide range of structures, and therefore of properties, is possible in this family, depending upon how the carbon is produced. The choices made by the different manufacturers are not all the same. Several good reviews of the materials science aspects of this topic can be found in the literature [6, 7].

The crystal structure of pure graphite is shown schematically in Fig. 7A.1. It consists of parallel sheets containing interconnected hexagons of carbon, called

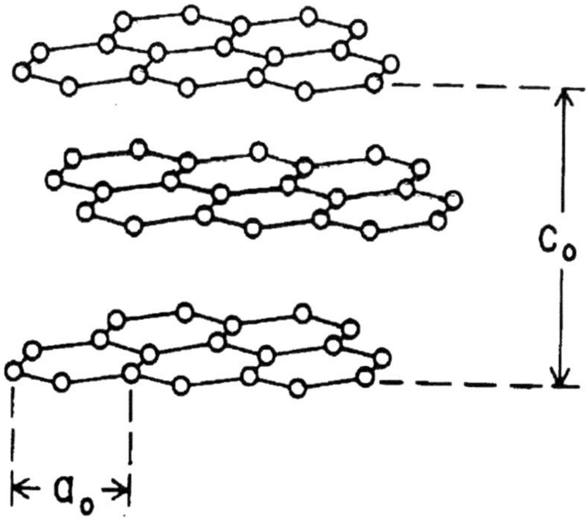

Fig. 7A.1 Model of a portion of the crystal structure of graphite

graphene layers or sheets. They are stacked with alternate layers on top of one another. This is described as A–B–A–B–A stacking.

Graphite is amphoteric, and either cations or anions can be inserted into it between the graphene layers. When cations are inserted, the host graphite structure takes on a negative charge. Cation examples are Li^+, K^+, Rb^+, and Cs^+. When anions are inserted, the host graphite structure takes on a positive charge, and anion examples are Br^-, SO_4^{2-}, or SbF_6^-.

The insertion of alkali metals into carbon was first demonstrated in 1926 [8], and the chemical synthesis of lithium–carbons was demonstrated in 1955 [9]. X-ray photoemission spectroscopy experiments showed that the inserted lithium gives up its electron to the carbon, and thus the structure can be viewed as Li^+ ions contained between the carbon layers of the graphite structure [10]. A general review of the early work on the insertion of species into graphite can be found in [11].

Insertion often is found to occur in *stages*, with nonrandom filling of positions between the layers of the host crystal structure. This ordering can occur in individual layers, and also in the filling of the stack of layers.

The possibility of the use of graphite as a reactant in the negative electrode of electrochemical cells containing lithium was first investigated some 30 years ago [12]. Those experiments were, however, unsuccessful. Swelling and defoliation occurred due to cointercalation of species from the organic solvent electrolytes that were used at that time. As discussed in Chap. 15B, this problem has been subsequently solved by the use of other liquid electrolytes.

Attention was again brought to this possibility by a conference paper presented in 1983 [13] that showed that lithium can be reversibly inserted into graphite at room temperatures when using a polymeric electrolyte. Although not publicly known at that time, two patents relating to the use of the insertion of lithium into graphite as a reversible negative electrode in lithium systems, at both elevated [14] and ambient [15] temperatures, had already been submitted by Bell Laboratories. Royalties paid for the use of these patents have become very large.

This situation changed abruptly as the result of the successful development by SONY in 1990 of commercial rechargeable batteries containing negative electrodes based upon materials of this family and their commercial introduction as the power source in camcorders [16, 17].

There has been a large amount of work on the understanding and development of graphites and related carbon-containing materials for use as negative electrode materials in lithium batteries since that time.

Lithium–carbon materials are, in principle, no different from other lithium-containing metallic alloys. However, since this topic is treated in more detail later, only a few points that specifically relate to carbonaceous materials will be discussed here.

One is that the behavior of these materials is very dependent upon the details of both the nanostructure and the microstructure. Therefore, the composition and the thermal and mechanical treatment of the electrode materials all play important roles in determining the resulting thermodynamic and kinetic properties. Materials with a more graphitic structure have properties that are much different from those with less well-organized structures. The materials that are used by the various commercial producers are not all the same, as they reflect the different choices that they have made for their specific products. However, the major producers of small consumer lithium batteries generally now use relatively graphitic carbons.

An important consideration in the use of carbonaceous materials as negative electrodes in lithium cells is the common observation of a considerable loss of capacity during the first charge–discharge cycle due to irreversible lithium absorption into the structure, as will be seen later. This has the distinct disadvantage that it requires that an additional amount of lithium be initially present in the cell. If this irreversible lithium is supplied from the positive electrode, an extra amount of the positive electrode reactant material must be put into the cell during its fabrication. As the positive electrode reactant materials often have relatively low specific capacities, e.g., around 140 mAh g^{-1}, this irreversible capacity in the negative electrode leads to a requirement for an appreciable amount of extra reactant material weight and volume in the total cell.

7A.2 Ideal Structure of Graphite Saturated with Lithium

Lithium can be inserted into the graphite structure up to a maximum concentration of one Li per six carbons, or LiC_6. One of the major influences of the presence of lithium in the graphite crystal structure is that the stacking of graphene layers is changed by the insertion of lithium. It changes from A–B–A–B–A stacking to A–A–A–A–A stacking. This is illustrated schematically in Fig. 7A.2.

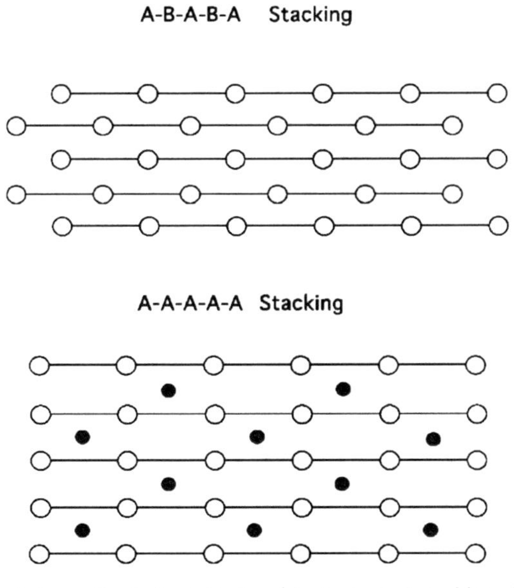

Fig. 7A.2 Difference between the A–B–A–B–A and A–A–A–A–A stacking of the graphene layers when lithium is inserted. The black circles are the lithium ions

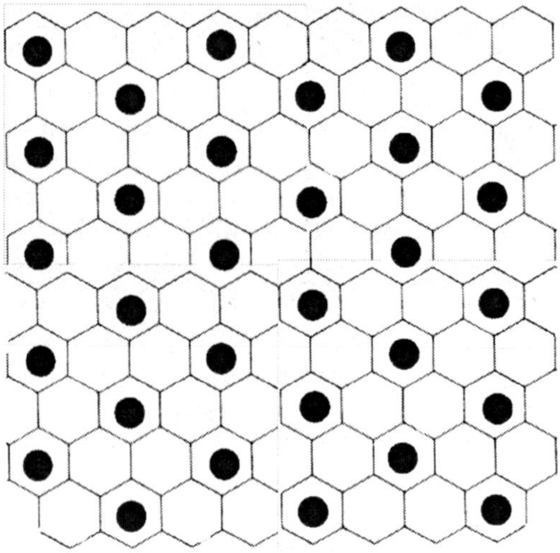

Fig. 7A.3 Schematic representation of the lithium distribution in the gallery space in relation to the carbon hexagonal network in the adjacent graphene layers

The distribution of lithium ions within the gallery space between the graphene layers is illustrated schematically in Fig. 7A.3.

7A.3 Variations in the Structure of Graphite

There is actually a wide range of lithium–carbon structures, and most such materials do not actually have the ideal graphite structure. The ones that are closest are made synthetically by vapor transport, and are called highly ordered pyrolytic graphite (HOPG). This is a slow and very expensive process. The graphites that are used commercially range from natural graphite to materials formed by the pyrolysis of various polymers or hydrocarbon precursors. They are often divided into two general types, designated as *soft, or graphitizing, carbons*, and *hard carbons* [18].

At modest temperatures and pressures there is a strong tendency for carbon atoms to be arranged in a planar graphene-type configuration, rather than a three-dimensional structure such as that in diamond.

Soft carbons are generally produced by the pyrolysis of liquid materials such as petroleum pitch, which is the residue from the distillation of petroleum fractions.

The carbon atoms in their structure are initially arranged in small graphene-type groups, but there is generally a significant amount of imperfection in their two-dimensional honeycomb networks, as well as randomness in the way that the layers are vertically stacked upon each other. In addition there is little coordination in the rotational orientation of nearby graphene layers. The term *turbostratic* is generally used to describe this general type of three-dimensional disorder in carbons [18].

Fig. 7A.4 Schematic drawing of the microstructure of graphite after heating to intermediate temperatures

The three types of initial disorder, in-plane defects, inter-plane stacking defects, and rotational misorientation, gradually become healed as the temperature is raised, the first two earlier than the rotational disorder between adjacent layers, for that requires more thermal energy.

The microstructure of such materials that have been heated to intermediate temperatures is shown schematically in Fig. 7A.4.

At this intermediate stage, the structure contains many small three-dimensional subgrains.

In addition to containing some internal imperfections, they differ from their neighbors in both vertical and horizontal orientations. They are separated by subgrain walls (boundaries) that have surface energy. This subgrain wall surface energy gradually gets reduced as the individual subgrains grow in size and the overall graphitic structure becomes more perfect.

The *hard carbons*, which are typically produced by the pyrolysis of solid materials, such as chars or glassy carbon, initially have a significant amount of initial cross-linking, related to the structure of their precursors. In addition, they can have a substantial amount of nanoporosity. As a result, it is more difficult to make these structural rearrangements and turbostratic disorder is more persistent. The result is the requirement for more thermal energy, i.e., higher temperatures.

The structure that results from the pyrolysis of carbonaceous precursors depends greatly upon the maximum temperature that is reached. Heating initially amorphous, or *soft*, carbons to the range of 1,000–2,000°C produces microstructures in which graphene sheets form and begin to grow, with diameters up to about 15 nm, and they become assembled into small stacks of 50–100 sheets. These subgrains initially have a turbostratic arrangement, but their alignment into larger ordered, i.e., graphitic, regions gradually takes place as the temperature is increased from 2,000 to 3,000°C.

7A.4 Structural Aspects of Lithium Insertion into Graphitic Carbons

One of the important features in the interaction of lithium with graphitic materials is the phenomenon of *staging*. Lithium that enters the graphite structure is not distributed uniformly between all the graphene layers at ambient temperatures. Instead, it resides in certain interlayer *galleries*, but not others, depending upon the total amount of lithium present.

The distribution is described by the number of graphene layers between those that have the lithium guest ions present. A stage 1 structure has lithium between all of the graphene layers; a stage 2 structure has an empty gallery between each occupied gallery, and a stage 4 structure has four graphene layers between each gallery containing lithium. This will be discussed a bit more later in this chapter.

This is obviously a simplification, for in any real material there will be regions with predominately one structure, and other regions with another.

The phenomenon of nonrandom gallery occupation is found in a number of other materials, and can be attributed to a catalytic effect, in which the ions that initially enter a gallery pry open the van der Waals-bonded interlayer space, making it easier for following ions to enter.

However, the situation is a bit more complicated, for there must be communication between nearby galleries in order for the structure to adopt the ordered stage structure. This is related to the intertunnel communication in the *hollandite* structure described in Chap. 6, but will not be further discussed here.

7A.5 Electrochemical Behavior of Lithium in Graphite

The electrochemical behavior of lithium in carbon materials is highly variable, depending upon the details of the graphitic structure. Materials with a more perfect graphitic structure react with lithium at more negative potentials, whereas those with less well organized structures typically operate over much wider potential ranges, resulting in cell voltages that are both lower and more state-of-charge dependent.

In a number of cases, the carbons that are used in commercial batteries have been heated to temperatures over about 2,400°C, where they become quite well graphitized. Capacities typically range from 300 to 350 mAh g^{-1}, whereas the maximum theoretical value (for LiC_6) is 372 mAh g^{-1}.

A typical discharge curve under operating conditions, with currents as large as 2–4 mA cm^{-2}, is shown in Fig. 7A.5.

This behavior is not far from what is found under near-equilibrium conditions, as shown in Fig. 7A.6. It can be seen that there is a difference between the data during charge, when lithium is being added, and discharge, when lithium is being deleted. This displacement (hysteresis) between the charge and discharge curves is at least partly due to the mechanical energy involved in the structural changes.

It can be seen that these data show plateaus, indicating the presence of three ranges of composition within which reconstitution reactions take place. As the composition changes along these plateaus the relative amounts of material with the two end compositions vary. This means that there will be regions, or domains, where the graphene layer stacking is of one type, and regions in which it has the other type.

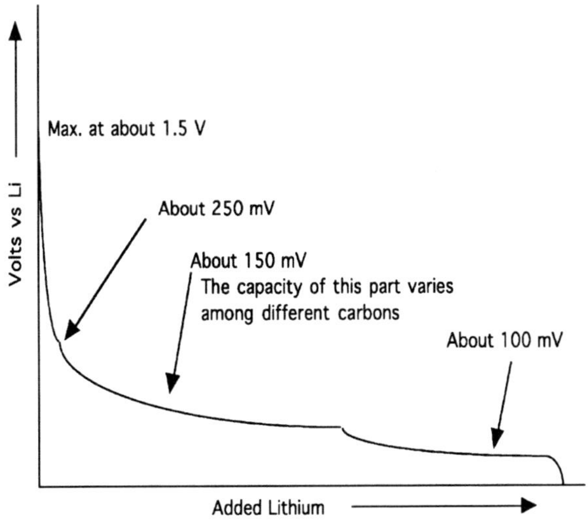

Fig. 7A.5 Typical discharge curve of a lithium battery negative electrode

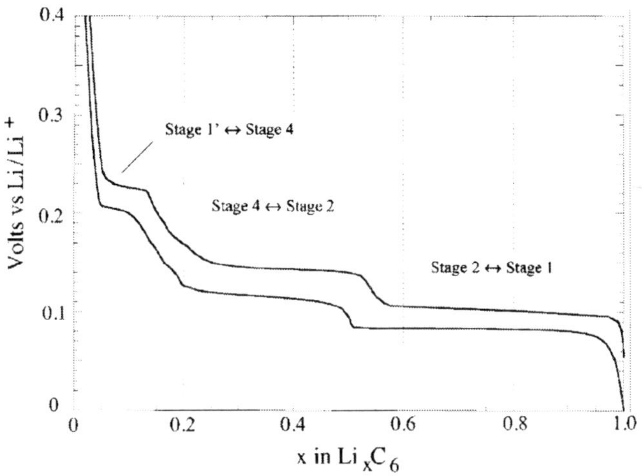

Fig. 7A.6 Potential vs. composition during lithiation and delithiation of a graphite electrode at the C/50 rate at ambient temperature (R. Yazami, personal communication)

The relative volumes of these two domains vary as the overall composition traverses these *two-phase regions*. The differences in stacking result in differences in interlayer spacing, and therefore considerable amount of distortion of the structure. Such a model was presented some time ago by Daumas and Herold [19].

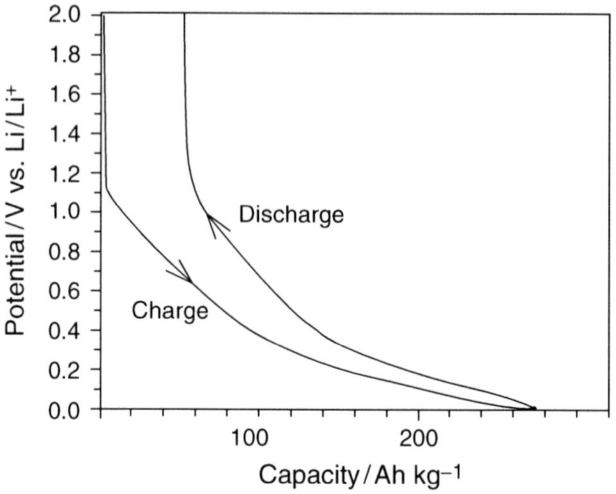

Fig. 7A.7 Typical data for the reaction of lithium with an amorphous carbon [6]

7A.6 Electrochemical Behavior of Lithium in Amorphous Graphite

The electrochemical behavior is quite different when the carbon has not been heated so high, and the structure is not so well ordered. There is a wide range of possible sites in which the lithium can reside, with different local structures, and therefore different energies. The result is that the potential varies gradually, rather than showing the steps characteristic of more ordered structures. This is shown in Fig. 7A.7. It can be seen that, in addition to varying with the state of charge, the potential is significantly greater than is found in the graphitic materials. This means that the cell voltages are correspondingly lower.

It can be seen that there was some capacity loss on the first cycle. The capacity upon the first charging (that is not useful) was greater than the capacity in the subsequent discharge cycle. The source of this phenomenon is not yet understood, but there must be some lithium that is *trapped* in the structure and does not come out during discharge. Because of this extra (useless) capacity during the initially charging of this negative electrode it is necessary to put extra capacity in the positive electrode. This is unfortunate, for the specific capacity of the positive electrodes in such systems is less than that in the negative electrodes. As a result, a significant amount of extra weight and volume is necessary.

7A.7 Lithium in Hydrogen-Containing Carbons

It is often found that there is a considerable amount of hydrogen initially present in various carbons, depending upon the nature of the precursor. This gradually disappears as the temperature is raised.

If the precursor is heated to only 500–700°C, there is still a lot of hydrogen present in the structure. It has been found experimentally that this can lead to a very large capacity for lithium that is proportional to the amount of hydrogen present [20–22]. There is a loss in this capacity upon cycling, perhaps due to the gradual loss of hydrogen in the structure.

The large capacity may be due to lithium binding to hydrogen-terminated edges of small graphene fragments. The local configuration would then be analogous to that in the organolithium molecule $C_2H_2Li_2$. This is consistent with the experimental observation of the dependence of the lithium capacity upon the amount of hydrogen present. This would also result in a change in the local bonding of the host carbon atom from sp^2 to sp^3.

In addition to a large capacity, experiments have shown a very large hysteresis with these materials [22]. Hysteresis is generally considered to be a disadvantage, as the discharge potential is raised, reducing the cell voltage.

Hysteresis is characteristic of reactions that involve a lot of mechanical energy as the result of shape and volume changes. However, in this case it is more likely due to the energy involved in the change of the bonding of the nearby carbon atoms [22].

The result of experiments performed on one example of a hydrogen-containing material are shown in Fig. 7A.8. It can be seen that there was a very large capacity loss on the first cycle. The capacity upon the first charging (that is not useful) was much greater than the capacities in subsequent cycles. As mentioned earlier, this extra lithium must be supplied by the positive electrode. The source of this phenomenon is not yet understood, but there must be a lot of lithium that is *trapped* in the structure and does not come out during the first, and subsequent, discharges.

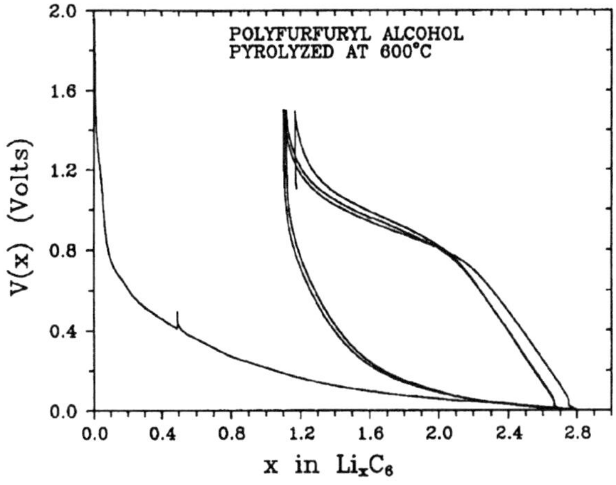

Fig. 7A.8 Charge–discharge curves for a material containing hydrogen [7]

7B Metallic Lithium Alloys

7B.1 Introduction

Attention has been given to the use of lithium alloys as an alternative to elemental lithium for some time. Groups working on batteries with molten salt electrolytes that operate at temperatures of 400–450°C, well above the melting point of lithium, were especially interested in this possibility. Two major directions evolved. One involved the use of lithium–aluminum alloys [23, 24], whereas another was concerned with lithium–silicon alloys [25–27].

Although this approach can avoid the problems related to lithium melting, as well as the others mentioned earlier, there are always at least two disadvantages related to the use of alloys. Because they reduce the activity of the lithium they necessarily reduce the cell voltage. In addition, the presence of additional species that are not directly involved in the electrochemical reaction always brings additional weight, and often, volume. Thus the maximum theoretical values of the specific energy are always reduced compared with what might be attained with pure lithium. The energy density is also often reduced. But lithium has a large specific volume, so that this is not always the case.

In practical cases, however, the excess weight and volume due to the use of alloys may not be very far from those required with pure lithium electrodes, for it is generally necessary to have a large amount of excess lithium in rechargeable cells in order to make up for the capacity loss related to the dendrite or filament growth problem upon cycling.

Lithium alloys have been used for a number of years in the high-temperature *thermal batteries* that are produced commercially for military purposes. These devices are designed to be stored for long times at ambient temperatures before use, where their self-discharge kinetic behavior is very slow. They must be heated to elevated temperatures when their energy output is desired. An example is the Li alloy/FeS_2 battery system that employs a chloride molten salt electrolyte. In order to operate, the temperature must be raised to over the melting point of the electrolyte. This type of cell typically uses either Li–Si or Li–Al alloys in the negative electrode.

The first use of lithium alloys as negative electrodes in commercial batteries to operate at ambient temperatures was the employment of Wood's metal alloys in lithium-conducting button-type cells by Matsushita in Japan. Development work on the use of these alloys started in 1983 [28], and the alloys became commercially available somewhat later.

7B.2 Equilibrium Thermodynamic Properties of Binary Lithium Alloys

Useful starting points when considering lithium alloys as electrode reactants are their phase diagrams and equilibrium thermodynamic properties. In some cases this

Table 7B.1 Plateau potentials and composition ranges of a number of binary lithium alloys at 400°C

Voltage vs. Li/Li^+	System	Range of y
0.910	Li_ySb	0–2.0
0.875	Li_ySb	2.0–3.0
0.760	Li_yBi	0.6–1.0
0.750	Li_yBi	1.0–2.82
0.570	Li_ySn	0.57–1.0
0.455	Li_ySn	1.0–2.33
0.430	Li_ySn	2.33–2.5
0.387	Li_ySn	2.5–2.6
0.283	Li_ySn	2.6–3.5
0.170	Li_ySn	3.5–4.4
0.565	Li_yGa	0.15–0.82
0.122	Li_yGa	1.28–1.48
0.090	Li_yGa	1.53–1.93
0.558	Li_yCd	0.12–0.21
0.373	Li_yCd	0.33–0.45
0.058	Li_yCd	1.65–2.33
0.507	Li_yPb	0–1.0
0.375	Li_yPb	1.1–2.67
0.271	Li_yPb	2.67–3.0
0.237	Li_yPb	3.0–3.5
0.089	Li_yPb	3.8–4.4
0.495	Li_yIn	0.22–0.86
0.145	Li_yIn	1.74–1.92
0.080	Li_yIn	2.08–2.67
0.332	Li_ySi	0–2.0
0.283	Li_ySi	2.0–2.67
0.156	Li_ySi	2.67–3.25
0.047	Li_ySi	3.25–4.4
0.300	Li_yAl	0.08–0.9

information is available, so that predictions can be made of their potentials and capacities. In other cases, experimental measurements are required. Relevant principles were discussed in Chap. 3, and will not be repeated here.

Elevated temperature data for a number of phases in the Li–Al, Li–Bi, Li–Cd, Li–Ga, Li–In, Li–Pb, Li–Sb, Li–Si, and Li–Sn binary lithium alloy systems, made using a LiCl–KCl molten salt electrolyte are listed in Table 7B.1.

7B.3 Experiments at Ambient Temperature

Experiments have also been performed to determine the equilibrium values of the electrochemical potentials and capacities in a smaller number of binary lithium systems at ambient temperatures [29, 30]. Because of slower kinetics at lower temperatures, these experiments took longer to perform. Data are presented in Table 7B.2.

Table 7B.2 Plateau potentials and composition ranges of lithium alloys at ambient temperatures under equilibrium conditions

Voltage vs. Li/Li^+	System	Range of y
0.956	Li_ySb	1.0–2.0
0.948	Li_ySb	2.0–3.0
0.828	Li_yBi	0–1.0
0.810	Li_yBi	1–3.0
0.680	Li_yCd	0–0.3
0.352	Li_yCd	0.3–0.6
0.055	Li_yCd	1.5–2.9
0.660	Li_ySn	0.4–0.7
0.530	Li_ySn	0.7–2.33
0.485	Li_ySn	2.33–2.63
0.420	Li_ySn	2.6–3.5
0.380	Li_ySn	3.5–4.4
0.601	Li_yPb	0–1.0
0.449	Li_yPb	1.0–3.0
0.374	Li_yPb	3.0–3.2
0.292	Li_yPb	3.2–4.5
0.256	Li_yZn	0.4–0.5
0.219	Li_yZn	0.5–0.67
0.157	Li_yZn	0.67–1.0
0.005	Li_yZn	1.0–1.5

7B.4 Liquid Binary Alloys

Although the discussion here has involved solid lithium alloys, similar considerations apply to those based on sodium or other species. In addition, it is not necessary that the active material be solid. The same principles hold for liquids.

An example was discussed in Chap. 3 relating to the so-called sodium–sulfur battery that operates at about 300°C. In this case, both of the electrodes are liquids, and the electrolyte is a solid sodium ion conductor. This configuration can thus be described as an L/S/L system. It is the inverse of conventional systems with solid electrodes and a liquid electrolyte, i.e., S/L/S systems.

7B.5 Mixed-Conductor Matrix Electrodes

To be able to achieve appreciable macroscopic current densities while maintaining low local microscopic charge and particle flux densities, many battery electrodes that are used in conjunction with liquid electrolytes are produced with porous microstructures containing very fine particles of the solid reactant materials. This porous structure with high reactant surface area is permeated with the electrolyte.

This porous fine-particle approach has several characteristic disadvantages, however. Among these are difficulties in producing uniform and reproducible microstructures and limited mechanical strength when the structure is highly porous.

In addition, they often suffer Ostwald ripening, sintering, or other time-dependent changes in both microstructure and properties during cyclic operation.

Furthermore, it is often necessary to have an additional material present to improve the electronic transport within an electrode. Various highly dispersed carbons are often used for this purpose.

A quite different approach was introduced some years ago [31–33] in which it was demonstrated that a rather dense solid electrode can be fabricated that has a composite microstructure in which particles of the reactant phase or phases are finely dispersed within a solid electronically conducting matrix in which the electroactive species is also mobile, i.e., within a mixed conductor. There is thus a large internal reactant/mixed-conducting matrix interfacial area. The electroactive species is transported through the solid matrix to this interfacial region, where it undergoes the chemical part of the electrode reaction. Since the matrix material is also an electronic conductor, it can also act as the electrode's current collector. The electrochemical part of the reaction takes place on the outer surface of the composite electrode.

When such an electrode is discharged by deletion of the electroactive species, the residual particles of the reactant phase remain as relics in the microstructure. This provides fixed permanent locations for the reaction to take place during following cycles, when the electroactive species again enters the structure. Thus this type of configuration has the additional advantage that it can provide a mechanism for the achievement of true microstructural reversibility.

For this concept to be applicable, the matrix and the reactant phases must be thermodynamically stable in contact with each other. One can evaluate this possibility if one has information about the relevant phase diagrams as well as the titration curves of the component binary systems. The stability window of the matrix phase must span the reaction potential of the reactant material. It has been shown that one can evaluate the possibility that these conditions are met from knowledge of the binary titration curves.

Since there is generally a common component, these two binaries can also be treated as a ternary system. Although ternary systems are not explicitly discussed here, it can be simply stated that the two materials must lie at corners of the same constant-potential tie triangle in the relevant isothermal ternary phase diagram in order to not interact. The potential of the tie triangle determines the electrode reaction potential, of course. An additional requirement is that the reactant material must have two phases present in the tie triangle, but the matrix phase only one.

The kinetic requirements for a successful application of this concept are readily understandable. The primary issue is the rate at which the electroactive species can reach the matrix/reactant interfaces. The critical parameter is the chemical diffusion coefficient of the electroactive species in the matrix phase. This can be determined by various techniques, as discussed in later chapters.

The first example that was demonstrated was the use of the phase with the nominal composition $Li_{13}Sn_5$ as the matrix, in conjunction with reactant phases in the lithium–silicon system at temperatures near 400°C. This is an especially favorable case, due to the very high chemical diffusion coefficient of lithium in the $Li_{13}Sn_5$ phase.

The relation between the potential–composition data for these two systems under equilibrium conditions is shown in Fig. 7B.1 [31]. It is seen that the phase $Li_{2.6}Sn$ ($Li_{13}Sn_5$) is stable over a potential range that includes the upper two-phase reconstitution reaction plateau in the lithium–silicon system. Therefore, lithium can react with Si to form the phase $Li_{1.71}Si$ ($Li_{12}Si_7$) inside an all-solid composite electrode containing the $Li_{2.6}Sn$ phase, which acts as a lithium-transporting, but electrochemically inert matrix.

Figure 7B.2 shows the relatively small polarization that is observed during the charge and discharge of this electrode, even at relatively high current densities [31]. It is seen that there is a potential overshoot due to the free energy involved in the nucleation of a new second phase if the reaction goes to completion in each direction. On the other hand, if the composition is not driven quite so far, so that there is some of the reactant phase remaining, this nucleation-related potential overshoot does not appear, as seen in Fig. 7B.3 [31].

This concept has also been demonstrated at ambient temperature in the case of the Li–Sn–Cd system [34, 35]. The composition dependence of the potentials in the two binary systems at ambient temperatures is shown in Fig. 7B.4, and the calculated phase stability diagram for this ternary system is shown in Fig. 7B.5. It was shown that the phase $Li_{4.4}Sn$, which has fast chemical diffusion for lithium [36], is stable at the potentials of two of the Li–Cd reconstitution reaction plateaus, and therefore can be used as a matrix phase. The behavior of this composite electrode, in which Li reacts with the Cd phases inside of the Li–Sn phase, is shown in Fig. 7B.6.

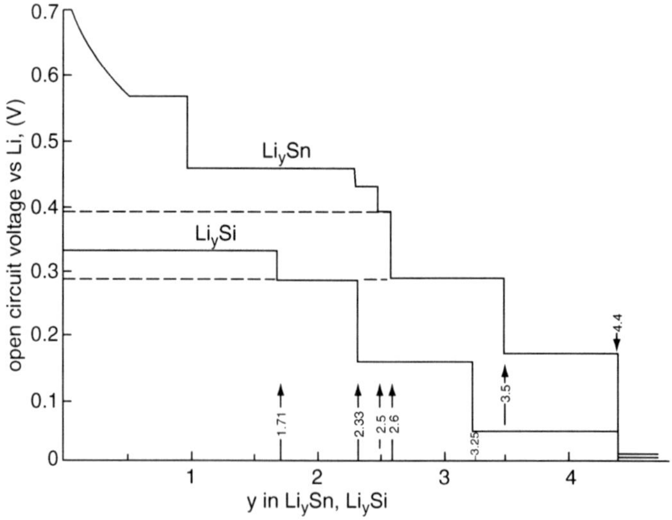

Fig. 7B.1 Composition dependence of the potential in the Li–Sn and Li–Si systems

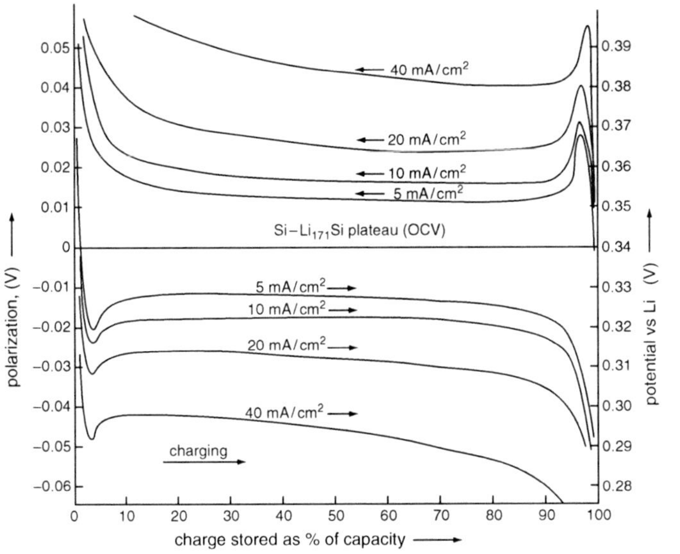

Fig. 7B.2 Charge and discharge curves of the Li–Si alloy in the matrix of the electrochemically inert mixed-conducting Li–Sn alloy at different current densities

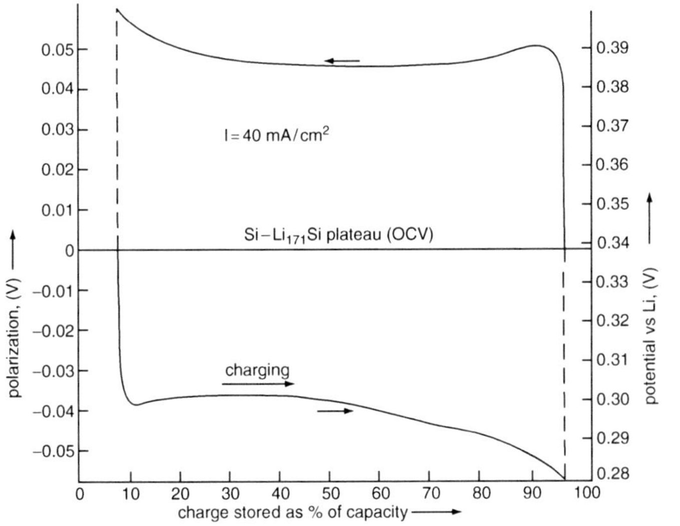

Fig. 7B.3 Charge and discharge curves of the Li–Si, Li–Sn composite if the capacity is limited so that the reaction does not go to completion in either direction. There is no large nucleation overshoot in this case

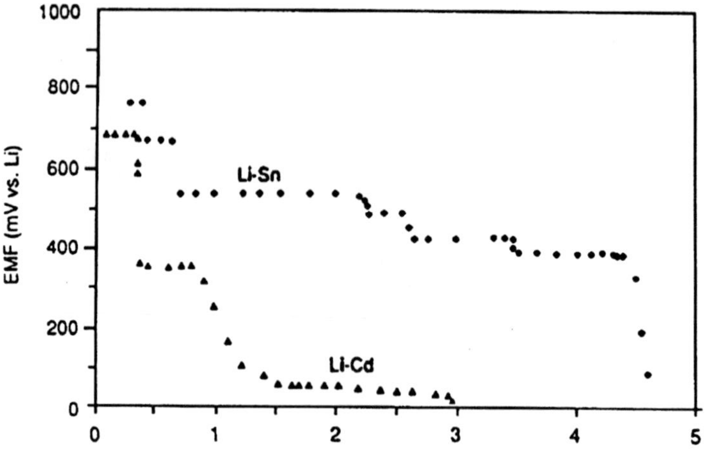

Fig. 7B.4 Potential vs. composition for Li–Sn and Li–Cd systems at ambient temperature

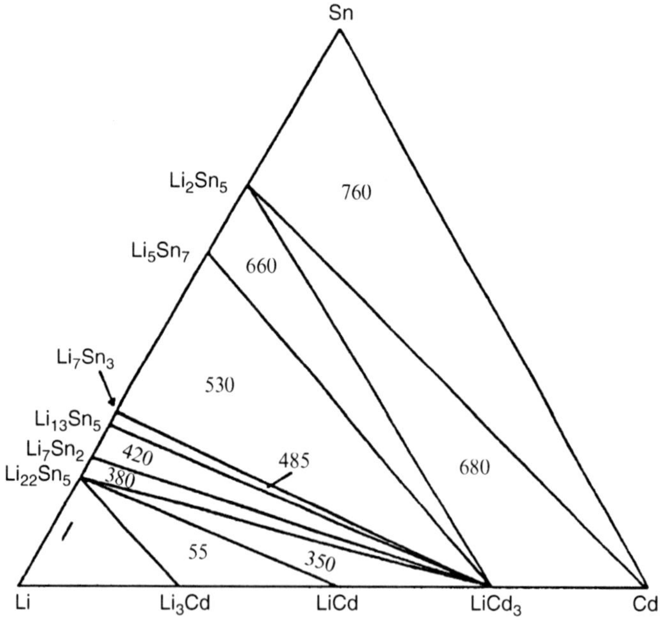

Fig. 7B.5 Calculated phase stability diagram for the Li–Cd–Sn system at ambient temperature. Numbers are voltages (millivolts) vs. Li

To achieve good reversibility, the composite electrode microstructure must have the ability to accommodate any volume changes that might result from the reaction that takes place internally. This can be taken care of by clever microstructural design and alloy fabrication techniques.

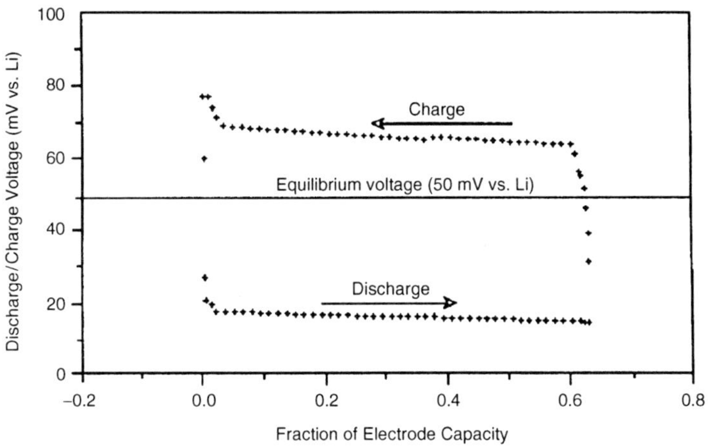

Fig. 7B.6 Charge–discharge curve of the Li–Cd system with a fast mixed-conducting phase in the lithium–tin system at ambient temperature

7B.6 Decrepitation

A phenomenon called *decrepitation*, that is also sometimes called *crumbling*, can occur in materials that undergo significant volume changes upon the insertion of guest species. These dimensional changes cause mechanical strain in the microstructure, often resulting in the fracture of particles in an electrode into smaller pieces.

This can be a striking, and sometimes disastrous, phenomenon, for it is not specifically related to fine particles, or even to electrochemical systems. As an example, it has been shown that some bulk solid metals can be caused to fracture, and can even be converted into powders by repeated exposure to hydrogen gas if they form metal hydrides under the particular thermodynamic conditions present. This is, of course, different from the hydrogen embrittlement problem in metals with body-centered cubic crystal structures, which involves the segregation of hydrogen to dislocations within the microstructure, influencing their mobility.

Decrepitation is often particularly evident during cycling of electrochemical systems. It can readily result in the loss of electronic contact between reactive constituents in the microstructure and the current collector. As a consequence, the reversible capacity decreases.

This phenomenon has long been recognized in some electrochemical systems in which metal hydrides are employed as negative electrode reactants, and is also mentioned in Chap. 11C.

Similar phenomena also occur in lithium systems employing alloy electrodes, some of which undergo very large changes in specific volume if the composition is varied over a wide range in order to achieve a large charge capacity.

Because of its potentially large capacity, a considerable amount of attention has been given recently to the Li–Sn system, which is a fine example of this phenomenon. The phase diagram of the Li–Sn system shows that there are six intermediate phases. The thermodynamic and kinetic properties of the different phases in this system were investigated some time ago at elevated temperatures [36, 37] and also at ambient temperatures [29, 30, 34, 35]. The volume changes that occur in connection with phase changes in this alloy system are large. The phase that forms at the highest lithium concentration, $Li_{4.4}Sn$, has a specific volume that is 283% of that of pure tin. Thus Li–Sn electrodes swell and shrink, or *breathe*, a lot as lithium is added or deleted.

Observations on metal hydrides that undergo larger volume changes have shown that this process does not continue indefinitely. Instead, it is found that there is a terminal particle size that is characteristic of a particular material. Particles with smaller sizes do not fracture further.

Experiments on lithium alloy electrodes have also shown that the electrochemical cycling behavior is significantly improved if the initial particle size is already very small [38], and it is reasonable to conclude that this is related to the terminal particle size phenomenon.

A theoretical study of the mechanism and the influence of the important parameters related to decrepitation utilized a simple one-dimensional model to calculate the conditions under which fracture will be caused to occur in a two-phase structure due to a specific volume mismatch [39]. This model predicts that there will be a terminal particle size below which further fracture will not occur. The value of this characteristic dimension is material-specific, depending upon two parameters, the magnitude of a strain parameter related to the volume mismatch and the fracture toughness of the lower specific-volume phase. For the same value of volume mismatch, the tendency to fracture will be reduced and the terminal particle size will be larger the greater the toughness of the material. The results of this model calculation are shown in Fig. 7B.7 [39].

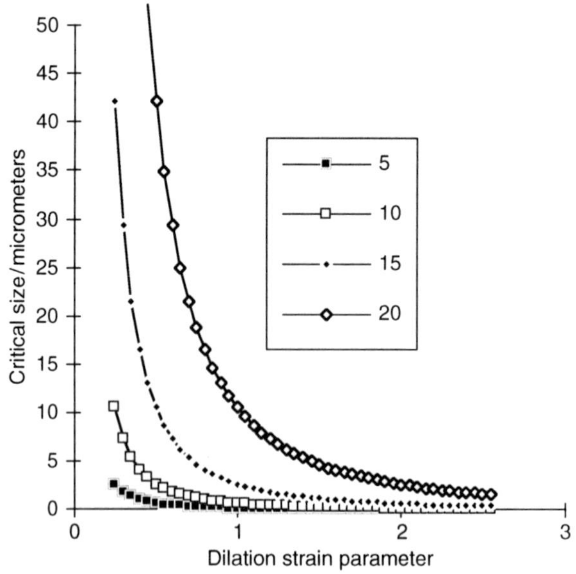

Fig. 7B.7 Variation of the critical particle size as a function of the dilation strain for several values of the fracture toughness of the phase in tension

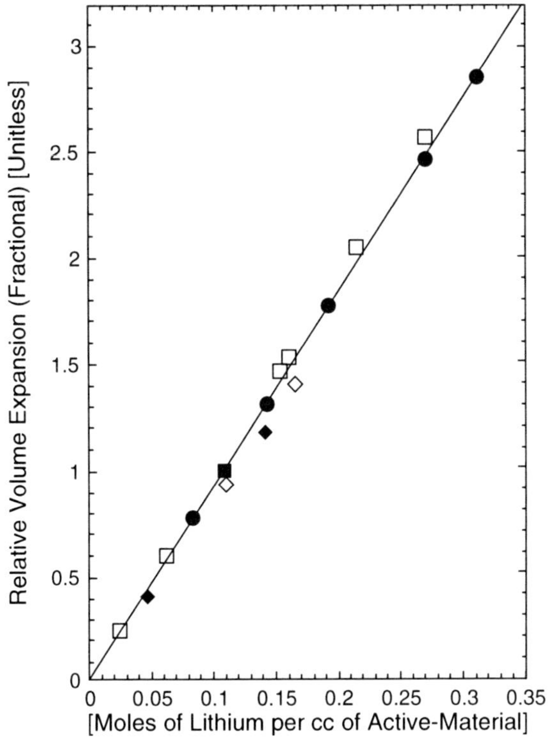

Fig. 7B.8 Relation between volume expansion and the amount of lithium introduced into lithium alloys

The magnitude of the volume change depends upon the amount of lithium that has entered the alloy crystal structure, and is essentially the same for all lithium alloys. This is shown in Fig. 7B.8 [40].

7B.7 Modification of the Micro and Nanostructure of the Electrode

Some innovative approaches have been employed to ameliorate the decrepitation problem due to the large volume changes inherent in the use of metal alloy and silicon negative electrodes in lithium systems. If that can be done, there is the possibility of a substantial improvement in the electrode capacity.

The general objective is to give the reactant particles room to *breathe*, so that they do not impinge upon each other. However, this has to be done so that they are maintained in electrical contact with the current collector system. Thus they cannot be physically isolated.

One interesting direction involves the modification of the shape of the surface upon which thin films of active material are deposited [41]. When the reactant film is dense, the volume changes and related stresses parallel to the surface cause separation from the substrate and loss of electronic contact. But if the surface is rough, there are high spots and low spots that have different local values of current density when the active material is electrodeposited. The deposition rate is greater at the higher locations, and less elsewhere. The result is that the active material is mostly deposited at the high spot locations, and grows in a generally columnar shape away from the substrate. This leaves some space between the columnar growths to allow for their volume changes during operation of the electrode. This is illustrated schematically in Fig. 7B.9.

Another alternative would be to make separated conductive spots on the surface, perhaps by the use of photolithography, which become the preferred locations for the deposition of reactant. By control of the spot arrangement, the electrodeposition can result in the formation of reactant material with limited impingement, thus allowing more *breathing room* when it undergoes charge and discharge.

It has been recently shown that a very attractive potential solution to this cycling problem is the use of reactant material in the form of nanowires. This is illustrated schematically in Fig. 7B.10.

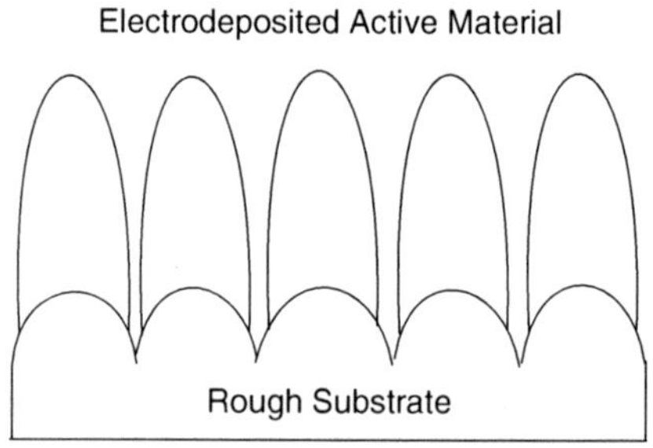

Fig. 7B.9 Schematic drawing of the preferential deposition of reactant material upon protrusions on the substrate surface

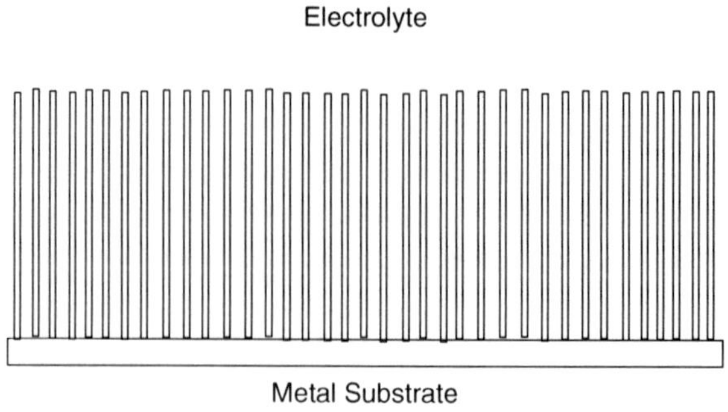

Fig. 7B.10 Schematic drawing of electrode with a large number of nanowires

The particular example has been silicon [42]. Such wires can be grown directly upon a metallic substrate, so that they are all in good electronic contact. Because there is some space between the individual wires, they can expand and contract as lithium is added or deleted without the constraints present in either thin film or powdered electrode structures. Experiments showed that such fine wires can attain essentially the theoretical capacity of the Li–Si system.

7B.8 Formation of Amorphous Products at Ambient Temperatures

This chapter has been primarily concerned with understanding the behavior of negative electrode materials under equilibrium or near-equilibrium conditions, from which the potential and capacity limits can be determined. Actual behavior in real applications always deviates from these limiting values, of course.

It was mentioned in Chap. 6 that repeated cycling can cause crystalline materials to become amorphous. The spectrum of materials in which amorphous phases have been formed under these conditions is now quite broad, and includes some materials of potential interest as positive electrode reactants, such as some vanadium-based materials with the general formula RVO_4, in which R is Al, Cr, Fe, In, or Y [43].

There have been a number of observations that the operation of negative electrode materials at very high lithium activities can result in the formation of amorphous, rather than crystalline, products. The properties of these amorphous materials are different from those of the corresponding crystalline materials. This is very different from the amorphization of positive electrode materials under cycling conditions.

One example is a group of nitride alloys with structures related to that of Li_3N, which is known to be a fast ionic conductor for lithium, but in which some of the lithium is replaced by a transition metal, such as Co, which have been found to become amorphous upon the first insertion of lithium [44–47].

Experimental evidence for the electrochemical amorphization of alloys in the Li–Si, Li–Sn, or Li–Ag systems was presented by Limthongkul [48]. In the latter two cases, this was only a transient phenomenon.

Especially interesting, however, have been experiments that gave evidence for the formation of amorphous silicon during the initial lithiation of a number of silicon-containing precursors, including SiB_3, SiO, $CaSi_2$, and $NiSi_2$ [49–51]. The electrochemical behavior of these materials after the initial lithiation cycle was essentially the same as that found in Si powder that was initially amorphous. There was, however, an appreciable amount of irreversible capacity in the first cycles of these precursors, about 1 mol of Li in the case of SiB_3 and the disilicides, which was evidently due to an irreversible displacement reaction with Li to form 1 mol of amorphous silicon. In the case of SiO the irreversible capacity amounted to about 2 mols of Li, which was surely related to the irreversible formation of Li_2O as well as the amorphous silicon.

Some of these materials with amorphous Si are of considerable potential interest as negative electrode reactants in lithium systems, as their charge/discharge curves are in an attractive potential range, they have reasonable kinetics, and their reversible capacities are quite high. The materials with silicon nanowire structure appear to be particularly attractive.

References

1. R.A. Huggins and D. Elwell, J. Cryst. Growth 37, 159 (1977)
2. C. Wagner, J. Electrochem. Soc. 101, 225 (1954)
3. C. Wagner, J. Electrochem. Soc. 103, 571 (1956)
4. G. Deublein and R.A. Huggins, Solid State Ionics 18/19, 1110 (1986)
5. U. von Sacken, E. Nodwell and J.R. Dahn, Solid State Ionics 69, 284 (1994)
6. M. Winter, K.-C. Moeller and J.O. Besenhard, Carbonaceous and Graphitic Anodes, in *Lithium Batteries, Science and Technology*, ed. by G.-A. Nazri and G. Pistoia, Kluwer Academic, Boston, MA (2004), p. 144
7. J.R. Dahn, A.K. Sleigh, H. Shi, B.M. Way, W.J. Weydanz, J.N. Reimers, Q. Zhong and U. von Sacken, Carbons and Graphites as Substitutes for the Lithium Anode, in *Lithium Batteries*, ed. by G. Pistoia, Elsevier, Amsterdam (1994), p. 1
8. K. Fredenhagen and G. Cadenbach, Z. Anorg. Allg. Chem. 158, 249 (1926)
9. D. Guerard, A. Herold, Carbon 13, 337 (1975)
10. G.K. Wertheim, P.M.Th.M. Van Attekum and S. Basu, Solid State Commun. 33, 1127 (1980)
11. L.B. Ebert, Intercalation Compounds of Graphite, in Annual Review of Materials Science, Vol. 6, ed. by R.A. Huggins, Annual Reviews, Palo Alto, CA (1976), p. 181
12. J.O. Besenhard and H.P. Fritz, J. Electroanal. Chem. 53, 329 (1974)
13. R. Yazami and P. Touzain, J. Power Sources 9, 365 (1983)
14. S. Basu, U.S. Patent No. 4,304,825, Dec. 8, 1981
15. S. Basu, U.S. Patent No. 4,423,125, Dec. 27, 1983
16. T. Nagaura and K. Tozawa, in *Progress in Batteries and Solar Cells*, Vol. 9, ed. by JEC Press and IBA, JEC Press, Brunswick, OH (1990), p. 209
17. T. Nagaura, in *Progress in Batteries and Solar Cells*, Vol. 10, JEC Press, Brunswick, OH (1991), p. 218
18. R.E. Franklin, Proc. R. Soc (Lond) A209, 196 (1951)
19. N. Daumas and A. Herold, C.R. Acad. Sci. C 286, 373 (1969)
20. T. Zheng, Y. Liu, E.W. Fuller, S. Tseng, U. von Sacken and J.R. Dahn, J. Electrochem. Soc. 142, 2581 (1995)
21. T. Zheng, J.S. Xue and J.R. Dahn, Chem. Mater. 8, 389 (1996)
22. T. Zheng, W.R. McKinnon and J.R. Dahn, Hysteresis During Lithium Insertion in Hydrogen-Containing Carbons, J. Electrochem. Soc. 143, 2137 (1996)
23. N.P Yao, L.A. Heredy and R.C. Saunders, J. Electrochem. Soc. 118, 1039 (1971)
24. E.C. Gay, et al., J. Electrochem. Soc. 123, 1591 (1976)
25. S.C. Lai, J. Electrochem. Soc. 123, 1196 (1976)
26. R.A. Sharma and R.N. Seefurth, J. Electrochem Soc. 123, 1763 (1976)
27. R.N. Seefurth and R.A. Sharma, J. Electrochem. Soc. 124, 1207 (1977)
28. H. Ogawa, Proceedings of Second International Meeting on Lithium Batteries, Elsevier Sequoia, Lausanne, Switzerland (1984), p. 259
29. J. Wang, P. King and R.A. Huggins, Solid State Ionics 20, 185 (1986)
30. J. Wang, I.D. Raistrick and R.A. Huggins, J. Electrochem. Soc. 133, 457 (1986)
31. B.A. Boukamp, G.C. Lesh and R.A. Huggins, J. Electrochem. Soc. 128, 725 (1981)
32. B.A. Boukamp, G.C. Lesh and R.A. Huggins, in Proc. Symp. on Lithium Batteries, ed. by H.V. Venkatasetty, Electrochem. Soc., Pennington, NJ (1981), p. 467

33. R.A. Huggins and B.A. Boukamp, US Patent 4,436,796
34. A. Anani, S. Crouch-Baker and R.A. Huggins, in Proc. Symp. on Lithium Batteries, ed. by A.N. Dey, Electrochem. Soc., Pennington, NJ (1987), p. 382
35. A. Anani, S. Crouch-Baker and R.A. Huggins, J. Electrochem. Soc. 135, 2103 (1988)
36. C.J. Wen and R.A. Huggins, J. Solid State Chem. 35, 376 (1980)
37. C.J. Wen and R.A. Huggins, J. Electrochem. Soc. 128, 1181 (1981)
38. J. Yang, M. Winter and J.O. Besenhard, Solid State Ionics 90, 281 (1996)
39. R.A. Huggins and W.D. Nix, Ionics 6, 57 (2000)
40. A. Timmons, PhD Dissertation, Dalhousie University, Canada (2007)
41. M. Fujimoto, S. Fujitani, M. Shima, et al., US Patent 7,195,842 (March 27, 2007)
42. C.K. Chan, H. Peng, G. Liu, K. McIlwrath, X. Feng Zhang, R.A. Huggins and Y. Cui, Nat. Nanotechnol.3, 31 (2008)
43. Y. Piffard, F. Leroux, D. Guyomard, J.-L. Mansot and M. Tournoux, J. Power Sources 68, 698 (1997)
44. M. Nishijima, T. Kagohashi, N. Imanishi, Y. Takeda, O. Yamamoto and S. Kondo, Solid State Ionics 83, 107 (1996)
45. T. Shodai, S. Okada, S-i. Tobishima, and J-i. Yamaki, Solid State Ionics 86–88, 785 (1996)
46. M. Nishijima, T. Kagohashi, Y. Takeda, N. Imanishi and O. Yamamoto, in Eighth International Meeting on Lithium Batteries, Nagoya, Japan (1996), p. 402
47. T. Shodai, S. Okada, S. Tobishima and J. Yamaki, in Eighth International Meeting on Lithium Batteries, Nagoya, Japan (1996), p. 404
48. P. Limthongkul, PhD Thesis, Massachussets Institute of Technology, Cambridge, MA (2002)
49. B. Klausnitzer, PhD Thesis, University of Ulm, Germany (2000)
50. A. Netz, PhD Thesis, University of Kiel, Germany (2001)
51. A. Netz, R.A. Huggins and W. Weppner, Presented at 11th International Meeting on Lithium Batteries, Monterey, CA (2002). Abstract No. 47

Chapter 8
Convertible Reactant Electrodes

8.1 Introduction

As mentioned earlier, most of the commercial lithium batteries have used carbonaceous materials as their negative electrode reactants for a number of years. The major impetus in this direction was the development and commercialization of the camcorder battery by SONY in 1991.

Nevertheless, there has been some continued consideration of the use of metallic and metal–metalloid alloys, due to the possibility of significant increases in specific capacity and capacity density, as was discussed in the previous chapter.

Another alternative suddenly appeared in 1996 as the result of the surprise announcement by Fuji Photo Film Co. of the development of a new generation of lithium batteries based on the use of an amorphous tin-based composite oxide in the negative electrode [1, 2]. In one case, the oxide was made by melting together SnO, B_2O_3, $Sn_2P_2O_7$, and Al_2O_3 powders. Elements such as boron, phosphorus, and aluminum are recognized to be glass formers in oxide systems, and they are electrochemically inactive. Because the resulting oxide is not electrically conductive, carbon was also present, as well as a polymer binder. The multicomponent oxide is converted to a mixture of a lithium–tin alloy and residual oxides during the first discharge cycle of the cell, and thereafter, its properties are essentially those of the resulting lithium alloy, as discussed in the previous chapter.

Fuji claimed that these electrodes had a volumetric capacity of 3,200 Ah L^{-1}, which is four times that commonly achieved with carbon-based negative electrodes, and a specific capacity of 800 mAh g^{-1}, more than twice that generally found in carbon-containing negative electrodes. This development caused a renewed interest in the potential of the use of noncarbonaceous lithium alloy electrodes.

R.A. Huggins, *Advanced Batteries*: Materials Science Aspects,

8.2 Electrochemical Formation of Metals and Alloys from Oxides

The principles involved in the formation of metals and alloys by the electrochemical conversion of oxides deserve some attention. The lithium–tin system is a relevant example.

If an electrode initially contains an oxide that is less stable than lithium oxide, there will be a thermodynamic driving force for a *displacement reaction* in which Li_2O will be formed at the expense of the prior oxide. The other product will contain the residual metallic species. An electrode initially containing SnO can be used as an example. If it is employed as the negative electrode of an electrochemical cell in which lithium is present in the positive electrode, the lithium activity will become larger as the potential is made more negative. This will tend to cause the *displacement reaction* in which Li_2O will be formed at the expense of the initial tin oxide.

$$2Li + SnO = Li_2O + Sn \tag{8.1}$$

The difference in the values of their Gibbs free energies of formation ($-562.1\,\text{kJ mol}^{-1}$ for Li_2O and $-256.8\,\text{kJ mol}^{-1}$ in the case of SnO at 25°C) provides the driving force. In this case it is quite strong, equivalent to 1.58 V.

The other product will be elemental Sn, and as additional *Li* is brought into the electrode this Sn will tend to react further to form the various Li–Sn alloys that exist in the lithium–tin phase diagram. This simplified picture is consistent with what has been found in experiments on oxides of this general type [3, 4].

If the formation of Li_2O is not reversible, the electrode will maintain a composite microstructure and behave as a binary lithium–tin alloy after the first cycle. This initial Li_2O formation represents a significant initial capacity loss, analogous to, but much larger than, what is generally found in lithium–carbon electrodes.

8.3 Lithium–Tin Alloys at Ambient Temperature

As was mentioned earlier, the lithium–tin system has been investigated experimentally at both elevated temperature and room temperature. The experimentally obtained electrochemical titration curve is shown in Fig. 8.1 [5].

It is seen that under equilibrium conditions there are a number of constant-potential plateaus at this temperature. These are listed in Table 8.1.

It has also been found that the kinetics on the longest plateau, at 0.53 V vs. Li and from $x = 0.7$–2.33 in Li_xSn, are quite favorable, even at quite high currents at ambient temperature [5]. This is consistent with the results of measurements of the chemical diffusion coefficient in the two adjacent phases, $Li_{0.7}Sn$ and $Li_{2.33}Sn$, which were found to be quite high, 6–8×10^{-8} and 3–$5 \times 10^{-7}\,\text{cm}^2\,\text{s}^{-1}$, respectively.

The composition dependence of both the potential and the chemical diffusion coefficient of the most lithium-rich phase, $Li_{4.4}Sn$ ($Li_{22}Sn_5$), has also been determined at ambient temperature [6]. The chemical diffusion coefficient in that phase reaches

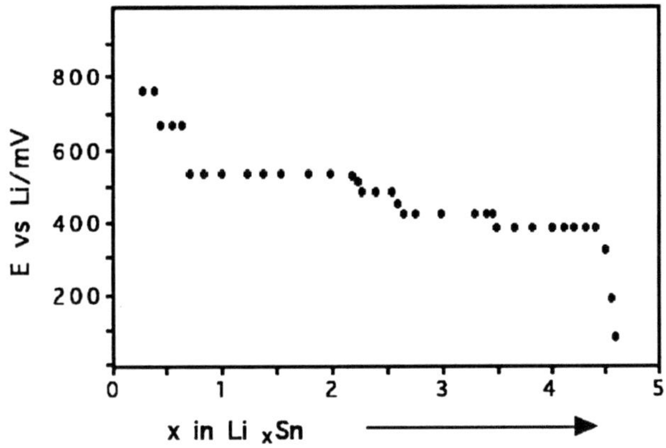

Fig. 8.1 Equilibrium titration curve for the reaction of lithium with tin at 25°C

Table 8.1 Plateau potentials and composition ranges of lithium–tin alloys at 25°C

Plateau Potential (voltage vs. Li)	Range of composition parameter x in Li_xSn
0.660	0.4–0.7
0.530	0.7–2.33
0.485	2.33–2.63
0.420	2.63–3.5
0.380	3.5–4.4

a peak value of about 6×10^{-7} cm^2 s^{-1} at the stoichiometric composition. Thus, it is clear that lithium has a very high mobility, and reactions occur quite rapidly in this alloy system, even at ambient temperature.

8.4 The Lithium–Tin Oxide System

The Fuji Photo Film development was said to involve the use of an amorphous tin-based composite oxide. It can be assumed that this is an example of a convertible oxide electrode, and that reaction with lithium on the first charge cycle results in the formation of a microstructure containing fine dispersions of both Li–Sn alloys and Li_2O. The latter is known to be a lithium-transporting solid electrolyte [7, 8], with a value of ionic conductivity at 25°C of 1.5×10^{-9} ohm^{-1} cm^{-1}. Thus, these electrodes can be thought of as having a composite microstructure with the reactant phase intimately mixed with a solid electrolyte.

It is useful to look at two simple cases in order to see the effect of starting with a convertible oxide. Consider the use of either SnO_2 or SnO as precursors for the in-situ formation of Li–Sn alloys. If the simplifying assumption is made that

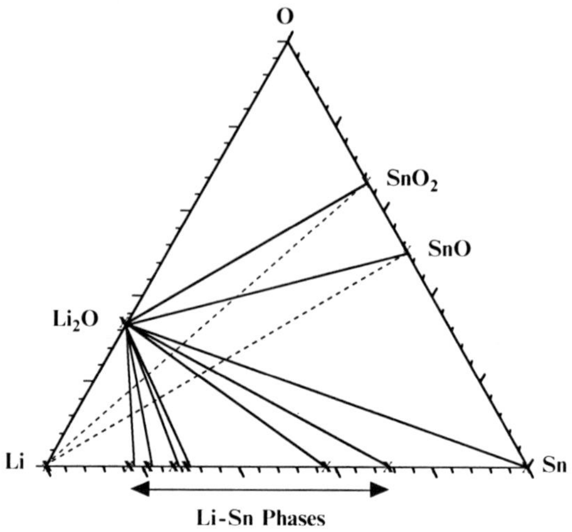

Fig. 8.2 Isothermal phase stability diagram for the Li–Sn–O system

there are no stable ternary phases, a simple isothermal phase stability diagram can be constructed to use as a thinking tool in working out the properties of this system. A simplified version of such a diagram is shown in Fig. 8.2.

If lithium reacts with one of the binary tin oxide phases, the overall composition will move toward the Li corner of the ternary diagram, and under conditions close to equilibrium, it will move along one of the dotted lines shown in that figure. It is seen that the overall composition will move through a series of three-phase triangles, crossing two-phase tie lines between them.

As discussed earlier, when three phases are present in a ternary system and the temperature and the overall pressure are held constant, there are no degrees of freedom, according to the Gibbs phase rule. This means that the electrical potential must be independent of composition, i.e., there will be a potential plateau, so long as the composition remains in that triangle. This potential can be calculated from the Gibbs free energy change involved in the virtual reaction that takes place within the relevant three-phase region, as discussed earlier.

If the oxide is initially SnO_2, the reaction of lithium with it causes the overall composition to move from SnO_2 into a constant potential triangle with SnO_2, Li_2O, and SnO at its corners. That reaction can be written as

$$2Li + SnO_2 = Li_2O + SnO. \tag{8.2}$$

From the values of the Gibbs free energy of formation of SnO_2, SnO, and Li_2O, it can be found that this reaction will take place at 1.88 V vs. Li under equilibrium conditions.

Further reaction with lithium moves the composition into the SnO, Li_2O, Sn triangle. The relevant reaction for this constant potential triangle is (8.1), given earlier.

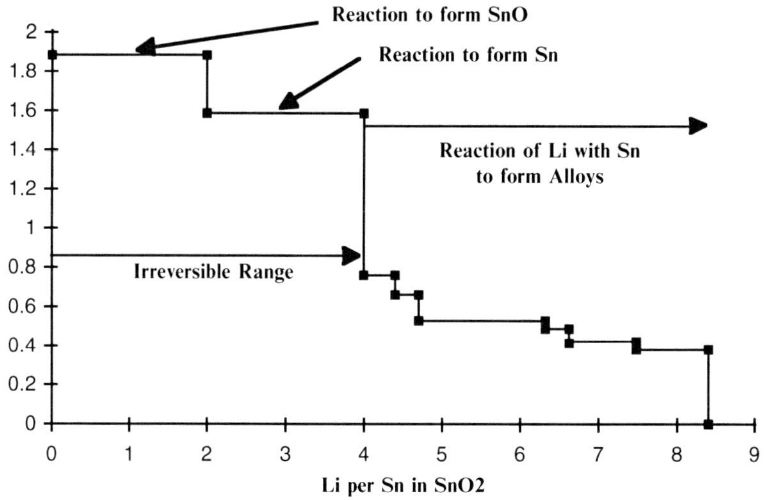

Fig. 8.3 Theoretical titration curve for the reaction of lithium with SnO_2 at 25°C

The equilibrium potential during this composition range is 1.58 vs. Li, as mentioned earlier.

Further reaction with lithium causes the composition to move through a series of constant potential triangles, as the lithium reacts with tin and its alloys.

If, on the other hand, the initial oxide is SnO instead of SnO_2, the overall composition line goes from SnO toward the lithium corner of the phase stability diagram. In this case it does not move through the 1.88-V triangle, but starts at 1.58 V. The later reaction stages are the same as when the initial oxide is SnO_2.

These results can be expressed in terms of theoretical electrochemical titration curves, in which the potential (V vs. Li) is plotted vs. the composition. There will be a series of constant-voltage plateaus as the composition moves across the three-phase regions of the diagram in Fig. 8.2. These are shown for two cases, starting with SnO_2, and starting with SnO, in Figs. 8.3 and 8.4.

Calculated values of the potentials of the various lithium–tin–oxygen plateaus at room temperature, based upon the thermodynamic data for the oxides, are included in Table 8.2.

8.5 Irreversible and Reversible Capacities

It can be seen that if one starts with SnO_2, two lithiums will initially react to form Li_2O and SnO at a potential of 1.88 V vs. Li. Following this, additional two lithiums will react with the SnO to form more Li_2O and Sn at a potential of 1.582 V vs. Li. If it can be assumed that these two reactions are irreversible, and that this lithium cannot be deleted to decompose the Li_2O and reform the tin oxides, four lithium atoms are irreversibly consumed if the initial oxide is SnO_2. Two lithium atoms

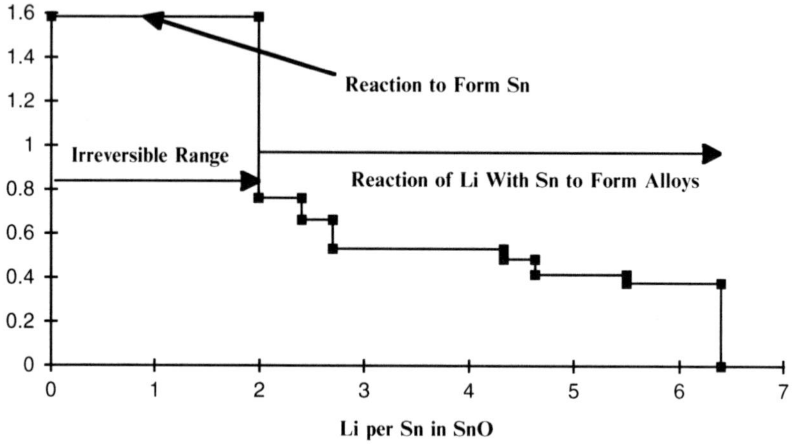

Fig. 8.4 Theoretical titration curve for the reaction of lithium with SnO at 25°C

Table 8.2 Equilibrium potentials of plateaus in three-phase regions in the Li–Sn–O ternary system at ambient temperature

Three-phase equilibrium	Potential
$Li_2O-O_2-SnO_2$	2.912
Li_2O-SnO_2-SnO	1.880
$Li_2O-SnO-Sn$	1.582
$Li_2O-Sn-Li_{0.4}Sn$	0.760
$Li_2O-Li_{0.4}Sn-Li_{0.714}Sn$	0.660
$Li_2O-Li_{0.714}Sn-Li_{2.33}Sn$	0.530
$Li_2O-Li_{2.33}Sn-Li_{2.6}Sn$	0.485
$Li_2O-Li_{2.6}Sn-Li_{3.5}Sn$	0.420
$Li_2O-Li_{3.5}Sn-Li_{4.4}Sn$	0.380

react irreversibly if the initial oxide is SnO, rather than SnO_2. This represents a large irreversible capacity, 711.3 mAh g^{-1} of SnO_2 if the electrode starts as SnO_2, and 397.9 mAh g^{-1} of SnO if the electrode is initially SnO.

This irreversible lithium must come from somewhere in the cell, i.e., from the positive electrode. Thus, a significant amount of extra lithium-containing positive electrode reactant must be present when the cell is fabricated in order to take advantage of the properties of the lithium–tin alloy in the negative electrode.

Now consider the active reversible capacity, rather than the irreversible capacity. This is the capacity that is obtained by the reaction of additional lithium with the tin that resulted from the earlier irreversible reactions.

From the theoretical titration diagrams in Figs. 8.3 and 8.4, it can be seen that an additional 4.4 lithium atoms can react in each case, assuming that the overall composition starts with pure tin after the conversion of the oxides. If it can be assumed that this is all reversible, and that the electrode starts as SnO_2, this amounts to a reversible capacity of 782 mAh g^{-1} of SnO_2. On the other hand, if the electrode starts as SnO, the reversible capacity is 875 mAh g^{-1} of SnO.

8.6 Other Possible Convertible Oxides

Similar irreversible and reversible capacities are theoretically found in other oxide systems. Theoretical data on both the tin oxides and several of these oxides are shown in Table 8.3.

In addition to the matter of capacities, the potential ranges must also be considered. Table 8.4 shows theoretical data for several oxides. The voltage necessary to decompose the oxide to its metal can be calculated simply from its Gibbs free energy of formation.

These calculated voltage values relate to the potential of 1 atm of oxygen, not lithium. To convert them to voltages vs. lithium, it is necessary to know the relationship between the potential of lithium and that of 1 atm of oxygen. This is calculated by the use of the Gibbs free energy of formation of lithium oxide. The result is that the potential of lithium is -2.91 V vs. oxygen at ambient temperature. The difference between the two voltages vs. oxygen indicates the relative potentials of the various metals and lithium. This relationship is illustrated in Fig. 8.5.

The voltage difference between the metals and lithium indicates the maximum voltage of any possible reaction plateaus (vs. lithium) in each of these possible lithium–metal systems. Information about the voltage stability of a number of oxides, and the maximum theoretical stability of their respective metals relative to lithium is given in Table 8.4.

In order to investigate the possibility of further capacity in such cases, one would have to have information about further possible lithium–metal reactions and their potentials and capacities, i.e., their titration curves. In addition, there is the matter of the kinetics of such reactions, and this would have to be determined experimentally.

Table 8.3 Theoretical irreversible and reversible capacities of several convertible oxides

Starting oxide	Reversible capacity (mAh/g oxide)	Irreversible capacity (mAh/g oxide)	Total capacity (mAh/g oxide)	Ratio (Rev./total)
SnO	875.36	398	1273	0.69
SnO_2	782.43	711	1494	0.52
ZnO	493.92	659	1152	0.43
CdO	605.25	417	1023	0.59
PbO	540.32	240	780	0.69

Table 8.4 Theoretical data on other possible convertible oxides

Oxide	Stability of oxide (V)	Maximum E vs. Lithium (V)
Al_2O_3	2.73	0.18
B_2O_3	2.06	0.85
CdO	1.19	1.72
PbO	0.98	1.93
SiO_2	2.22	0.69
SnO	1.33	1.58
ZnO	1.66	1.25

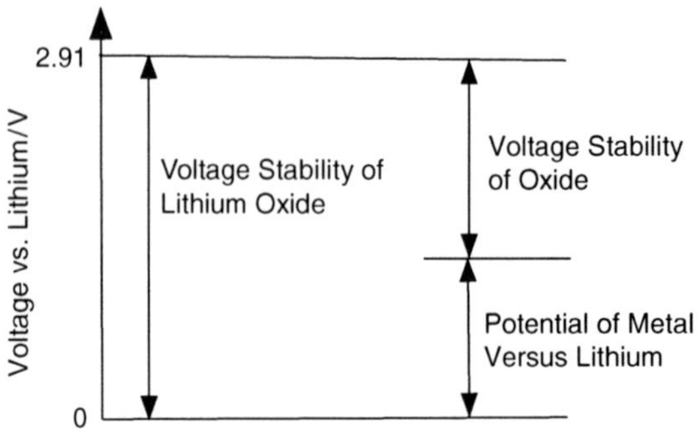

Fig. 8.5 Relative potentials in lithium–metal–oxygen systems

8.7 Final Comments

Although the possibility of making alloy electrodes with large capacities by the reaction of lithium with corresponding oxides initially appears to be attractive, it can be seen that there can also be significant disadvantages. The primary ones are the inherent large initial irreversible capacity and the excess weight and volume of the oxides present.

References

1. Fujifilm, http://www.fujifilm.co.jp/eng/news_e/nr079.html (1996)
2. Y. Idota, et al., Science 276, 1395 (1997)
3. I. A. Courtney and J. R. Dahn, J. Electrochem. Soc. 144, 2045 (1997)
4. I. A. Courtney and J. R. Dahn, J. Electrochem. Soc. 144, 2943 (1997)
5. J. Wang, I. D. Raistrick and R. A. Huggins, J. Electrochem. Soc. 133, 457 (1986)
6. A. Anani, S. Crouch-Baker and R. A. Huggins, "Measurement of Lithium Diffusion in Several Binary Lithium Alloys at Ambient Temperature," in *Proc. Symp. on Lithium Batteries*, ed. by A. N. Dey, The Electrochemical Society, Pennington, NJ (1987), p. 365
7. R. A. Huggins, Electrochimica Acta 22, 773 (1977)
8. R. M. Biefeld and R. T. Johnson, Jr., J. Electrochem. Soc. 126, 1 (1979)

Chapter 9
Positive Electrodes in Lithium Systems

9.1 Introduction

Several types of lithium batteries are used in a variety of commercial products, and are produced in very large numbers. According to various reports, the sales volume in 2008 is approximately 10 billion dollars per year, and it is growing rapidly. Most of these products are now used in relatively small electronic devices, but there is also an extremely large potential market if lithium systems can be developed sufficiently to meet the requirements for hybrid, or even plug-in hybrid, vehicles.

As might be expected, there is currently a great deal of interest in the possibility of the development of improved lithium batteries in both the scientific and technological communities. An important part of this activity is aimed at the improvement of the positive electrode component of lithium cells, where improvements can have large impacts upon the overall cell performance.

However, before giving attention to some of the details of positive electrodes for use in lithium systems, some comments will be made about the evolution of lithium battery systems in recent years.

Modern advanced battery technology actually began with the discovery of the high ionic conductivity of the solid phase $NaAl_{11}O_{17}$, called sodium beta alumina, by Kummer and coworkers at the Ford Motor Co. laboratory [1]. This led to the realization that ionic transport in solids can actually be very fast, and that it might lead to a variety of new technologies. Shortly thereafter, workers at Ford showed that one can use a highly conducting solid electrolyte to produce an entirely new type of battery, using molten sodium at the negative electrode and a molten solution of sodium in sulfur as the positive electrode, with the sodium-conducting solid electrolyte in between [2].

This attracted a lot of attention, and scientists and engineers from a variety of other fields began to get interested in this area, which is so different from conventional aqueous electrochemistry, in the late 1960s. This concept of a liquid electrode, solid electrolyte (L/S/L) system was quite different from conventional S/L/S batteries. The development of the $Na/NiCl_2$ *Zebra* battery system, which has since turned out to be more attractive than the Na/Na_xS version, came along somewhat later [3–5]. This is discussed elsewhere in this text.

As might be expected, consideration was soon given to the possibility of analogous lithium systems, for it was recognized that an otherwise equivalent lithium cell should produce higher voltages than a sodium cell. In addition, lithium has a lower weight than sodium, another potential plus. There was a difficulty, however, for no lithium-conducting solid electrolyte was known that had a sufficiently high ionic conductivity to be used for this purpose.

Instead, a concept employing a lithium-conducting molten salt electrolyte, a eutectic solution of LiCl and KCl that has a melting point of 356°C, seemed to be an attractive alternative. However, because a molten salt electrolyte is a liquid, the electrode materials had to be solids. That is, the lithium system had to be of the S/L/S type.

Elemental lithium could not be used, because of its low melting point. Instead, solid lithium alloys, primarily the Li/Si and Li/Al systems, were investigated [6], as discussed elsewhere in this text.

A number of materials were investigated as positive electrode reactants at that time, with most attention given to the use of either FeS or FeS_2. Upon reaction with lithium, these materials undergo *reconstitution reactions*, with the disappearance of the initial phases and the formation of new ones [7].

9.2 Insertion Reaction, Instead of Reconstitution Reaction, Electrodes

An important next step was the introduction of the concept that one can reversibly *insert* lithium into solids to produce electrodes with useful potentials and capacities. This was first demonstrated by Whittingham in 1976, who investigated the addition of lithium to the layer-structured TiS_2 to form Li_xTiS_2, where x went from 0 to 1 [8,9].

Evidence that this insertion-driven solid solution redox process is quite reversible, even over many cycles, is shown in Fig. 9.1, where the charge and discharge behavior of a Li/TiS_2 cell is shown after 76 cycles [10].

Subsequently, the insertion of lithium into a significant number of other materials, including V_2O_5, LiV_3O_8, and V_6O_{13} was investigated in many laboratories. In all of these cases, this involved the assumption that one should assemble a battery with pure lithium negative electrodes and positive electrodes with small amounts of, or no, lithium initially. That is, the electrochemical cell is assembled in the charged state.

The fabrication method generally involved the use of glove boxes and a molten salt or organic liquid electrolyte. This precluded operation at high potentials, and the related oxidizing conditions, as discussed elsewhere.

That work involved the study of materials by the addition of lithium, and thus scanned their behavior at potentials lower than about 3 V vs. Li, for this is the

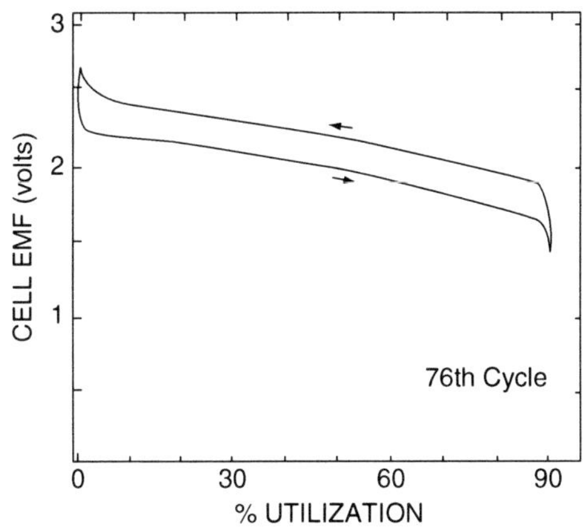

Fig. 9.1 Charge/discharge behavior of a Li/TiS_2 cell after 76 cycles

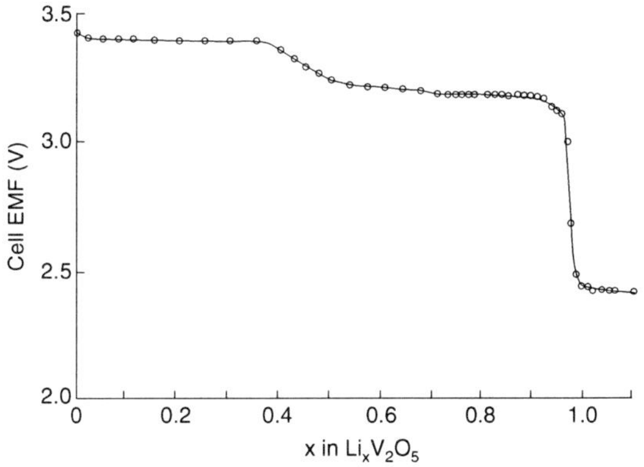

Fig. 9.2 Variation of the potential with the concentration of lithium guest species in the V_2O_5 host structure

starting potential for most electrode materials that are synthesized in air. As lithium is added and the cell is discharged, the potential of the positive electrode goes down toward that of pure lithium.

9.3 More than One Type of Interstitial Site or More than One Type of Redox Species

The variation of the potential depends upon the distribution of available interstitial places that can be occupied by the Li guest ions. If all sites are not the same in a given crystal structure, the result can be the presence of more than one plateau in the voltage–composition curve. An example of this is the equilibrium titration curve for the insertion of lithium into the V_2O_5 structure shown in Fig. 9.2 [11].

It will be seen later that similar voltage/composition behavior can result from the presence of more than one species that can undergo a redox reaction as the amount of inserted lithium is varied.

9.4 Cells Assembled in the Discharged State

On the other hand, if a positive electrode material initially contains lithium, and some or all of the lithium is deleted, the potential goes up, rather than down, as it does upon the insertion of lithium. Therefore, it is possible to have positive electrode materials that react with lithium at potentials above about 3 V, if they already contain lithium, and this lithium can be electrochemically extracted.

This concept is shown schematically in Fig. 9.3 for a hypothetical material that is *amphoteric*, and can react at both high and low potentials.

This approach, involving the use of materials in which lithium is already present, was first demonstrated in Prof. Goodenough's laboratory in Oxford. The first examples of materials initially containing lithium, and electrochemically deleting lithium from them, were the work on $Li_{1-x}CoO_2$ [12] and $Li_{1-x}NiO_2$ [13] in 1980. They showed that it is possible in this way to obtain high reaction potentials, up to over 4 V.

It was not attractive to use such materials in cells with metallic Li negative electrodes, however, and this approach did not attract any substantial interest at that time. This abruptly changed as the result of the surprise development by SONY of a lithium battery containing a carbon negative electrode and a $LiCoO_2$ positive electrode that became commercially available in 1990. These cells were initially

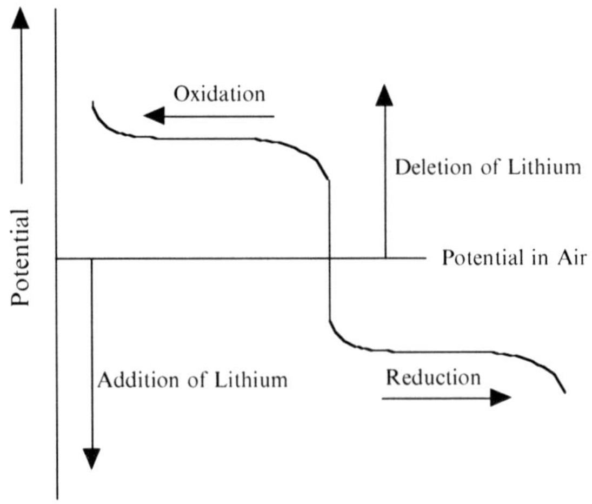

Fig. 9.3 Schematic of the behavior of a material that is amphoteric, i.e, that can be both electrochemically oxidized at high potentials by the deletion of lithium, and electrochemically reduced at lower potentials by the addition of lithium

assembled in the discharged state. They were activated by charging, whereby lithium left the positive electrode material, raising its potential, and moved to the carbon negative electrode, whose potential was concurrently reduced.

This cell can be represented as

$$\mathrm{Li}_x\mathrm{C}/\text{organic solvent electrolyte}/\mathrm{Li}_{1-x}\mathrm{CoO}_2,$$

and the cell reaction can be written as

$$C + LiCoO_2 = Li_xC + Li_{1-x}CoO_2. \tag{9.1}$$

This general type of cell and related reaction are most common in commercial cells at the present time. Some of the recent progress made in this area is discussed in Sect. 9A. Sections 9B and 9C will discuss other topics relating to positive electrodes in lithium battery systems.

9A Solid Positive Electrodes in Lithium Systems

9A.1 Introduction

In almost every case, the materials that are now used as positive electrode reactants in reversible lithium batteries operate by the use of insertion reactions. This general concept has been discussed several times in this text already. As pointed out in Chap.9, the early ambient temperature lithium battery developments were based upon the observation that lithium could be readily inserted into solids with crystal structures containing available interstitial space. A number of such materials were found, the most notable being TiS_2 and V_6O_{13}. These cells utilized elemental lithium, or lithium alloys, in the negative electrode.

Although precautions had to be taken in preparing and handling the negative materials, due to their propensity to oxidize, the positive electrode materials were typically stable in air.

As the insertion of lithium causes the potential to decrease, and those positive electrodes necessarily operated at potentials lower than that of air, the voltage of such cells was limited to about 3 V.

The shift in concept to the use of air-stable positive electrode materials that already contained lithium, and their operation by the deletion of lithium, led to the possibility of batteries with significantly higher voltages. But this also required a different strategy for the negative electrodes, for they must be initially devoid of lithium. Cells can be assembled in air in the discharged state. To be put into operation, they must be charged, the lithium initially in the positive electrode being transferred to the negative electrode.

This different approach did not attract any substantial interest until the surprise development by SONY Energytec [14, 15] of a commercial lithium cell that was produced with a $LiCoO_2$ positive electrode, an organic solvent electrolyte, and a carbon negative electrode, i.e., in the discharged state. Upon charging, lithium is transferred from the positive electrode to the carbon negative electrode. Such a cell can be represented simply as

$$Li_xC/\text{organic solvent electrolyte}/Li_{1-x}CoO_2.$$

It is interesting that the most commonly used positive electrode in small consumer electronics batteries now also is $LiCoO_2$, although a considerable amount of research is underway in the quest for a more desirable material.

A charge/discharge curve showing the reversible extraction of lithium from $LiCoO_2$ is shown in Fig. 9A.1. It is seen that approximately 0.5 Li per mole of $LiCoO_2$ can be reversibly deleted and reinserted. The charge involved in the transfer of lithium ions is balanced by the Co^{3+}/Co^{4+} redox reaction. This process cannot go further, because the layered crystal structure becomes unstable, and there is a transformation into another structure.

Quite a number of materials are now known from which it is possible to delete lithium at high potentials. Some of these will be described briefly later, but it is important to realize that this is a very active research area at the present time, and no such discussion can be expected to be complete.

There are a number of interesting materials that have a *face-centered cubic packing* of oxide ions, including both those with the *spinel* structure, e.g., $LiMn_2O_4$, variants containing more than one redox ion, and those with *ordered cation distributions*, which are often described as having *layered structures* (e.g., $LiCoO_2$ and $LiNiO_2$). There are also materials with *hexagonal close-packed oxide ion packing*, including some with ordered *olivine-related structures* (e.g., $LiFePO_4$).

In addition, there are a number of interesting materials that have more open crystal structures, sometimes called *framework*, or *skeleton* structures. These are sometimes described as containing *polyanions*. Examples are some sulfates,

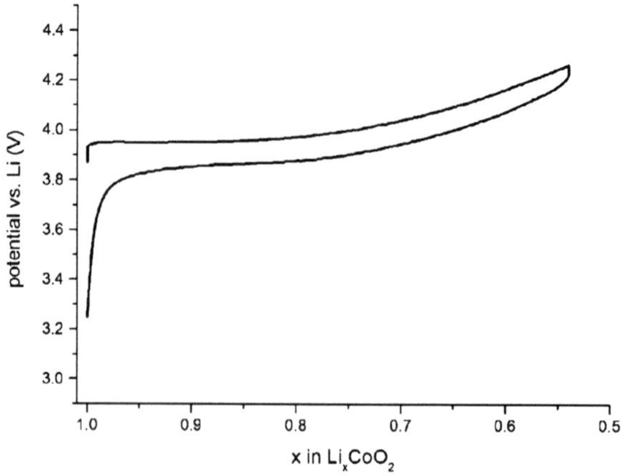

Fig. 9A.1 Charge/discharge behavior of Li_xCoO_2

molybdates, tungstates, and phosphates, as well as Nasicon, and Nasicon-related materials (e.g., $Li_3V_2(PO_4)_3$ and $LiFe_2(SO_4)_3$). In these materials lithium ions can occupy more than one type of interstitial position. Especially interesting are materials with more than one type of polyanion. In some cases the reaction potentials are related to the potentials of the redox reactions of ions in octahedral sites, which are influenced by the charge and crystallographic location of other highly charged ions on tetrahedral sites in their vicinity.

Since the reaction potentials of these positive electrode materials are related to the redox reactions that take place within them, consideration should be given to this matter.

The common values of the formal valence of a number of redox species in solids are given in Table 9A.1. In some cases, the capacity of a material can be enhanced by the use of more than one redox reaction. In such cases, an issue is whether this can be done without a major change in the crystal structure.

An example of the reaction of lithium with an electrode material containing two redox ions, a Li–Mn–Fe phosphate with the olivine structure, is shown in Fig. 9A.2 [16].

Table 9A.1 Common valences of redox ions in solids

Element	Valences	Valence range	Comments
Ti	2, 3, 4	2	
V	2, 3, 4, 5	3	
Cr	2, 3, 6	1	6 is poisonous
Mn	2, 3, 4, 6, 7	2	6, 7 usable?
Fe	2, 3	1	
Co	2, 3	1	
Ni	2, 3, 4	2	
Cu	1, 2	1	

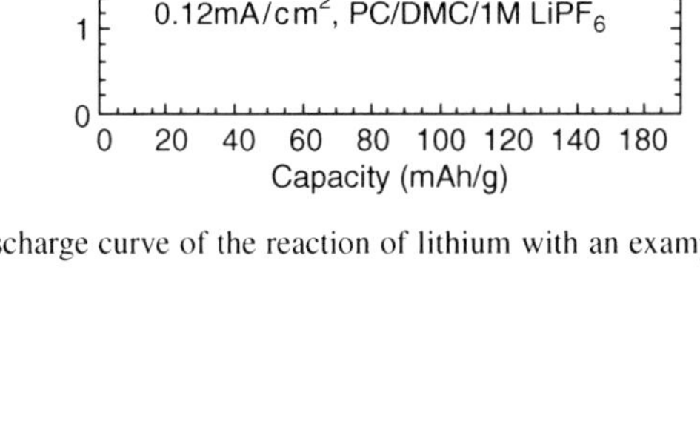

Fig. 9A.2 Charge–discharge curve of the reaction of lithium with an example of a double-cation olivine material

Table 9A.2 Potentials of redox reactions in a number of host materials/volts vs. lithium

Redox system	Nasicon framework phosphates	Layered close-packed oxides	Cubic close-packed spinels	Hexagonal close-packed olivines
V^{2+}/V^{3+}	1.70–1.75			
Nb^{3+}/Nb^{4+}	1.7–1.8			
Nb^{4+}/Nb^{5+}	2.2–2.5			
Ti^{3+}/Ti^{4+}	2.5–2.7		1.6	
Fe/Fe^{2+}	2.65			
Fe^{2+}/Fe^{3+}	2.7–3.0			3.4
V^{3+}/V^{4+}	3.7–3.8			
Mn^{2+}/Mn^{3+}		4.0	1.7	>4.3
Co^{2+}/Co^{3+}		4.2	1.85	>4.3
Ni^{2+}/Ni^{3+}		4.8		>4.3
Mn^{3+}/Mn^{4+}			4.0	
Fe^{3+}/Fe^{4+}	4.4			
Co^{3+}/Co^{4+}			5.0	

Not all redox reactions are of practical value in electrode materials, and in some cases, their potentials depend upon their environments within the crystal structure. Some experimental data are presented in Table 9A.2.

When lithium or other charged mobile guest ions are inserted into the crystal structure, their electrostatic charge is balanced by a change in the oxidation state of one or more of the redox ions contained in the structure of the host material. The reaction potential of the material is determined by the potential at which this oxidation or reduction of these ions occurs in the host material. In some cases, this redox potential is rather narrowly defined, whereas in others redox occurs over a range of potential, due to the variation of the configurational entropy with the guest species concentration, as well as the site distribution.

9A.2 Influence of the Crystallographic Environment on the Potential

It has been shown that the environment in which a given redox reaction takes place can affect the value of its potential. This matter has been investigated by comparing the potentials of the same redox reactions in a number of oxides with different polyanions, but with the same type of crystal structure. Some of the early references to this topic are [17, 18].

These materials all have crystal structures in which the redox ion is octahedrally surrounded by oxide ions, and the oxide ions also have cations with a different charge in tetrahedral environments on the other side. The electron clouds around the oxide ions are displaced by the presence of adjacent cations with different charges. This is shown schematically in Fig. 9A.3.

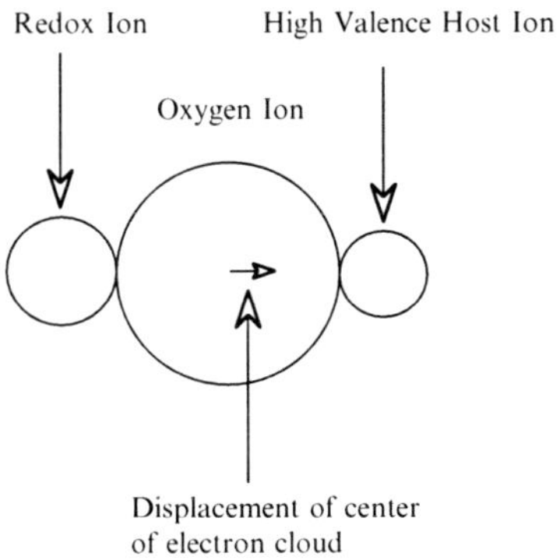

Fig. 9A.3 Schematic of the displacement of the electron cloud around an oxide ion by the charge upon nearby cations

One of the first cathode materials with a polyanion structure to be investigated was $Fe_2(SO_4)_3$. It can apparently reversibly incorporate up to two Li's per formula unit, and has a very flat discharge curve, indicating a reconstitution reaction at 3.6 V vs. Li/Li^+ [19, 20].

9A.3 Oxides with Structures in Which the Oxygen Anions Are in a Face-Centered Cubic Array

9A.3.1 Materials with Layered Structures

As mentioned earlier, the positive electrode reactant in the SONY cells was Li_xCoO_2, whose properties were first investigated at Oxford [21]. It can be synthesized so that it is stable in air, with $x = 1$. Its crystal structure can be described in terms of a close-packed face-centered cubic arrangement of oxide ions, with the Li^+ and Co^{3+} cations occupying octahedrally coordinated positions in between layers of oxide ions. The cation positions are ordered such that the lithium ions and the transition metal ions occupy alternate layers between close-packed (111) planes of oxide ions. As a result, these materials are described as having layered, rather than simple cubic, structures. This is shown schematically in Fig. 9A.4. However, there is a slight distortion of the cubic oxide stacking because of the difference between the bonding of the monovalent and trivalent cations.

When lithium ions move between octahedral sites within the layers of this structure they must go through nearby tetrahedral sites that lie along the jump path. A simple minimum energy path model for this process is discussed in Sect. 15B.

Li_xCoO_2 can be cycled many times over the range $1 > x > 0.5$, but there is a change in the structure and a loss of capacity if more Li^+ ions are deleted.

Because it has an inherently lower cost and is somewhat less poisonous, it would be preferable to use $LiNiO_2$ instead of $LiCoO_2$. However, it has been found that

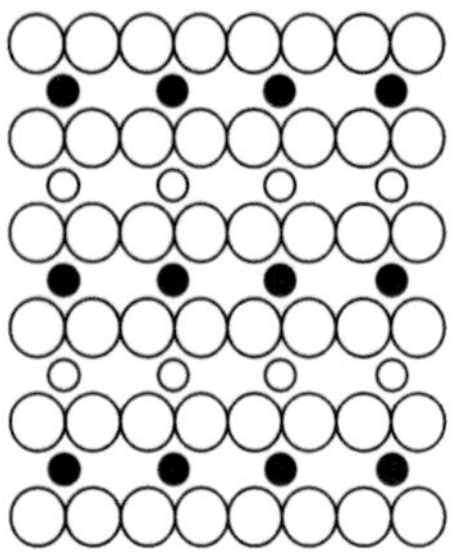

Fig. 9A.4 Simplified schematic of a layered structure in which there is alternate occupation of the cation layers between the close-packed oxide ion layers. The solid and open small circles represent two different types of cations. The larger circles are oxide ions

Li_xNiO_2 is difficult to prepare with the right stoichiometry, as there is a tendency for nickel ions to reside on the lithium layers. This results in a loss of capacity. It was also found that $LiNiO_2$ readily loses oxygen at high potentials, destroying its layer structure, and tending to lead to safety problems because of an exothermic reaction with the organic solvent electrolyte.

There have been a number of investigations of the modification of Li_xNiO_2 by the substitution of other cations for some of the Ni^{3+} ions. It has been found that the replacement of 20–30% of the Ni^{3+} ions by Co^{3+} ions will impart sufficient stability [22]. Other aliovalent alternatives have also been explored, including the introduction of Mg^{2+}, Al^{3+} or Ti^{4+} ions.

In the case of $LiMnO_2$, that also has the alpha $NaFeO_2$ structure, it has been found that if more than 50% of the lithium ions are removed during charging, conversion to the spinel structure tends to occur. About 25% of the Mn ions move from octahedral sites in their normal layers into the alkali metal layers, and lithium is displaced into tetrahedral sites [23]. But this conversion to the spinel structure can be avoided by the replacement of half of the Mn ions by chromium [24]. In this case, the capacity (190 mAh g^{-1}) is greater than the capacity that can be accounted for by a single redox reaction, such as Mn^{3+} to Mn^{4+}. This implies that the chromium ions are involved, whose oxidation state can go from Cr^{3+} to Cr^{6+}. Unfortunately, the use of chromium is not considered desirable because of the toxicity of Cr^{6+}.

The replacement of some of the manganese ions in $LiMnO_2$ by several other ions in order to prevent the conversion to the spinel structure has been investigated [25].

A number of other layer-structure materials have also been investigated. Some of them contain two or more transition metal ions at fixed ratios, often including Ni, Mn, Co, and Al. In some cases, there is evidence of ordered structures at specific compositions and well-defined reaction plateaus, at least under equilibrium or near-equilibrium conditions. This indicates reconstitution reactions between adjacent phases.

There have been several investigations of layer phases with manganese and other transition metals present. A number of these, including $LiMn_{1-y}Co_yO$, have been found to not be interesting, as they convert to the spinel structure rather readily.

However, the manganese–nickel materials, $Li_xMn_{0.5}Ni_{0.5}O_2$ and related compositions, have been found to have very good electrochemical properties, with indications of a solid solution insertion reaction in the potential range 3.5–4.5 V vs. Li [26–29]. It appears as though the redox reaction involves a change from Ni^{2+} to Ni^{4+}, whereas the Mn remains as Mn^{4+}. This means that there is no problem with Jahn-Teller distortions, which are related to the presence of Mn^{3+}. The presence of the manganese ions is apparently useful in stabilizing this structure.

At higher manganese concentrations these materials adopt the spinel structure and apparently react by reconstitution reactions, as will be discussed later in this chapter.

Success with this cation combination apparently led to considerations of compositions containing three cations, such as Mn, Ni, and Co. One of these is $LiMn_{1/3}Ni_{1/3}Co_{1/3}O_2$ [30, 31]. The presence of the cobalt ions evidently stabilizes the layer structure against conversion to the spinel structure. These materials have good electrochemical behavior, and have been studied in many laboratories, but one concern is that they evidently have limited electronic conductivity.

In these materials, as well, when they are fully lithiated, the nickel is evidently predominantly divalent, the cobalt trivalent, and the manganese tetravalent. Thus, the major electrochemically active species is nickel, with the cobalt playing an active role only at high potentials. The manganese evidently does not play an active role. It does reduce the overall cost, however.

An extensive discussion of the various approaches to the optimization of the layer structure materials can be found in [32].

9A.3.2 Materials with the Spinel Structure

The spinel class of materials, with the nominal formula AB_2O_4, has a related structure that also has a close-packed face-centered cubic arrangement of oxide ions. Although this structure is generally pictured in cubic coordinates, it also has parallel layers of oxide ions on (111) planes, and there are both octahedrally coordinated sites and tetrahedrally coordinated sites between the oxide ion planes. The number of octahedral sites is equal to the number of oxide ions, but there are twice as many tetrahedral sites. The octahedral sites reside in a plane intermediate between every two oxide ion planes. The tetrahedral sites are in parallel planes slightly above and below the octahedral site planes between the oxide ion planes.

In *normal spinels*, the A (typically monovalent or divalent) cations occupy 1/8 of the available tetrahedral sites, and the B (typically trivalent or quadrivalent) cations 1/2 of the B sites. In *inverse spinels*, the distribution is reversed.

The spinel structure is quite common in nature, indicating a large degree of stability. As mentioned earlier, there is a tendency for the materials with the layer structures to convert to the closely related spinel structure. This structure is shown schematically in Fig. 9A.5.

A wide range of materials with different A and B ions can have this structure, and some of them are quite interesting for use in lithium systems. An especially

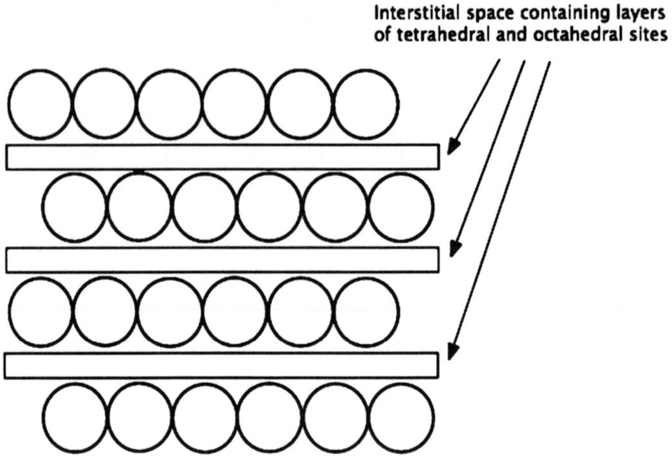

Fig. 9A.5 Schematic of the spinel structure in which the cations between the close-packed (111) planes of oxide ions are distributed among both tetrahedral and octahedral sites

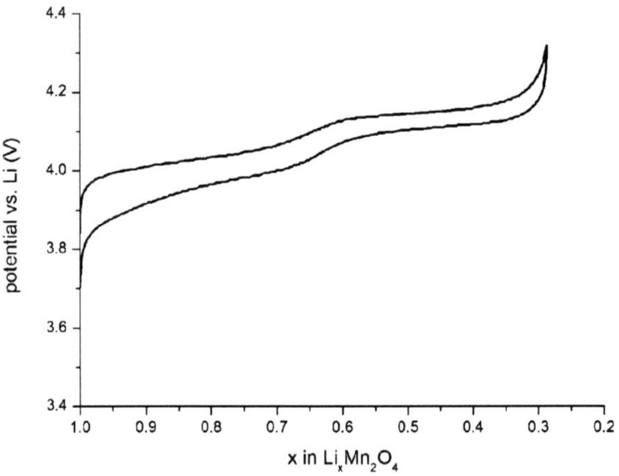

Fig. 9A.6 Charge/discharge behavior of $Li_xMn_2O_4$

important example is $Li_xMn_2O_4$. There can be both lithium insertion and deletion from the nominal composition in which $x = 1$. This material has about 10% less capacity than Li_xCoO_2, but it has somewhat better kinetics and does not have as great a tendency to evolve oxygen.

$Li_xMn_2O_4$ can be readily synthesized with x equal to unity, and this composition can be used as a positive electrode reactant in lithium batteries. A typical charge/discharge curve is shown in Fig. 9A.6.

It is seen that there are two plateaus. This is related to an ordering reaction of the lithium ions on the tetrahedral sites when x is about 0.5.

Although the $Li_xMn_2O_4$ system, first investigated by Thackeray et al. [33, 34], has the inherent advantages of low cost, good kinetics, and being nonpoisonous, it has been found to have some problems that can result in a gradual loss of capacity [35]. Thorough discussions of early work to optimize this material can be found in [36, 37].

One of the problems with this material is the loss of Mn^{2+} into the organic solvent electrolyte as the result of a disproportionation reaction when the potential is low near the end of discharge.

$$2Mn^{3+} = Mn^{4+} + Mn^{2+}. \quad (9A.1)$$

These ions travel to the carbon negative electrode, with the result that a layer of manganese metal is deposited that acts to block lithium ion transport.

Another problem that can occur at low potentials is the local onset of Jahn-Teller distortion that can cause mechanical damage to the crystal structure. On the other hand, if the electrode potential becomes too high as the result of the extraction of too much lithium, oxygen can escape and react with the organic solvent electrolyte.

These problems are reduced by modification of the composition of the electrode by the presence of additional lithium and a reduction of the manganese [38]. This increase in stability comes at the expense of the capacity. Although the theoretical capacity of $LiMn_2O_4$ is 148 mAh g^{-1}, this modification results in a capacity of only 128 mAh g^{-1}.

There have also been a number of investigations in which various other cations have been substituted for part of the manganese ions. But in order to avoid the loss of a substantial amount of the normal capacity, it was generally thought at that time that the extent of this substitution must be limited to relatively small concentrations.

At that time the tendency was to perform experiments only up to a voltage about 4.2 V above the Li/Li^+ potential, as had been done for safety reasons when using Li_xCoO_2. But it was soon shown that it is possible to reach potentials up to 5.4 V vs. Li/Li^+ using some organic solvent electrolytes [39, 40].

Experiments on the substitution of some of the Mn^{2+} ions in Li_xMnO_2 by Cr^{3+} ions [41] showed that the capacity upon the 3.8-V plateau was decreased in proportion to the concentration of the replaced Mn ions. But when the potential was raised to higher values, it was found that this missing capacity at about 4 V reappeared at potentials about 4.9 V that was obviously related to the oxidation of the Cr ions that had replaced the manganese ions in the structure. This particular option, replacing inexpensive and nontoxic manganese with more expensive and toxic chromium is, of course, not favorable.

In both the cases of chromium substitution and nickel substitution the sum of the capacities of the higher potential plateau and the lower plateau is constant. This implies that there is a one-to-one substitution, and thus that the oxidation that occurs in connection with the chromium and nickel ions is a one-electron process. This is in contradiction to the normal expectation that these ions undergo a three-electron (Cr^{3+} to Cr^{6+}) or a two-electron (Ni^{2+} to Ni^{4+}) oxidation step.

Another example is the work on lithium manganese spinels in which some of the manganese ions have been replaced by copper ions. One of these is $LiCu_xMn_{2-x}O_4$ [42–44]. Investigations of materials in which up to a quarter of the manganese ions are replaced by Cu ions have shown that a second plateau appears at 4.8–5.0 V vs.

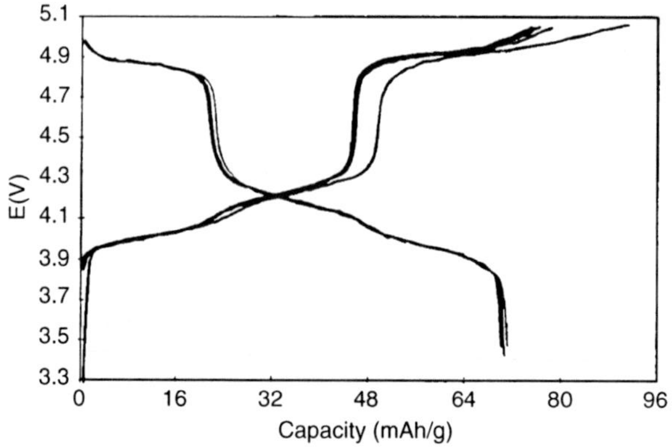

Fig. 9A.7 Potential–composition curves for $LiCu_{0.5}Mn_{1.5}O4$

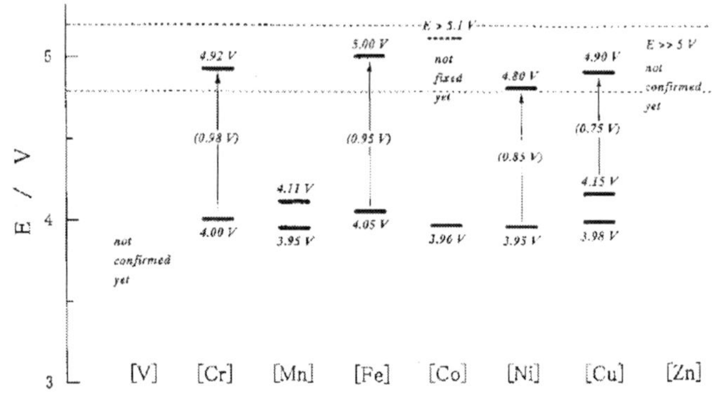

Fig. 9A.8 Potential ranges, vs. Li, of redox potentials found as the result of the introduction of a number of cations into lithium manganese spinels. The operating potential range of lithium manganese spinel itself is also shown

Li/Li^+ that is due to a Cu^{2+}/Cu^{3+} reaction, in addition to the normal behavior of the Li–Mn spinel in the range 3.9–4.3 V vs. Li/Li^+ that is related to the Mn^{3+}/Mn^{4+} reaction. Data for this case are shown in Fig. 9A.7 [42]. Unfortunately, the overall capacity seems to be reduced when there is a substantial amount of copper present in this material [43]. When x is 0.5 the total capacity is about 70 mAh g^{-1}, with only about 25 mAh g^{-1} obtainable in the higher potential region.

The redox potentials that are observed when a number of elements are substituted into lithium manganese spinel structure materials are shown in Fig. 9A.8 [45].

An especially interesting example is the spinel structure material with a composition $Li_xNi_{0.5}Mn_{1.5}O_4$. Its electrochemical behavior is different from the others, showing evidence of two reconstitution reactions, rather than solid solution behavior [46].

The constant potential charge/discharge curve for this material in the high potential range is shown in Fig. 9A.9 [46]. Careful coulometric titration experiments showed that this apparent plateau is actually composed of two reactions with a potential separation of only 20 mV.

In addition to this high-potential reaction, this material also has a reconstitution reaction with a capacity of one Li per mole at 2.8 V vs. Li, as well as further lithium uptake via a single phase reaction below 1.9 V. These features are shown in Fig. 9A.10 [46]. It is not fully known what redox reactions are involved in this

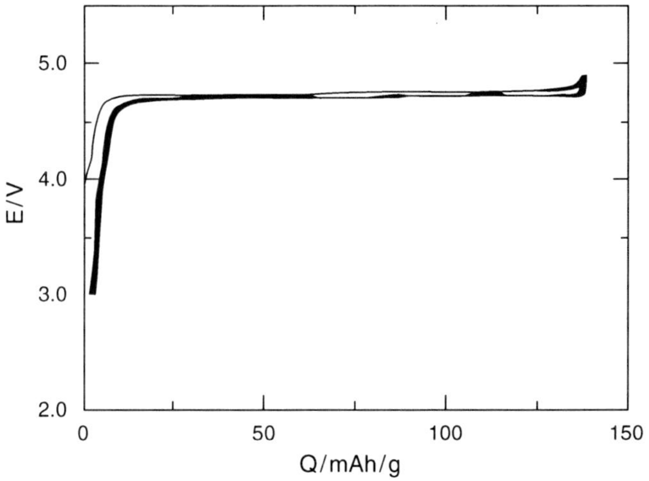

Fig. 9A.9 Charge/discharge curves for $Li_xNi_{0.5}Mn_{1.5}O_4$

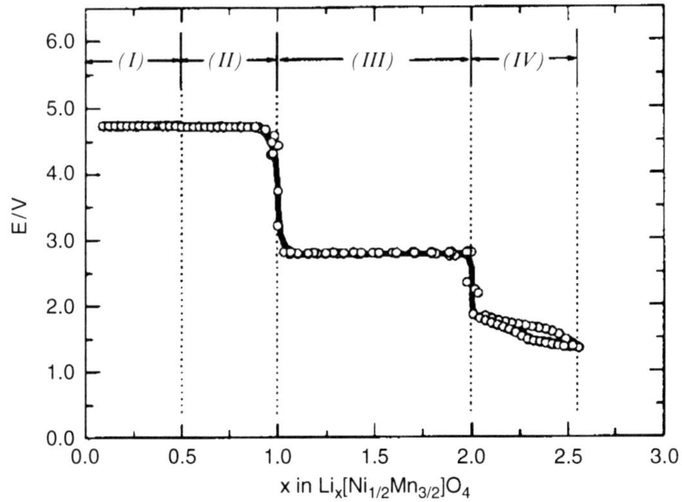

Fig. 9A.10 Coulometric titration curve for the reaction of lithium with $Li_xNi_{0.5}Mn_{1.5}O_4$

behavior, but it is believed that those at the higher potentials relate to nickel, and the lower ones to manganese.

9A.3.3 Lower Potential Spinel Materials with Reconstitution Reactions

Although this discussion here has centered about lithium-containing materials that exhibit high-potential reactions, and thus are useful as reactants in the positive electrode, attention should also be given to another related spinel-structure material that has a reconstitution reaction at 1.55 V vs. Li [47,48]. This is $Li_{1.33}Ti_{1.67}O_4$ that can also be written as $Li_x[Li_{0.33}Ti_{1.67}O_4]$, for some of the lithium ions share the octahedral sites in an ordered arrangement with the titanium ions. It also sometimes appears in the literature as $Li_4Ti_5O_{12}$.

This spinel-structure material is unusual in that there is essentially no change in the lattice dimensions with variation of the amount of lithium in the crystal structure, and it has been described as undergoing a *zero-strain insertion reaction* [49]. This is an advantage in that there is almost no volume change-related hysteresis, leading to very good reversibility upon cycling.

As was mentioned in Chap. 7, this material can also be used on the negative electrode side of a battery. Although there is a substantial voltage loss compared to the use of carbons, the good kinetic behavior can make this option attractive for high-power applications, where the lithium-carbons can be dangerous because their reaction potential is rather close to that of elemental lithium.

A charge/discharge curve for this interesting material is shown in Fig. 9A.11.

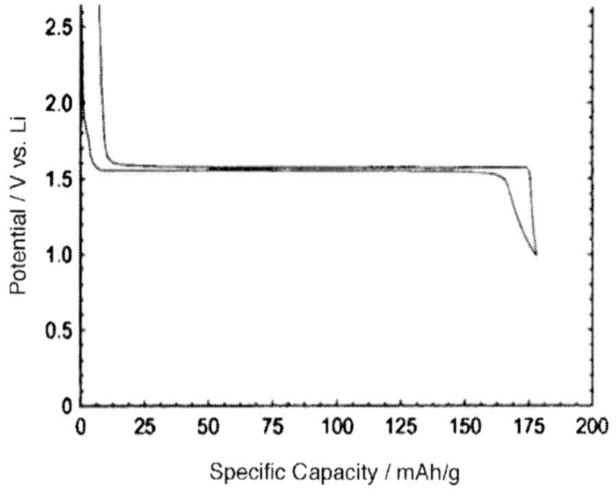

Fig. 9A.11 Charge/discharge curve for $Li_4Ti_5O_{12}$

9A.4 Materials in Which the Oxide Ions Are in a Close-Packed Hexagonal Array

Whereas in the spinel and the related layered materials such as Li_xCoO_2, Li_xNiO_2, and Li_xMnO_2 the oxide ions are in a cubic close-packed array, there are also many materials in which the oxide ions are in a hexagonal close-packed configuration. Some of these are currently of great interest for use as positive electrode reactants in lithium batteries, but are generally described as having *framework structures*. They are sometimes also called *scaffold, skeleton, network*, or *polyanion* structures.

9A.4.1 The Nasicon Structure

The *Nasicon structure* first attracted attention within the solid state ionics community because some materials with this structure were found to be very good solid electrolytes for sodium ions. One such composition was $Na_3Zr_2Si_2PO_{12}$.

This structure has monoclinic symmetry, and can be considered as consisting of MO_6 octahedra sharing corner oxide ions with adjacent XO_4 tetrahedra. Each octahedron is surrounded by six tetrahedra, and each tetrahedron by four octahedra. These are assembled as a three-dimensional network of M_2X_3 groups. Between these units is three-dimensional interconnected interstitial space, through which small cations can readily move. This structure is shown schematically in Fig. 9A.12.

Unfortunately, Nasicon was found to be not thermodynamically stable vs. elemental sodium, so that it did not find use as an electrolyte in the Na/Na_xS and $Na/NiCl_2$ Zebra cells, that are discussed elsewhere in this text, at that time.

Fig. 9A.12 Schematic representation of the Nasicon structure

However, by using M cations whose ionic charge can be varied, it is possible to make materials with this same structure that undergo redox reactions upon the insertion or deletion of lithium within the interstitial space. The result is that although Nasicon materials may not be useful for the function for which they were first investigated, they may be found to be useful for a different type of application.

As mentioned earlier, it has been found that the identity of the X ions in the tetrahedral parts of the structure influences the redox potential of the M ions in the adjacent octahedra [50, 51]. This has been called an *induction effect*.

A number of compositions with this structure have been investigated for their potential use as positive electrode reactants in lithium cells [50–53]. An example is $Li_3V_2(PO_4)_3$, whose potential vs. composition data are shown in Fig. 9A.13 [53]. The related differential capacity plot is shown in Fig. 9A.14.

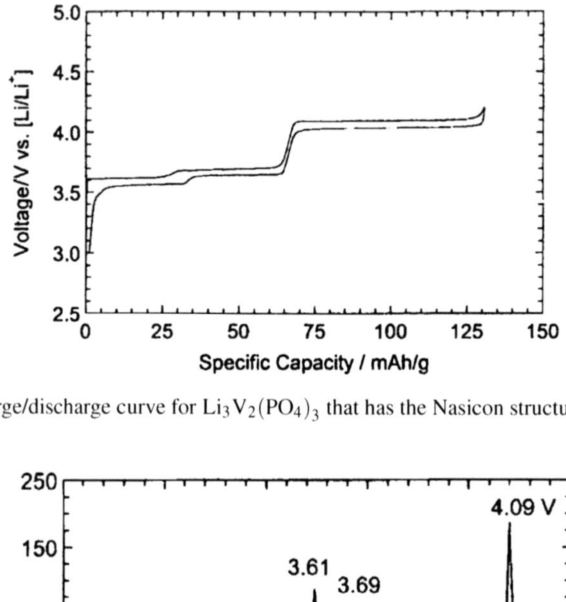

Fig. 9A.13 Charge/discharge curve for $Li_3V_2(PO_4)_3$ that has the Nasicon structure

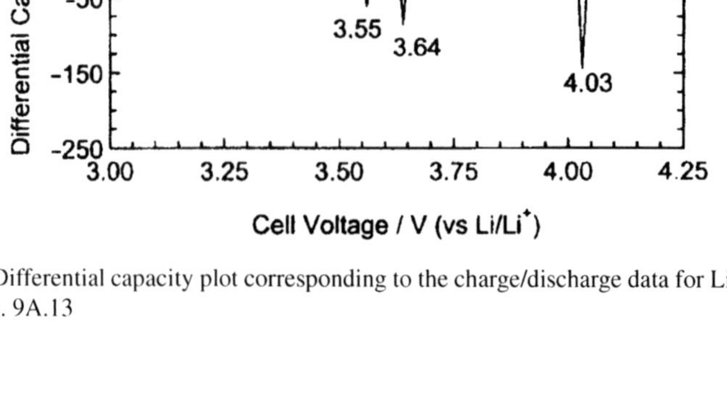

Fig. 9A.14 Differential capacity plot corresponding to the charge/discharge data for $Li_3V_2(PO_4)_3$ shown in Fig. 9A.13

It is seen that the titration curve shows three two-phase plateaus, corresponding to the extraction of two of the lithium ions in the initial structure. The first two plateaus indicate that there are two slightly different configurations for one of the two lithium ions. The potential must be increased substantially, to over 4 V, for the deletion of the second. Experiments showed that it is possible to extract the third lithium from this material by going up to about 5 V, but that this process is not readily reversible, whereas the insertion/extraction of the first two lithium ions is highly reversible.

These phosphate materials all show significantly more thermal stability than is found in some of the other, e.g., layer- and spinel-structure, positive electrode reactants. This is becoming ever more important as concerns about the safety aspects of high-energy batteries mount.

9A.4.2 Materials with the Olivine Structure

Another group of materials that have a hexagonal stacking of oxide ions are those with the Olivine structure. These materials have caused a great deal of excitement, as well as controversy, in the research community since it was first shown that they can reversibly react with lithium at ambient temperature [54]. The most interesting of these materials is $LiFePO_4$ that has the obvious advantage of being composed of safe and inexpensive materials.

The olivine structure can be described as M_2XO_4, in which the M ions are in half of the available sites of the close-packed hexagonal oxygen array. The more highly charged X ions occupy one-eighth of the tetrahedral sites. Thus, it is a hexagonal analog of the cubic spinel structure discussed earlier. However, unlike spinel, the two octahedral sites in olivine are crystallographically distinct, and have different sizes. This results in a preferential ordering if there are two M ions of different sizes and/or charges. Thus, $LiFePO_4$ and related materials containing lithium and transition metal cations have an ordered cation distribution. The M_1 sites containing lithium are in linear chains of edge-shared octahedral that are parallel to the c-axis in the hexagonal structure in alternate a–c planes. The other (M_2) sites are in a zigzag arrangement of corner-shared octahedral parallel to the c-axis in the other a–c planes. The result is that lithium transport is highly directional in this structure.

Experiments showed that extraction of lithium did not readily occur with olivines containing the Mn, Co, or Ni, but proceeded readily in the case of $LiFePO_4$. Deletion of lithium from $LiFePO_4$ occurs by a reconstitution reaction with a moving two-phase interface in which $FePO_4$ is formed at a potential of 3.43 V vs. Li. Although the initial experiments only showed the electrochemical removal of about 0.6 Li ions per mole, subsequent work has shown that greater values can be attained. A reaction with one lithium ion per mole would give a theoretical specific capacity of 170 mAh g^{-1}, which is higher than that obtained with $LiCoO_2$. It has been found that the extraction/insertion of lithium in this material can be quite reversible over many cycles.

These phases have the mineralogical names triphylite and heterosite, although the latter was given to a mineral that also contains manganese. Although this reaction potential is significantly lower than that of many of the materials discussed earlier in this chapter, other properties of this class of materials make them attractive for application in lithium-ion cells. There is active commercialization activity, as well as a measure of conflict over various related patent matters.

These materials do not tend to lose oxygen and react with the organic solvent electrolyte nearly so much as the layer-structure materials, and they are evidently much safer at elevated temperatures. As a result, they are being considered for larger format applications, such as in vehicles or load leveling, where there are safety questions with some of the other positive electrode reactant materials.

It appeared that the low electronic conductivity of these materials might limit their application, so work was undertaken in a number of laboratories aimed at the development of two-phase microstructures in which electronic conduction within the electrode structure could be enhanced by the presence of an electronic conduction, such as carbon [55]. Various versions of this process quickly became competitive and proprietary.

A different approach is to dope the material with highly charged (supervalent) metal ions, such as niobium, that could replace some of the lithium ions on the small M_1 sites in the structure, increasing the *n*-type electronic conductivity [56]. On the other hand, experimental evidence seems to indicate that the electronic conduction in the doped Li_xFePO_4 is *p*-type, not *n*-type [56, 57]. This could be possible if the cation doping is accompanied by a deficiency of lithium.

This interpretation has been challenged, however, based upon observations of the presence of a highly conductive iron phosphide phase, Fe_2P, under certain conditions [58]. Subsequent studies of phase equilibria in the Li–Fe–P–O quaternary system [59] seem to contradict that interpretation.

Regardless of the interpretation, it has been found that the apparent electronic conductivity in these Li_xFePO_4 materials can be increased by a factor of 10^8, reaching values above 10^{-2} S cm^{-1} in this manner. These are higher than those found in some of the other positive electrode reactants, such as $LiCoO_2$ (10^{-3} S cm^{-1}) and $LiMn_2O_4$ ($2–5 \times 10^{-5}$ S cm^{-1}).

An interesting observation is that very fine scale cation-doped Li_xFePO4 has a restricted range of composition at which the two phases *LiFePO4* and *FePO4* are in equilibrium, compared with undoped and larger particle-size material [60]. Thus, there is more solid solubility in each of the two end phases. This may play an important role in their increased kinetics, for in order for the moving interface reconstitution phase transformation involved in the operation of the electrode to proceed there must be diffusion of lithium through the outer phase to the interface. The rate of diffusional transport is proportional to the concentration gradient. A wider compositional range allows a greater concentration gradient, and thus faster kinetics.

These materials have been found to be able to react with lithium at very high power levels, greater than those that are typical of common hydride/H_xNiO_2 cells, and commercial applications of this material are being vigorously pursued.

9A.5 Materials Containing Fluoride Ions

Another interesting variant has also been explored somewhat. This involves the replacement of some of the oxide ions in lithium transition metal oxides by fluoride ions. An example of this is the lithium vanadium fluorophosphate, $LiVPO_4F$, that was found to have a triclinic structure analagous to the mineral tavorite, $LiFePO_4.OH$ [61]. As in the case of the Nasicon materials mentioned earlier, the relevant redox reaction in this material involves the V^{3+}/V^{4+} couple. The charge/discharge behavior of this material is shown in Fig. 9A.15 [62], and the related differential capacity results are presented in Fig. 9A.16.

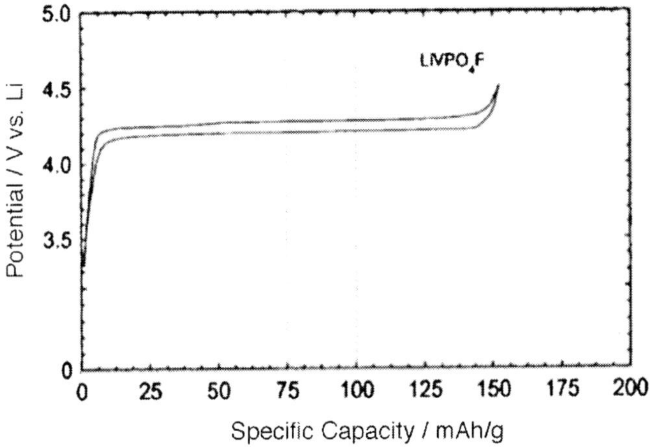

Fig. 9A.15 Charge/discharge behavior of $LiVPO_4F$

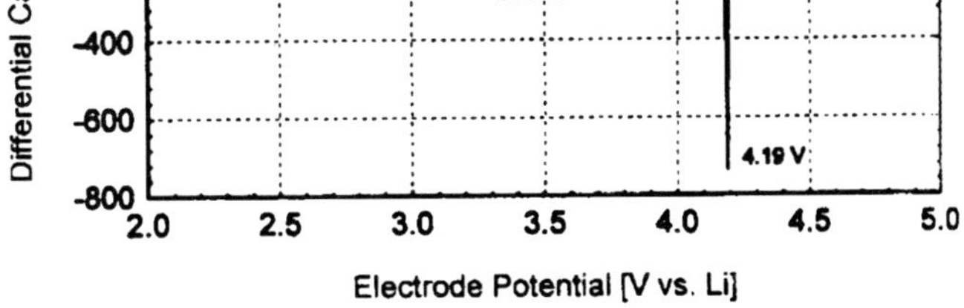

Fig. 9A.16 Differential capacity plot corresponding to the charge/discharge data for $LiVPO_4F$ shown in Fig. 9A.15

9A.6 Hybrid Ion Cells

An additional variant involves the use of positive electrode reactants that contain other mobile cations. An example of this was the reports of the use of $Na_3V_2(PO_4)F_3$ as the positive electrode reactant and either graphite [63] or $Li_{4/3}Ti_{5/3}O_4$ [64] as the negative reactant in lithium-conducting electrolyte cells. It appears as though the mobile insertion species in the positive electrodes gradually shifts from Na^+ to Li^+. Consideration of this type of mixed-ion materials may lead to a number of interesting new materials.

9A.7 Amorphization

It was pointed out in Chap. 6 that crystal structures can become amorphous as the result of multiple insertion/extraction reactions. A simple explanation for this phenomenon can be based upon the dimensional changes that accompany the variation in the composition. These dimensional changes are typically not uniform throughout the material, so quite significant local shear stresses can result that disturb the regularity of the atomic arrangements in the crystal structure, resulting in regions with amorphous structures. The degree of amorphization should increase with cycling, as is found experimentally.

There is also another possible cause of this effect that has to do with the particle size. As particles become very small, a significant fraction of their atoms actually resides on the surface. Thus, the surface energy present becomes a more significant fraction of the total Gibbs free energy. Amorphous structures tend to have lower values of surface energy than their crystalline counterparts. As a result, it is easy to understand that there will be an increasing tendency for amorphization as particles become smaller.

9A.8 The Oxygen Evolution Problem

It is generally considered that a high cell voltage is desirable, and the more the better, since the energy stored is proportional to the voltage, and the power is proportional to the square of the voltage. However, there are other matters to consider, as well. One of these is the evolution of oxygen from a number of higher potential positive electrode materials.

There is a direct relationship between the electrical potential and the chemical potential of oxygen in materials containing lithium. In this connection it is useful to remember that the chemical potential has been called the *escaping tendency* in the well-known book on thermodynamics by Pitzer and Brewer.

Experiments have shown that a number of the high-potential positive electrode reactant materials lose oxygen into the electrolyte. It is also generally thought that

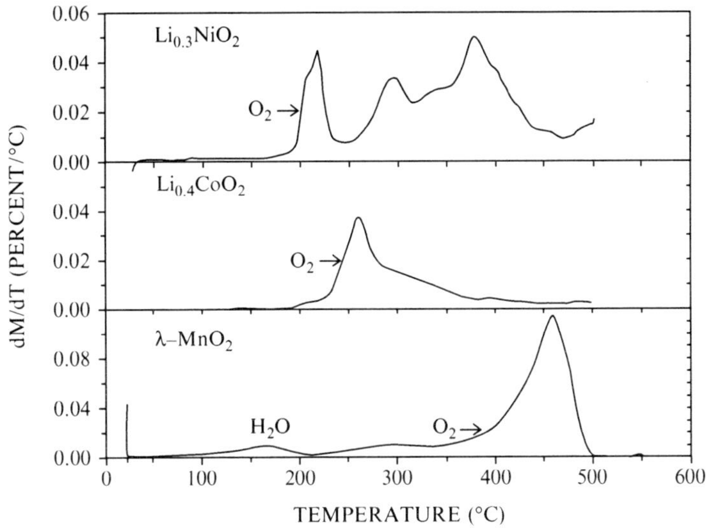

Fig. 9A.17 The influence of temperature upon the derivative of the sample weight vs. temperature for three different layer-structure materials [65]

the presence of oxygen in the organic solvent electrolytes is related to thermal runaway and the safety problems that are sometimes encountered in lithium cells. An example of experimental measurements that clearly show oxygen evolution is shown in Fig. 9A.17.

The relationship between the potential and the chemical potential of oxygen in electrode materials was investigated a number of years ago, but under conditions that are somewhat different from those in current ambient temperature lithium cells. Nevertheless, the principles are the same, and thus it is useful to review what was found about the thermodynamics of such systems at that time [66].

As discussed in this chapter, many of the positive electrode materials in lithium batteries are ternary lithium transition metal oxides. Since there are three kinds of atoms, i.e., three components, present, compositions in these systems can be represented on an isothermal Gibbs triangle. As discussed in Chaps. 2 and 4, the Gibbs phase rule can be written as

$$F = C - P + 2, \tag{9A.2}$$

where F is the number of degrees of freedom, C the number of components, and P the number of phases present. At constant temperature and overall pressure $F = 0$ when $C = P = 3$. This means that all of the intensive variables have fixed values when three phases are present in such three-component systems. Since the electrical potential is an intensive property, this means that the potential has the same value, independent of how much of each of the three phases is present.

It has already been pointed out that the *isothermal phase stability diagram*, an approximation of the Gibbs triangle in which the phases are treated as though they have fixed, and very narrow, compositions, is a very useful thinking tool to use when considering ternary materials.

The compositions of all of the relevant phases are located on the triangular coordinates, and the possible two-phase tie lines identified. Tie lines cannot cross, and the stable ones can readily be determined from the energy balance of the appropriate reactions. The stable tie lines divide the total triangle into subtriangles that have two phases at the ends of the tie lines along their boundaries. There are different amounts of the three corner phases at different locations inside the subtriangles. All of these compositions have the same values of the intensive properties, including the electrical potential.

The potentials within the subtriangles can be calculated from thermodynamic data on the electrically neutral phases at their corners. From this information it is possible to calculate the voltages vs. any of the components. This means that one can also calculate the equilibrium oxygen activities and pressures for the phases in equilibrium with each other in each of the subtriangles. As was shown in Chap. 4, one can also do the reverse, and measure the equilibrium potential at selected compositions in order to determine the thermodynamic data, including the oxygen pressure. The relation between the potential and the oxygen pressure is of special interest because of its practical implications for high-voltage battery systems.

The experimental data that are available for ternary lithium–transition metal oxide systems are, however, limited to only three systems and one temperature. The Li–Mn–O, Li–Fe–O, and Li–Co–O systems were studied quantitatively using molten salt electrolytes at 400°C [66]. Because of the sensitivity of lithium to both oxygen and water, they were conducted in a helium-filled glove box. The maximum oxygen pressure that could be tolerated was limited by the formation of Li_2O in the molten salt electrolyte, which was determined to occur at an oxygen partial pressure of 10^{-25} atm at 400°C. This is equivalent to 1.82 V vs. lithium at that temperature. Thus, it was not possible to study materials with potentials above 1.82 V vs. lithium at that temperature.

As an example, the results obtained for the Li–Co–O system under those conditions are shown in Fig. 9A.18 [66], which was also included in Chap. 4.

The general equilibrium equation for a ternary subtriangle that has two binary transition metal oxides (MO_y and MO_{y-x}) and lithium oxide (Li_2O) at its corners can be written as

$$2xLi + MO_y = xLi_2O + MO_{y-x}. \tag{9A.3}$$

According to Hess's law, this can be divided into two binary reactions, and the Gibbs free energy change ΔG_r is the sum of the two.

$$\Delta G_r = \Delta G_r^1 + \Delta G_r^2. \tag{9A.4}$$

One is the reaction

$$MO_y = (x/2)O_2 + MO_{y-x}. \tag{9A.5}$$

The related Gibbs free energy change is given by

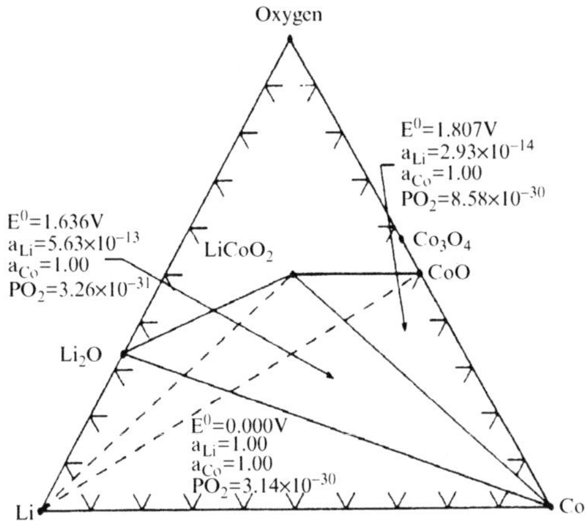

Fig. 9A.18 Equilibrium data for the Li–Co–O ternary system at 400°C

$$\Delta G_r^1 = -RT \ln K, \tag{9A.6}$$

where K is the equilibrium constant.

The other is the formation of Li_2O that can be written as

$$2xLi + (x/2)O_2 = xLi_2O \tag{9A.7}$$

for which the Gibbs free energy change is the standard Gibbs free energy of formation of Li_2O.

$$\Delta G_r^2 = x\Delta G_f{}^{\circ}(Li_2O). \tag{9A.8}$$

The potential is related to ΔG_r by

$$E = -\Delta G_r/zF. \tag{9A.9}$$

that can also be written as

$$E = RT/(4F)\ln(pO_2) - \Delta G_f{}^{\circ}(Li_2O)/(2F). \tag{9A.10}$$

This can be simplified to become a linear relation between the potential E and ln $p(O_2)$, with a slope of $RT/(4F)$ and an intercept related to the Gibbs free energy of formation of Li_2O at the temperature of interest.

Experimental data were obtained on the polyphase equilibria within the subtriangles in the Li–Mn–O, Li–Fe–O, and Li–Co–O systems by electrochemical titration of lithium into various Li_xMO_y materials to determine the equilibrium potentials and compositional ranges. The results are plotted in Fig. 9A.19 [66].

It is seen that there is a clear correlation between the potentials and the oxygen pressure in all cases. The equation for the line through the data is

$$E = 3.34 \times 10^{-2} \log p(O_2) + 2.65V. \tag{9A.11}$$

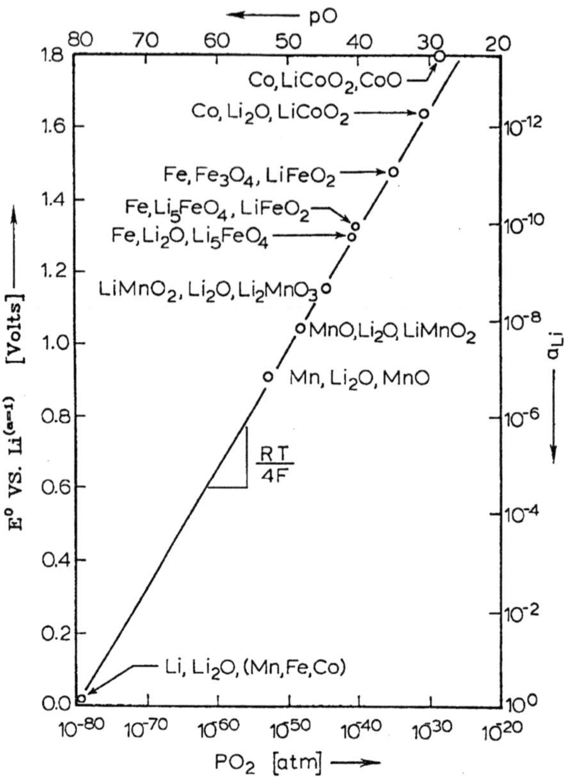

Fig. 9A.19 Experimental data on the relation between the potential and the oxygen pressure in phase combinations in the Li–Mn–O, Li–Fe–O, and Li–Co–O systems at 400°C

The data fit this line very well, even though the materials involved had a variety of compositions and crystal structures. Thus, the relation between the potential and the oxygen pressure is obviously independent of the identity and structures of the materials involved.

Extrapolation of the data in Fig. 9A.19 shows that the oxygen pressure would be 1 atm at a potential of 2.65 V vs. Li/Li^+ at 400°C.

At 25°C the Gibbs free energy of formation of Li_2O is $-562.1\,kJ\ mol^{-1}$, so the potential at 1 atm oxygen is 2.91 V vs. Li/Li^+. This is about what is observed as the initial open circuit potential in measurements on many transition metal oxide materials when they are fabricated in air.

Evaluating (9A.10) for a temperature of 298 K, it becomes

$$E = 1.476x10^{-2} \log p(O_2) + 2.91V. \tag{9A.12}$$

At this temperature the slope of the potential vs. oxygen pressure curve is somewhat less than at the higher temperature. But considering it the other way round, the pressure increases more rapidly as the potential is raised.

Table 9A.3 Values of the equilibrium oxygen pressure over oxide phases in Li–M–O systems at 298 K

E vs. Li/Li^+ (V)	Logarithm of equilibrium oxygen pressure (atm)
1	−129
1.5	−95
2	−62
2.5	−28
3	6
3.5	40
4	73.7
4.5	107.6
5	141.4

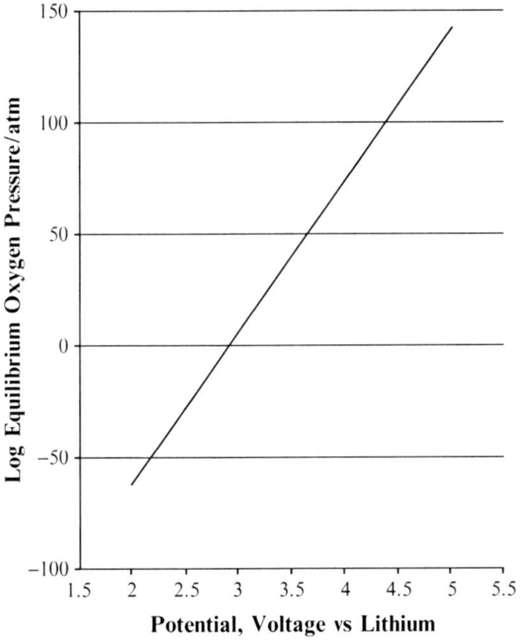

Fig. 9A.20 Dependence of the logarithm of the equilibrium oxygen pressure upon the potential in lithium–transition metal-oxide systems

This result shows that the equilibrium oxygen pressures in the Li–M–O oxide phases increase greatly as the potential is raised. Values of the equilibrium oxygen pressure as a function of the potential are shown in Table 9A.3. These data are plotted in Fig. 9A.20.

It can be seen that these values become very large at high electrode potentials, and from the experimental data taken under less extreme conditions, it is obvious that the critical issue is the potential, not the identity of the electrode reactant material or the crystal structure.

One can understand the tendency toward the evolution of oxygen from oxides at high potentials from a different standpoint. Considerations of the influence of the potential on the point defect structure of oxide solid electrolytes has shown that electronic holes tend to be formed at higher potentials, and excess electrons at lower potentials. The presence of holes means that some of the oxide ions have a charge of 1-, rather than 2-. That is, they become peroxide ions, O^-. This is an intermediate state on the way to neutral oxygens, as in the neutral oxygen gas molecule O_2.

Such ions have been found experimentally on the oxygen electrode surface where the transition between neutral oxygen molecules and oxide ions takes place at the positive electrodes of fuel cells.

9A.9 Closing Comments

It is evident that this is a very active research area, with a number of different avenues being explored in the pursuit of higher potentials, greater capacity, longer cycle life, greater safety, and lower cost. It will be interesting to see which of these new materials, if any, actually come into commercial application.

9B Liquid Positive Electrode Reactants

9B.1 Introduction

The discussion of positive electrodes in lithium batteries thus far has assumed that the reactants are solids. However, this is not necessary, and there are two types of primary batteries that have been available commercially for a number of years, in which the reactant is a liquid: the Li/SO_2 and $Li/SOCl_2$ (thionyl chloride) systems. They both have very high specific energies. But because of safety considerations these batteries are not in general use, and are being produced for military and space purposes.

9B.2 The Li/SO_2 System

These cells are generally constructed with large surface area carbon electrodes on the positive side, and X-ray experiments have shown that $Li_2S_2O_4$ is formed there upon discharge.

The discharge curve is very flat, at 3.0 V, as shown in Fig. 9B.1.

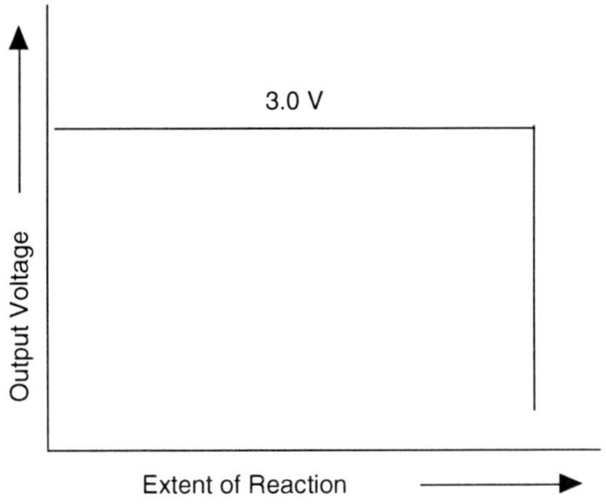

Fig. 9B.1 Discharge curve for a Li/SO_2 cell

Table 9B.1 Gibbs free energies of formation of phases in the Li–S–O system at 25°C

Phase	Gibbs free energy of formation (kJ mol^{-1})
Li_2O	−562.1
SO_2	−300.1
Li_2S	−439.1
$Li_2S_2O_4$	−1,179.2

As discussed earlier, this type of behavior indicates that the cell operates by a reconstitution reaction. It should be possible to calculate the voltage by consideration of the thermodynamic properties of the phases involved in this system at ambient temperature. These are shown in Table 9B.1.

From this information the stable tie lines in the ternary phase stability diagram for this system can be determined, as discussed earlier. The reaction equations relevant to each of the subtriangles can also be identified, and their potentials calculated. The resulting diagram is shown in Fig. 9B.2.

It can be seen that the $Li_2S_2O_4$–SO_2–O subtriangle has a potential of 3.0 V vs. lithium. Since the SO_2–$Li_2S_2O_4$ tie line on the edge of that triangle points at the lithium corner, no oxygen is formed by the reaction of lithium with SO_2 to produce $Li_2S_2O_4$.

The formal reaction for this cell is therefore

$$2Li + 2SO_2 = Li_2S_2O_4. \tag{9B.1}$$

The theoretical specific energy of this cell can be calculated to be 4,080 kWh kg^{-1}, a high value.

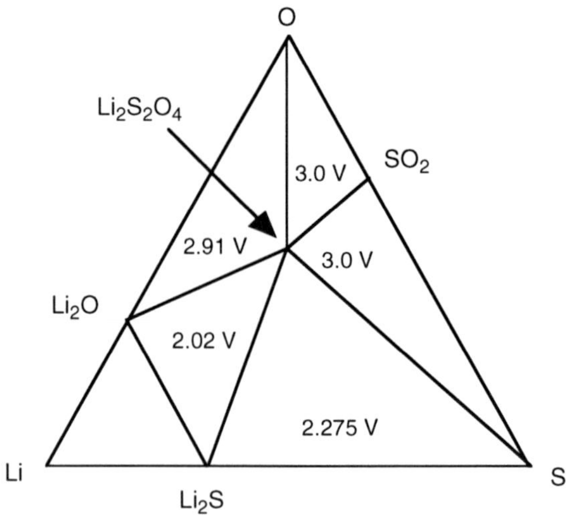

Fig. 9B.2 Phase stability diagram for the ternary Li–S–O system at ambient temperature

9B.3 The Li/$SOCl_2$ System

The lithium/thionyl batteries react at a somewhat higher constant voltage plateau, at 3.66 V.

The formal reaction is known to be

$$4Li + 2SOCl_2 = 4LiCl + S + SO_2. \tag{9B.2}$$

This involves the Li–S–Cl–O quaternary system. In order to visualize the behavior of this system in a manner similar to that for the Li/SO_2 cell above, a tetrahedral figure would have to be drawn, and the constant voltage plateaus related to each of the subtetrahedra calculated. While straightforward, this is a bit too complicated to be included here, however.

The theoretical specific energy of this cell can be calculated to be 7,250 kWh kg^{-1}, which is a very high value.

9C Hydrogen and Water in Positive Electrode Materials

9C.1 Introduction

The electrochemical insertion and deletion of hydrogen is a major feature in some important types of aqueous batteries. The use of metal hydrides as negative electrode reactants in aqueous systems is discussed in Sect. 10C, and the hydrogen-driven $H_2NiO_2/HNiO_2$ phase transformation is the major reaction in the positive electrode of a number of nickel cells, as described in Sect. 11B.

It is generally known that alkali metals react vigorously with water, with the evolution of hydrogen. In addition, a number of materials containing lithium are sensitive to air and/or water, and thus have to be handled in dry rooms or glove boxes. Yet, most of the lithium-containing oxides now used as positive electrode reactants in lithium battery systems are synthesized in air, often with little heed given to this problem.

It has long been known that hydrogen (protons) can be present in oxides, including some that contain lithium, and that water (a combination of protons and extra oxide ions) can be absorbed into some selected cases. There are several different mechanisms whereby these can happen.

9C.2 Ion Exchange

It is possible to simply exchange one type of cationic species for another of equal charge without changing the ratio of cations to anions or introducing other defects in oxides. For example, the replacement of some or all of the sodium cations present in oxides by lithium cations is discussed in several places in this text.

Especially interesting is the exchange of lithium ions by protons. One method is chemically driven ion exchange, in which there is interdiffusion in the solid state between native ionic species and ionic species from an adjacent liquid phase. An example of this is the replacement of lithium ions in an oxide solid electrolyte or mixed conductor by protons as the result of immersion in an acidic aqueous solution. Protons from the solution diffuse into the oxide, replacing lithium ions, which move back into the solution. The presence of anions in the solution that react with lithium ions to form stable products, such as LiCl, can provide a strong driving force. An example could be a lithium transition metal oxide, $LiMO_2$, placed in an aqueous solution of HCl. In this case the ion exchange process can be written as a simple chemical reaction

$$HCl + LiMO_2 = HMO_2 + LiCl. \tag{9C.1}$$

The LiCl product can either remain in solution or precipitate as a solid product.

One can also use electrochemical methods to induce ion exchange. That is, one species inside a solid electrode can be replaced in the crystalline lattice by a different species from the electrolyte electrochemically. The species that is displaced leaves the solid and moves into the electrolyte or into another phase. This electrochemically driven displacement process is now sometimes called *extrusion* by some investigators.

9C.3 Simple Addition Methods

Instead of exchanging with lithium, hydrogen can be simply added to a solid in the form of interstitial protons. The charge balance requirement can be accomplished by the coaddition of either electronic or ionic species, i.e., either by the introduction of extra electrons or the introduction of negatively charged ionic species, such as O^{2-} ions. If electrons are introduced, the electrical potential of the material will become more negative, with a tendency toward *n*-type conductivity.

Similarly, oxygen, as oxide ions, can be introduced into solids, either directly from an adjacent gas phase or by reaction with water, with the concurrent formation of gaseous hydrogen molecules. Oxide ions can generally not reside upon interstitial sites in dense oxides because of their size, and thus their introduction requires the presence of oxygen vacancies in the crystal lattice. If only negatively charged oxide ions are introduced, electroneutrality requires the simultaneous introduction of electron holes. Thus, the electrical potential of the solid becomes more positive, with a tendency toward *p*-type conductivity.

There is another possibility, first discussed by Stotz and Wagner [67, 68]. This is the simultaneous introduction of species related to both the hydrogen component and the oxygen component of water, i.e., both protons and oxide ions. This requires, of course, mechanisms for the transport of both hydrogen and oxygen species within the crystal structure. As mentioned already, hydrogen can enter the crystal structure of many oxides as mobile interstitial protons. The transport of oxide ions that move by vacancy motion requires the preexistence of oxide ion vacancies. As will be discussed in Sect. 16C, this typically involves cation doping. In this dual mechanism the electrical charge is balanced. Neither electrons nor holes are involved, so the electrical potential of the solid is not changed. The concurrent introduction of both protons and oxide ions is, of course, compositionally equivalent to the addition of water to the solid, although the species H_2O does not actually exist in the crystal structure.

9C.4 Thermodynamics of the Lithium–Hydrogen–Oxygen System

A number of features of the interaction between lithium, hydrogen, and oxygen in solids can be understood in terms of the thermodynamics of the ternary Li–H–O system. A useful thinking tool that can be used for this purpose is the *ternary phase stability diagram* with these three elements at the corners. This was discussed in some detail in Chap. 4.

The ternary phase stability diagram for the Li–H–O system at ambient temperature was determined [69] by using chemical thermodynamic data from Barin [70], and assuming that all relevant phases are in their standard states. An updated version is shown in Fig. 9C.1.

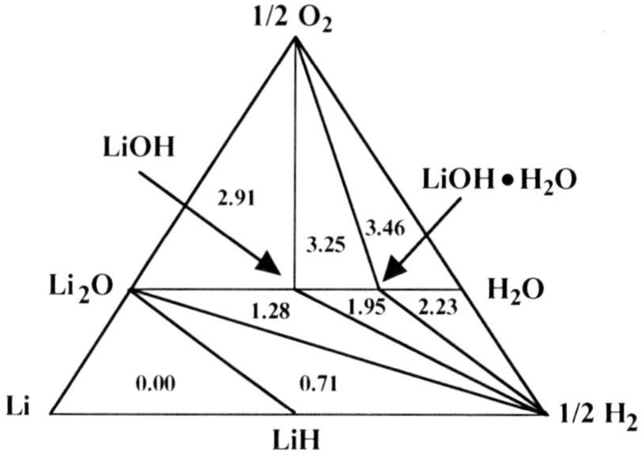

Fig. 9C.1 Calculated phase stability diagram for the Li–H–O system at 298 K, assuming unit activities of all phases. The numbers within the triangles are their respective potentials vs. pure lithium

Using the methods discussed in Chap. 4, the calculated voltages for the potentials of all compositions in the subtriangles are shown relative to pure lithium.

If one considers an electrochemical cell with pure lithium at the negative electrode, the potential of water that is saturated with $LiOH \cdot H_2O$ will be 2.23 V when hydrogen is present at 1 atm. On the other hand, water saturated with $LiOH \cdot H_2O$ will have a potential of 3.46 V vs. Li if 1 atm of oxygen is present. It can be seen that under these conditions water has a stability window of 1.23 V, as is the case in the binary hydrogen–oxygen system.

These results may seem to be in conflict with the general conclusion in the literature that the potential of lithium is −3.05 V relative to that of the standard hydrogen electrode (SHE) potential in aqueous electrochemical systems. This can be reconciled by recognizing that the values calculated here are for the case that water is in equilibrium with $LiOH \cdot H_2O$, which is very basic, with a pH of 14. The potentials of both hydrogen and pure oxygen, as well as all other zero-degree-of-freedom equilibria, decrease by 0.059 V per pH unit. Thus, in order to be compared with the potential of the SHE, these calculated values have to be corrected by 14×0.59, or 0.826 V. Then, the voltage between lithium and the SHE that is calculated in this way becomes 3.056 V, corresponding to the data in electrochemical tables.

These matters will be discussed more thoroughly in Sect. 13B.

9C.5 Examples of Phases Containing Lithium That are Stable in Water

A number of examples can be found in the literature that are consistent with, and illustrate these considerations. Particularly appropriate are several experimental results that were published by the group of J.R. Dahn some years ago.

They performed experiments on the addition of lithium to $LiMn_2O_4$ in a LiOH-containing aqueous electrolyte using a carbon negative electrode [71] and showed that the two-phase system $LiMn_2O_4-Li_2Mn_2O_4$, that is known to have a potential of 2.97 V vs. Li in nonaqueous cells [72] is stable in water containing LiOH. They used a Ag/AgCl reference electrode, referred their measurements to the SHE, and then converted to the lithium scale, assuming that the potential of the lithium electrode is −3.05 V vs. the SHE. They found that lithium began reacting with the $LiMn_2O_4$ at a potential of −0.1 V vs. the SHE, which is consistent with the value of 2.97 V vs. Li mentioned earlier.

As lithium was added beyond the two-phase composition limit the potential fell to that for hydrogen evolution. Their data showed hydrogen evolution at a potential 2.2 V vs. pure Li, and found oxygen evolution on a carbon negative electrode at 3.4 V vs. pure Li. It can readily be seen that these experimental results are consistent with the results of the Gibbs triangle calculations shown in Fig. 9C.1.

It was also found that the phase $VO_2(B)$ reacts with lithium at potentials within the stability range of water [73]. Electrochemical cell experiments were performed in which $Li_xVO_2(B)$ acted as the negative electrode, and $Li_xMn_2O_4$ as the positive electrode. These aqueous electrolyte cells gave comparable results to those with the same electrodes in organic solvent electrolyte cells.

9C.6 Materials That Have Potentials Above the Stability Window of Water

At normal pressures materials with potentials more positive than that of pure oxygen will tend to oxidize water to cause the evolution of electrically neutral molecular oxygen gas. For this to happen there must be a concurrent reduction process. One possibility is the insertion of positively charged ionic species, along with their charge-balancing electrons, into the material in question. The insertion of protons or lithium ions and electrons into high-potential oxides is one possible example of such a reduction process. When this happens, the potential of the material goes down toward that of pure oxygen.

9C.7 Absorption of Protons from Water Vapor in the Atmosphere

A number of materials that are used as positive electrode reactants in lithium battery systems have operating potentials well above the stability range of water. Cells containing these materials and carbon negative electrodes are typically assembled in air in the uncharged state. It is generally found that the open circuit cell voltage at the start of the first charge is consistent with lithium–air equilibrium, i.e., along the Li_2O/O_2 edge of the ternary phase stability diagram in Fig. 9C.1. This can be calculated to be 2.91 V vs. pure lithium. This can be explained by the reaction of these materials with water vapor in the atmosphere. Protons and electrons enter the crystal structures of these high potential materials, reducing their potentials to that value. This is accompanied by the concurrent evolution of molecular oxygen.

9C.8 Extraction of Lithium from Aqueous Solutions

An analogous situation can occur if a material that can readily insert lithium, rather than protons, has a potential above the stability range of water. If lithium ions and electrons enter the material's structure the potential will decrease until the value in equilibrium with oxygen is reached. Such a material can thus be used to extract lithium from aqueous solutions. This was demonstrated by experiments on the use of the λ-MnO_2 spinel phase that absorbed lithium when it was immersed in aqueous chloride solutions [74].

References

1. Y.F.Y. Yao and J.T. Kummer, J. Inorg. Nucl. Chem. 29, 2453 (1967)
2. N. Weber and J.T. Kummer, Proc. Annu. Power Sources Conf. 21, 37 (1967)
3. J. Coetzer, J. Power Sources 18, 377 (1986)
4. R.C. Galloway, J. Electrochem. Soc. 134, 256 (1987)
5. R.J. Bones, J. Coetzer, R.C. Galloway, D.A. Teagle, J. Electrochem. Soc. 134, 2379 (1987)
6. R.A. Huggins, J. Power Sources 81–82, 13 (1999)
7. D.R. Vissers, Z. Tomczuk and R.K. Steunenberg, J. Electrochem. Soc. 121, 665 (1974)
8. M.S. Whittingham, Science 192, 1126 (1976)
9. M.S. Whittingham, J. Electrochem. Soc. 123, 315 (1976)
10. M.S. Whittingham, Intercalation Compounds, in *Fast Ion Transport*, ed. by B. Scrosati, A. Magistris, C.M. Mari and G. Mariotto, KluwerAcademic, Dordrecht, (1993), p. 69
11. P.G. Dickens, S.J. French, A.T. Hight and M.F. Pye, Mater. Res. Bull. 14, 1295 (1979)
12. K. Mizushima, P.C. Jones, P.J. Wiseman and J.B. Goodenough, Mater. Res. Bull. 15, 783 (1980)
13. J.B. Goodenough, K. Mizushima and T. Takada, Jpn. J. Appl. Phys. 19 (Suppl. 19-3), 305 (1980)
14. T. Nagaura and K. Tozawa, in *Progress in Batteries and Solar Cells*, Vol. 9, JEC Press, Brunswick, OH (1990), p. 209
15. T. Nagaura, in *Progress in Batteries and Solar Cells*, JEC Press, Vol. 10, Brunswick, OH (1991), p. 218
16. A. Yamada, M. Hosoya, S.C. Chung, Y. Kudo and K.Y. Liu, *Concepts in Design of Olivine-Type Cathodes*, Abstract No. 205, Electrochemical Society Meeting, San Francisco (2001)
17. K.S. Nanjundaswamy, A.K. Padhi, J.B. Goodenough, S. Okada, H. Ohtsuka, H. Arai, and J. Yamaki, Solid State Ionics 92, 1 (1996) **[Nanjundaswamy, 1996 #42]
18. A.K. Padhi, K.S. Nanjundaswamy, C. Masquelier and J.B. Goodenough, J. Electrochem. Soc. 144, 2581 (1997) **[Padhi, 1997 #9]
19. S. Okada, H. Ohtsuka, H. Arai and M. Ichimura, "Characteristics of New Low-Cost High-Voltage Cathode, Fe2(SO4)3," ECS Proceedings, Hawaii (1993) **[Okada, #43]

20. S. Okada, T. Takada, M. Egashira, J. Yamaki, M. Tabuchi, H. Kageyama, T. Kodama and R. Kanno, "Characteristics of 3D Cathodes with Polyanions for Lithium Batteries," presented at *Second Hawaii Battery Conference*, Hawaii (1999) **[Okada, 1999 #44]
21. K. Mizushima, P.C. Jones, P.J. Wiseman and J.B. Goodenough, Mater. Res. Bull. 15, 783 (1980)
22. I. Sadadone and C. Delmas, J. Mater. Chem. 6, 193 (1996)
23. P.G. Bruce, A.R. Armstrong and R. Gitzendanner, J. Mater. Chem. 9, 193 (1999)
24. Y. Grincourt, C. Storey and I.J. Davidson, J. Power Sources 97–98, 711 (2001)
25. J.M. Paulson, R.A. Donaberger and J.R. Dahn, Chem. Mater. 12, 2257 (2000)
26. M.E. Spahr, P. Novak, B. Schneider, O. Haas, R.J. Nesper, J. Electrochem. Soc. 145, 1113 (1998)
27. T. Ohzuku and Y. Makimura, Chem. Lett. 744 (2001)
28. Z. Lu, D.D. MacNeil and J. R. Dahn, Electrochem. Solid State Lett. 4, A191 (2001)
29. K. Kang, Y.S. Meng, J. Breger, C.P. Grey and G. Ceder, Science 311, 977 (2006)
30. Z. Liu, A. Yu and J.Y. Lee, J. Power Sources 81–82, 416 (1999)
31. M. Yoshio, H. Noguchi, J.-I. Itoh, M. Okada and T. Mouri, J. Power Sources 90, 176 (2000)
32. M.S. Whittingham, Chem. Rev. 104, 4271 (2004)
33. M.M. Thackeray, W.I.F. David, P.G. Bruce and J.B. Goodenough, Mater. Res. Bull. 18, 461 (1983)
34. M.M. Thackeray, P.J. Johnson, L.A. de Picciotto, P.G. Bruce, and J.B. Goodenough, Mater. Res. Bull. 19, 179 (1984)
35. M.M. Thackeray, "The Structural Stability of Transition Metal Oxide Insertion Electrodes for Lithium Batteries," in *Handbook of Battery Materials*, ed. by J.O. Besenhard, Wiley-VCH, New York (1999), p. 293
36. D. Guyomard and J.M. Tarascon, Solid State Ionics 69, 222 (1994)
37. G. Amatucci and J.-M. Tarascon, J. Electrochem. Soc. 149, K31 (2002)
38. R.J. Gummow, A. De Kock and M.M. Thackeray, Solid State Ionics 69, 59 (1994)
39. D. Guyomard and J.-M. Tarascon, US Patent 5,192,629 (1993)
40. D. Guyomard and J.-M. Tarascon, Solid State Ionics 69, 293 (1994)
41. C. Sigala, D. Guyomard, A. Verbaere, Y. Piffard and M. Tournoux, Solid State Ionics 81, 167 (1995)
42. Y. Ein-Eli and W.F. Howard, J. Electrochem. Soc. 144, L205 (1997)
43. Y. Ein-Eli, W.F. Howard, S.H. Lu, S. Mukerjee, J. McBreen, J.T. Vaughey and M.M. Thackeray, J. Electrochem. Soc. 145, 1238 (1998)
44. Y. Ein-Eli, S.H. Lu, M.A. Rzeznik, S. Mukerjee, X.Q. Yang and J. McBreen, J. Electrochem. Soc. 145, 3383 (1998)
45. T. Ohzuku, S. Takeda and M. Iwanaga, J. Power Sources 81–82, 90 (1999)
46. K. Ariyoshi, Y. Iwakoshi, N. Nakayama and T. Ohzuku, J. Electrochem. Soc. 151, A296 (2004)
47. K.M. Colbow, J.R. Dahn and R.R. Haering, J. Power Sources 26, 397 (1989)
48. T. Ohzuku, A. Ueda and N. Yamamoto, J. Electrochem. Soc. 142, 1431 (1995)
49. J.B. Goodenough, H.Y.P. Hong and J.A. Kafalas, Mater. Res. Bull. 11, 203 (1976)
50. K.S. Nanjundaswamy, A.K. Padhi, J.B. Goodenough, S. Okada, H. Ohtsuka, H. Arai, J. Yamaki, Solid State Ionics 92, 1 (1996)
51. A.K. Padhi, K.S. Nanjundaswamy, C. Masquelier, S. Okada and J.B. Goodenough, J. Electrochem. Soc. 144, 1609 (1997)
52. J. Barker and M.Y. Saidi, US Patent 5,871,866 (1999)
53. M.Y. Saidi, J. Barker, H. Huang, J.L. Swoyer and G. Adamson, Electrochem. Solid State Lett., 5, A149 (2002)
54. A.K. Padhi, K.S. Nanjundaswamy and J.B. Goodenough, J. Electrochem. Soc. 144, 1188 (1997)
55. N. Ravet, J.B. Goodenough, S. Besner, M. Simoneau, P. Hovington and M. Armand, Electrochem. Soc. Meeting Abstract 99-2, 127 (1999)
56. S.Y. Chung, J.T. Bloking and Y.-M. Chiang, Nat. Mater. 1, 123 (2002)

57. R. Amin and J. Maier, Solid State Ionics 178, 1831 (2008)
58. P.S. Herle, B. Ellis, N. Coombs and L.F. Nazar, Nat. Mater. 3, 147 (2004)
59. S.P. Ong, L. Wang, B. Kang and G. Ceder, presented at the Materials Research Society Meeting in San Francisco (2007).
60. N. Meethong, H.-Y.S. Huang, S.A. Speakman, W.C. Carter and Y.-M. Chiang, Adv. Funct. Mater. 17, 1115 (2007)
61. J. Barker, M.Y. Saidi and J.L. Swoyer, J. Electrochem. Soc. 151, A1670 (2004)
62. J. Barker, R.K.B. Gover, P. Burns and A.J. Bryan, Electrochem. Solid State Lett. 10, A130 (2007)
63. J. Barker, R.K.B. Gover, P. Burns and A.J. Bryan, Electrochem Solid State Lett. 9, A190 (2006)
64. J. Barker, R.K.B. Gover, P. Burns and A.J. Bryan, J. Electrochem. Soc. 154, A882 (2007)
65. J.R. Dahn, E.W. Fuller, M. Obrovac and U. von Sacken, Solid State Ionics 69, 265 (1994)
66. N.A. Godshall, I.D. Raistrick and R.A. Huggins, J. Electrochem. Soc. 131, 543 (1984)
67. S. Stotz and C. Wagner, Ber. Bunsenges. Phys. Chem. 70, 781 (1966)
68. C. Wagner, Ber. Bunsenges. Phys. Chem. 72, 778 (1968)
69. R.A. Huggins, Solid State Ionics 136–137, 1321 (2000)
70. I. Barin, *Thermochemical Data of Pure Substances*, VCH, Weinheim (1989)
71. W. Li, W.R. McKinnon and J.R. Dahn, J. Electrochem. Soc. 141, 2310 (1994)
72. J.M. Tarascon and D. Guyomard, J. Electrochem. Soc. 138, 2864 (1993)
73. W. Li and J.R. Dahn, J. Electrochem. Soc. 142, 1742 (1995)
74. H. Kanoh, K. Ooi, Y. Miyai and S. Katoh, Sep. Sci. Technol. 28, 643 (1993)

Chapter 10
Negative Electrodes in Aqueous Systems

10.1 Introduction

The following sections of this chapter will discuss three examples of negative electrodes that are used in aqueous electrolyte battery systems, the zinc electrode, the "cadmium" electrode, and metal hydride electrodes.

It will be seen that these operate in quite different ways. In the first case, there is an exchange between solid zinc and zinc (zincate) ions in the electrolyte.

In the second, the "cadmium" electrode is actually a two-phase system, with elemental cadmium in equilibrium with another solid, its hydroxide.

In the third, hydrogen is exchanged between a solid metal hydride and hydrogen-containing ionic species in the electrolyte.

10A The Zinc Electrode in Aqueous Systems

10A.1 Introduction

Elemental zinc is used as the negative electrode in a number of aqueous electrolyte batteries. The most prominent example is the very common Zn/MnO_2 primary "alkaline cell" that is used in a wide variety of small electronic devices. As will be discussed in Sect. 11A, the positive electrode reaction involves the insertion of hydrogen into the MnO_2 crystal structure. A discussion of Zn/MnO_2 technology can be found in [1].

The initial open circuit voltage of these cells is in the range 1.5–1.6 V. This is greater than the decomposition voltage of water, which can be calculated from its Gibbs free energy of formation, 237.1 kJ mol^{-1}, to be 1.23 V at ambient temperatures from

$$\Delta E = -\left(\frac{\Delta G_f{}^{\circ}(H_2O)}{2F}\right). \qquad (10A.1)$$

It will be shown here that this is possible because the zinc negative electrode is covered by a thin layer of ionically conducting ZnO, and the thermodynamic result is that its potential is several hundred millivolts lower than the potential at which gaseous hydrogen is normally expected to evolve if an unoxidized metal electrode were to be in contact with water.

10A.2 Thermodynamic Relationships in the H–Zn–O System

The potential and stability of the zinc electrode can be understood by consideration of the thermodynamics of the ternary H–Zn–O system, and its representation in a ternary phase stability diagram.

In addition to the elements and water, the only other relevant phase in this system is ZnO, and the value of its Gibbs free energy of formation is $-320.5\,\text{kJ mol}^{-1}$ at 25°C.

As discussed earlier, one can use the values of the Gibbs free energy of formation of the different phases to determine which tie lines are stable in a ternary phase stability diagram. In this case, the only possibilities would be either a line between Zn and H_2O or a line between ZnO and hydrogen. Because the Gibbs free energy of formation of ZnO is more negative than that of water, the second of these possible tie lines must be the more stable. The simple result in this case is shown in Fig. 10A.1. It shows that a subtriangle is formed that has Zn, ZnO, and H_2 at its corners. Another has water, ZnO, and hydrogen at its corners. The potentials of all compositions in the first triangle with respect to oxygen can be calculated from the Gibbs free energy change related to the simple binary reaction along its edge,

$$Zn + {}^1/_2 O_2 = ZnO \qquad (10A.2)$$

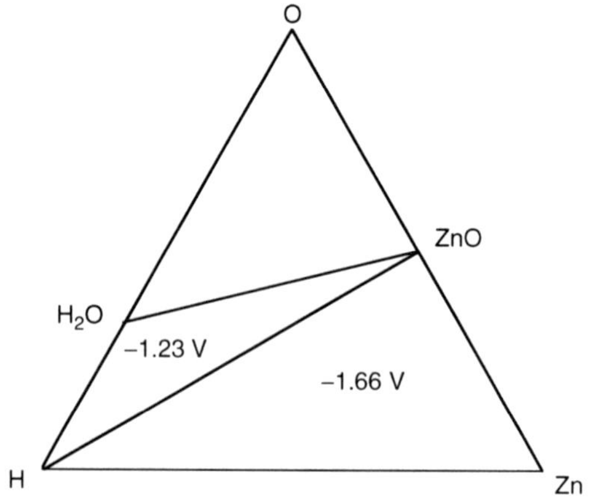

Fig. 10A.1 Ternary phase stability diagram for the H–Zn–O system. The numbers within the ternary subtriangles are the potentials relative to pure oxygen

and the result is $-1.66\,\text{V}$. The potential of all compositions in the second triangle is likewise related to the Gibbs free energy of formation of water, or $-1.23\,\text{V}$ relative to pure oxygen. That means that zinc has a potential that is 0.43 V more negative than the potential of pure hydrogen in aqueous electrolytes.

Because of the presence of the thin ionically conducting, but electronically insulating, layer of ZnO, water is not present at the electrochemical interface, the location of the transition between ionic conduction and electronic conduction, and hydrogen gas is not formed on the zinc. Thus the effective stability range of the electrolyte is extended, as discussed in Sect. 16A.

10A.3 Problems with the Zinc Electrode

Although its low potential is very attractive, there are two negative features of the use of zinc electrodes in aqueous systems. Both relate to its rechargeability in basic aqueous electrolytes.

One of these is that ZnO dissolves in KOH electrolytes, producing an appreciable concentration of zincate ions, $Zn(OH)_4{}^{2-}$, in which the Zn^{2+} cations are tetrahedrally surrounded by four OH^- groups. Nonuniform zincate composition gradients during recharging, as well as the ZnO on the surface, lead to the formation of *protrusions, filaments*, and *dendrites* during the redeposition of zinc from the electrolyte at appreciable currents.

The other is that the zinc has a tendency to not redeposit upon the electrode at the same locations during charging of the cell as those from which it was removed during discharge. Gravitational demixing causes the concentration of zincate ions to increase at lower locations, leading to slight differences in the electrolyte conductivity. The result is that there is a gradual redistribution of the zinc, so that the lower portions of the electrode become somewhat thicker or denser as it is discharged and recharged. This effect is often called *shape-change*.

10B The "Cadmium" Electrode

10B.1 Introduction

Cadmium/nickel, Ni/Cd, or NiCad, cells have been important products for many years. They have alkaline electrolytes and use the "nickel"-positive electrode, H_xNiO_2, which is discussed in Sect. 11B. Because they have both higher capacity and a reduced problem with environmental pollution, cadmium is considered to be environmentally hazardous – batteries with metal hydride, rather than cadmium, negative electrodes are gradually taking a larger part of the market. They are discussed in Sect. 10C.

10B.2 Thermodynamic Relationships in the H−Cd−O System

As in the case of the H−Zn−O system described in the last section, the first step in understanding what determines the potential of the cadmium electrode involves the use of available thermodynamic data to determine the relevant ternary phase stability diagram, for the driving forces of electrochemical reactions are the related reactions between electrically neutral species.

In this case, the key issue is the value of the Gibbs free energy of formation of CdO, which has been found to be −229.3 kJ mol^{-1} at 25°C. This is less than the value for the formation of water, so a tie line between water and cadmium must be more stable than a tie line between CdO and hydrogen. Also, $Cd(OH)_2$ is a stable phase between water and CdO, because its Gibbs free energy of formation at 25°C is −473.8 kJ mol^{-1}, whereas the sum of the others is −466.4 kJ mol^{-1}. From these data, the ternary phase stability diagram shown in Fig. 10B.1 can be drawn. It is clear that it is different from the one for the H−Zn−O system in Sect. 11A.

It is seen that $Cd(OH)_2$ is also stable when Cd is in contact with water. The potential of the cadmium electrode is determined by the potential of the subtriangle that has water, $Cd(OH)_2$, and cadmium at its corners.

Since there are three phases as well as three components, Cd, hydrogen, and oxygen, present, there are no degrees of freedom, according to the Gibbs phase rule, as discussed earlier. Therefore, the cadmium reaction should occur at a constant potential, independent of the state of charge. This is what is experimentally found.

The potential of all compositions in this triangle is determined by the reaction

$$^{1}/_{2}O_2 + Cd + H_2O = Cd(OH)_2 \tag{10B.1}$$

and from the Gibbs free energies of formation of the relevant phases it is found that its value is −1.226 V relative to that of pure oxygen, as is shown in Fig. 10B.1.

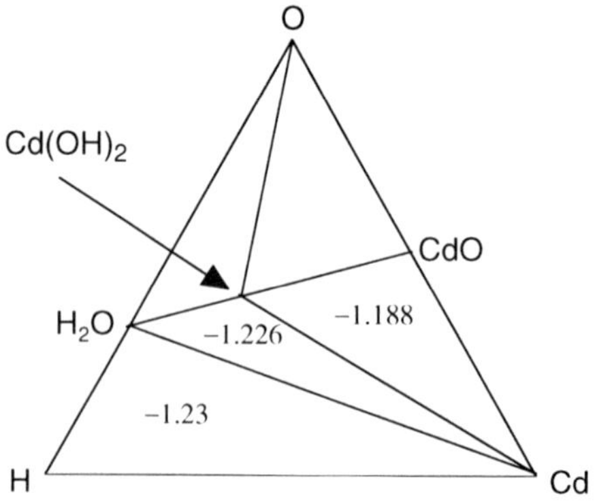

Fig. 10B.1 The H−Cd−O ternary phase stability diagram showing the potentials of the compositions in the subtriangles vs. pure oxygen, in volts

The discharge of the cadmium electrode can be written as an electrochemical reaction

$$Cd + 2(OH)^- = Cd(OH)_2 + 2e^-. \quad (10B.2)$$

This shows that there is consumption of water from the electrolyte during discharge, as can also be seen in the neutral species reaction in (10B.1). This consumption of water must be considered in the determination of the electrolyte composition.

This is different from the zinc electrode discussed in Sect. 10A, where the equivalent discharge reaction is

$$Zn + 2(OH)^- = ZnO + H_2O + 2e^- \quad (10B.3)$$

and there is no net change in the amount of water.

10B.3 Comments on the Mechanism of Operation of the Cadmium Electrode

There is another matter to be considered in the behavior of the cadmium electrode, since discharge involves the formation of a layer of $Cd(OH)_2$ on top of the Cd. This would require a mechanism to either transport Cd^{2+} or OH^- ions through the growing $Cd(OH)_2$ layer, both of which seem unlikely. This reaction is generally thought to involve the formation of an intermediate species that is soluble in the KOH electrolyte. The most likely intermediate species is evidently $Cd(OH)_3^-$.

The kinetics of the cadmium electrode are sufficiently rapid that the potential changes relatively little on either charge or discharge. Typical values are a deviation of 60 mV during charge, and 15 mV during discharge at the C/2, or 2-h, rate. In addition, there are small potential overshoots at the beginning in both directions if the full capacity had been employed in the previous step. This is, of course, what would be expected if the microstructure started with only one phase, and the second phase has to be nucleated.

This is shown in Fig. 10B.2 [2].

One of the questions that had arisen in earlier considerations of the mechanism of this electrode was the possibility of the formation of CdO. X-ray investigations have found no evidence for its presence. Thus if this phase were present it would have to be either as extremely thin layers or be amorphous.

However, this question can be readily answered by consideration of the potential of the reaction in the triangle with Cd, CdO, and $Cd(OH)_2$ at its corners. This can be determined simply by the reaction along its edge

$$^1/_2 O_2 + Cd = CdO. \quad (10B.4)$$

From the Gibbs free energy of formation of CdO this is found to be -1.188 V relative to the potential of oxygen. This is 38 mV positive of the equilibrium potential of the main reaction. Since it is not expected that the electrode potential would deviate so far during operation of these electrodes, the formation of CdO is unlikely.

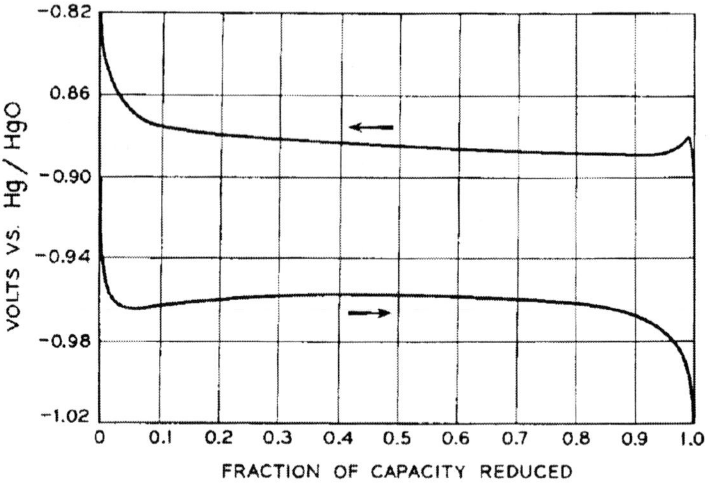

Fig. 10B.2 The charge–discharge behavior of a sintered-plate cadmium electrode, measured at about the C/2 rate in 2M KOH at 25°C

10C Metal Hydride Electrodes

10C.1 Introduction

Metal hydrides are currently used as the negative electrode reactant in large numbers of reversible commercial batteries with aqueous electrolytes, generally in combination with "nickel" positive electrodes.

There are several families of metal hydrides, and the electrochemical properties of some of these materials are comparable to those of cadmium. Developmental efforts have led to the production of small consumer batteries with comparable kinetics, but with up to twice the energy content per unit volume of comparable small *normal* Cd/Ni cells. Typical values for AA size cells are shown in Table 10C.1. For this reason, as well as because of the poisonous nature of cadmium, hydride cells are taking a larger and larger portion of this market.

Table 10C.1 Typical capacities of AA size cells used in many small electronic devices

Type of cell	mWh cm^{-3}
Normal Cd/Ni	110
High-capacity Cd/Ni	150
Hydride/Ni	200

10C.2 Comments on the Development of Commercial Metal Hydride Electrode Batteries

Although there had been research activities earlier in several laboratories, work on the commercialization of small metal hydride electrode cells began in Japan's Government Industry Research Institute (GIRIO) laboratory in Osaka in 1975. By 1991 there were a number of major producers in Japan, and the annual production rate had reached about 1 million cells. Activities were also underway in other countries. Those early cells had specific capacity values of about 54 Wh kg^{-1} and specific powers of about 200 W kg^{-1}.

The production rate grew rapidly, reaching an annual rate of about 100 million in 1993, and over 1 billion cells in 2005. The properties of these small consumer cells also improved greatly. By 2006 the specific capacity had reached 100 Wh kg^{-1}, and the specific power 1,200 W kg^{-1}. The energy density values also improved, so that they are now up to 420 Wh L^{-1}.

They are generally designed with excess negative electrode capacity, i.e., $N/P > 1$. This is increased for higher power applications.

The metal hydrides used in small consumer cells are multicomponent metallic alloys, typically containing about 30% rare earths. Prior to this development, the largest commercial use of rare earth materials was for specialty magnets. The major source of these materials is in China, where they are very abundant. Rare earths are also available in large quantities in the USA and South Africa.

In addition to the large current production of small consumer batteries, development efforts have been aimed at the production of larger cells with capacities of 30–100 Ah at 12 V. The primary force that is driving this move toward larger cells is their use in hybrid electric vehicles. In order to meet the high-power requirements, the specific capacity of these cells has to be sacrificed somewhat, down to about 45–60 Wh kg^{-1}.

10C.3 Hydride Materials Currently Being Used

There are two major families of hydrides currently being produced that can be roughly identified as AB_5 and AB_2 alloys.

The AB_5 alloys are based upon the pioneering work in the Philips laboratory that started with the serendipitous discovery of the reaction of gaseous hydrogen with $LaNi_5$. The basic crystal structure is of the layered hexagonal $CaCu_5$ type. Alternate layers contain both lanthanum and nickel, and only nickel. This structure is illustrated in Fig. 10C.1.

The reaction of hydrogen with materials of this type can be written as

$$3H_2 + LaNi_5 = LaNi_5H_6. \qquad (10C.1)$$

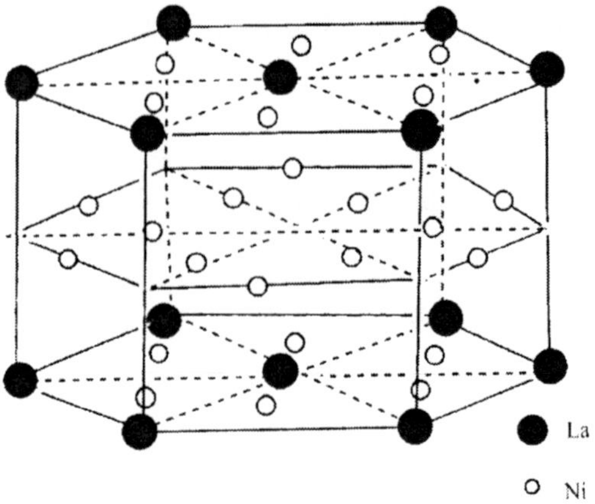

Fig. 10C.1 Schematic of the layered hexagonal lattice of $LaNi_5$ [3]

The hydrogen atoms reside in tetrahedral interstitial sites between the host atoms. It has been found that hydrogen can occupy interstitial sites in suitable alloys in which the *holes* have spherical radii of at least 0.4 Å. In larger interstitial positions hydrogen atoms are often *off center*.

Developmental work has involved the partial or complete replacement of the lanthanum with other metals, predominantly with mischmetall (Mm), a mixture of rare earths, and zirconium. A typical composition of the relatively inexpensive mischmetall is 45–58% Ce, 20–27% La, 13–20% Nd, and 3–8% Pr.

In addition, it has been found advantageous to replace some or all of the nickel with other elements, such as aluminum, manganese, and cobalt. Furthermore, it is possible to change the A/B ratio. One major producer uses a composition that has a higher A/B ratio than 5, in the direction of A_2B_7, for example. These materials show relatively flat two-phase discharge voltage plateaus, indicating a reconstitution reaction. Various compositional factors influence the pressure (cell voltage) and the hydrogen (charge) capacity of the electrode, as well as the cycle life. There has also been a lot of developmental work on preparative methods and the influence of microstructure upon the kinetic and cycle life properties of small cells with these materials.

10C.4 Disproportionation and Activation

Another reaction between hydrogen and these alloys can also take place, particularly at elevated temperatures. It can be written as

$$H_2 + LaNi_5 = LaH_2 + 5Ni \tag{10C.2}$$

and is called *disproportionation.* At 298 K the Gibbs free energies of formation of $LaNi_5$ and LaH_2 are −67 and −171 kJ mol^{-1}, respectively, so there is a significant driving force for this to occur, at least on the surface. Experiments have shown that the surface tends to contain regions that are rich in lanthanum, combined with oxygen. In addition, there are clusters of nickel. Because of the presence of these nickel islands, which are permeable to hydrogen, hydrogen can get into the interior of the alloy.

It is often found that a cyclic activation process is necessary in order to get full reaction of hydrogen with the total alloy. As hydrogen works its way into the interior there is a local volume expansion that often causes cracking and the formation of new fresh surface that is not covered with oxygen. This cracking can cause the bulk material to be converted into a powder, and is called *decrepitation.*

10C.5 Pressure–Composition Relation

If the particle size is small and there are no surface contamination or activation problems, the $LaNi_5$ alloy reacts readily with hydrogen at a few atmospheres pressure. This is illustrated in Fig. 10C.2.

This flat curve is an indication that this is a reconstitution, rather than insertion, reaction.

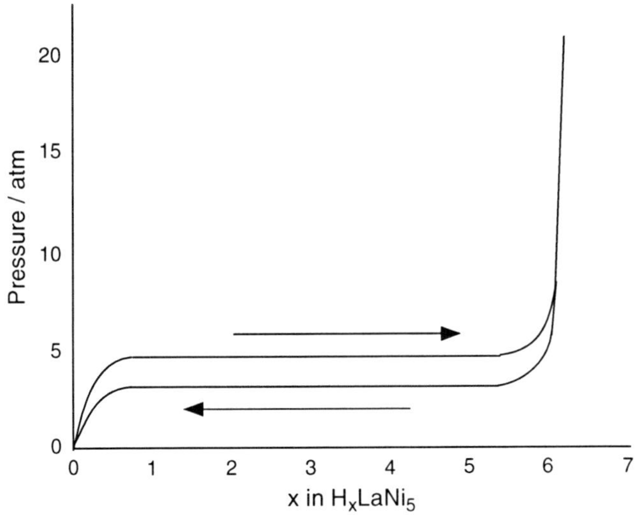

Fig. 10C.2 Pressure–composition isotherm for the reaction of $LaNi_5$ with hydrogen

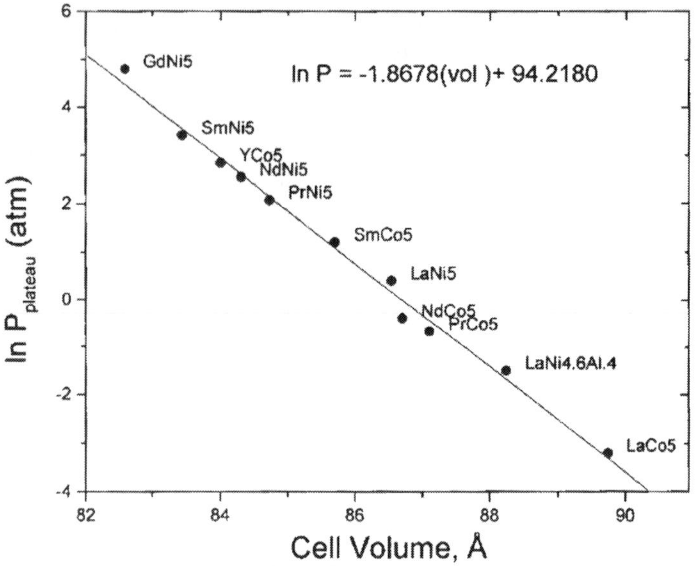

Fig. 10C.3 Relation between the logarithm of the plateau pressure and the volume of the crystal structure's unit cell [3]

There is a slight difference in the potential when hydrogen is added from that when hydrogen is removed. This hysteresis is probably related to the mechanical work that must occur due to the volume change in the reaction.

The pressure plateaus for the alloys that are used in batteries are a bit lower, so that the electrochemical potential remains somewhat positive of that for the evolution of hydrogen on the negative electrode.

It has been found that the logarithm of the potential at which this reaction occurs depends linearly upon the lattice parameter of the host material for this family of alloys. This is shown in Fig. 10C.3.

In order to reduce the blocking of the surface by oxygen, as well as to help hold the particles together, thin layers of either copper or nickel are sometimes put on their surfaces by the use of electroless plating methods [3]. PVDF or a similar material is also often used as a binder.

10C.6 The Influence of Temperature

The equilibrium pressure over all metal hydride materials increases at higher temperatures. This is shown schematically in Fig. 10C.4.

The relation between the potential plateau pressure and the temperature is generally expressed in terms of the Van't Hoff equation

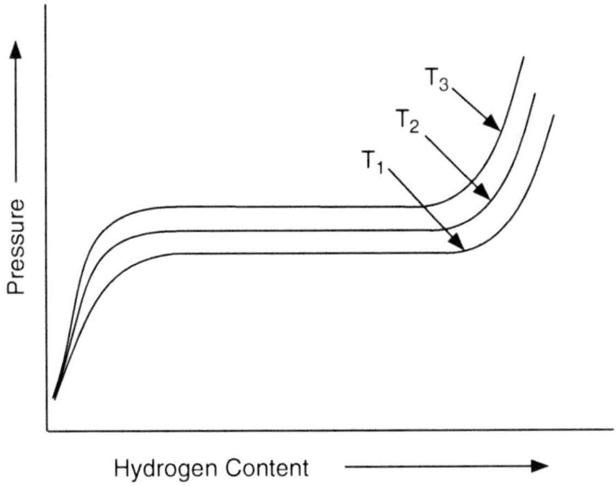

Fig. 10C.4 Schematic variation of the equilibrium pressure of a metal hydride system with temperature

$$\ln\, p(H_2) = \left({}^{\Delta H}/_{RT}\right) - \left({}^{\Delta S}/_{R}\right). \tag{10C.3}$$

This can readily be derived from the general relation

$$\Delta G = RT \ln\, p(H_2) \tag{10C.4}$$

and

$$\Delta G = \Delta H - T\Delta S. \tag{10C.5}$$

This relationship is shown in Fig. 10C.5 for $LaNi_5$ and a commercial mischmetal-containing alloy [3]. It can be seen that the pressure is lower, and thus the electrical potential is higher, in the case of the practical alloy.

This type of representation is often used to compare metal hydride systems that are of interest for the storage of hydrogen from the gas phase. Figure 10C.6 is an example of such a plot [4].

It can be seen that the range of temperature and pressure that can be considered for the storage of hydrogen gas is much greater than that which is of interest for the use in aqueous electrolyte battery systems.

Higher pressure in gas systems is equivalent to a lower potential in an electrochemical cell, as can be readily seen from the Nernst equation

$$E = -\left({}^{RT}/_{zF}\right) \ln\, p(H_2). \tag{10C.6}$$

The reaction potential must be above that for the evolution of hydrogen, and if it is too high, the cell voltage is reduced. As a result, the range of materials is quite constrained, and a considerable amount of effort has been invested in making minor modifications by changes in the alloy composition.

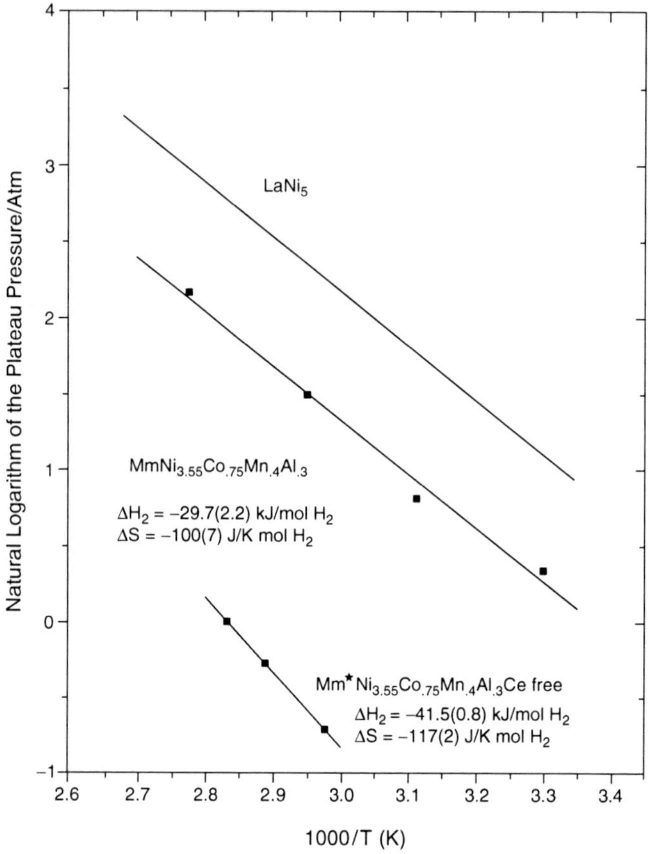

Fig. 10C.5 Van't Hoff plot for $LaNi_5H_x$ and two compositions of a $MmNi_{3.55}Co_{0.75}Mn_{0.4}Al_{0.3}H_x$ alloy

10C.7 AB_2 Alloys

The other major group of materials that are now being used as battery electrodes are the AB_2 alloys. There are two general types of AB_2 structures, sometimes called Friauf-Laves phases: the C14, or $MgZn_2$, type in which the B atoms are in a close-packed hexagonal array, and the C15, or $MgCu_2$, type in which the B atoms are arranged in a close-packed cubic array. Many materials can be prepared with these, or closely related, structures. The A atoms are generally either Ti or Zr. The B elements can be V, Ni, Cr, Mn, Fe, Co, Mo, Cu, and Zn. Some examples are listed in Table 10C.2.

It has generally been found that the C14-type structure is more suitable for hydrogen storage applications. A typical composition can be written as $(Ti,Zr)(V,Ni,Cr)_2$.

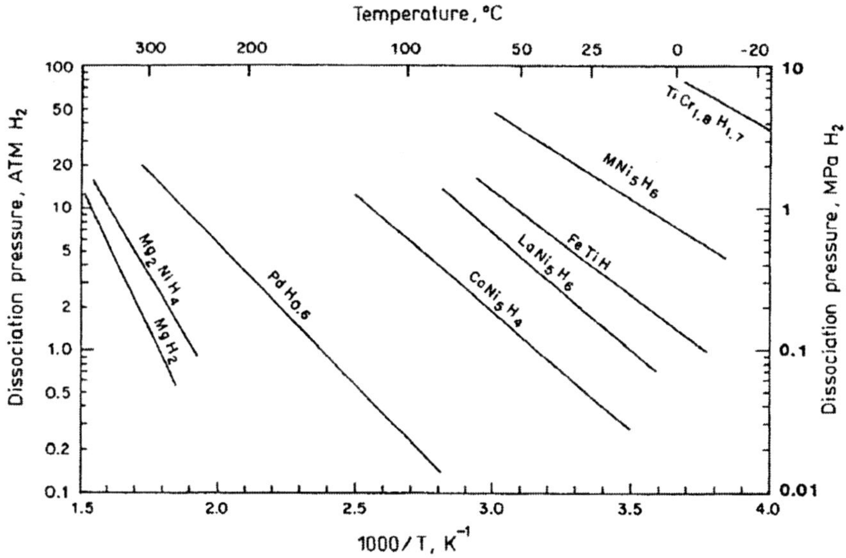

Fig. 10C.6 Example of Van't Hoff plot showing data for a wide range of materials

Table 10C.2 Structures of AB_2 Phase Materials

Material	C14 structure (hexagonal)	C15 structure (cubic)
$TiMn_2$	X	
$ZrMn_2$	X	
ZrV_2		X
$TiCr_2$	X	X
$ZrCr_2$	X	X
$ZrMo_2$		X

10C.8 General Comparison of These Two Structural Types

Both of these systems can provide a high charge storage density. Present data indicate that this can be slightly (5–10%) higher for some of the AB_2 materials than for the AB_5 materials.

However, there is a significant difference in the electrochemical characteristics of these two families of alloys. As illustrated in Fig. 10C.2, the hydrogen pressure is essentially independent of the composition over a wide range in the AB_5 case. Thus, the cell potential is essentially independent of the state of charge, characteristic of a reconstitution reaction. On the other hand, the hydrogen activity generally varies appreciably with the state of charge in the AB_2 alloys, giving charge-dependent cell voltages. The fact that the cell voltage decreases substantially during discharge of cells with the AB_2 alloy hydrides can be considered to be a significant disadvantage for the use of these materials in batteries.

A serious issue, particularly in the AB_2 materials, is the question of oxidation, and subsequent corrosion, particularly of the B metals. This can lead to drastic reductions in the capacity and cycle life, as well as causing a time-dependent increase in gaseous hydrogen pressure in the cell. Because of this, special preetching treatments have been developed to reduce this problem. The vanadium content of the surface is evidently particularly important.

10C.9 Other Alloys That Have Not Been Used in Commercial Batteries

An alloy of the AB_2 type based upon Ti and Mn, with some V, is currently being used for the storage of gaseous hydrogen, rather than in batteries. One commercial application is in fuel cell-propelled submarines manufactured in Germany. However, its hydrogen activity range is too high to be applicable for use in batteries.

Some years ago there was a development program in the Battelle laboratory in Switzerland funded by Daimler Benz aimed at the use of titanium–nickel materials of the general compositions A_2B (Ti_2Ni) and AB (TiNi) for use in automobile starter batteries. This early work showed that if electrodes have the right composition and microstructure and are properly prepared, they can perform quite well for many cycles. However, this development was never commercialized.

An interesting side issue is the fact that the TiNi phase, which is very stable in KOH, is one of the materials that is known to be ferroelastic, and to have mechanical memory characteristics. Its mechanical deformation takes place by the formation and translation of twin boundaries, rather than by dislocation motion. As a result, it is highly ductile, yet extremely resistant to fracture. Thus, it would be interesting to consider the use of minor amounts of this phase as a metallic binder in hydride, or other, electrodes. It should be able to accommodate the repeated microscopic mechanical deformation that typically occurs within the electrode structure upon cycling without fracturing.

10C.10 Microencapsulation of Hydride Particles

A method was developed some years ago in which a metallic coating of either copper or nickel is deposited by electroless methods upon the hydride particles before the mechanical formation of the hydride electrode [5]. This ductile layer helps the formation of electrodes by pressing, acts as a binder, contributes to the electronic conductivity, and thus improves electrode kinetics, and helps against overcharge. It evidently also increases the cycle life. Since both copper and nickel do not corrode in KOH, this layer also acts to prevent oxidation and corrosion.

10C.11 Other Binders

In addition to the copper or nickel metallic binders, some Japanese cells use PTFE, silicon rubber, or SEBS rubber as a binder. It has been found that this can greatly influence the utilization at high (up to 5C) rates. With the rubber binders (e.g., 3-wt% PTFE), flexible thin sheet electrodes can be made that make fabrication of the small spiral cells easier.

10C.12 Inclusion of a Solid Electrolyte in the Negative Electrode of Hydride Cells

An interesting development was the work in Japan on the use of a proton-conducting solid electrolyte in the negative electrodes of hydride cells. This material is tetramethyl ammonium hydroxide pentahydrate, $(CH_3)_4NOH.5H_2O$, which has been called TMAH5. It is a clathrate hydrate, and melts (at about 70°C) rather than decomposes, when it is heated. Thus, it can be melted to impregnate a preformed porous electrode to act as an internal electrolyte. This is typically not true for other solid electrolytes, and can be advantageous in increasing the electrode–electrolyte contact area.

TMAH5 has a conductivity of about $5 \times 10^{-3}\,\mathrm{S\,cm^{-2}}$ at ambient temperatures. While this value is higher than the conductivity of almost all other known proton-conducting solid electrolytes, it is less than that of the normal KOH aqueous electrolyte. Thus, if this solid electrolyte were to be used, one would have to be concerned with the development of fine scale geometries. This could surely be done, but it would probably involve the use of screen printing or tape casting fabrication methods, rather than conventional electrode fabrication procedures.

Both hydride/H_xNiO_2 cells and hydride/MnO_2 cells have been produced using this solid electrolyte. Because of the lower potential of the MnO_2-positive electrode relative to the *nickel* electrode, the latter cells have lower voltages.

10C.13 Maximum Theoretical Capacities of Various Metal Hydrides

The maximum theoretical specific capacities of various hydride negative electrode materials are listed in Table 10C.3. Values are shown for both the hydrogen charged and uncharged weight bases. They include two AB_5-type alloys that are being used by major producers, as well as the basic $LaNi_5$ alloy and two AB_2 materials.

Table 10C.3 Specific capacities of several AB_5 and AB_2 alloys

Material	Uncharged ($mAh\ g^{-1}$)	H_2 Charged ($mAhg^{-1}$)
$LaNi_5H_6$	371.90	366.81
$MmNi_{3.5}Co_{0.7}Al_{0.8}H_6$	393.93	388.23
$(LaNd)(NiCoSi)_5H_4$	248.80	246.51
$TiMn_2$	509.67	500.16
$(Ti,Zr)(V,Ni)_2$	448.78	441.39

As would be expected, small commercial cells have practical values that are less than the theoretical maxima presented in the last few pages. Hydride electrodes generally have specific capacities of 320–385 $mAh\ g^{-1}$. For comparison, the H_xNiO_2 *nickel* positive electrodes typically have practical capacities about 240 $mAh\ g^{-1}$.

References

1. R.F. Scarr, J.C. Hunter and P.J. Slezak, Alkaline Manganese Dioxide Batteries, in *Handbook of Batteries*, 3rd Edition, ed. by D. Linden and T.B. Reddy, McGraw-Hill, New York (2002), p. 10.1
2. P.C. Milner and U.B. Thomas, The Nickel-Cadmium Cell, in *Advances in Electrochemistry and Electrochemical Engineering*, Vol. 5, ed. by C.W. Tobias, Wiley-Interscience, New York (1967), p. 1
3. J.J. Reilly, Metal Hydride Electrodes in *Handbook of Battery Materials*, ed. by J.O. Besenhard, Wiley-VCH, New York (1999), p. 209
4. G.D. Sandrock and E.L. Huston, Chemtech 11, 754 (1981)
5. T. Sakai, H. Yoshinaga, H. Miyamura, N. Kuriyama and H. Ishikawa, J. Alloys Compounds 180, 37 (1992)

Chapter 11
Positive Electrodes in Aqueous Systems

11.1 Introduction

The following sections of this chapter will discuss three topics relating to positive electrodes in aqueous electrolyte battery systems, the manganese dioxide electrode, the "nickel" electrode, and the so-called *memory effect* that is found in batteries that have "nickel" positive electrodes.

The first of these deals with a very common material, MnO_2, which is used in the familiar "alkaline" cells that are found in a very large number of small portable electronic devices. This electrode operates by a simple proton insertion reaction.

MnO_2 can have a number of different crystal structures, and it has been known for many years that they exhibit very different electrochemical behavior. It is now recognized that the properties of the most useful version can be explained by the presence of excess protons in the structure, whose charge compensates for that of the Mn^{4+} cation vacancies that result from the electrolytic synthesis method.

The so-called "nickel" electrode is discussed in the following section. This electrode is also ubiquitous, as it is used in several types of common batteries. Actually, this electrode is not metallic nickel at all, but a two-phase mixture of nickel hydroxide and nickel oxy-hydroxide. It is reversible, and also operates by the insertion and deletion of protons. The mechanism involves proton transport through one of the phases that acts as a solid electrolyte. The result is the translation of a two-phase interface at essentially constant potential.

The third topic in this group is a discussion of what has been a vexing problem for consumers. It occurs in batteries that have "nickel" positive electrodes. The mechanism that results in the appearance of this problem is now understood, and explains the reason for the success of the commonly used solution.

R.A. Huggins, *Advanced Batteries*: Materials Science Aspects,

11A Manganese Dioxide Electrodes in Aqueous Systems

11A.1 Introduction

Manganese dioxide, MnO_2, is the reactant that is used on the positive side of the very common *alkaline* cells that have zinc as the negative electrode material. There are several versions of MnO_2, some of which are much better for this purpose than others. Thus this matter is more complicated than it might seem at first.

MnO_2 is polymorphic, with several different crystal structures. The form found in mineral deposits has the rutile (beta) structure, and is called *pyrolusite*. It is relatively inactive as a positive electrode reactant in KOH electrolytes. However, it can be given various chemical treatments to make it more reactive. One of these produces a modification containing some additional cations that is called *birnessite*. Manganese dioxide can also be produced chemically, and then generally has the delta structure. The material that is currently much more widely used in batteries is produced electrolytically, and is called *EMD*. It has the gamma (*ramsdellite*) structure.

The reason for the differences in the electrochemical behavior of the several morphological forms of manganese dioxide presented a quandary for a number of years. It was known, however, that the electrochemically active materials contain about 4% water in their structures that can be removed by heating to elevated temperatures (100–400°C), but the location and form of that water remained a mystery. This problem was solved by Ruetschi, who introduced a cation vacancy model for MnO_2 [1,2].

The basic crystal structure of the various forms of MnO_2 contains Mn^{4+} ions in octahedral holes within hexagonally (almost) close-packed layers of oxide ions. That means that each Mn^{4+} ion has six oxygen neighbors, and these MnO_6 octahedra are arranged in the structure to share edges and corners. Differences in the edge and corner-sharing arrangements result in the various polymorphic structures.

If some of the Mn^{4+} ions are missing (cation vacancies), their missing positive charge has to be compensated by something else in the crystal structure. The Ruetschi model proposed that this charge balance is accomplished by the local presence of four protons. These protons would be bound to the neighboring oxide ions, forming a set of four OH^- ions. This local configuration is sometimes called a *Ruetschi defect*. There is very little volume change, as OH^- ions have essentially the same size as O^{2-} ions, and these species play the central role in determining the size of the crystal structure.

Reduction of the MnO_2 occurs by the introduction of additional protons during discharge, as first proposed by Coleman [3], and does produce a volume change. The charge of these added mobile protons is balanced by a reduction in the charge of some of the manganese ions present from Mn^{4+} to Mn^{3+}. Mn^{3+} ions are larger than Mn^{4+} ions, and this change in volume during reduction has been observed experimentally.

The presence of protons (or OH^- ions) related to the manganese ion vacancies facilitates the transport of additional protons as the material is discharged. This is why these materials are very electrochemically reactive.

11A.2 The Open Circuit Potential

The EMD is produced by oxidation of an aqueous solution of manganous sulfate at the positive electrode of an electrolytic cell. This means that the MnO_2 that is produced is in contact with water.

The phase relations, and the related ternary phase stability diagram, for the H–Mn–O system can be determined by use of available thermodynamic information [4,5], as discussed in previous chapters. From this information, it becomes obvious which neutral species reactions determine the potential ranges of the various phases present, and their values.

Following this approach, it is found that the lower end of the stability range of MnO_2 is at a potential that is 1.014 V vs. one atmosphere of H_2. The upper end is well above the potential at which oxygen evolves by the decomposition of water.

Under equilibrium conditions all oxides exist over a range of chemical composition, being more metal-rich at lower potentials and more oxygen-rich at higher potentials. In the higher potential case, an increased oxygen content can result from either the presence of cation (Mn) vacancies or oxygen interstitials. In materials with the rutile, and related, structures that have close-packed oxygen lattices, the excess energy involved in the formation of interstitial oxygens is much greater than that for the formation of cation vacancies. As a result, it is quite reasonable to assume that cation vacancies are present in the EMD MnO_2 that is formed at the positive electrode during electrolysis.

Because of the current that flows during the electrolytic process, the potential of the MnO_2 that is formed is actually higher than the equilibrium potential for the decomposition of water. A number of other oxides with potentials above the stability range of water have been shown to oxidize water. Oxygen gas is evolved, and they become reduced by the insertion or protons. Therefore, it is quite reasonable to expect that EMD MnO_2 would have Mn vacancies, and that there would also be protons present, as discussed by Ruetschi [1,2].

When such positive oxides oxidize water and absorb hydrogen as protons and electrons, their potentials decrease to the oxidation limit of water, 1.23 V vs. H_2 at 25°C. This is the value of the open circuit potential of MnO_2 electrodes in Zn/MnO_2 cells.

This water oxidation phenomenon that results in the insertion of protons into MnO_2 is different from the insertion of protons by the absorption of water into the crystal structure of materials that initially contain oxygen vacancies, originally discussed by Stotz and Wagner [6]. It has been shown that both mechanisms can be present in some materials [7,8].

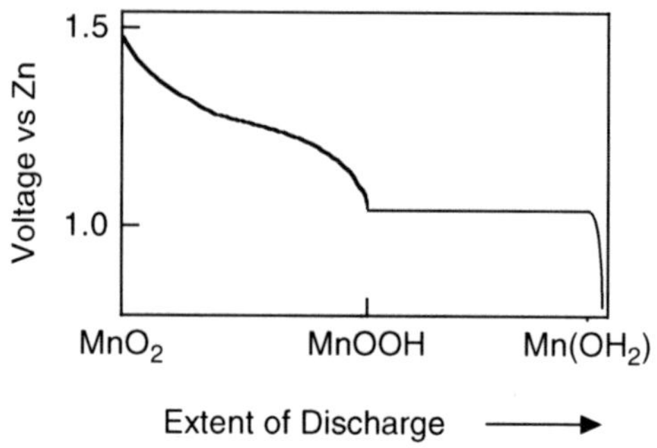

Fig. 11A.1 Schematic discharge curve of Zn/MnO_2 Cell

11A.3 Variation of the Potential During Discharge

As mentioned earlier, this electrode operates by the addition of protons into its crystal structure. This is a single-phase insertion reaction, and therefore the potential varies with the composition, as discussed in Chap. 6.

If all of the initially present Mn^{4+} ions are converted to Mn^{3+} ions, the overall composition can be expressed as $HMnO_2$ or MnOOH.

It is also possible to introduce further protons, so that the composition moves in the direction of $Mn(OH)_2$. In this case, however, there is a significant change in the crystal structure, so that the mechanism involves the translation of a two-phase interface between MnOOH and $Mn(OH)_2$. This is analogous to the main reaction involved in the operation of the "nickel" electrode, as discussed in Sect. 11B.

The sequence of these two types of reactions during discharge of a MnO_2 electrode is illustrated in Fig. 11A.1.

The second, two-phase, reaction occurs at such a low cell voltage, that the energy that is available is generally not used. Such cells are normally considered to only be useful down to about 1.2 V.

11B The "Nickel" Electrode

11B.1 Introduction

The "nickel" electrode is widely used in battery technology, e.g. on the positive side of so-called Cd/Ni, Zn/Ni, Fe/Ni, H_2/Ni, and metal hydride/Ni cells, in some cases for a very long time. It has relatively rapid kinetics and exhibits unusually good cycling behavior. This is directly related to its mechanism of operation, which involves a solid-state insertion reaction involving two ternary phases, $Ni(OH)_2$ and *NiOOH*, with no soluble product. Although the attractive properties of this electrode have led to many investigations, there are still a number of aspects of its operation that are not fully understood. This chapter will focus primarily upon the microstructural mechanism of this two-phase insertion reaction and the thermodynamic features of the ternary H–Ni–O system that determine the observed potentials.

11B.2 Structural Aspects of the $Ni(OH)_2$ and NiOOH Phases

The nanostructure of this electrode can be most simply described as a layer type configuration in which slabs of NiO_2 are separated by *galleries* in which various mobile guest species can reside. The structure of the NiO_2 layers consists of parallel sheets of hexagonally close-packed O^{2-} ions between which nickel ions occupy essentially all of the octahedral positions.

As will be described later, the mechanism of operation of this electrode involves the transition between $Ni(OH)_2$ and NiOOH upon oxidation, and the reverse upon reduction. Both of these phases are vario-stoichiometric (have ranges of stoichiometry). One can thus also describe their compositions in terms of the value of x in the general formula H_xNiO_2.

In the case of stoichiometric β $Ni(OH)_2$, the equilibrium crystal structure, which is isomorphous with *brucite*, $Mg(OH)_2$, has galleries that contain a proton concentration such that one can consider it as consisting of nickel-bonded layers of OH^- ions instead of O^{2-} ions. The nominal stoichiometry could thus be written as H_2NiO_2. Stoichiometric NiOOH has half as many protons in the galleries, and thus can be thought of as having an ordered mixture of O^{2-} and OH^- ions. Its nominal composition would then be H_1NiO_2.

When it is initially prepared, $Ni(OH)_2$ is often in the α modification, with a substantial amount of hydrogen-bonded water in the galleries. This structure is, however, not stable, and it gradually loses this water and converts to the equilibrium β $Ni(OH)_2$ structure, in which the galleries are free of water and contain only protons.

The equilibrium form of NiOOH, likewise called the β form, also has only protons in the galleries. However, there is also a γ modification of the NiOOH phase that contains water, as well as other species from the electrolyte, in the galleries. This γ modification forms at high charge rates or during prolonged overcharge in the alkali electrolyte. In both cases, the potential is quite positive. It can also be formed by electrochemical oxidation of the α $Ni(OH)_2$ phase.

One can understand the transition of the β NiOOH to the γ modification at high potentials under overcharge conditions qualitatively in terms of the structural instability of the H_xNiO_2 type phase when the proton concentration is reduced substantially. Under those conditions, the bonding between adjacent slabs will be primarily of the relatively weak van der Waals type. This allows the entry of species from the electrolyte into the gallery space. This type of behavior is commonly found in other insertion reaction materials, such as TiS_2, mentioned in Chap. 9, if the interslab forces are weak and the electrolyte species are compatible.

The general relation between these various phases is generally described in terms of the scheme presented by Bode and coworkers [9].

Table 11B.1 Interslab distances for a number of phases related to the "Nickel" electrode

Phase	Spacing (Å)
β-$Ni(OH)_2$	4.6
β-$NiOOH$	4.7
$NaNiO_2$	5.2
$Na_y(H_2O)_zNiO_2$	5.5
$Na_y(H_2O)_zCoO_2$	5.5
γ-$H_xNa_y(H_2O)_zNiO_2$	7.0
γ-$H_xK_y(H_2O)_zNiO_2$	7.0
γ'-$H_xNa_y(H_2O)_{2z}NiO_2$	9.9

A number of very good papers were published by the Delmas group in Bordeaux [10–14] that were aimed at the stabilization of the α $Ni(OH)_2$ phase by the presence of cobalt, so that one might be able to cycle between the α $Ni(OH)_2$ and γ NiOOH phases. Since both of these phases have water, as well as other species, in the galleries, they have faster kinetics than the proton-conducting γ phases, although the potential is less positive. An important feature of their work has been the synthesis of sodium analogs by solid-state preparation methods and the use of solid-state ion exchange techniques (*chimie douce* or *soft chemistry*) to replace the sodium with other species [15].

The available information concerning the interslab spacing, the critical feature of the crystallographic structure of these phases in the "nickel" electrode, is presented in Table 11B.1. It is readily seen that the crystallographic changes involved in the β $Ni(OH)_2$–β NiOOH reaction are very small, as they have almost the same value of interslab spacing. This is surely an important consideration in connection with the very good cycle life that is generally experienced with these electrodes. It can also be seen that the structural change involved in the α $Ni(OH)_2$–γ NiOOH transformation is somewhat larger. There are also differences in the slab stacking sequence in these various phases, but that factor will not be considered here.

On the one hand, both the α and β versions of the $Ni(OH)_2$ phase are predominantly ionic, rather than electronic conductors, and have a pale green color. On the other hand, the NiOOH phase is a good electronic conductor, and both the β and γ versions are black.

11B.3 Mechanism of Operation

The normal cycling reaction of commercial cells containing this electrode involves back and forth conversion between the β $Ni(OH)_2$ structure and the β NiOOH structure. It has been well established that these are separate, although vario-stoichiometric, phases, rather than end members of a continuous solid solution. The experimental evidence for this conclusion involves both X-ray measurements that show no gradual variation in lattice parameters with the extent of reaction [16], as well as similar IR observations [17] that indicate only changes in the amounts of the two separate phases as the electrode is charged or discharged.

Although the electrode potential when this two-phase structure is present is appreciably above the potential at which water is oxidized to form oxygen gas, as recognized long ago by Conway [18], gaseous oxygen evolution cannot happen if the solid electrolyte $Ni(OH)_2$ separates the water from the electronic conductor NiOOH. Oxygen evolution can only occur when the electronically conducting NiOOH phase is present on the surface in contact with the aqueous electrolyte.

Therefore, as a first approximation, one can describe the microstructural changes occurring in the electrode in terms of the translation of the $Ni(OH)_2/NiOOH$ interface. When the electrode is fully reduced, its structure consists of only $Ni(OH)_2$, whereas upon full oxidation, only NiOOH is present. This is shown schematically in Fig. 11B.1. The crystallographic transition between the $Ni(OH)_2$ and NiOOH structures, with their different proton concentrations in the galleries, is shown schematically in Fig. 11B.2.

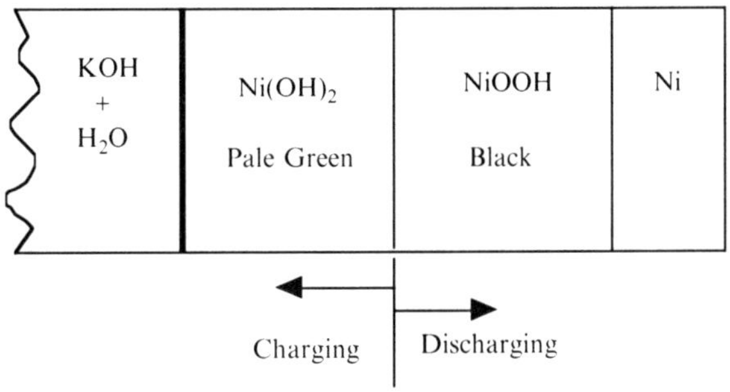

Fig. 11B.1 The microstructure of the *nickel* electrode. The major phases present are $Ni(OH)_2$, which is a proton-conducting solid electrolyte, and NiOOH, a proton-conducting mixed conductor. The electrochemical reaction takes place by the translation of the $Ni(OH)_2/NiOOH$ interface and the transport of protons through the $Ni(OH)_2$ phase

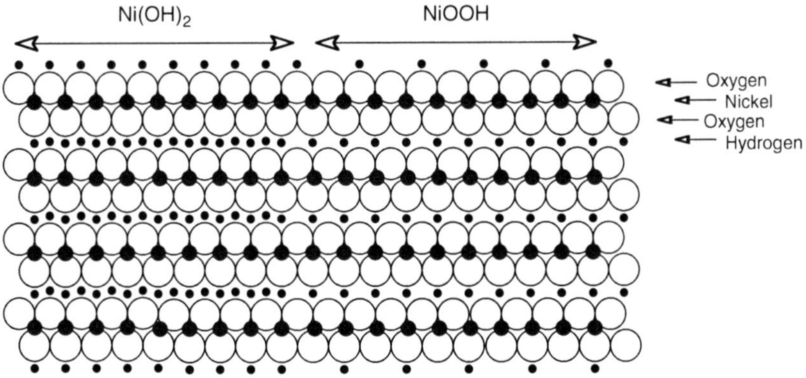

Fig. 11B.2 The crystallographic transition between the $Ni(OH)_2$ and NiOOH structures, showing the step in the proton concentration in the galleries

It has long been known [19] that the NiOOH forms first at the interface between the $Ni(OH)_2$ and the underlying electronic conductor, rather than at the electrolyte/$Ni(OH)_2$ interface. Other authors [20, 21] have observed the motion of the color boundary during charge and discharge of such electrodes.

11B.4 Relations Between Electrochemical and Structural Features

It is useful to consider the operation of this electrode in terms of the net reaction in which hydrogen is either added to or deleted from the layer structure. In the case of oxidation, this can be written as a neutral chemical reaction:

$$Ni(OH)_2 - 1/2H_2 = NiOOH. \tag{11B.1}$$

However, in electrochemical cells, this oxidation reaction takes place electrochemically, and since this normally involves an alkaline electrolyte, it is generally written in the electrochemical literature as

$$Ni(OH)_2 + OH^- = NiOOH + H_2O + e^-. \tag{11B.2}$$

However, the general rule is that electrochemical reactions take place at the boundary where there is a transition between ionic conduction and electronic conduction. Since $Ni(OH)_2$ is predominantly an ionic conductor (a solid electrolyte), the electrochemical reaction occurs at the $Ni(OH)_2$/NiOOH interface, where neither H_2O nor OH^- are present. The electrochemical reaction should therefore more properly be written as

$$Ni(OH)_2 - H^+ = NiOOH + e^-. \tag{11B.3}$$

In order for the reaction to proceed, the protons must be transported away from the interface through the galleries in the $Ni(OH)_2$ phase and into the electrolyte. However, in the alkaline aqueous electrolyte environment hydrogen is not present as either H^+ or H_2. Instead, hydrogen is transferred between the electrolyte and the $Ni(OH)_2$ phase by the interaction of neutral H_2O molecules and OH^- ions in the electrolyte with the H^+ ions at the electrolyte/$Ni(OH)_2$ interface. Thus the reaction at the electrolyte/$Ni(OH)_2$ interface must be electrically neutral and can be written as:

$$OH^-{}_{(electrolyte)} + H^+{}_{(Ni(OH)_2)} = H_2O. \tag{11B.4}$$

The equilibrium coulometric titration curve shows that under highly reducing conditions, when only the pale green $Ni(OH)_2$ phase is present throughout, there is a relatively steep potential – composition dependence. However, the fact that this part of the titration curve is not vertical indicates that there is a range of composition in this phase. It was shown some time ago that up to about 0.25 electrons (and thus 0.25 protons) per mole can be deleted from the $Ni(OH)_2$ phase before the onset of the two-phase $Ni(OH)_2/NiOOH$ equilibrium [21]. Translated to the crystallographic picture, this means that the proton concentration in the phase nominally called $Ni(OH)_2$ can deviate significantly from the stoichiometric value, up to a proton vacancy fraction of some 12.5%. The proton-deficient composition limit for the $Ni(OH)_2$ phase can thus be expressed as $H_{1.75}NiO_2$.

When both phases are present, there is a relatively long constant-potential plateau in the limit of negligible current density. This extends from the proton-deficient concentration limit in the $Ni(OH)_2$ phase ($H_{1.75}NiO_2$) to the maximum proton concentration in the NiOOH phase. According to Barnard et al. [21] this is when about 0.75 electrons (or protons) per mole are deleted from the electrode. This is equivalent to a composition of $H_{1.25}NiO_2$. Under more oxidizing conditions, when further protons are deleted, the potential of the NiOOH phase becomes more positive.

The apparent length of the constant potential two-phase plateau that is observed experimentally depends upon when the NiOOH phase reaches the electrolyte/electrode interface, and thus upon the thickness of the $Ni(OH)_2$ phase and the geometrical shape of the $Ni(OH)_2/NiOOH$ interface. The morphology of this interface, which is often not flat [22], is dependent upon several factors. As will be discussed subsequently, a flat interface is inherently unstable during the oxidation reaction. But it will tend toward a smooth shape when it translates in the reduction direction. In both cases, it will be shown that the current density is a critical parameter.

Under more oxidizing conditions, when only the NiOOH phase is present, the electrode is black and electronically conducting. This phase has wide ranges of both composition and potential. As mentioned earlier, the upper limit of proton concentration has been found to be approximately $H_{1.25}NiO_2$ for the β modification. Upon further oxidation in the NiOOH single-phase regime, the gallery proton concentration is reduced. It is generally found that the proton concentration can be substantially lower for the γ modification than in the β case. These can thus be far from the nominal composition of NiOOH.

11B.5 Self-discharge

Since the NiOOH phase is a good mixed-conductor, with a high mobility of both ionic and electronic species, equilibrium with the adjacent electrolyte is readily attained. In the absence of current through the external circuit, there will be a chemical reaction at the NiOOH surface with water in the electrolyte that results in the addition of hydrogen to the electrode. This causes a shift in the direction of a less positive potential. This increase in the hydrogen content and decrease of the potential thus results effectively in a gradual self-discharge of the electrode.

The electrochemical literature generally assumes that this self-discharge reaction involves the generation of oxygen, since the potential of the electrode is more positive than that necessary for the evolution of oxygen by the decomposition of water, as mentioned earlier.

There are two possible oxygen evolution reactions involving species in the electrolyte:

$$2NiOOH + H_2O = 2Ni(OH)_2 + 1/2O_2, \tag{11B.5}$$

which can also be written as

$$H_2O = 1/2O_2\ (into\ electrolyte\ or\ gas) + H_2\ (into\ electrode) \tag{11B.6}$$

and

$$2OH^- = H_2O + 1/2O_2 + 2e^-. \tag{11B.7}$$

However, the latter does not provide any hydrogen to the electrode, and thus cannot contribute to self-discharge. Instead, it is the electrochemical oxygen evolution reaction, involving passage of current through the outer circuit, as mentioned later.

The rate of the self-discharge reaction can be simply measured for any value of electrode potential in the single-phase NiOOH regime, where the potential is state-of-charge dependent by using a potentiostat to hold the potential at a constant value, and measuring the anodic current through the external circuit that is required to maintain that value of the potential (and thus also the corresponding proton concentration in the electrode). This is the opposite of the self-discharge process, and can be written as

$$2Ni(OH)_2 + OH^- = 2NiOOH + H_2O + e^-. \tag{11B.8}$$

Measurements of the self-discharge current as a function of potential in the NiOOH regime for the case of electrodes produced by two different commercial manufacturers are shown in Fig. 11B.3. The differences between the two curves are not important, as they are related to differences between the microstructures of the two electrodes.

If anodic current is passed through the NiOOH electrode, part will be used to counteract the self-discharge mentioned earlier. If the magnitude of the current is

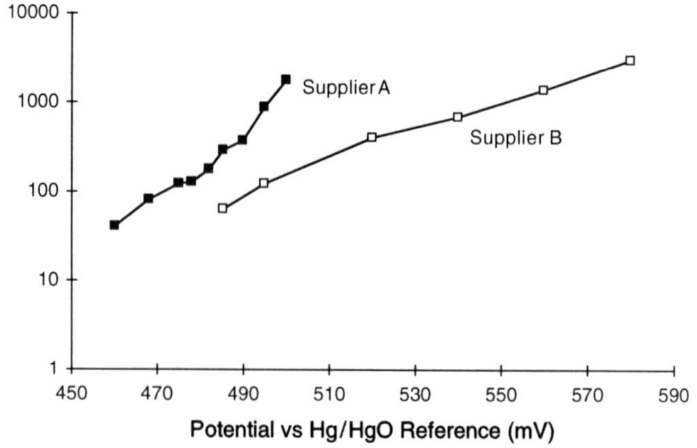

Fig. 11B.3 Self-discharge current as a function of potential in the NiOOH regime measured on electrodes produced by two different commercial manufacturers

greater than the self-discharge current, additional protons will be removed from the electrode's crystal structure, making the potential more positive. This results in an increased rate of self-discharge. Thus a steady-state will evolve in which the applied current will be just balanced by the rate of self-discharge and the proton concentration in the galleries will reach a new steady (lower) value.

11B.6 Overcharge

If the applied current density exceeds that which can be accommodated by the kinetics of the compositional change and the self-discharge process, another mechanism must come into play. This is the direct generation of oxygen gas at the electrolyte/NiOOH interface by the decomposition of water in the electrolyte. This can be described by the reaction

$$2OH^- = H_2O + 1/2O_2 + 2e^- \tag{11B.9}$$

in which the electrons go into the current collector.

The relationship between the potential of the "nickel" electrode and the amount of hydrogen that is deleted when it is charged (oxidized) is shown schematically in Fig. 11B.4.

11B.7 Relation to Thermodynamic Information

The available thermodynamic data relating to the various phases in the H–Ni–O system can be used to produce a ternary phase stability diagram. From this information, one can also readily calculate the potentials of the various possible stable phase combinations. This general methodology [23–26] has been used with great success to understand the stability windows of a number of electrolytes, as well as the potential-composition behavior of many electrode materials in lithium, sodium, oxide, and other systems.

With this information, the microstructural model discussed earlier, and the available information about the stoichiometric ranges of the important phases, one should be able to explain the observed electrochemical behavior of the "nickel" electrode.

Unfortunately, reliable thermodynamic information for this system is rather scarce. The data that have been used are included in Table 11B.2, taken mostly from the compilation in [27]. Unfortunately, no recognition was given to the question of stoichiometry or the ordered/disorder state of crystal perfection, or even to the differences between the α and β structures of $Ni(OH)_2$ and the β and γ structures of NiOOH. Therefore, the calculated potentials can only be considered semiquantitative at present.

The results of these calculations are shown in the partial ternary phase stability diagram of Fig. 11B.5, in which all phases are assumed to have their nominal

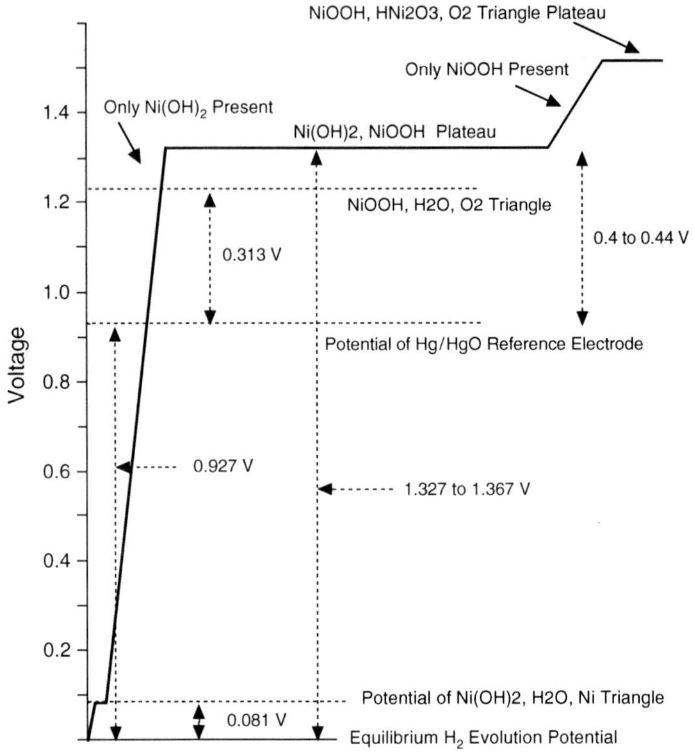

Fig. 11B.4 Variation of the potential as hydrogen is removed from the nickel electrode during oxidation when the cell is being charged

Table 11B.2 Thermodynamic data

Phase	$\Delta G_f{}^\circ$ (25 °C)
NiO	−211.5
NiOOH	−329.4
$Ni(OH)_2$	−458.6
H_2O	−237.14

compositions. The two-phase tie lines and three-phase triangles indicate the phases that are stable in the presence of each other at ambient temperatures. Also shown are the potentials of the various relevant three-phase equilibria vs. the hydrogen evolution potential at one atmosphere. The composition of the electrode during operation on the main plateau follows along the heavy line that points toward the hydrogen corner of the diagram and lies on the edge of the triangle in which all compositions have a potential of 1.339 V vs. hydrogen.

Parts of this figure are incomplete, for the data available did not indicate what happens when additional hydrogen is removed from $HNiO_2$ and the potential exceeds 1.339 V at the time that it was first written [28]. It was obvious, of course,

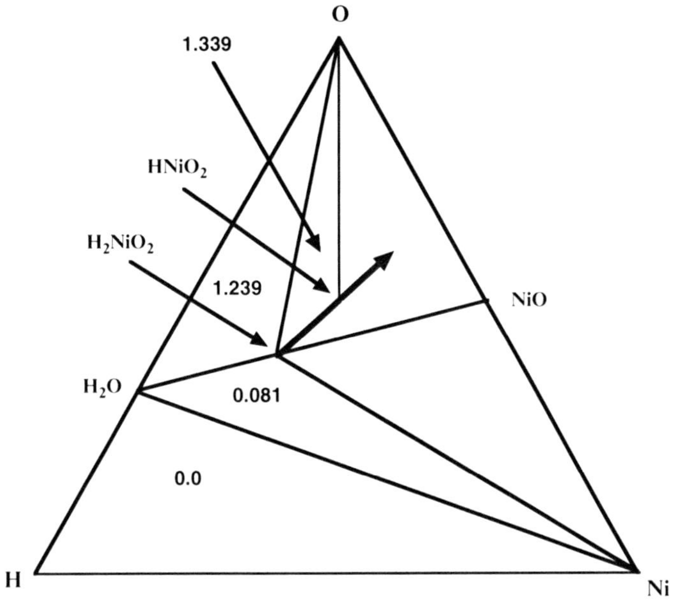

Fig. 11B.5 Partial Gibbs triangle. The main charge–discharge reaction takes place along the *thick line* at 1.339 V vs. H_2. The overall composition moves along that line upon further charging

that the composition follows further along the arrow, but the species at the corners of the subtriangle that has the observed higher potential could not be identified. Subsequent information that led to an explanation of the so-called *memory effect* is included in Sect. 11C.

One important question is whether there is a stable tie line between NiOOH and H_2O. The alternative is a tie line between $Ni(OH)_2$ and oxygen. Only one of these can be stable, as tie lines cannot cross. The Gibbs free energy change involved in the determining reaction can be calculated

$$2NiOOH + H_2O = 2Ni(OH)_2 + 1/2O_2. \qquad (11B.10)$$

This is found to be -21.26 kJ at 25 °C, which means that the NiOOH$-H_2O$ tie line is not stable, and that there is a stable three-phase equilibrium involving $Ni(OH)_2$, NiOOH, and oxygen. A situation in which both $Ni(OH)_2$ and NiOOH are in contact with water can only be metastable.

It is possible to calculate the potential of the $Ni(OH)_2$, NiOOH, O_2 triangle relative to the one atmosphere hydrogen evolution potential from the reaction

$$1/2H_2 + NiOOH = Ni(OH)_2 \qquad (11B.11)$$

that is the primary reaction of the "nickel" electrode, as discussed earlier, since this reaction occurs along one of the sides of this three-phase equilibrium triangle. From the data in Table 11B.2, it can be determined that the Gibbs free energy change $\Delta G_r{}^\circ$ accompanying this reaction is -129.2 kJ mol^{-1}. From the relation

$$\Delta G_r^\circ = -zFE \tag{11B.12}$$

since z is unity for this reaction, the equilibrium potential is 1.34 V positive of the hydrogen evolution potential.

Since the potential of a Hg/HgO reference electrode is 0.93 V positive of the reversible hydrogen evolution potential (RHE), this calculation predicts that the equilibrium value of the two-phase constant potential plateau for the main reaction of the nickel electrode should occur at about 0.41 V positive of the Hg/HgO reference. This is also about 0.11 V more positive than the equilibrium potential of oxygen evolution from water.

This result can be compared with the experimental information from Barnard et al. [21] on the potentials of both the *activated* (highly disordered) and *deactivated* (more highly ordered) β $Ni(OH)_2$–β NiOOH reaction. Their data fell in the range 0.44–0.47 V positive of the Hg/HgO electrode potential. They found the comparable values for the a $Ni(OH)_2$–γ NiOOH reaction to be in the range 0.39–0.44 V relative to the Hg/HgO reference. Despite the lack of definition of the structures to which the thermodynamic data relate, this should be considered to be a quite good correlation.

Further oxidation causes the electrode composition to move along the arrow further away from the hydrogen corner of the ternary diagram, and leads to an electrode structure in which the $Ni(OH)_2$ phase is no longer stable, as is found experimentally. The potential moves to more positive values as the stoichiometry of the NiOOH phase changes, and if no other reaction interferes, should eventually arrive at another, higher, plateau in which the lower proton concentration limit NiOOH is in equilibrium with some other phase or phases.

Another complicating fact is that electrolyte enters the β NiOOH at high potentials, converting it to the γ modification. As mentioned earlier, the water-containing α $Ni(OH)_2$ and γ NiOOH phases are not stable, and during normal cycling are gradually converted to the corresponding β phases that have only protons in their galleries. When these metastable phases are present the electrode potential of the reaction plateau is less positive, as is characteristic of insertion structures with larger interslab spacings. Correspondingly, the apparent capacity of the electrode prior to rapid oxygen evolution is greater. These several factors are discussed further in the next section.

11C Cause of the Memory Effect in "Nickel" Electrodes

11C.1 Introduction

It is often found that batteries with "nickel" positive electrodes, e.g. Cd/Ni, hydride/Ni, Zn/Ni, Fe/Ni, and H_2/Ni cells, have a so-called *memory effect*, in which the available capacity apparently decreases if they are used under conditions in which they are repeatedly only partially discharged before recharging. In many cases, these batteries are kept connected to their chargers for long periods of time. It is also widely known that this problem can be "cured" by subjecting them to a slow, deep discharge.

The phenomena that take place in such electrodes have been studied by many investigators over many years, but no rational and consistent explanation of the *memory effect* related to "nickel" electrodes emerged until recently. Although it has important implications for the practical use of such cells, some of the major reviews in this area do not even mention this problem, and others give it little attention and/or no explanation.

In studying this apparent loss of capacity, a number of investigators have shown that a second plateau appears at a lower potential during discharge of "nickel" electrodes [29–44]. Importantly, it is found that under low current conditions the total length of the two plateaus remains constant. As the capacity on the lower one, sometimes called *residual capacity*, becomes greater, the capacity of the higher one shrinks. The relative lengths of the two plateaus vary with the conditions of prior charging. This is shown schematically in Fig. 11C.1.

Since the capacity of the lower plateau is at about 0.78 V positive of the reversible hydrogen electrode potential, it is generally not useful for most of the applications for which "nickel" electrode batteries are employed. The user does not see this capacity, but instead, sees only the dwindling capacity on the upper plateau upon discharge. Thus, it is quite obvious that the appearance of this lower plateau and reduction in the length of the upper plateau is an important component of the memory effect.

It is also found that this lower plateau and the *memory effect* both disappear if the cell is deeply discharged. Thus the existence of the lower plateau and its disappearance are both obviously related to the *curing* of the memory effect.

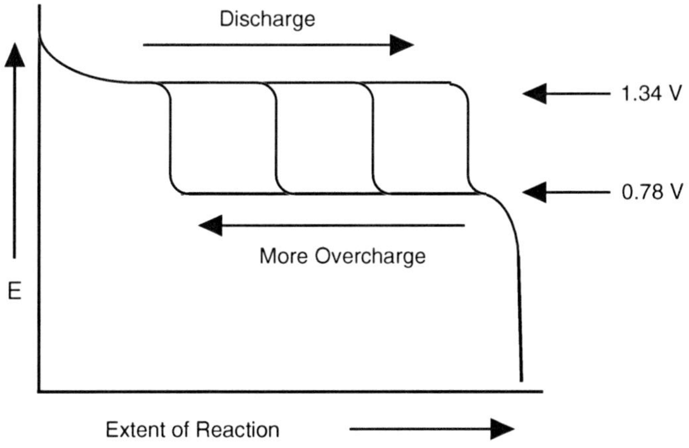

Fig. 11C.1 Schematic representation of the two discharge plateaus. With increasing overcharge the length of the upper one decreases and the lower one increases

These phenomena can now be explained on the basis of available thermodynamic and structural information by using the ternary Gibbs phase stability diagram for the H–Ni–O system as a thinking tool [45, 46].

11C.2 Mechanistic Features of the Operation of the "Nickel" Electrode

The microscopic mechanism of the basic operation of these electrodes was discussed in Sect. 11B. However, it is important for this discussion, and will be briefly reviewed here. It involves an insertion reaction that results in the translation of a two-phase interface between H_2NiO_2 and $HNiO_2$, both of which are vario-stoichiometric (have ranges of stoichiometry). The H_2NiO_2 is in contact with the alkaline electrolyte, and the $HNiO_2$ is in contact with the metallic current collector. The outer layer of the H_2NiO_2 phase is pale green and is predominantly an ionic conductor, allowing the transport of protons to and from the two-phase $H_2NiO_2/HNiO_2$ boundary. $HNiO_2$ is a good electronic conductor, and is black. The electrochemical reaction takes place at that ionic/electronic two-phase interface. This boundary is displaced as the reaction proceeds, and the motion of the color boundary has been experimentally observed. When the electrode is fully reduced, its structure consists of only H_2NiO_2, whereas oxidation causes the interface to translate in the opposite direction until only $HNiO_2$ is present. Although these are both ternary phases, the only compositional change involves the amount of hydrogen present, and the structure of the host "NiO_2" does not change. Thus this is a pseudo-binary reaction, although it takes place in a ternary system, and the potential is independent of the overall composition; i.e. the state of charge.

Once the H_2NiO_2 has been completely consumed, and the $HNiO_2$ phase comes into contact with the aqueous electrolyte it is possible to obtain further oxidation. This involves a change in the hydrogen content of the $HNiO_2$ phase. The variation of the composition of this single phase results in an increase in the potential from this two-phase plateau to higher values, as is expected from the Gibbs phase rule.

After the low-hydrogen limit of the composition of the $HNiO_2$ phase is reached, further oxidation can still take place. Another potential plateau is observed, and oxygen evolution occurs. This is often called *overcharging*, and obviously involves another process.

A number of authors have shown that the length of the lower plateau observed upon discharging is a function of the amount of the γ NiOOH phase formed during overcharging [38]. However, other authors [40] have shown that it is possible to prevent the formation of the γ phase during overcharging by using a dilute electrolyte. Yet the lower discharge potential plateau still appears. There is also evidence that the γ phase can disappear upon extensive overcharging, but the lower discharge plateau is still observed [38].

Neutron diffraction studies [43], which see only crystalline structures, showed a gradual transition between the γ and β NiOOH structures upon discharge, with no discontinuity at the transition between the upper and lower discharge plateaus. There was no evidence of a change in the compositions of either of the two phases, just a variation in their amounts, which changed continuously along both discharge plateaus. These authors attributed the presence of the lower plateau to undefined "technical parameters".

Several other authors have explained the presence of the lower discharge plateau in terms of the formation of some type of barrier layer [30,36], and there is evidence for the formation of β H_2NiO_2, which is not electronically conducting, on the lower plateau [41]. This can, of course, be interpreted as a barrier.

These studies all seem to assume that the oxygen that is formed during operation upon the upper plateau during charging comes only from decomposition of the aqueous electrolyte. However, something else is obviously happening that leads to the formation of the lower plateau that is observed upon discharge. It must also relate to a change in the amounts, compositions, or structure of the solid phase, or phases, present.

Although the electrochemical behavior of the "nickel" electrode upon the lower potential plateau can be understood in terms of a pseudo-binary insertion/extraction hydrogen reaction, the evolution of oxygen and the formation of the second discharge plateau indicate that the assumption that the oxygen comes from (only) the electrolysis of the aqueous electrolyte during overcharge cannot be correct. To understand this behavior, recognition must be given to the fact that the evolution of oxygen indicates that at this potential this electrode should be treated in terms of the ternary H–Ni–O system, rather than as a simple binary phase reaction.

Use of the Gibbs triangle as a *thinking tool* to understand the basic reactions in the H–Ni–O system has been discussed in several places [47–49]. The major features of the lower-potential portion of this system can be readily determined from available thermodynamic information. A major part of the Gibbs phase stability triangle for this system is shown in Fig. 11C.2, copied from Sect. 11B.

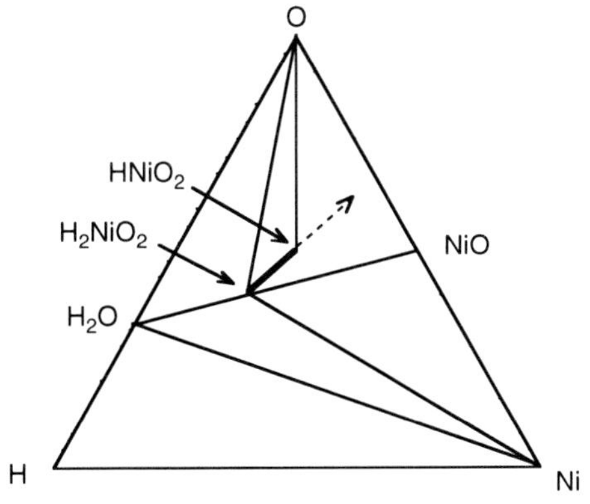

Fig. 11C.2 Partial Gibbs triangle. The main charge–discharge reaction takes place along the *thick line*. The overall composition moves along the *dashed line* upon further charging

Since the two phases H_2NiO_2 and $HNiO_2$ are on a tie line that points to the hydrogen corner, neither hydrogen insertion nor deletion involves any change in the Ni/O_2 ratio, and this can be considered to be a pseudo-binary reaction. The tie line between those two phases is one side of a triangle that has pure oxygen at its other corner. This means that both of these phases are stable in oxygen, as is well known.

As the result of the Gibbs phase rule, movement of the overall composition along this tie line occurs at a constant potential plateau. It was shown in Sect. 11B that the potential of this plateau is 1.34 V vs. pure hydrogen at 25 °C.

Thus the equilibrium electrode potential of the basic $H_2NiO_2-HNiO_2$ reaction is not only composition-independent, but also more positive than the potential of the decomposition of water, as is experimentally observed. Also, because the H_2NiO_2 that is between the $HNiO_2$ and the water is a solid electrolyte, there is little or no oxygen evolution.

As additional hydrogen is removed the potential moves up the curve where only $HNiO_2$ is present. When the overall composition exceeds the stability range of that phase it moves further from the hydrogen corner and enters another region in the phase diagram, as indicated by the dashed line in Fig. 11C.2.

11C.3 Overcharging Phenomena

The potential then moves along the upper charging (or overcharging) plateau. Since all of the area within a Gibbs triangle must be divided into subtriangles, the overall composition must be moving into a new subtriangle. One corner of this new triangle must be $HNiO_2$, and another must be oxygen. This is consistent with the observation that oxygen is evolved at this higher charging potential. The question is then, what is the composition of the phase that is at the third corner?

If gaseous oxygen is evolved from the electrode, not just from decomposition of the water, the third-corner composition must be below (i.e. have less oxygen) than all compositions along the dashed line.

One possibility might be the phase Ni_3O_4, another could be NiO. However, neither of these phases, which readily crystallize, has been observed. There must be another phase with a reduced ratio of oxygen to nickel.

Although evidently not generally recognized by workers interested in the "nickel" electrode, it has been found [50–52] that a phase with a composition close to HNi_2O_3 can be formed under conditions comparable to those during charging the electrode on the upper voltage plateau. This phase can form as an amorphous product during the oxidation of $HNiO_2$. Its crystal structure and composition were determined after hydrothermal crystallization. In addition, the mean nickel oxidation state was found by active oxygen analysis to be only 2.65.

The composition HNi_2O_3 lies on a line connecting $HNiO_2$ and NiO. This would then lead to a subtriangle as shown in Fig. 11C.3, which meets the requirement that there be another phase in equilibrium with both $HNiO_2$ and oxygen that has a reduced ratio of oxygen to nickel.

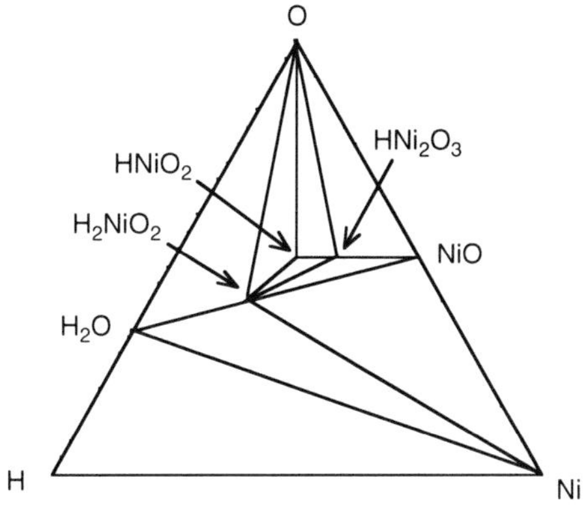

Fig. 11C.3 Gibbs triangle showing the presence of the HNi_2O_3 phase

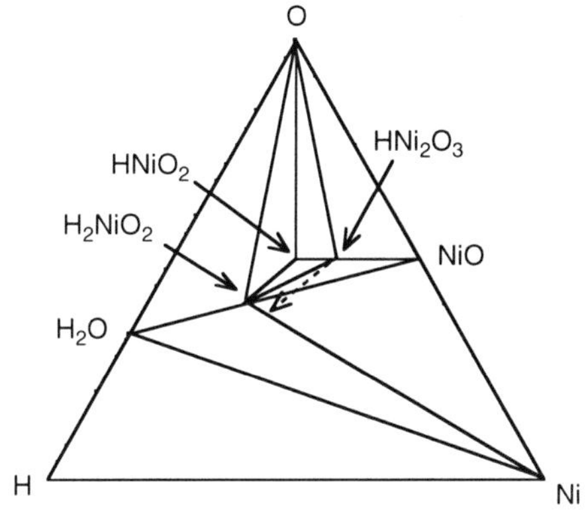

Fig. 11C.4 Composition path during the discharge of the HNi_2O_3 formed during overcharge

The gradual formation of amorphous HNi_2O_3 during oxygen evolution upon the upper overcharging plateau, and its influence upon behavior during discharge, is the key element in the *memory effect* puzzle.

As overcharge continues, oxygen is evolved, and more and more of the HNi_2O_3 phase forms. Thus the overall composition of the solid gradually shifts along the line connecting $HNiO_2$ and HNi_2O_3.

Upon discharge, the overall composition moves in the direction of the hydrogen corner of the Gibbs triangle. This is indicated by the dashed line in Fig. 11C.4.

It is seen that the HNi_2O_3 portion of the total solid moves into a different subtriangle that has H_2NiO_2, HNi_2O_3, and NiO at its corners. From the available thermodynamic data, one can calculate that the potential in this subtriangle is 0.78 V vs. hydrogen. That is essentially the same as experimentally found for the lower discharge plateau. The larger the amount of HNi_2O_3 that has been formed during overcharging, the longer the corresponding lower discharge plateau will be. The upper discharge plateau becomes correspondingly shorter.

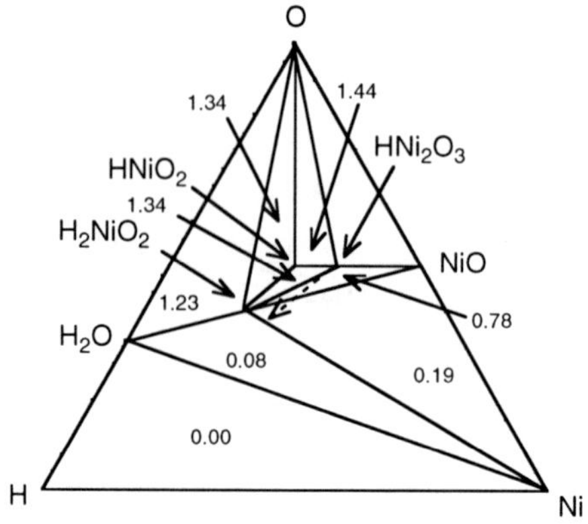

Fig. 11C.5 Calculated values of the potential in the various subtriangles in the H−Ni−O ternary system vs. hydrogen. It is seen how the composition path during discharge of HNi_2O_3 leads to the observation of the lower discharge plateau at about 0.78 V, and the disappearance of that phase when the potential moves to a much lower value

After traversing this triangle, the overall composition of what had been HNi_2O_3 moves into another subtriangle that has H_2NiO_2, NiO, and Ni at its corners. The HNi_2O_3 disappears, and the major product is H_2NiO_2. The potential in this subtriangle can be calculated to be 0.19 V vs. hydrogen.

If the electrode is now recharged, its potential does not go back up to the 0.8 V plateau, since HNi_2O_3 is no longer present, but goes to the potential for the oxidation of its major component, H_2NiO_2. The overall composition again moves away from the hydrogen corner, and the H_2NiO_2 loses hydrogen and gets converted to $HNiO_2$. This is the standard charging cycle low potential plateau. This also means that the lower reduction plateau is no longer active, for the HNi_2O_3 has disappeared, and the *memory effect* has been *cured*.

The calculated potentials in the various subtriangles of the H−Ni−O system are shown in Fig. 11C.5.

The reactions in the H−Ni−O system obviously have very rapid kinetics, for this electrode can be both charged and discharged at high rates. Therefore, it is quite reasonable to expect the phases present to be at or near their equilibrium amounts and compositions. This is indicated by the very good correlation between experimental results and the information obtained by the use of ternary phase stability diagrams based upon the available thermodynamic data.

11C.4 Conclusions

The basic mechanisms that are involved in causing the *memory effect* have been identified. The key element is the formation of an amorphous HNi_2O_3 phase upon overcharging into the potential range where oxygen is evolved. Upon subsequent reduction, the presence of this phase produces the potential plateau at about 0.8 V vs. hydrogen, reducing the available capacity at the normal higher reduction potential. The more the overcharge, the more HNi_2O_3 that is formed, and the longer the lower plateau. If the electrode undergoes further reduction this phase disappears, and the potential drops to a much lower value. Subsequent charging of the electrode brings the composition back to the initial state, and the *memory effect is cured.*

This model provides an understanding of the main features of the *memory effect*, and also explains the several confusing and apparently contradictory observations in the literature. It is expected that further experimental work will address this matter. Additional confirmation of the presence of HNi_2O_3 in the microstructures of overcharged electrodes would be especially useful.

The implication from this mechanism is that the major reason for the *memory effect*, a decrease in the capacity at the normal discharge potential, is related to extensive overcharging, rather than to the use of shallow discharge cycles.

References

1. P. Ruetschi, J. Electrochem. Soc. 131, 2737 (1984)
2. P. Ruetschi and R. Giovanoli, J. Electrochem. Soc. 135, 2663 (1988)
3. J.J. Coleman, Trans. Electrochem. Soc. 90, 545 (1946)
4. I. Barin, *Thermochemical Data of Pure Substances*, VCH, New York (1995)
5. M. Pourbaix, *Atlas of Electrochemical Equilibria*, Pergamon Press, New York (1966)
6. S. Stotz and C. Wagner, Ber. Bunsenges. Phys. Chem. 70, 781 (1966)
7. A. Netz, W.F. Chu, V. Thangadurai, R.A. Huggins and W. Weppner, Ionics 5, 426 (1999)
8. R.A. Huggins, J. Power Sources 153, 365 (2006)
9. H. Bode, K. Dehmelt and J. Witte, Electrochim. Acta 11, 1079 (1966)
10. P. Oliva, et al., J. Power Sources 8, 229 (1982)
11. C. Faure, et al., J. Power Sources 35, 249 (1991)
12. C. Faure, C. Delmas and P. Willmann, J. Power Sources 35, 263 (1991)
13. C. Faure, C. Delmas and M. Fouassier, J. Power Sources 35, 279 (1991)
14. C. Delmas, in *Solid State Ionics II*, G.-A. Nazri, D. F. Shriver, R. A. Huggins and M. Balkanski, Eds., Materials Research Society, Warrendale, PA (1991), p. 335
15. C. Delmas, et al., Solid State Ionics 28–30, 1132 (1988)
16. G.W.D. Briggs, E. Jones and W.F.K. Wynne-Jones, Trans. Faraday Soc. 51, 394 (1955)
17. F.P. Kober and J. Electrochem. Soc. 112, 1064 (1965)
18. B.E. Conway and P.L. Bourgault, Can. J. Chem. 37, 292 (1959)
19. E.M. Kuchinskii and B.V. Erschler, J. Phys. Chem. (USSR) 14, 985 (1940)
20. G.W.D. Briggs and M. Fleischmann, Trans. Faraday Soc. 67, 2397 (1971)
21. R. Barnard, C.F. Randell and F.L. Tye, J. Appl. Electrochem. 10, 109 (1980)
22. R.W. Crocker and R.H. Muller, Presented at the Meeting of the Electrochemical Society, Toronto (1992)
23. R.A. Huggins, in *Fast Ion Transport in Solids*, P.P. Vashishta, J.N. Mundy and G. K. Shenoy, Eds., North-Holland, Amsterdam (1979), p. 53
24. W. Weppner and R.A. Huggins, Solid State Ionics 1, 3 (1980)
25. N.A. Godshall, I.D. Raistrick and R.A. Huggins, Mater. Res. Bull. 15, 561 (1980)
26. N.A. Godshall, I.D. Raistrick and R.A. Huggins, J. Electrochem. Soc. 131, 543 (1984)
27. J. Balej and J. Divisek, Presented at the Meeting of the Bunsengesellschaft, Wien (1992).
28. R.A. Huggins, M. Wohlfahrt-Mehrens and L. Jörissen, Presented at Symposium on Intercalation Chemistry and Intercalation Electrodes, Meeting of the Electrochemical Society, Hawaii (1993)

29. P.C. Milner and U.B. Thomas, in *Advances in Electrochemistry and Electrochemical Engineering*, C.W. Tobias, Ed., Wiley-Interscience, New York (1967), p. 1
30. R. Barnard, G.T. Crickmore, J.A. Lee and F.L. Tye, J. Appl. Electrochem. 10, 61 (1980)
31. B. Klapste, K. Mickja, J. Mrha and J. Vondrak, J. Power Sources 8, 351 (1982)
32. A.H. Zimmerman and P.K. Effa, J. Electrochem. Soc. 131, 709 (1984)
33. H.S. Lim and S.A. Verzwyvelt, J. Power Sources 22, 213 (1988)
34. H. Vaidyanathan, J. Power Sources 22, 221 (1988)
35. J. McBreen, Mod. Aspects Electrochem. 21, 29 (1990)
36. A.H. Zimmerman, in *Nickel Hydroxide Electrode*, Vol. 90–94, D.A. Corrigan and A.H. Zimmerman, Eds., Electrochem. Soc. Proc., The Electrochemical Society, Pennington, NJ (1990), p. 311
37. A.H. Zimmerman, Proc. IECEC 4, 63 (1994)
38. P. Wilde, Ph.D. Thesis, University of Ulm, Germany (1996)
39. N. Sac-Epee, M.R. Palacín, B. Beaudoin, A. Delahaye-Vidal, T. Jamin, Y. Chabre and J.-M. Tarascon, J. Electrochem. Soc. 144, 3896 (1997)
40. N. Sac-Epee, M.R. Palacìn, A. Delahaye-Vidal, Y. Chabre and J.-M. Tarâscon, J. Electrochem. Soc. 145, 1434 (1998)
41. C. Leger, C. Tessier, M. Ménétrier, C. Denage and C. Delmas, J. Electrochem. Soc. 146, 924 (1999)
42. F. Fourgeot, S. Deabate, F. Henn and M. Costa, Ionics 6, 364 (2000)
43. S. Deabate, F. Fourgeot and F. Henn, Ionics 6, 415 (2000)
44. F. Barde, M.R. Palacin, Y. Chabre, O. Isnard and J.-M. Tarascon, Chem. Mater. 16, 3936 (2004)
45. R.A. Huggins, Solid State Ionics 177, 2643 (2006)
46. R.A. Huggins, J. Power Sources 165, 640 (2007)
47. R.A. Huggins, M. Wohlfahrt-Mehrens and L. Jörissen, Presented at Meeting of the Electrochemical Society, Hawaii (1992)
48. R.A. Huggins, M. Wohlfahrt-Mehrens and L. Jörissen, in *Solid State Ionics III*, G.-A. Nazri, J.-M. Tarascon and M. Armand, Eds., Materials Research Society Proceedings 293, Pittsburgh, PA (1993), p. 57
49. R.A. Huggins, H. Prinz, M. Wohlfahrt-Mehrens, L. Jörissen and W. Witschel, Solid State Ionics 70/71, 417 (1994)
50. C. Greaves, M.A. Thomas and M. Turner, Power Sources 9, 163 (1983)
51. C. Greaves, A.M. Malsbury and M.A. Thomas, Solid State Ionics 18/19, 763 (1986)
52. A.M. Malsbury and C. Greaves, J. Solid State Chem. 71, 418 (1987)

Chapter 12
Other Topics Related to Electrodes

12.1 Introduction

This chapter will include discussions of three additional topics related to electrodes: Electrodes into Which Both Cations and Anions Can Be Inserted (Sect. 12A) Flow Batteries with Liquid Electrodes (Sect. 12B), and Reactions in Fine Particle Electrodes (Sect. 12C).

12A Mixed-Conducting Host Structures into Which Either Cations or Anions Can Be Inserted

12A.1 Introduction

As mentioned in Chap. 6, it has been known for some 25 years that charge can be stored in some important battery electrodes by the insertion of ionic species from the electrolyte. Insertion reactions play an especially important role in current versions of lithium batteries, where lithium cations are typically the inserted species in both electrodes. Hydrogen cations (protons) are the guest species during the operation of both the $Ni(OH)_2/NiOOH$ electrode and the "MnO_2" electrode in aqueous systems.

Whereas most attention has been given to materials in which the guest species are cations, it is also possible to have anion insertion into some crystal structures. Materials in which the structure can accommodate *either* cations or anions are especially interesting.

There are also some cases in which *both* cations and anions can be inserted into a crystal structure. One example of this type will be briefly discussed in this chapter, ternary materials with the hexagonal transition metal bronze structure.

Most of the attention will be given to the *hexacyanometallates*, however. These materials have structures which are variants of the cubic ReO_3 type structure which have rather large intercell windows. They can accommodate a wide variety of guest ions of both charges. Cations can be inserted into the structure at relatively low potentials, and anions can be inserted at more positive potentials. This can lead to a number of interesting features and properties.

12A.2 Insertion of Species into Materials with Transition Metal Oxide Bronze Structures

It has also long been known that a number of ions can be readily inserted into the structures of ternary oxides such as the tungsten, molybdenum and vanadium bronzes. These bronze families can exist in several different crystal structures, depending upon the identity and concentration of the lower-charge cations present. If the inserted ions are relatively small, these materials often have the cubic ReO_3 structure.

Especially interesting, however, are materials with the hexagonal tungsten bronze structure with the general formula M_xWO_3, in which there are two types of crystallographic tunnels. There is a hexagonal array of rather large linear tunnels that penetrate through the structure parallel to the *c*-axis. This structure is only obtained when M is a large cation, such as K^+, Rb^+, or Cs^+. An example is K_xWO_3, where $x = 0.3$ and the K^+ ions partly occupy the positions in the large tunnels. It can be readily prepared by either solid state or electrochemical methods at elevated temperatures. This structure is shown schematically in Fig. 12A.1.

This material is dark blue-black, due to the presence of both W^{5+} and W^{6+} species. If it is heated to intermediate temperatures (e.g., 400°C) in air, O^{2-} anions are introduced into the large tunnels. These balance the charge of the K^+ ions,

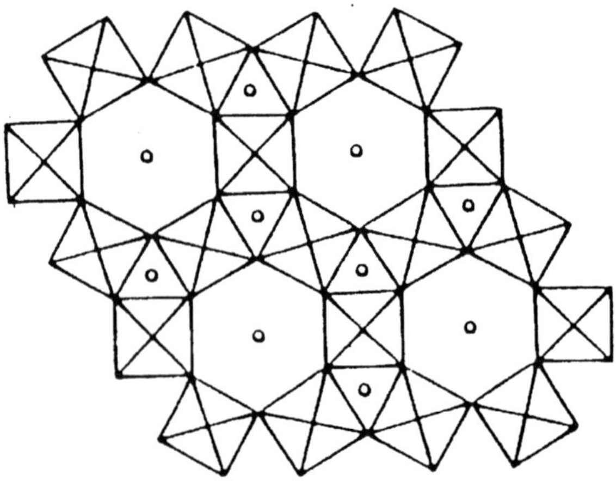

Fig. 12A.1 Representation of the hexagonal tungsten bronze structure in the *c*-direction, *small circles* show the presence of both large and small tunnel sites

causing all the tungsten ions to become W^{6+}. The material is thus bleached, becoming white.

Li^+ cations can subsequently be inserted into this structure at room temperature. They go into the set of smaller tunnels that are oriented in the cross-direction, rather than into the large tunnels. The presence of the lithium ions causes the reduction of some of W^{6+} ions to W^{5+}, and the material becomes dark. Thus the low temperature insertion of Li^+ cations, which is both very rapid and reversible, can be employed to make this an interesting electrochromic material [1–4].

12A.3 Materials with Cubic Structures Related to Rhenium Trioxide

Another crystal structure that can have interesting properties is the ReO_3 (or BX_3) structure, which has cubic symmetry. This structure can be thought of as a simple cubic arrangement of corner-shared octahedral BX_6 groups. There is empty space in the center of each of these cubes that is interconnected by a three-dimensional set of tunnels through the centers of the cube faces. It is possible for this cube-center space to be either empty, partly, or fully occupied by cations, assuming that the charges of the other species present are adjusted so as to maintain overall charge neutrality.

If there is a cation in the center of every cube, the nominal formula is then ABX_3, and this is the well-known perovskite structure, which is adopted by many oxides. In order for it to be stable, the A ions must be relatively large. The more highly charged B ions are quite small.

It has been shown that a variety of ions can reside on the A sites of the perovskite structure. In addition, there may be mixed occupation by more than one type of ion. An example of this is the family of Li–La titanates, in which Li ions, lanthanum ions, and vacancies are distributed among the available A sites. It has been shown that these materials can have a relatively high lithium ionic conductivity at positive potentials [5–9], and that they are also interesting fast mixed-conductors at more negative potentials [10].

12A.4 Hexacyanometallates

There is a family of materials with crystal structures that are analagous to the *BX_3 rhenium trioxide* and *ABX_3 perovskite* materials, but in which the X positions are occupied by cyanide anions, which are appreciably larger than oxide ions. These materials are sometimes called hexacyanometallates, and the B positions are often occupied by transition metal ions. The transition metal hexacyanometallates are examples of the large family of insoluble mixed-valence compounds. An overview of these materials can be found in the extensive review by Robin and Day [11].

The prototype material is "Prussian blue," which is also sometimes called "Berlin blue." Its nominal formula is $KFe_2(CN)_6$, or $K_{0.5}Fe(CN)_3$. It has a dark blue-black color, has been known for a very long time, and has been widely used as a dyestuff. It was evidently the first coordination compound reported in the scientific literature [12]. An account of the early work on the preparation and chemical composition of materials in the Prussian blue family can be found in [13]. They have been studied extensively because of their electrochromic properties, and there has been renewed interest in them in recent years in connection with their use in "modified electrode surfaces" that are interesting for catalytic purposes.

The general formula for materials in this family is A_xP^{3+} R^{2+} $(CN)_6$, where the P^{3+} and R^{2+} species are distributed in an ordered arrangement upon the B sites of the A_xBX_3 structure. The value of x, which specifies the amount of A, can often be varied from 0 to 2. When $x = 0$ the material has the ReO_3 (BX_3) structure, and when $x = 2$, the structure is analogous to the ABX_3 perovskites. In the case of Prussian blue, the A sites are half full, and x is nominally equal to 1.

The structure of Prussian blue was first discussed by Keggin and Miles in 1936 [14], on the basis of powder X-ray diffraction results. They found it to be cubic, like the ReO_3 and perovskite materials, with a simple cube edge length about 5.1 Å. In normal Prussian blue K^+ ions fill half of the A positions, and Fe ions are in the B positions. In order to keep overall charge balance, half of the Fe ions have a formal charge of 3+ (and thus can be described as P^{3+} ions), and half have a formal charge of 2+ (and can be described as R^{2+} ions). The carbon ends of the CN^- ions point toward the Fe^{2+} ions and the nitrogen ends toward the Fe^{3+} ions. Thus one can think of the P^{3+} ions being in a nitrogen hole, and the R^{2+} being in a carbon hole. This orientation of the CN^- ions has been confirmed by Mossbauer and infrared studies [11, 15, 16]. This structure is shown schematically in Figs. 12A.2 and 12A.3.

All is not quite so simple, however, as Keggin and Miles also introduced the concept that there are two different versions, "*soluble*" Prussian blue, represented by the general formula $KFeFe(CN)_6$, as discussed above, and "*insoluble*" Prussian blue, with the general formula $Fe_4[Fe(CN)_6]_3$. The terms "soluble" and "insoluble" are actually misnomers, for neither is appreciably soluble. The term "soluble" had been employed by dye makers for many years to indicate the relative ease by which the potassium version can be peptized to form a clear sol by washing in water. We can assume that this is the result of the removal of the potassium from the structure. If there is no potassium in the structure, this cannot occur. Thus the use of the "insoluble" name.

It can readily be seen that these two general formulas lead to different structural interpretations. If the composition of the "insoluble" alternative, $Fe_4[Fe(CN)_6]_3$, is correct, it can be written as $Fe^{3+}{}_4Fe^{2+}{}_3(CN)_{18}$ or $(Fe^{3+})_{1/3}(Fe^{3+}Fe^{2+})(CN)_6$. This would indicate that 1/6 of the A positions are occupied by Fe^{3+} ions, instead of their being half-filled with K^+ ions in the "soluble" version. The presence of hydrated Fe^{3+} ions at the A sites has been reported by [17].

Although incorporated water is not included in these compositional formulas, all of these materials are generally found to contain substantial amounts of water when they are prepared by the use of aqueous methods. When some ethanol is present

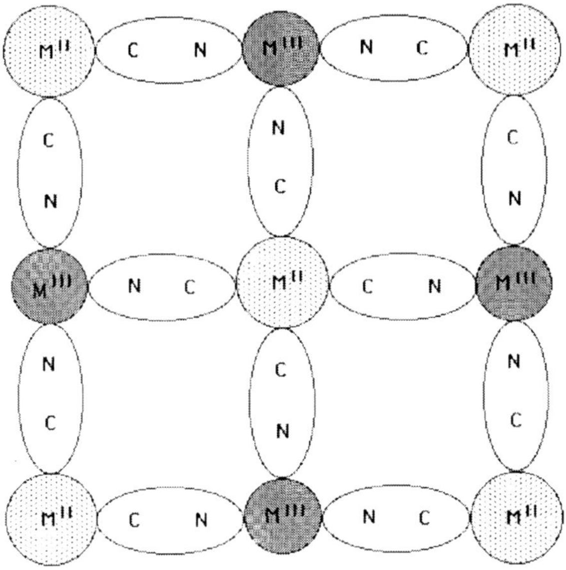

Fig. 12A.2 Schematic representation of one plane in the structure of the hexacyanometallate host lattice

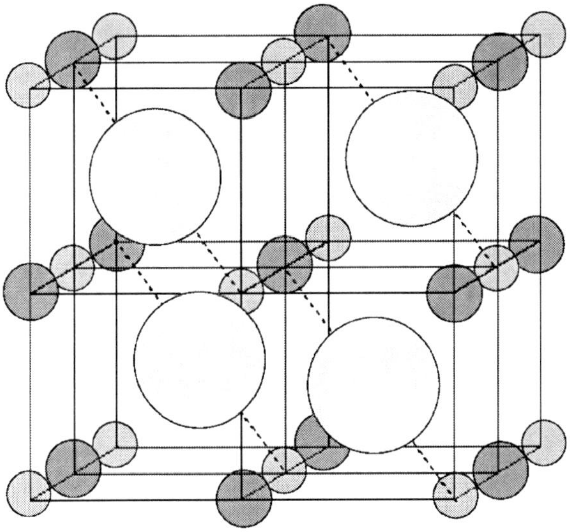

Fig. 12A.3 Schematic representation of the structure of Prussian blue, in which half of the A sites are filled

in the water, the behavior changes somewhat, indicating an influence of the co-absorbed species [18]. It has also been found that the behavior is very different in nonaqueous electrolytes, such as propylene carbonate [19]. The presence of even a relatively small amount of water (e.g. 1 vol%) in such nonaqueous electrolytes leads to behavior quite similar to that found in aqueous electrolytes. This provides strong evidence for the importance of the hydration sheath around the insertable cations in the A position in the structure.

On the other hand, water-free materials can also be produced using solid-state methods [20]. The dry-produced materials with high values of x are very sensitive to moisture and oxidation in air, turning blue and undergoing decrepitation.

Moreover, the Keggin and Miles interpretation of the crystal structure of the "insoluble" material has been challenged by more recent single crystal X-ray diffraction work by Ludi et al. [21], who postulate the presence of water on some of the B sites. Thus this matter remains unresolved. However, it is reasonable to assume that even in the case of materials that initially have the "insoluble" composition, the structure changes in the direction of the "soluble" $K_{1/2}Fe(CN)_3$ cation insertion composition during cycling in KOH solutions into potential regions in which the potassium activity is high [22]. Thus this simpler "soluble" structural description will be used in the discussion in this paper.

Because of the possibility that the ions on the two types of B sites may have variable valences these materials can often be either reduced or oxidized, or both. The mechanism whereby this occurs involves the insertion or removal of charged species on the A sites.

Reduction can occur by increasing the concentration of A^+ ions in the structure. In the case that both P and R species are Fe ions, the additional positive charge in the A sites is balanced by the reduction of some of the Fe^{3+} species on the P sites to Fe^{2+}. When the A^+ concentration reaches 2, essentially all of the Fe^{3+} ions have been reduced, so that there are Fe^{2+} ions on both the P and R sites. Thus the composition can be written as $A_2P^{2+}R^{2+}$ $(CN)_6$. This interpretation has been verified by in situ Mossbauer experiments [23] (see Fig. 12A.4).

Prussian blue and its analogs can also be oxidized by the removal of A^+ species. In this case the decreased positive charge on the A sites is balanced by the oxidation of Fe^{2+} species on the R positions to Fe^{3+}, and the nominal composition becomes $P^{3+}R^{3+}(CN)_6$. Thus Fe ions can participate in both reduction and oxidation processes as the concentration of positively charged A^+ ions is varied.

In addition to the (complete) removal of the A^+ species, the structure can be further oxidized by the insertion of negatively charged anionic species (B^-) into the A sites. In this case the composition can be nominally described as A_xP^{3+} R^{3+} $(CN)_6B_y$, where x is 0.

Thus these materials can have insertion of either cations or anions into the A sites, depending upon the potential, as discussed below. The insertion kinetics depends greatly upon the identity of the inserted species. In some cases this can be remarkably rapid. Due to the large size of the CN^- anions, the openings between adjacent unit cells are quite large. Thus it is possible for relatively large ions to move throughout the structure by jumping through these windows. Species in the A sites are often highly mobile, and can be accompanied by an appreciable amount of water of hydration.

These hexacyanoferrates are generally stable in water at low to moderate values of pH. In acidic electrolytes containing K^+ ions they can be reduced and re-oxidized

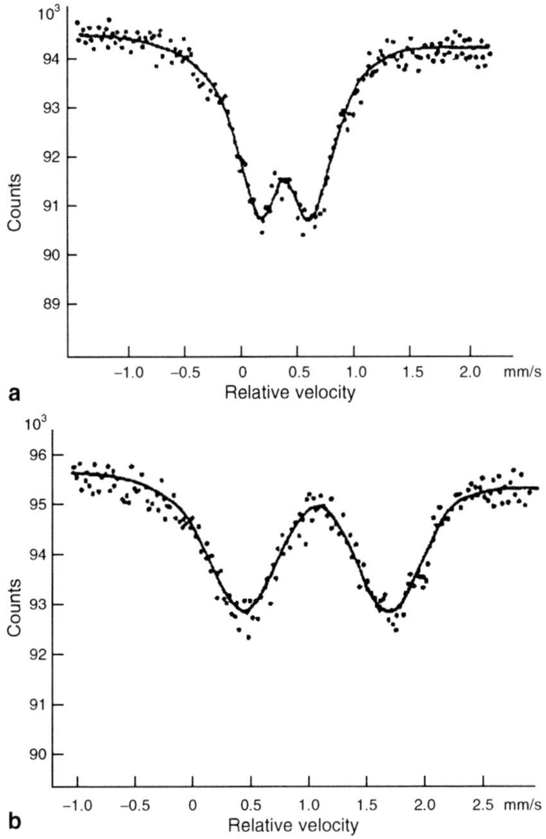

Fig. 12A.4 In-situ Mössbauer spectra obtained at two different values of electrode potential. **a** 0.6 V vs. SCE and **b** −0.2 V vs. SCE [23]

many times, and show excellent reversibility. Over 10^7 cycles have been demonstrated in some cases [24].

12A.5 Electrochemical Behavior of Prussian Blue

Prussian blue and its analogs can be readily reduced and oxidized electrochemically. Much of the experimental work in the literature on the electrochemical behavior of Prussian blue has been performed using potassium – containing aqueous electrolytes with a pH value of about 4, and potentials are generally referenced to the *standard calomel electrode* (*SCE*), which is about 478 mV positive of the reversible hydrogen evolution electrode potential at that value of pH. Although very common, this choice of reference electrode is unfortunate, as the definition of its potential relative to neutral chemical species requires knowledge of the pH, in accordance with the *Gibbs Phase Rule*, as discussed in Chap. 13.

Prussian blue, which has a dark blue-black color, can be both reduced and oxidized electrochemically. Reduction occurs at a potential about 195 mV positive of the SCE potential, and thus about 678 mV positive of the reversible hydrogen potential. The reduction product is white, and is generally called "Everitt's salt," although it is sometimes also designated as "Prussian White." As already mentioned, its composition can nominally be described as $K_2FeFe(CN)_6$ when K^+ ions are in the A sites.

The interpretation of the reaction that occurs in the iron hexametallocyanates at about 0.2 V positive of the SCE is the reduction of high-spin Fe^{3+} to low-spin Fe^{2+} as the potential is reduced, and the corresponding oxidation as the potential is increased has been established on the basis of in situ Mossbauer experiments [23].

Oxidation of Prussian blue occurs when the potential is made more positive. This occurs at about 870 mV vs. the SCE reference, and thus 1.348 mV positive of the reversible hydrogen potential. The product is only lightly colored, and is generally called "Berlin Green". Its composition can be nominally described as $FeFe(CN)_6$.

A second oxidation reaction, involving the insertion of anions into the A position, is sometimes found at about 1,100 mV vs. the SCE potential, where a yellow product called "Prussian Yellow" is formed. This material is unstable in water, as would be expected from its potential, some 1.8 V positive of the reversible hydrogen potential. Its nominal compositions can be written as $FeFe(CN)_6A_x$.

All four of these materials have essentially the same basic cubic unit cell, with a lattice parameter of about 10.2 Å. Although incorporated water is not included in these nominal compositions, these materials are generally found to contain substantial amounts of water of hydration around the A species.

Electrochemical experiments have often been made using cyclic voltammetry. A typical example is shown in Fig. 12A.5. It is seen that there are reversible current peaks in two quite different potential regions, relating to the reduction and oxidation

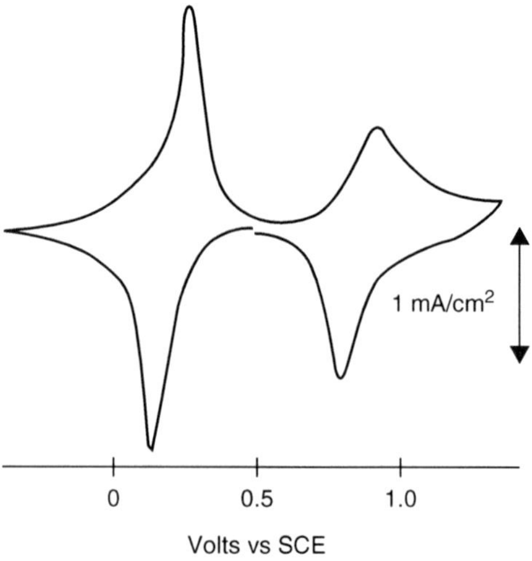

Fig. 12A.5 Typical voltammogram of Prussian blue, after [25]

reactions described above. Another oxidation reaction at even more positive potentials is not shown in that figure because the scan rate was rather low and the potential was not increased sufficiently to see it. Another voltammogram is shown in Fig. 12A.6 under conditions that made it possible to see the formation of the chemically unstable Prussian yellow at more positive potentials.

The critical potentials in the Prussian blue system are shown schematically in Fig. 12A.7. One can translate the semi-quantitative dynamic data obtained

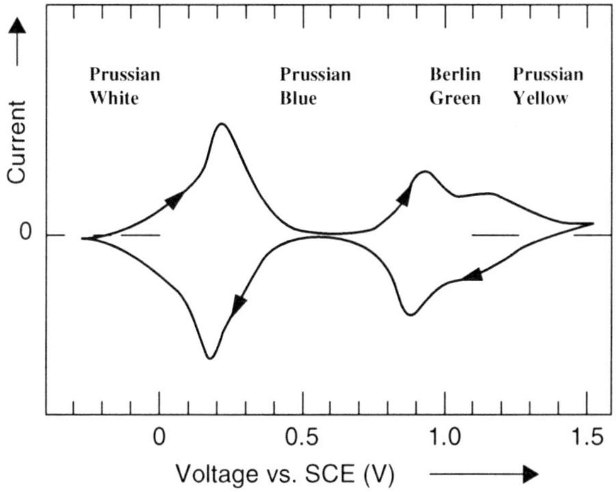

Fig. 12A.6 Voltammogram that also shows the reaction to form Prussian yellow at more positive potentials

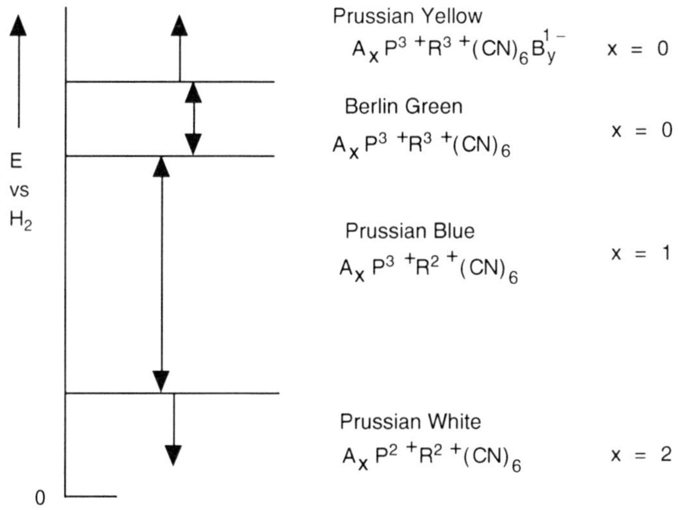

Fig. 12A.7 Schematic representation of the potentials at which the several reactions occur in Prussian blue

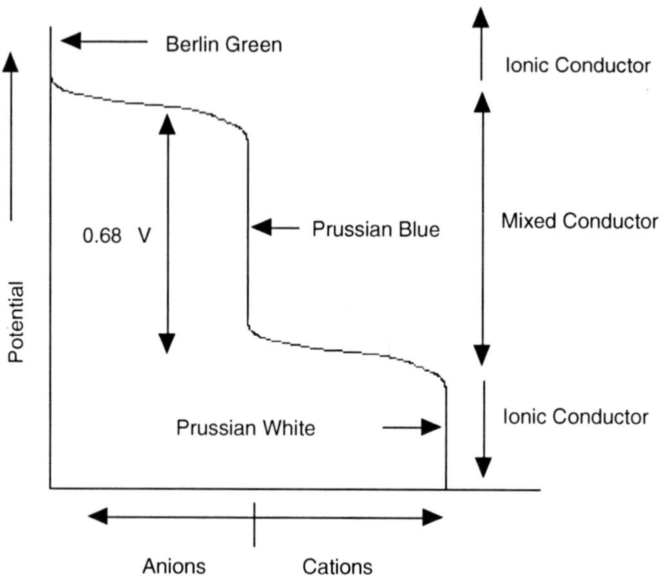

Fig. 12A.8 Schematic equilibrium electrochemical titration curve for the Prussian blue system

from cyclic voltammetry experiments into the results that would be expected if electrochemical potential spectroscopy experiments were performed. Likewise, they could be expressed as an equilibrium titration curve, as indicated schematically in Fig. 12A.8. In this case the formation of Prussian yellow is not included, as it is not stable in water, as indicated earlier.

Experiments of this type would provide more quantitative information about the potentials at which the reduction and oxidation reactions take place than can be obtained from the more common dynamic cyclic voltammetry experiments. However, in the case of Prussian blue, the kinetics of the relevant phenomena are so fast that there is not a lot of difference between the information obtained from dynamic and static experiments.

12A.6 Various Cations Can Occupy the A Sites in the Prussian Blue Structure

A number of different cations can be present in the A positions in the hexacyanometallate structure. Monovalent examples include Li^+, Na^+, K^+, NH_4^+, Rb^+, and Cs^+ ions. The ability of these various ions to reversibly enter the structure has been interpreted in terms of their size when hydrated. These are shown in Table 12A.1, which also includes the hydrated divalent Ba^{2+} ion.

Table 12A.1 Assumed radii of hydrated cations

Cation	Radii (Å)
Li^+	2.37
Na^+	1.83
K^+	1.25
$NH_4{}^+$	1.25
Rb^+	1.18
Cs^+	1.19
Ba^{2+}	2.88

The hard-sphere model of the Prussian blue structure has an intercell window radius of 1.6 Å. Thus it is expected that it would be difficult for cations with hydrated radii greater than this value to readily move into and out of the structure. This corresponds to what is found experimentally from dynamic electrochemical measurements [26].

Hydrated $NH_4{}^+$ and Rb^+ can be cycled many times, although their insertion/extraction kinetics are significantly slower than hydrated K^+ ions. Both smaller (Cs^+) and larger (Li^+ and Na^+) hydrates showed much more restricted behavior.

In nonaqueous electrolytes where the mobile cations are not hydrated these insertion reactions occur much more slowly and are not so interesting from a practical point of view. This is typical of the ionic transport behavior when the mobile species is relatively small in comparison to the host structure through which it moves.

12A.7 The Substitution of Other Species for the Fe^{3+} and the Fe^{2+} in the P and R Positions in the Prussian Blue Structure

There have been a number of studies of analogues to Prussian blue and the related reduced and oxidized phases, Prussian white, Berlin green, and Prussian yellow, all of which contain Fe ions. Whereas most of the published work has involved the substitution of Fe ions by other transition metal ions, it is also possible to synthesize analagous materials in which the A positions are filled ($x = 2$) and divalent alkaline earth cations are in the P positions [27]. In these cases, the crystal structure becomes distorted relative to the cubic structure commonly found in the materials with only transition metal cations. Experiments were performed on compositions containing monovalent cations with different sizes in the A positions, and different alkaline earth cations in the P positions. It was found that the amount and type of distortion correlates rather well with the calculated value of a tolerance factor that quantifies the departure of the relative ionic sizes from the ideal values of the related cubic (perovskite) structure. The largest A^+ ion (Cs^+) and the smallest alkaline earth ion (Mg^+) give the largest value of the tolerance factor, and thus the least distorted structure. The covalent binding in the transition metal – only materials makes the cubic configuration very stable, so that these materials can be thought of as having stiff framework structures, rather than as close packed ionic compounds.

12A.8 Other Materials with $x = 2$ That Have the Perovskite Structure

There has been work that shows that materials can be made in which other divalent ions can replace reduced Fe ions (Fe^{2+}) on the P sites when $x = 2$ so that the structure is equivalent to that of a perovskite. Some of these, including Co and Cu [28, 29], are shown in Table 12A.2.

12A.9 The Electronic Properties of Members of the Prussian Blue Family

Despite its color, Prussian blue is an electronic semiconductor with a band gap of about 0.5 V. Because of the mobility of the K^+ ions among the A sites within its structure, it is also a fast ionic conductor.

Experiments on dry thin films have shown that the voltage/current relation in the case of potassium – containing Prussian blue is non-ohmic, in that electronic conduction is very low below a threshold voltage of about 0.5 V [30]. The apparent electronic conductivity can vary over a number of orders of magnitude depending upon the extent of hydration [31].

On the other hand, both the reduced material, Everitt's salt, and the oxidized material, Prussian yellow, show ohmic electronic properties. This behavior is shown in Fig. 12A.9.

Table 12A.2 Examples in which Fe ions have been replaced by other species

Substitution for Fe^{3+}	Substitution for Fe^{2+}
Mn^{3+}	Cu^{2+}
Co^{2+}, M^+	Os^{2+}
Ni^{3+}	Ru^{2+}
In^{3+}	Alkaline earth ions
Cu^{2+}, M^+	

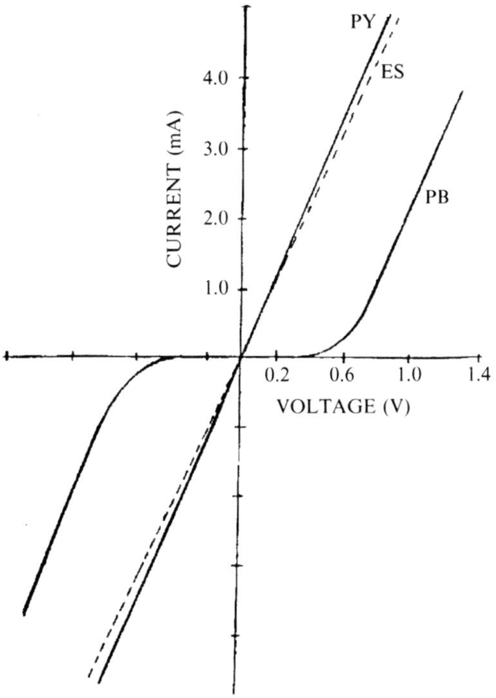

Fig. 12A.9 Electronic conductivity of members of the Prussian blue family. PB is Prussian blue, ES is Everitt's salt (Prussian white), and PY is Prussian yellow

12A.10 Batteries with Members of the Prussian Blue Family on Both Sides

It has been shown that one can make Galvanic cells with an open circuit voltage of 0.68 V by using materials of this family upon both electrodes. Upon charging, Everitt's salt and Berlin green form on the negative and positive electrodes, respectively. Upon discharging, they both change to Prussian blue, and the voltage falls to zero. It has also been shown that a simple battery can be made by putting Prussian blue on both sides of a film of Nafion [32].

The higher oxidation plateau, at which Prussian Yellow is formed, gives an initial voltage of 0.905 V, but this decays to 0.68 V because Prussian Yellow is unstable in water [33].

12A.11 Catalytic Behavior

Layers of the materials in the Prussian blue family have been explored for a number of catalyst applications. As in other cases, there is often confusion about the mechanisms underlying the observed behavior, and whether they involve surface or bulk phenomena. As an example, it has been shown [34] that CO can be reduced to methanol on Prussian blue electrodes. This may be important for cleaning hydrogen gas for use in low-temperature fuel cells.

12A.12 Electrochromic Behavior

The color change between the different phases can be caused to occur electrochemically, so that Prussian blue can be used for electrochromic applications, and this matter has been pursued in a number of laboratories. The color change kinetics can be faster in Prussian blue electrodes than in the common A_xWO_3 electrochromic materials [24].

The optical absorption spectrum depends upon the identity of the cations present in the P and R positions. Figures 12A.10 and 12A.11 show two examples from [26], Prussian blue and Ruthenium purple. In both cases the disappearance of the absorption peak at potentials below that at which the insertion of the second cation into the A sites in the structure can be seen.

It has also been shown that a single-layer electrochromic device can be made by applying transparent conductors to both sides of a film of Prussian blue. When a sufficiently high voltage is applied, the A^+ ions are driven to the negative side, forming white Everitt's salt. At the same time, Berlin green (or Prussian yellow) is formed on the cation-deficient positive side. The relative thicknesses of these layers and the dark Prussian blue layer in the middle are dependent upon the magnitude of the applied voltage [35]. Thus the transparency of the configuration can be varied by changing the voltage.

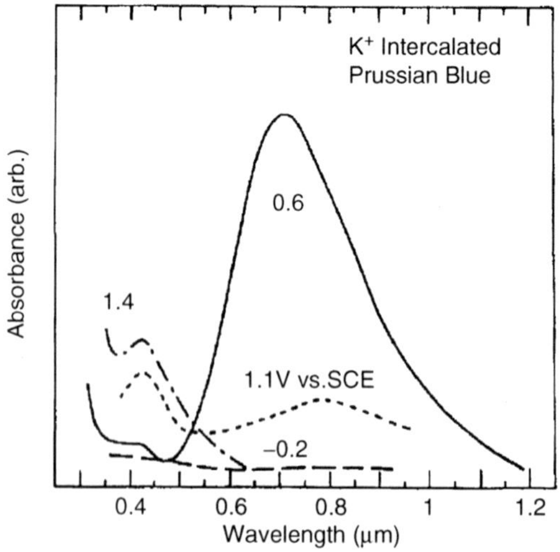

Fig. 12A.10 Optical absorption of Prussian blue at different potentials

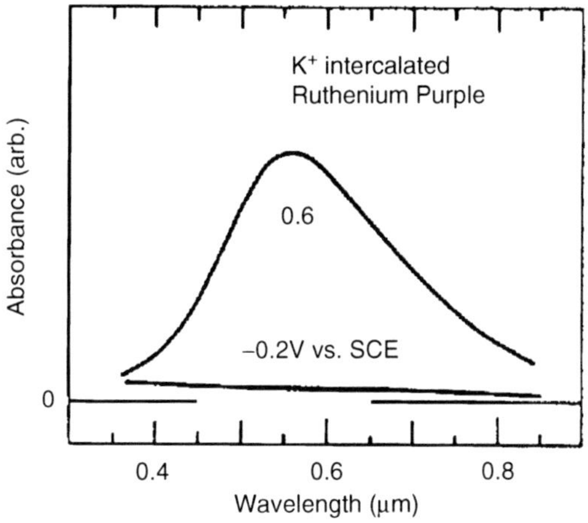

Fig. 12A.11 Optical absorption of Ruthenium purple, in which Ru is on the R sites in the structure

Thus it can be seen that the Prussian blue materials are different from those that are typically considered for use in electrochromic systems. One often finds electrochromic materials divided into two classes. One includes those which become dark when the potential is reduced, such as WO_3 and mixed-conducting titanium oxides. This is generally related to the insertion of cationic species into the structure and the reduction of ions in the host structure, and is sometimes called anodic coloration. The other group, including $Ni(OH)_2$, IrO_2, and YH_x, become dark when species, e.g. protons, are removed from their structures. They exhibit cathodic coloration. Prussian blue exhibits both these features, becoming dark as the potential is reduced from a high value, and becoming dark when the potential is increased from a low value. These features are indicated schematically in Fig. 12A.12, in which the coloration potentials are put on a general potential scale.

12A.13 Insertion of Species into Graphite

It has long been known that a variety of species can be rather easily inserted between the graphene sheets of the graphite structure. A discussion of much of the early work can be found in [36]. Insertion often is found to occur in "stages," with nonrandom filling of positions between the layers of the host crystal structure.

The graphite structure is amphoteric, so that both negatively charged and positively charged species can be inserted into it. These typically are at least partially ionized, exchanging electrons with the electronic structure of the graphite host, whose charge is adjusted so as to attain the requisite overall charge neutrality.

When cations are inserted, the host graphite structure takes on a negative charge. A well-known example is the insertion of potassium into graphite. The K^+ ions

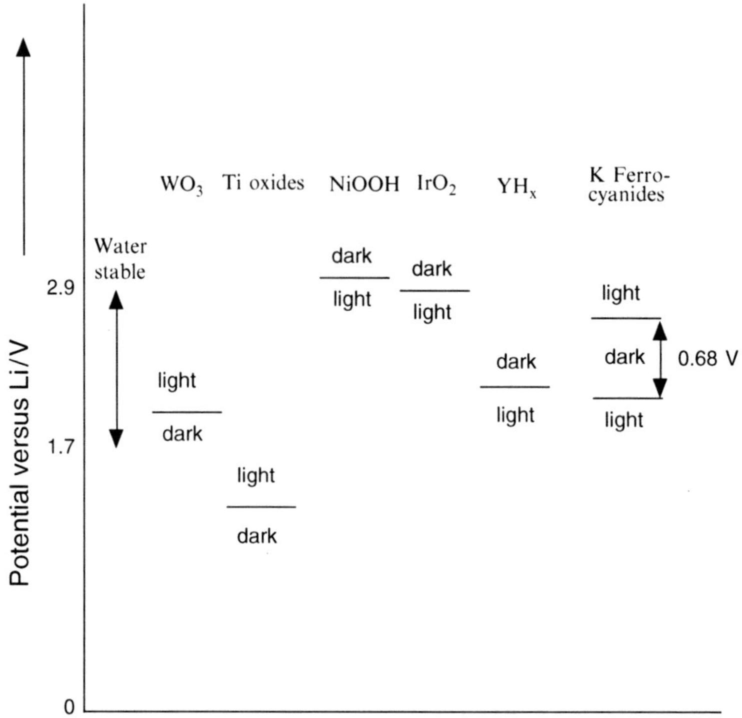

Fig. 12A.12 Color changes in a number of electrochromic materials as a function of potential (volts vs. lithium)

occupy ordered sites within the carbon planes of the graphite. The maximum composition that can theoretically be reached in the crystal structure is KC_8.

It was reported in 1983 [37] that lithium can be reversibly inserted into graphite at room temperature when using a polymeric electrolyte. The cation arrangement is different from that in the case of potassium, and the maximum composition that can be reached is LiC_6.

Prior experiments with liquid electrolytes had been unsuccessful due to co-intercalation of species from the organic electrolytes that were used at that time. This problem has been subsequently solved by the use of other electrolytes, as discussed in Chap. 15. Chapter 7 included a discussion of graphites and related carbon-containing materials that have been investigated for use as negative electrode materials in lithium batteries in recent years. The impetus for this was the successful development in 1990 by Sony Energytec, Inc. of commercial rechargeable batteries containing negative electrodes based upon materials of this family [38].

When anions are inserted into graphite, acid salts are formed, and the host carbon structure takes on a positive charge. Examples of insertable anions are Br^-, SO_4^{2-}, and SbF_6^-. Neutral species or combinations can also be inserted into the graphite structure, and examples include HF, $MnCl_2$, $NiCl_2$, $ZnCl_2$, $CuCl_2$, $CuBr_2$, $AlCl_3$, $FeCl_3$, $TiCl_3$, and $ZrCl_4$.

As discussed in Chap. 10, it has also been shown that mobile cationic species can enter the graphite structure and react with relatively immobile anion species that had been previously inserted [39].

12A.14 Insertion of Guest Species into Polymers

There was a considerable amount of interest some years ago in the possibility of the electrochemical insertion of species into polymers, with the expectation that one could use these materials as electrodes in battery systems. One of the concepts driving this interest was the idea of using inexpensive large-scale polymer preparation technology to produce thin-film electrochemical systems.

Both anionic and cationic species can be inserted into some polymers. One of the earliest and most prominent examples was polyacetylene, where it was shown that both cations and a number of anions, including I^-, ClO_4^-, AsF_6^-, and SO_4^{2-} can be electrochemically inserted into the structure [40–42].

Whereas the use of materials such as polyacetylene as battery electrodes has not become practical for a variety of reasons, other polymeric materials are still of interest for other potential applications, such as for electrochromic devices.

Especially interesting are insertion reactions involving members of the polythiophene family, such as PFPT. In this case, electrochemical experiments indicate the presence to two different reactions at appreciably different potentials [43]. This is shown in Fig. 12A.13.

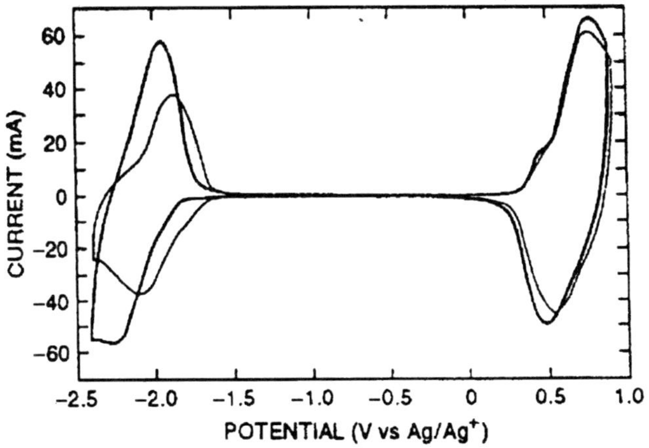

Fig. 12A.13 Cyclic voltammogram of PFPT in two different organic electrolytes [43]. PFPT is poly 3-(4-fluorophenyl)-thiophene

Whereas the low potential reaction surely involves the insertion of protons into the structure, the identity of the species that reversibly reacts at more positive potentials is not yet clear.

12A.15 Summary

There are a number of insertion reaction materials in which the host structure can be amphoteric, and thus can accept either cationic or anionic species as guests. It has long been known that graphite has this property, and a number of polymeric materials have been found that can also act as amphoteric hosts. Two of the latter have been briefly mentioned here, polyacetylene and the polythiophene family.

Especially interesting are inorganic materials with partially filled guest ion substructures and ions in the host substructure that can be both reduced and oxidized. One example of this type is the hexacyanometallate family, of which Prussian blue is the best-known example. In this case hydrated potassium ions can enter and leave the structure reversibly and with very fast kinetics. Some other ions can also do this, but their kinetic properties are not as favorable.

Materials with these characteristics can have a number of interesting electronic, electrochemical, and optical properties.

12B Cells with Liquid Electrodes: Flow Batteries

12B.1 Introduction

Except for the Zebra cell discussed in Chap. 4, all of the electrochemical cells that have been discussed thus far have electrodes that are solids. In the *Zebra cell* case liquid electrodes could be used because the electrolyte is solid, resulting in a L/S/L configuration. There is another group of cells that have liquid electrode reactants, although their electrode structures contain solid current collectors. These are generally called *Flow Batteries*, since the liquid reactant is stored in tanks and is pumped (flows) through the cell part of the electrochemical system. Thus such systems can be considered to be rechargeable batteries.

A number of such systems have been explored, and in some cases commercialized. However, most of them have not really been commercially successful to date. As will be seen below, this could well change in the near future.

The general physical arrangement is shown in Fig. 12B.1, whereas the configuration of the cell portion of the system is shown schematically in Fig. 12B.2.

It can be seen that this is also a type of L/S/L configuration. The electrolyte is a proton-conducting *solid polymer*, and the electrode reactants are liquids on its two sides. In the Zebra cell the reactants are both electronically conducting, whereas in

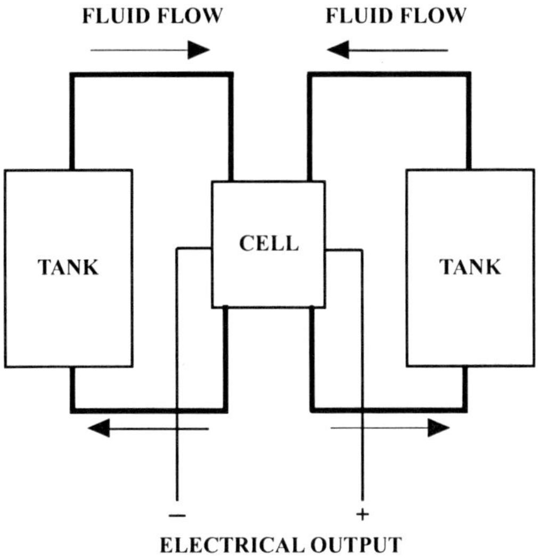

Fig. 12B.1 General physical arrangement of a flow battery

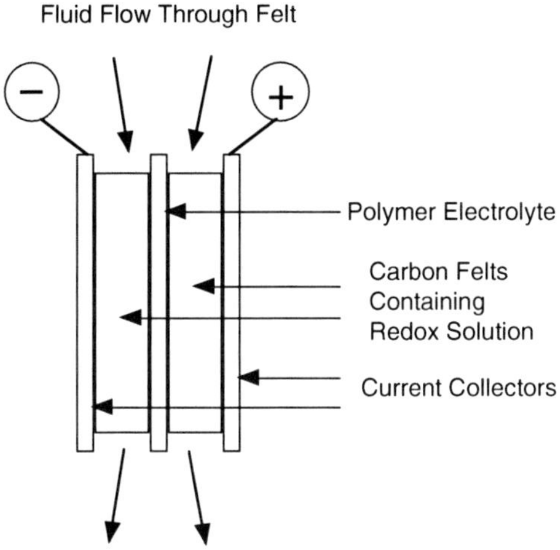

Fig. 12B.2 The cell portion of the system. In some cases there are multiple bipolar cell configurations

the flow cells the electrode reactants are ionic aqueous solutions that are electronic insulators. In order to get around this problem and provide electronic contact to an external electrical circuit, the liquid reactants permeate an *electronically conducting graphite felt*. This felt provides contact, both to the polymer electrolyte and to a *graphite current collector*.

The electrode reactants are typically acidic, e.g., 2 M H_2SO_4 aqueous solutions of ions that can undergo *redox reactions*. The function of the polymer electrolyte is to transport protons from one side to the other, thus changing the pH and charges on the dissolved *redox ions*.

An important difference from the Zebra cell is that the reactant materials, the redox ion solutions, can be pumped into and out of the electrode compartments. This means that the capacity is not fixed by the cell dimensions, but is determined by the size of the liquid electrode reactant tanks. This can produce very large capacities. Thus flow batteries deserve consideration for relatively large stationary applications, such as remote solar or wind installations, whose outputs are dependent upon the time of day and/or the weather.

The open circuit voltage across the electrolyte is determined by the difference in the chemical potentials across the electrolyte. As current passes through the cell protons are transferred, changing the pH, so that the ionic compositions of the two electrode reactant fluids gradually change. Thus the cell potential varies with the state of charge. The change in the voltage with the amount of charge passed depends, of course, upon the size of the tanks. Thus it is easy to see that electrochemical systems of this type can have very large charge capacities if the electrode reactant tanks are large. This is one of the potential advantages of flow battery systems.

A number of different redox systems have been explored, as indicated in Table 12B.1.

A general discussion of these various systems can be found in [44].

There is some confusion in the terminology used to describe these systems, for the liquid reactants on the two sides are sometimes called *electrolytes*, even though they do not function as electrolytes in the battery sense. Another terminology that is also sometimes used is to call the liquid reactant on the negative side of the cell the *anolyte*, and that on the positive side of the cell the *catholyte*.

One of the most attractive flow systems involves the vanadium redox system [45–50]. In this case the negative electrode reactant solution contains a mixture of V^{2+} and V^{3+} ions, whereas the positive electrode reactant solution contains a mixture of V^{4+} and V^{5+} ions. Charge neutrality requirements mean that when protons (H^+ ions) are added or deleted from such liquids by passage through the polymer electrolyte in the cell, the ratio of the charges on the redox species is varied.

These systems are generally assembled in the uncharged state, in which the chemical compositions of the two liquid reactants are the same. In the vanadium system this is done by adding vanadyl sulfate to 2 M H_2SO_4, which gives an equal

Table 12B.1 Various redox systems used in flow batteries

System	Negative electrode reactant	Positive electrode reactant	Nominal voltage
V/Br	V	Bromine	1.0
Cr/Fe	Cr	Fe	1.03
V/V	V	V	1.3
Sulfide/Br	Polysulfide	Bromine	1.54
Zn/Br_2	Elemental Zinc	Bromine	1.75
Ce/Zn	Zn	Ce	<2

mixture of V^{3+} and V^{4+} ions. They are then charged by passing current, causing the transport of protons through the polymer electrolyte, so that the ion contents on the two sides become different.

12B.2 Redox Reactions in the Vanadium/Vanadium System

One can write the reactions in the electrode solutions of the vanadium system as

$$VO_2{}^+ + 2H^+ + e^- = VO^{2+} + H_2O \tag{12B.1}$$

or in terms of the vanadium ions

$$V^{5+} + e^- = V^{4+} \tag{12B.2}$$

in the positive electrode reactant solution, and

$$V^{2+} = V^{3+} + e^- \tag{12B.3}$$

in the negative electrode reactant solution.

So that the overall reaction is

$$VO_2{}^+ + 2H^+ + V^{2+} = VO^{2+} + H_2O + V^{3+} \tag{12B.4}$$

or

$$V^{5+} + V^{2+} = V^{4+} + V^{3+} \tag{12B.5}$$

12B.3 Resultant Electrical Output

The variation of the open circuit cell potential with the state of charge in the case of the V/V system with concentrations of 2 mol l^{-1} of each V species is shown in Fig. 12B.3. Typical operation would involve cycling between 20 and 80% of capacity, and thus at voltages between 1.3 and 1.58 V.

12B.4 Further Comments on the Vanadium/Vanadium Redox System

Since each cell produces a relatively low voltage, such batteries generally contain a number of cells arranged in series in order to produce a greater overall output voltage. Parallel configurations can be used to provide higher currents. Depending upon the application, it may be desirable to permit relatively rapid charging, which

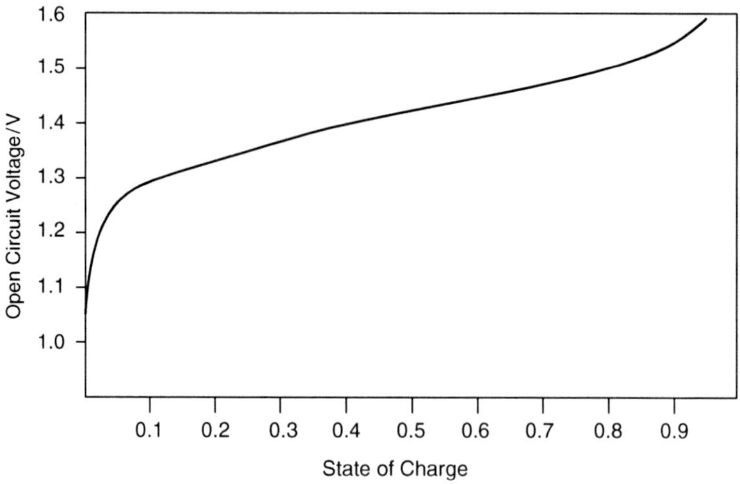

Fig. 12B.3 Variation of the open circuit potential vs. state of charge for the case of V/V flow cell at 298 K

may not be necessary during discharge. Thus it may be desirable to include a mechanism to change the number of cells and their series/parallel arrangements during different operating conditions.

If all of the cells are fed from common liquid supplies, this can result in a large voltage applied across the liquid reactants and this can result in the passage of a considerable amount of current, this is a form of self-discharge, and is sometimes called shunt or bypass current. This is different, however, from the self-discharge that results from neutral species, or neutral combinations of species, traveling through or around the electrolyte in other types of electrochemical systems.

Because there are no solid-state volume changes during charging and discharging, as are typical for electrochemical cells with solid electrodes, the components of the cells, as well as the total system, can have long lives. Thus long cycle life is not a problem, even with repeated deep charges and discharges.

The vanadium redox system can be used over a temperature range from 10 to 35°C, and typically operates at or near ambient temperature. At higher temperatures the current density increases, but the cell voltage is reduced somewhat. The overall result is that the power is greater at somewhat elevated temperatures.

The electrode kinetics are good, and additional catalysts are not required. The coulometric and voltage efficiencies are high, except for the self-discharge mechanism mentioned above.

The specific energy and energy density are determined primarily by the electrode reactants themselves, which are the major components in these systems. Typical values are $15\,W\,h\,kg^{-1}$ and $18\,W\,h\,l^{-1}$. Round trip efficiencies are typically 70–75%.

Since the electrode reactants both consist of vanadium sulfate solutions in aqueous sulfuric acid, only differing by the oxidation states of the vanadium ions, contamination by leakage across the electrochemical cell membranes only results in some capacity loss, and is fully reversible.

Because the cell voltage is a function of the state of charge, it is possible to determine the state of charge of such systems remotely, which may be an advantage in some cases. The cell design also makes monitoring of the voltage across each cell possible.

Since the cells can be configured in a variety of different series/parallel arrangements, the charging and discharging cycles can operate at different voltages. As a result, such as system can be used as a DC/DC converter.

12C Reactions in Fine Particle Electrodes

12C.1 Introduction

The discussion of the mechanism of compositional changes in electrodes in this text has thus far been related to the behavior of individual particles, or of bulk reactant phases.

Two different cases have been discussed: insertion reactions, in which the composition of a single phase varies by the insertion or extraction of a solute species from the electrolyte, and their redistribution via chemical diffusion, and poly-phase reconstitution reactions, in which the microstructure consists of two phases in equilibrium that have different compositions. In this latter case, the overall composition varies by a change in the relative amounts of the two phases via the translation of a two-phase reaction front as the result of chemical diffusion through the outer phase.

There are also other possibilities. The translation of the two-phase interface in reconstitution reactions can occur by a different mechanism, and the composition of individual particles in a fine particle electrode can be different in the same general location.

A related interface transport mechanism is also relevant to the behavior of lithium–carbons in the negative electrodes of lithium batteries.

12C.2 Translation of Two-Phase Interface by Chemical Diffusion

The discussion of poly-phase reconstitution reactions generally assumes that a change in the overall composition of the electrode reactant occurs by the translation of a two-phase interface at which two phases of different compositions are in equilibrium with each other, consistent with the relevant phase diagram.

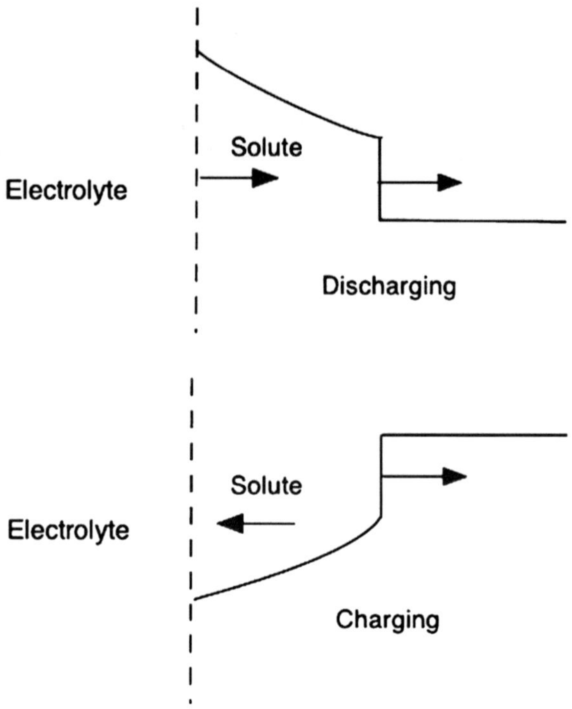

Fig. 12C.1 Composition profile in a lithium battery positive electrode when the progress of a two-phase interface is controlled by diffusional transport through the outer phase that is in contact with the electrolyte

The normal model for this process involves transport of solute species to or from the two-phase interface by chemical diffusion through the outer phase that is in contact with the electrolyte.

This can be represented schematically as illustrated in Fig. 12C.1.

During discharge of such electrodes, the amount of lithium in the electrode increases, whereas during charging the amount of lithium decreases. In both cases the reaction front moves away from the electrolyte and into the electrode structure.

When this is the case, the kinetic behavior of the electrode depends upon the composition profile and the chemical diffusion coefficient in the outer phase.

12C.3 Alternative Mechanism for the Translation of Poly-Phase Interfaces

Another mechanism for reconstitution reactions was proposed by Delmas [51], in connection with Li_xFePO_4 electrodes, who called it a "domino-cascade model". He pointed out that the compositional change accompanying the translation of a

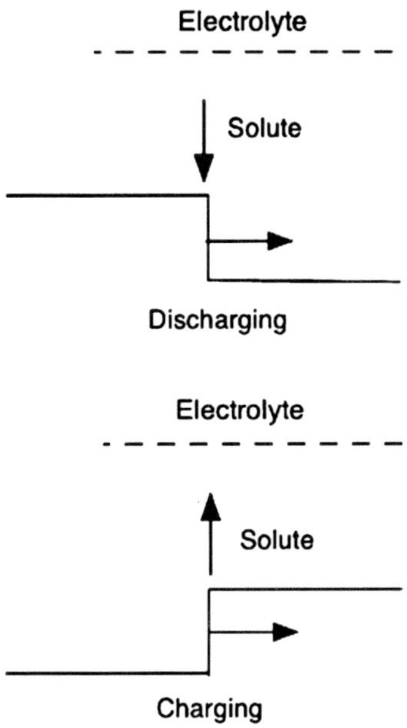

Fig. 12C.2 Composition profiles in a positive electrode in a lithium battery when the progress of a two-phase interface involves transverse transport within the reaction front

two-phase interface can also take place by the insertion or deletion of solute species by their transport along the two-phase interface itself, rather than through one of the adjacent phases.

This mechanism is illustrated schematically in Fig. 12C.2.

When this is the case, no composition gradient needs to exist in either of the adjacent phases, and chemical diffusion kinetics in them, a topic that has received considerable attention in the research community, is of no importance. Instead, the solute enters and leaves the microstructure by transport along the two-phase interface. In solids, diffusion along grain boundaries and poly-phase interfaces is generally much more rapid than bulk diffusion.

In addition to kinetic advantages, the dimensional changes involved in the phase transformation occur at the transverse interface, rather than within either of the two phases. This reduces the amount of mechanical strain energy involved in the process. Since mechanical work is a major source of energy loss and hysteresis in cycling electrodes, this can be a significant advantage.

This mechanism will be most important when the reactant particles are small enough that they act as single crystals, and this transverse mechanism can readily receive and deliver solute species from and to the electrolyte. If the particles are polycrystalline, this process will be mechanically and kinetically hindered. This is thus an important particle-size effect, with monocrystalline nanoparticles behaving quite differently from larger particles.

12C.4 Reactions in Electrodes Containing Many Small Particles

Reconstitution reactions occur in fine particle electrodes by the translation of reaction fronts, with reacted particles behind, and unreacted particles ahead. It has been observed in several studies, however, that there can be microscopic regions in which individual particles have either one structure and composition or the other. This implies that the kinetics of the transformation of individual particles is rapid compared to the nucleation process in the reaction front. The region in which there are particles of both phases can have an appreciable width. In such a case the overall composition in the front will gradually vary from that characteristic of one phase to that characteristic of the other. This can be represented as shown in Fig. 12C.3.

12C.5 Mechanism Involved in Changing the Composition of Lithium–Carbons

A similar mechanism probably occurs when there is a change in the composition of lithium–carbons. As was discussed in Sect. 7A, lithium–carbons have a stage structure, and changes in the amount of lithium present occur as the result of changes in the stage structure, i.e., the spacing between graphene layer pairs between which lithium is present. It is difficult to understand how this can take place by lithium

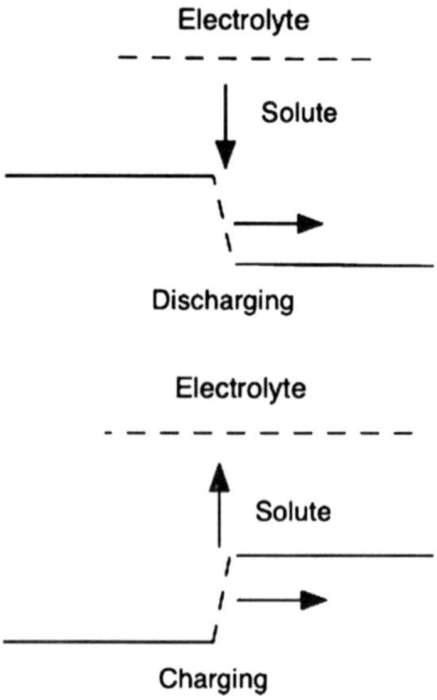

Fig. 12C.3 Schematic of the variation of the overall composition with location when there is an intermediate transition region in which some of the particles have been converted to the new phase, and others have not

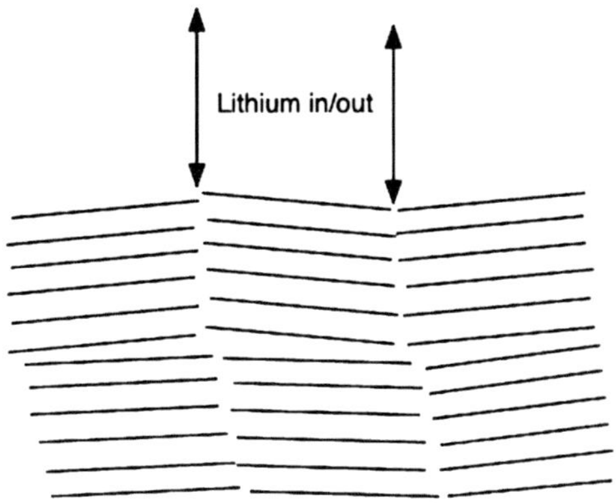

Fig. 12C.4 Mechanism for the transport of lithium into and out of the structure of lithium–graphite. The stage structure within the subgrains is not shown

entering or leaving the structure over a significant distance parallel to the graphene sheets, for example, how can a transition between stage 2 and stage 3 structures take place by long-distance lithium transport parallel to the graphene sheets.

However, the graphite structure is generally divided into small subgrains that have minor differences in orientation, and lithium transport along the subgrain boundaries transverse to the graphene sheets provides a mechanism whereby the stage structure can change locally. This is indicated schematically in Fig. 12C.4.

References

1. S.-K. Joo, I. D. Raistrick, R. A. Huggins, Materials Research Bulletin *20*, 897 (1985)
2. S.-K. Joo, I. D. Raistrick, R. A. Huggins, Materials Research Bulletin *20*, 1265 (1985)
3. S.-K. Joo, I. D. Raistrick, R. A. Huggins, Solid State Ionics *17*, 313 (1985)
4. S.-K. Joo, I. D. Raistrick, R. A. Huggins, Solid State Ionics *18/19*, 592 (1986)
5. Y. Inaguma, et al., Solid State Communication *86*, 689 (1993)
6. H. Kawai, J. Kuwano, Journal of the Electrochemical Society *141*, L78 (1994)
7. Y. Inaguma, L. Chen, M. Itho, T. Nakamura, Solid State Ionics *70/71*, 196, 203 (1994)
8. Y. Inaguma, J. Yu, Y.-J. Shan, M. Itho, T. Nakamura, Journal of the Electrochemical Society *142*, L8 (1995)
9. A. D. Robertson, S. Garcia Martin, A. Coats, A. R. West, Journal of Materials Chemistry *5*, 1405 (1995)
10. P. Birke, S. Scharner, R. A. Huggins, W. Weppner, Journal of the Electrochemical Society *144*, L167 (1997)
11. M. B. Robin, P. Day, Advances in Inorganic Chemistry and Radiochemistry *10*, 247 (1967)
12. J. Brown, Philosophical Transcation *33*, 17 (1724)
13. H. B. Weiser, Inorganic colloid chemistry, Vol. 3, *Colloidal Salts* (Wiley, New York, NY, 1938), p. 343 ff
14. J. F. Keggin, F. D. Miles, Nature, 577 (1936)
15. J. F. Duncan, P. W. R. Wrigley, Journal of the Chemical Society 1120 (1963)
16. D. Knapp, Kent State University (personal communication)
17. R. E. Wilde, S. N. Ghosh, B. J. Marshall, Inorganic Chemistry *9*, 2512 (1970)

18. L. M. Siperko, T. Kuwana, Journal of the Electrochemical Society *130*, 396 (1983)
19. A. L. Crumblis, P. S. Lugg, N. Morosoff, Inorganic Chemistry *23*, 4701 (1984)
20. M. B. Armand, M. S. Whittingham, R. A. Huggins, Materials Research Bulletin *7*, 101 (1972)
21. A. Ludi, Chemie in unserer Zeit *22*, 123 (1988)
22. H. Kellawi, D. R. Rosseinsky, Journal of Electroanalytical Chemistry *131*, 373 (1982)
23. K. Itaya, T. Ataka, S. Toshima, T. Shinohara, Journal of Physical Chemistry *86*, 2415 (1982)
24. T. Oi, in: R. A. Huggins (Ed.), Vol. 16, *Annual Review of Materials Science* (Annual reviews, Palo Alto, CA, 1986), p. 185
25. K. Itaya, H. Akahoshi and S. Toshima, Journal of the Electrochemical Society *129*, 1498 (1982)
26. K. Itaya, T. Ataka, S. Toshima, Journal of the American Chemical Society *104*, 4767 (1982)
27. I. D. Raistrick, N. Endow, S. Lewkowitz, R. A. Huggins, Journal of Inorganic and Nuclear Chemistry *39*, 1779 (1977)
28. R. Rigamonti, Gazzetta Chimica Italiana *67*, 137, 146 (1937)
29. R. Rigamonti, Gazzetta Chimica Italiana *68*, 803 (1938)
30. A. Xidix, V. D. Neff, Journal of the Electrochemical Society *138*, 3637 (1991)
31. K. Tennakone, W. G. O. Dharmaratne, Journal of Physics C, Solid State Physics *16*, 5633 (1983)
32. K. Honda, H. Hayashi, Journal of the Electrochemical Society *134*, 1330 (1987)
33. V. D. Neff, Journal of the Electrochemical Society *132*, 1382 (1985)
34. K. Ogura, S. Yamasaki, Journal of the Chemical Society Faraday Transactions *81*, 267 (1985)
35. M. K. Carpenter, R. S. Conell, Journal of the Electrochemical Society *137*, 2464 (1990)
36. L. B. Ebert, in: R. A. Huggins (Ed.), Vol. 6, *Annual Review of Materials Science* (Annual Reviews, Inc., Palo Alto, CA, 1976), p. 181
37. R. Yazami, P. Touzain, Journal of Power Sources *9*, 365 (1983)
38. T. Nagaura, K. Tozawa, Progress in Batteries and Solar Cells *9*, 209 (1990)
39. M. B. Armand, in: W. van Gool (Ed.), *Fast Ion Transport in Solids* (North-Holland Publishing, Amsterdam, 1973), p. 665
40. P. J. Nigrey, A. G. MacDiarmid, A. J. Heeger, Journal of the Chemical Society Chemical Communications 578 (1979)
41. A. G. MacDiarmid, A. J. Heeger, Synthetic Metals *1*, 101 (1979/80)
42. P. J. Nigrey, D. MacInnes, D. P. Nairns, A. G. MacDiarmid, A. J. Heeger, Journal of the Electrochemical Society *128*, 1651 (1981)
43. A. Rudge, J. Davey, S. Gottesfeld, J. P. Ferraris, in: B. M. Barnett, E. Dowgiallo, G. Halpert, Y. Matsuda and Z. Takehara (Eds.), *Proceedings of Symposium on New Sealed Rechargeable Batteries and Supercapacitors* (The Electrochemical Society, Pennington, NJ, 1993), p. 74
44. C. Ponce de Leon, A. Frias-Ferrer, J. Gonzalez-Garcia, D.A. Szanto and F.C. Walsh, J. Power Sources *160*, 716 (2006)
45. E. Sum and M. Skyllas-Kazacos, J. Power Sources *15*, 179 (1985)
46. E. Sum, M. Rychcik and M. Skyllas-Kazacos, J. Power Sources *16*, 85 (1985)
47. M. Skyllas-Kazacos, M. Rychcik, R. Robins, A. Fane and M. Green, J. Electrochem. Soc. *133*, 1057 (1985)
48. M. Rychcik and M. Skyllas-Kazacos, J. Power Sources *19*, 45 (1987)
49. M. Rychcik and M. Skyllas-Kazacos, J. Power Sources *22*, 59 (1988)
50. M. Skyllas-Kazacos and F. Grossmith, J. Electrochem. Soc. *134*, 2950 (1987)
51. C. Delmas, L. Croguennec, M. Menetrier and D. Carlier, Presented at the 14th International Meeting on Lithium Batteries, Tianjin, China, June 2008

Chapter 13
Potentials

13.1 Introduction

Potentials and potential gradients are important in battery systems. The difference in the potentials of the two electrodes determines the voltage of electrochemical cells, being larger when they are charged, and smaller when they are discharged. On the other hand, potential gradients are the driving forces for the transport of species within electrodes.

All potentials (potential energies) are relative, rather than having absolute values.

Since they cannot be measured on an absolute scale it is desirable to establish useful references against which they can be measured. This is not the case in electrochemistry alone, but is true for all disciplines. For example, when dealing with the potential energies of electrons in solids the solid-state physics community uses two different references, depending upon the problem being addressed. One is the potential energy of an electron at the bottom of the *valence band* in a solid, and the other is the so-called *vacuum level*, the energy of an isolated electron at an infinite distance from the solid in question. There is no universal relation between these two reference potentials, as the first is dependent upon the identity of the material involved, while the latter is not. This matter is discussed in Sect. 13.1.

In electrochemical systems potential differences are measured electrically as voltages between some reference electrode system and an electrode of interest. The voltage that is measured is a measure of the difference in the electrochemical potentials of the electrons in the two electrodes.

The approaches to this matter are different between the conventional electrochemical community, whose interests have traditionally been mostly concerned with aqueous systems, and the solid-state electrochemical community, many of whose members have come from solid-state materials background. This is despite the fact that some of the electrochemical systems of interest to the latter group also often include liquid electrolytes. As will be seen in Sect. 13A.9, a characteristic difference is in the focus on the properties of neutral species in the solid-state electrochemical community, and upon ionic species in the aqueous electrochemical community.

R.A. Huggins, *Advanced Batteries*: Materials Science Aspects,

Chemical reactions are sometimes used to impose specific potentials upon materials under study. This is discussed in Sect. 13B.13.

The matter of the distribution of the different electrical and chemical potentials within electrochemical cells is often misunderstood. This often depends upon the experimental conditions, and is discussed in Sect. 13C.4.

13A Potentials in and Near Solids

13A.1 Introduction

The term *potential* is often used for both a single potential and a potential difference. The standard practice in electrochemistry is to use certain reactions to provide reference electrode potentials against which other potentials can be measured. In aqueous systems a standard procedure is to use the reaction

$$2H^{+} + 2e^{-} = H_2 \tag{13A.1}$$

as the reference potential, and electrodes which involve this reaction are often called standard hydrogen electrodes (SHE), as discussed in Sect. 13A.9. On the same scale the potential of a lithium electrode at which the reaction is

$$Li^{+} + e^{-} = Li. \tag{13A.2}$$

occurs at a potential that is $-3.045V$ with regard to the potential of the SHE. Further, a fluorine electrode operating at a pressure of 1 atm of fluorine gas and for which the reaction can be written as

$$2F^{-} - 2e^{-} = F_2 \tag{13A.3}$$

has a potential of $+2.87$ V relative to the SHE at ambient temperature.

These potentials will be modified somewhat in other electrolytes because of differences in the *solvation energies*. If the solvation energy is not considered, the difference in electrode potential is always equal to that of the Nernst equation voltage for neutral species outside the electrolyte and one can always write

$$\Delta G_r{}^{\circ} = -zFE, \tag{13A.4}$$

where $\Delta G_r{}^{\circ}$ is the Gibbs free energy change of the relevant reaction, z is the number of electrons transferred in the reaction to which $\Delta G_r{}^{\circ}$ refers, F is the Faraday constant, and E is the cell voltage, which is equal to the difference in the electrode potentials on the two sides.

13A.2 Potential Scales

Another alternative way of looking at electrode potentials involves the use of a general potential scale based upon a particular reaction equilibrium. In molten salts, for example, it may be useful to use the chlorine or fluorine evolution electrode reaction as a reference against which other electrode potentials are measured.

In the subsequent chapters, reference will be made to chemical and electrostatic *macropotential differences* across a solid, as well as to gradients in those potentials within the solid. The chemical and electrostatic potentials are only two of a number of thermodynamic potentials. Since there is often a considerable amount of confusion relating to the different types of potentials inside solids and near their surfaces, this question will now be considered.

Understanding of the spatial distribution of the various thermodynamic potentials within a solid is important because of the relationship between the values of specific potentials and the local structure. Since many of the properties of solids are dependent upon the local structure, variations in properties with position are both possible and commonplace. Furthermore, under proper conditions, they can be controlled to advantage.

This relationship between local potentials, structure, and properties leads to two general types of application. Local values of some potentials may be experimentally observed by the proper use of appropriate probe techniques. This can lead to valuable information about the structure, and can therefore be used as an analytical techniques for a host of purposes. In addition, however, the situation can be reversed, and specific values of certain potentials can be imposed upon a material in order to change or control its structure and properties.

The total thermodynamic potential for a given species i at any point can arbitrarily be composed of several factors. Each of these factors has the dimensions of energy, as does the total thermodynamic potential. The total thermodynamic potential of a particular species has the properties of potential energy, and gradients in it produce forces tending to cause the superposition of a long-range drift motion upon the local random motion of that species within the solid.

13A.3 Electrical, Chemical, and Electrochemical Potentials in Metals

First, consider the matter of electrical potentials and the various related electrostatic potentials of individual species. To compare electrical potentials as well as the electrostatic energies of charged particles within and near different solids it must be recognized that neither of these quantities has an absolute value. Therefore, it is desirable to establish some sort of reference level electric potential. For this purpose it is useful to compare the thermodynamic potential of a charged particle i within a solid phase with the potential of an isolated particle of the same chemical composition in a vacuum at an infinite distance from all other charges. The value of the electrical potential at this charge-free infinity is defined arbitrarily as 0. This fixed reference value is called the *reference vacuum level*, E^{∞}. Unfortunately, the term *vacuum level* is also used in some of the current semiconductor literature for a different potential, as will be described later.

Consider a hypothetical experiment in which this charged particle is transferred from infinity to a position inside a solid. In the absence of any other potential gradients, the work that would be done can be divided into two parts. One of these is due to the interaction of the particle with the other particles within the bulk solid phase. This will typically include electrostatic, polarization, and repulsive interactions, and thus is dependent upon the identity of the particle, as well as the constitution of the bulk solid. It represents the chemical binding energy of the species in the solid, and is called the *chemical potential* of particle i, μ_i.

The second part of the work involved in transferring the particle from infinity to the interior of the solid is purely electrostatic, and is thus $z_i q(\Phi - E^{\infty})$, where z_i is the charge number of the particle (and represents both the sign and the number of elementary charges carried), q is the magnitude of the elementary charge (the value of the charge of the proton), and Φ is the local value of the *electrostatic macropotential* within the solid, which is called the *inner potential*.

The total work involved in this hypothetical experiment is called *the electrochemical potential* of the particle of species i, η_i, within the solid, and since $E^{\infty} = 0$, it can thus be written as

$$\eta_i = \mu_i + z_i q \Phi. \tag{13A.5}$$

If the interior of the solid can be considered to be compositionally homogeneous, it will thus have a uniform value of the inner potential Φ. However, the solid phase must have an exterior surface that separates it from its surroundings whether vacuum, gas, liquid, or another solid phase.

For simplicity, consider the case of an isolated solid phase surrounded by vacuum. Because of the structural discontinuity at the surface, there must be local redistributions of both particles and electrical charge compared with the configurational structure within the bulk solid. There can also be differences in the concentrations of intrinsic species between the surface and the interior, as well as adsorbed foreign species upon the surface.

It is useful to utilize a simple model in which the solid is divided into two parts, a uniform interior and a separate surface region. The latter is sometimes called the *selvedge*, the term used for the edge region of a piece of cloth, which is often woven differently from the interior to prevent it from unraveling. The selvedge thus contains all of the various local redistributions and compositional effects, which can be described as producing an electrical double layer. In addition, this surface region contains all the excess charge if the solid has a net electrostatic charge different from E^{∞}.

Therefore, the value of the inner potential Φ can be divided into two terms, one called the *surface potential* X, which is related to the dipolar effects of the electrical double layer in the selvedge. The second is known as the *outer potential* Ψ, and is the net externally measurable electrostatic potential of the solid.

The value of outer potential Ψ is dependent upon the amount of excess charge Q and the dimensions of the solid. For the case of a sphere of radius a,

$$\Psi = Q/a. \tag{13A.6}$$

The surface potential can be interpreted in terms of a simple model consisting of a uniform distribution of dipoles of moment M perpendicular to the surface with a concentration of N per cm^2 within the selvedge. If the positively charged ends of the dipoles are on the outside,

$$X = -4\pi NM. \tag{13A.7}$$

The relationship between these potentials for the case of two chemically identical solids with different amounts of excess charge, and thus different values of the outer potential, is shown in Fig. 13A.1.

Changes in the charge on a solid body are actually accomplished by the transfer of charged particles, e.g., electrons, so that the composition is actually slightly changed when the net charge is varied. However, this involves such minor changes in the concentrations of the particles present in the solid that they can be neglected.

Now consider the question of the *energies* (also sometimes called *potentials*) of charged species. As mentioned already, the electrostatic part of the total potential energy of a particle i of charge $z_i q$ inside a solid with an inner potential Φ is $z_i q\Phi$. In the case of electrons, $z_i = -1$. Thus, this part of the total potential energy of an

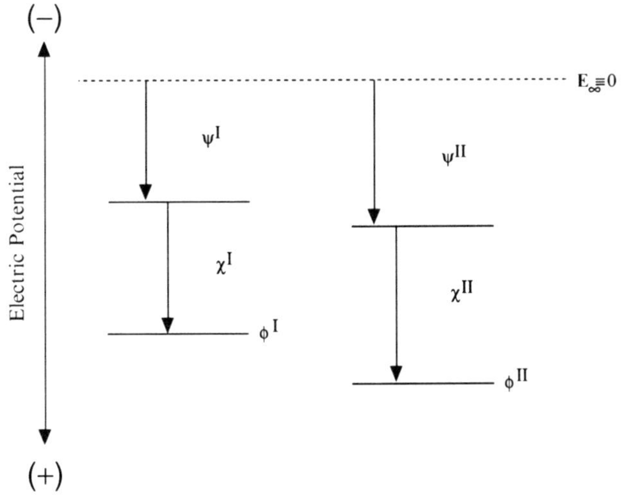

Fig. 13A.1 Relationship between potentials related to two chemically identical solids with different values of outer potential

electron has an absolute magnitude that is greater and more positive the lower the value of Φ, since Φ is negative relative to the zero reference E^{∞}.

Therefore, the total energy of an electron within a solid is its electrochemical potential η'_{e}, which has two components. One is related to the fact that the electron is within the solid and has the characteristics of a *chemical binding energy*, and the other is *purely electrostatic*. Thus

$$\eta_{e^-} = \mu_{e^-} + z_{e^-} q\Phi \tag{13A.8}$$

and

$$\eta_{e^-} = \mu_{e^-} + RT \ln a_{e^-} + z_{e^-} q(\Psi + \mathrm{X}). \tag{13A.9}$$

These energy relations for a single electron in the interior of a metal are shown in Fig. 13A.2. Note that the value of the chemical potential for the electron is also negative.

To reinforce the understanding of these matters, a hypothetical experiment can be considered in which a single electron exists between two parallel plates of the same metal which are maintained in a vacuum, but are connected to the opposite terminals of a battery. This is illustrated in Fig. 13A.3, in which the right hand plate in connected to the positive pole, and the left hand plate to the negative pole of the battery.

There will be a force acting upon the electron that is proportional to the negative value of the gradient in its potential energy in the vacuum $ze^{-}qd\psi/dx$. This will cause it to accelerate from left to right. This is consistent with our expectation from general electrostatics that a negatively charged particle will be attracted to a positively charged electrode.

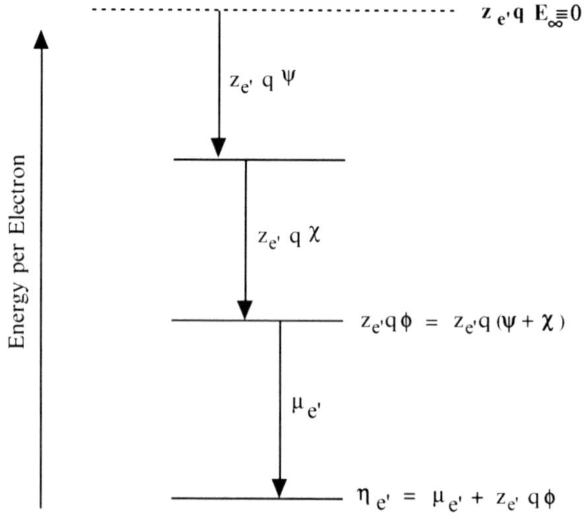

Fig. 13A.2 Energy relations for a single electron in a metal

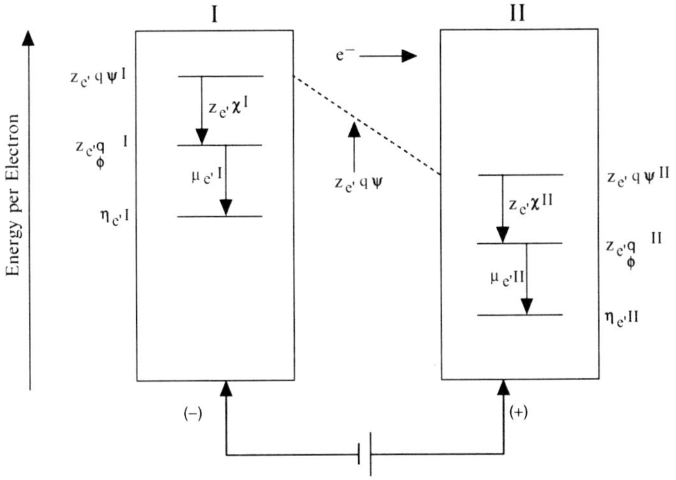

Fig. 13A.3 Relationship between the potentials in two identical materials connected to the terminals of a battery

It can be seen from this example that the values of $z_i q\Psi$, μ_i, and η_i all change as the externally measurable electrostatic potential of the solid Ψ is varied. However, if the chemical constitution is not altered, the values of X and μ_i remain the same, so that the relative values of $z_i q\Psi$, $z_i q\Phi$, and η_i do not change.

It will be seen later that the quantity Ψ can be varied externally, as in the earlier example, and also experimentally measured. It has therefore been found useful to define another quantity, the *real potential* α_i, which is independent of the value of Ψ. This is given as

$$a_i = \mu_i + z_i q \mathrm{X}. \tag{13A.10}$$

For the case of an uncharged solid, where $\Psi = 0$ and $\alpha_i = \mu_i, \alpha_i$, which generally has a negative value, is the work done in transferring a particle of species i from infinity to the interior of the uncharged solid.

The real potential of an electron $\alpha_{e'}$ thus has the same magnitude, but opposite sign, as the *electronic work function* $W_{e'}$, which is defined as the Gibbs free energy necessary to extract an electron from an uncharged solid into an exterior vacuum (at infinite distance). That is,

$$\alpha_{e^-} = W_{e^-}. \tag{13A.11}$$

The *binding energy* or chemical potential of species i can be written as

$$\mu_i = \mu_i^0 + RT \ln a_i, \tag{13A.12}$$

where μ_i^0 is the chemical potential in some standard state. In the case of electrons in a metal, $\mu_{e^-}^0$ is chosen as the chemical potential of an electron in the chemically pure metal. The activity of the electron in the pure metal is also defined as unity, so that in this case

$$\mu_{e^-} = \mu^0_{e^-}. \tag{13A.13}$$

The activity of any species i is related to its concentration $[i]$ by

$$a_i = \gamma_i[i], \tag{13A.14}$$

where γ_i is the activity coefficient, expressed in appropriate units. Thus, both the chemical potential μ_i and electrochemical potential η_i are composition-dependent.

In metals the concentration of electrons is typically very high, so that minor changes in composition due to impurities or doping produce negligible effects upon μ_{e^-} and η_{e^-}. However, this factor should be taken into consideration in more heavily alloyed metals as well as in nonmetals. The two contributions to μ_i, and thus to η_i, are shown in Fig. 13A.4.

Because the real potential and the work function include the term z_iqX that relates to the electrical double layer effects in the selvedge, these values are dependent upon the details of the structure in that region. Experiments have shown that this includes both the crystal face from which electrons are emitted and the presence of any impurities upon the surface. Some experimental values are given in Table 13A.1 for polycrystalline metals. Table 13A.2 shows the variation with crystal face on single crystals of several metals.

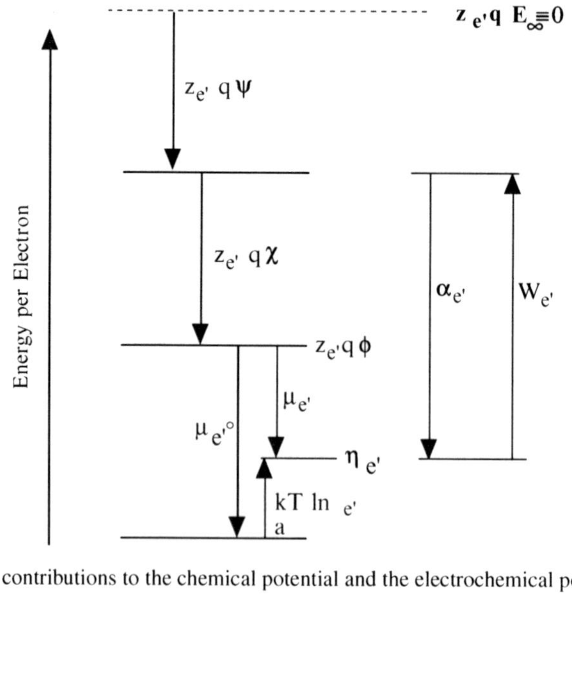

Fig. 13A.4 The contributions to the chemical potential and the electrochemical potential

Table 13A.1 Values of the electronic work function measured on polycrystals

Metal	Work function (eV)
Ag	4.33
Ba	2.39
Be	3.92
Ca	2.71
Co	4.41
Cs	1.87
Fe	4.48
K	2.26
Li	2.28
Mg	3.67
Mo	4.20
Na	2.28
Ni	4.61
Rb	2.16
Ta	4.19
U	3.27
W	4.49
Zn	4.28

Table 13A.2 Crystallographic orientation dependence of the work function – single crystal

Metal	Normal to surface	Work function (eV)
Cu	111	4.39
Cu	100	5.64
Ag	111	4.75
Ag	100	4.81
W	111	4.39
W	112	4.69
W	001	4.56
W	110	4.68

13A.4 Relation to the Band Model of Electrons in Solids

A combination of knowledge about the variation of the density of energy states available for electron occupation with energy and the state occupation probability, which is expressed in terms of the Fermi-Dirac relation, provides information about the distribution of the electrons within a solid among their allowed energy states. This type of information is often displayed in simplified form by use of an *energy band model*, in which the energy per electron is plotted vs. distance. An example of a simple band model of a metal is shown in Fig. 13A.5. This figure also shows the relationship to the various thermodynamic potentials discussed here.

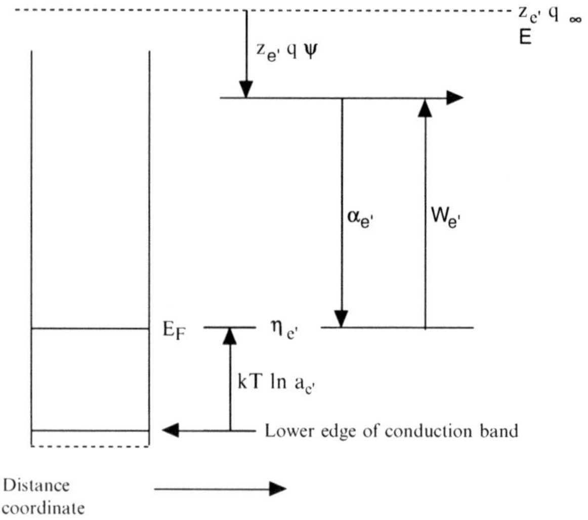

Fig. 13A.5 Simple band model of a metal

13A.5 Potentials in Semiconductors

While the discussion thus far has centered upon metals, similar conclusions are found for semiconductors. In (undoped) intrinsic semiconductors, $\mu_{e^-} = \mu^0_{e^-}$. and the electrochemical potential is the same as the *Fermi level* E_F. The Fermi level is midway between the electron energies at the top of the valence band E_V and the bottom of the conduction band E_C, regardless of the temperature. This situation is shown in Fig. 13A.6.

An important difference between metals and semiconductors, however, is that the electrostatic contribution to the total energy of an electron $z_{e^-}q\Phi$ is generally not the same throughout the solid in semiconductors. It often increases or decreases significantly near the surface, or at locations where the chemical composition varies, such as at *p–n junctions*.

What if the semiconductor is doped with an *altervalent*, or *aliovalent*, species, an atom that carries a different amount of electrical charge from those that are normally present? The *electroneutrality requirement* causes the ratio of conduction electrons to holes to change to compensate for the charge of this *foreign species*, or *dopant*. As an example, if *donors* are present the concentration of itinerant conduction band electrons is increased. The activity of the electrons is thus greater in such an *extrinsic* semiconductor than in the corresponding *intrinsic* (undoped) material. This raises the values of $kT\ln a_e^-$, thus reducing μ_e^- (which is negative), and raising η_e^-. In semiconductor band model language this raises E_F toward E_C. This also, of course, decreases the work function W_e^- since one can assume that X is not changed.

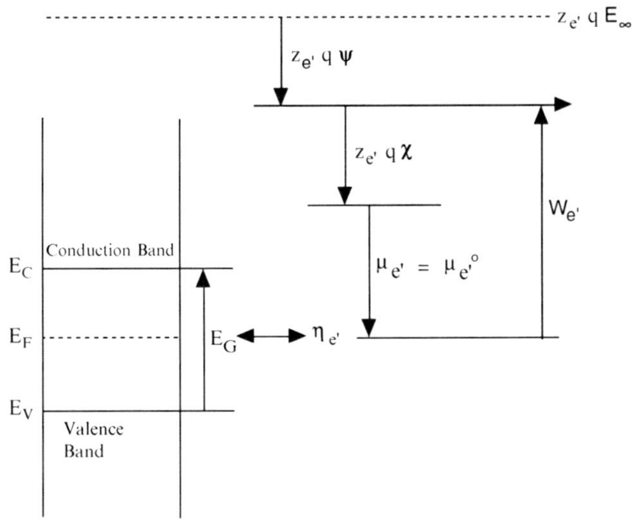

Fig. 13A.6 Relation between thermodynamic potentials and potentials commonly used in the energy band model of an intrinsic semiconductor. E_G is the band gap (E_C-E_V). The potential of the Fermi level E_F is equal to the electrochemical potential of the electrons, η_e^-

13A.6 Interactions Between Different Materials

Many applications of these concepts involve interactions between the various potentials and energies that have been discussed here. To understand such matters a simple case will be considered.

Consider two chemically different metals. If they are physically separated and not in equilibrium with each other, their potentials can be portrayed as illustrated in Fig. 13A.7.

13A.7 Junctions Between Two Metals

If these two metals are brought into contact, so that thermal equilibrium can be established between them, electrons will flow from one to the other until the total energy per electron is the same in both. This means that after equilibrium is attained, $\mu_{e'}^{I}$ must be equal to $\mu_{e'}^{II}$, or in band model language, $E_F^I = E_F^{II}$. The question is how this is achieved. The values of $\mu_{e'}^{\infty I}$ and $\mu_{e'}^{\infty II}$ are chemical binding energies of electrons in the lowest levels of the respective conduction bands. These are thus fixed by the chemical compositions of the two metals. The values of $kT \ln a_e'$ are determined by the electron concentrations in the metals. Since these concentrations are typically very high in metals, the relatively small number of electrons that pass from one metal to the other upon contact will make relatively small changes in the activities of the electrons. Thus, this term will also not change significantly.

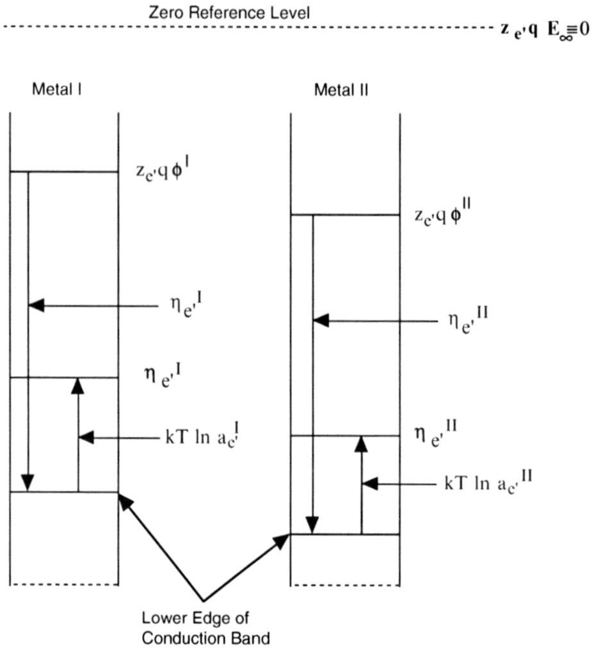

Fig. 13A.7 Potentials of two chemically different metals separated by a vacuum, and not in equilibrium with each other. The relative positions of the Fermi levels (electrochemical potentials) are arbitrary, depending upon prior history

This means that the equilibration of the Fermi levels occurs primarily by the adjustment of the electrostatic energy term $z_{e'}q\Phi$ or $z_{e'}q(\Psi+\mathrm{X})$.

Since in each case $\eta_{e'}=\mu_{e'}+z_{e'}q\Phi$, upon equilibration of the $\eta_{e'}$ values (Fermi levels) a fixed value of the difference in the internal electrostatic potentials will be established, directly related to the difference in the chemical potentials of electrons in the two metals. That is

$$z_{e'}q(\Phi^{I}-\Phi^{II})=\mu_{e'}^{II}-\mu_{e'}^{I}. \tag{13A.15}$$

The value of $z_{e'}q(\Phi^{I}-\Phi^{II})$ is called the *Galvanic voltage*, or *Galvanic potential difference*, and is characteristic of the two metals in question. It cannot be measured, however, because it is not possible to separate the two contributions to the value of $\Delta\Phi$, the differences in the outer potential $\Delta\Psi$ and in the surface double layer potential $\Delta\mathrm{X}$. The transfer of only a relatively few electrons from one metal to the other is expected to modify the X values significantly. As a result, $(\Phi^{I}-\Phi^{II})$ is not equal to $(\Psi^{I}-\Psi^{II})$.

However, it is possible to measure experimentally $(\Psi^{I}-\Psi^{II})$, since these are externally observable values. The energy difference $z_{e'}q(\Psi^{I}-\Psi^{II})$ is called the *Volta potential difference*. It is also sometimes called the *contact potential*. The use of this latter term is unfortunate, for it actually relates to a difference in electric potential between two free surfaces that are *not* in physical contact with each other.

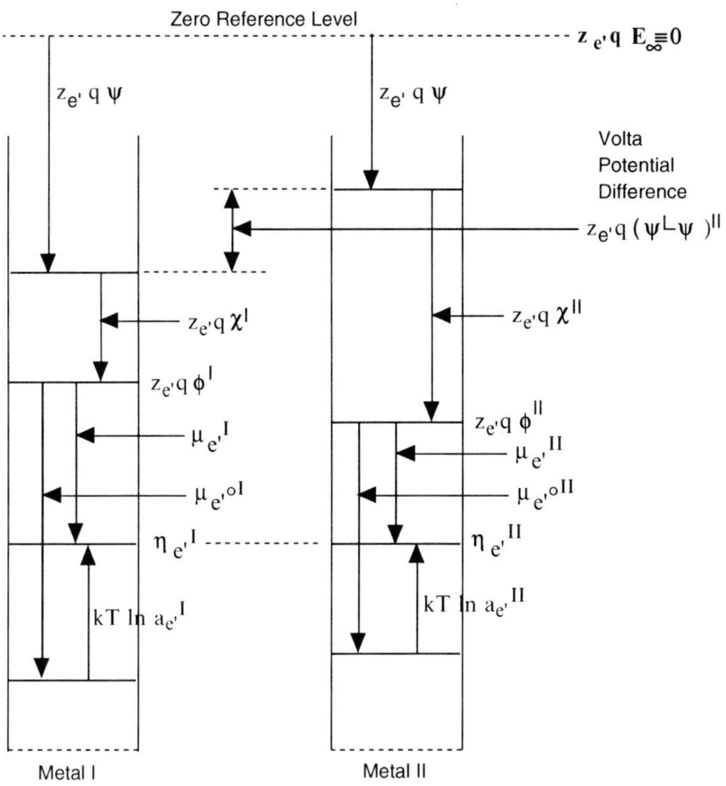

Fig. 13A.8 Relationship between various potentials when two chemically different metals are brought into electronic equilibrium, so that their Fermi energies become equal. For purposes of illustration, the two metals are shown physically separated

The relations between the various potentials when two chemically different metals are brought into equilibrium are illustrated in Fig. 13A.8.

It is seen that the Volta potential difference is equal to the difference in electron work functions

$$z_{e'}q(\Psi^{I} - \Psi^{II}) = W_{e'}^{I} - W_{e'}^{II} \tag{13A.16}$$

or

$$z_{e'}q(\Psi^{I} - \Psi^{II}) = z_{e'}q[(\Psi^{I} - \mu^{I}) - (\Psi^{II} - \mu^{II})]. \tag{13A.17}$$

13A.8 Junctions Between Metals and Semiconductors

Similar considerations are important in the case of equilibration between a metal and a semiconductor. Again, the important feature is that the Fermi levels must be equal under equilibrium conditions. This will not be discussed here, however. The principles are the same as have been elucidated in the last few pages, and this topic is addressed in many other places in the literature.

13A.9 Selective Equilibrium

This discussion has assumed that local equilibrium can be maintained within solids. In most practical cases this is reasonable at elevated temperatures. However, it is often not true for all species at lower temperatures. Indeed, it is often found that *selective equilibrium* occurs at low (e.g., ambient) temperatures. The concentrations of less mobile species can be established during processing at high temperatures, and *frozen in* by cooling to lower temperatures, where they may not be in accord with low-temperature equilibria. More mobile species can react to the influence of various forces, and reach appropriate equilibria at lower temperatures. The frozen-in less-mobile species do, however, influence the local electrostatic charge balance, and thus can play a major role in determining the concentrations of the more mobile defects, as all species participate in the charge balance.

13B Reference Electrodes

13B.1 Introduction

Reference electrodes play an important role in the study of many aspects of electrochemical systems. Experimental work reported in the literature can involve the use of different reference systems, and it is sometimes difficult to translate between measurements made with one from those made using another.

This situation is made even worse by the fact that reference electrodes that are used in solid state electrochemical systems are based upon the potentials of electrically neutral chemical species that can be understood by the use of normal chemical thermodynamics. On the other hand, the general practice in aqueous electrochemistry is to use reference electrodes that involve the properties of ions, and the pH of the electrolyte becomes important in some cases, but not in others.

These matters are discussed in terms of the Gibbs phase rule, showing the difference between zero-degree-of-freedom (ZDF) electrodes, and those in which an additional parameter, such as the electrolyte pH, must be specified.

The interrelationship between these two types will be illustrated using potential–pH plots, or *Pourbaix diagrams*. Use of this *thinking tool* provides a simple understanding of the glass electrode systems that are used to measure the pH of electrolytes, for example.

It will also be shown that in electrodes with a mixed-conducting matrix and an internal ZDF reaction, the potential is determined by the internal chemical reaction, rather than the external electrochemical reaction.

13B.2 Reference Electrodes in Nonaqueous Lithium Systems

Much of the current interest in batteries involves lithium-based systems with nonaqueous electrolytes. Thus, attention should first be directed to the matter of reference electrodes in lithium systems.

13B.2.1 Use of Elemental Lithium

Pure metallic lithium is typically used as a reference electrode in experimental activities to investigate the properties of individual electrode components, both those of interest as negative electrodes and those that act as positive electrodes, at ambient and near-ambient temperatures.

Because it is so extremely reactive, it is very difficult to maintain the surface of lithium free of oxide or other layers in even the cleanest gaseous and liquid environments. It is also important to realize that the organic electrolytes that are often used with lithium reference electrodes are typically not stable in the presence of elemental lithium. Reaction product layers are commonly present on the surface of the lithium, and separate the lithium from the bulk electrolyte. This topic is discussed in Sects. 14A.6 and 16A.6.

Therefore, the reaction that takes place at the electrochemical interface is typically not really known. It is also important that the identity of the electrolyte is not important, so long as it acts to transport Li ions, and not electrons. Despite these factors, elemental lithium is a widely used and highly reliable primary potential reference in a wide range of lithium-based electrochemical systems.

13B.2.2 Use of Two-phase Lithium Alloys

Some years ago there were substantial efforts to develop elevated temperature lithium-based batteries. Since they operated above the melting point of metallic lithium, metallic lithium could not be used. One of the reference electrodes that was often employed was a two-phase mixture of aluminum and the phase LiAl.

In auxiliary experiments the potential of that electrode could be compared to that of pure lithium, which was considered to be the primary reference, and to which all potentials were referred. Because of the entropy change involved in the formation of LiAl by the reaction of lithium with aluminum, this difference is temperature-dependent. Experiments [1] from 375 to 600°C showed that the potential of a two-phase mixture of lithium and LiAl is more positive than that of pure lithium, and that this potential difference ΔE can be expressed by the following relation:

$$\Delta E = 451 - 0.220T(K), \tag{13B.1}$$

where ΔE is in millivolts.

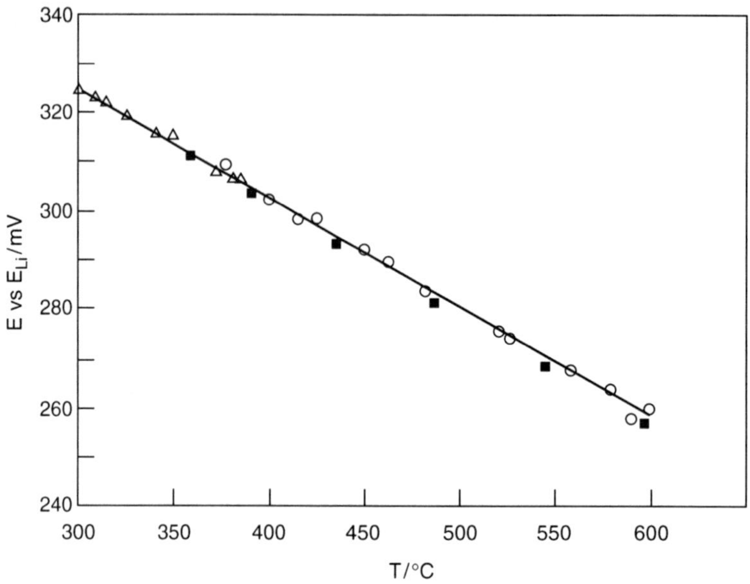

Fig. 13B.1 Temperature dependence of the voltage between two-phase Li–LiAl electrode and pure lithium

This is shown in Fig. 13B.1 [1]. The reason for the use of a two-phase mixture in this case, and why it is suitable, will become obvious from the discussion below.

Because of their high lithium activity, these two-phase electrodes also can have reaction product layers on their surfaces in the presence of some electrolytes. As in the case of pure lithium reference electrodes, neither the presence of such layers nor the details of the interfacial reactions have any significant influence upon their potential.

13B.3 Reference Electrodes in Elevated Temperature Oxide-Based Systems

Electrochemical systems with solid electrolytes are employed in high-temperature fuel cells, oxygen sensors, and related applications. In such cases there are also two general types of reference electrodes used.

13B.3.1 Gas Electrodes

An inert metal such as platinum in contact with pure oxygen gas is often used as a primary reference in these systems. Alternatively, air or some other gas with a known oxygen activity can be used. Gases with lower oxygen partial pressures will have less positive potentials. The potential difference between pure oxygen and a gas with a lower oxygen partial pressure is typically expressed in terms of the Nernst equation:

$$\Delta E = -RT/zF \ln p(O_2), \tag{13B.2}$$

where $p(O_2)$ is the oxygen partial pressure of the gas in question, R is the gas constant, z the charge carried by the electrons involved in the assumed reaction (-4), and F the Faraday constant. Air is often used as a reference instead of pure oxygen. Using (13B.2) it is possible to compare the potential of an air reference with that of pure oxygen. If it is assumed that air has an oxygen partial pressure of 0.79 atm, ΔE is equal to 6.09 mV at 1,200 K, or 927°C. Thus, the air reference potential is 6.09 mV lower than that of pure oxygen at that particular temperature.

13B.3.2 Polyphase Solid Reference Electrodes

An alternative to the use of a gas reference electrode is the use of solid electrodes. One example is a two-phase mixture of Ni and Ni_xO. If conditions are such that an equilibrium between Ni and its oxide can be attained, this combination will have a fixed value of oxygen activity, equal to that in Ni_xO at its Ni-rich compositional limit. Thus, this two-phase mixture can be used as a secondary reference instead of pure oxygen. The oxygen activity and the potential relative to oxygen can be calculated if the Gibbs free energy of formation of NiO is known. The formation reaction is as follows:

$$Ni + 1/2O_2 = NiO. \tag{13B.3}$$

The potential of this material combination is also less positive than that of pure oxygen. The difference can be calculated from the simple relation:

$$\Delta E = -(\Delta G_r{}^{\circ}/zF), \tag{13B.4}$$

where $\Delta G_r{}^{\circ}$ is the Gibbs free energy change involved in the formation reaction. In this case, $z = -2$, as only one oxygen atom is involved. Because the Gibbs free energy contains an entropy term, the value of ΔE will be temperature-dependent. At 925°C the value of $\Delta G_r{}^{\circ}$ is -132.16 kJ mol^{-1}. Thus, the potential of the two-phase Ni, Ni_xO system is 0.685 V less positive than the potential of pure oxygen at that temperature.

One should note that, as was the case in the lithium systems, the potentials and potential differences in these oxide-related cases are independent of the details of the interfacial reactions. The identity of the electrolyte is not important, so long as it effectively transports oxygen ions and has a relatively low electronic conductivity.

The open circuit voltage of an H_2/O_2 fuel cell is also independent of the details of the interfacial electrochemical reactions as well as the identity and detailed properties of the electrolyte. Regardless of whether the electrolyte transports hydrogen ions or oxygen ions, the voltage is always determined by the thermodynamics of the reaction in which water is formed from hydrogen and oxygen. The electrolyte does not need to be solid, it can also be liquid, and either acidic or basic. It can also have a composite structure, such as when a liquid electrolyte is contained within a solid polymer, such as Nafion.

13B.3.3 Metal Hydride Systems

Similarly, the potentials of metal hydride electrodes and the voltages of electrochemical cells employing them are independent of the interfacial reactions and the electrolyte, depending only upon the difference between the hydrogen activities of the electrode materials.

13B.4 Relations Between Binary Potential Scales

One can determine the relation between different potential scales if they both refer to a common reference. As an example, consider the relation between a scale based upon the potential of pure lithium, or one based on sodium. Lithium and sodium both react with oxygen to form their respective oxides. If it can be assumed that those reactions were to occur with oxygen at unit activity (1 atm), the difference between the potentials of Li and Na and that of pure oxygen can be calculated from their respective oxide formation reactions.

At 25°C, the Gibbs free energy of formation values [2] are $-562.104\,\mathrm{kJ\,mol^{-1}}$ for Li_2O, and $-379.090\,\mathrm{kJ\,mol^{-1}}$ for Na_2O. Using the relation between the voltages and these Gibbs free energies of formation, it is found that the potential of pure Li is 2.91 V, and that of sodium 1.96 V, negative of pure oxygen at 298 K. Those values are the ranges of stability of their respective oxides.

13B.5 Potentials in the Ternary Lithium–Hydrogen–Oxygen System

This situation is modified in the case of a ternary system. As an example, if lithium is also present in addition to hydrogen and oxygen, the potential limits of the stability of water are shifted. To understand this, the isothermal Li–H–O ternary phase diagram must be considered.

This ternary system was discussed briefly in Sect. 9B.3. There are several phases in this system in addition to the elements: Li_2O, LiH, LiOH, $LiOH.H_2O$, and H_2O. The values of the standard Gibbs free energy of formation of these phases are given in Table 13B.1.

Table 13B.1 Values of the standard Gibbs free energy of formation of phases in the Li–H–O System at 25°C

Phase	$\Delta G_f{}^\circ$ (kJ mol^{-1})
Li_2O	−562.1
LiOH	−439.0
$LiOH.H_2O$	−689.5
H_2O	−237.1
LiH	−68.5

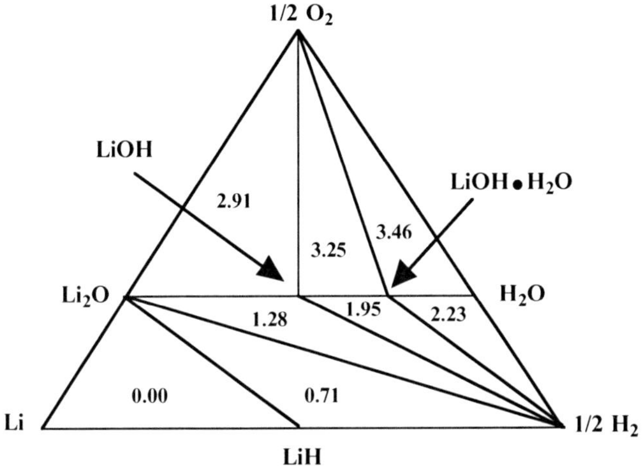

Fig. 13B.2 Isothermal Gibbs triangle for the Li–H–O system at 25°C. The numbers within the subtriangles are the calculated values of their respective potentials vs. pure lithium

The locations of these phases are shown in the isothermal ternary Gibbs triangle in Fig. 13B.2, which was also included in Sect. 9B.3. Assuming that all of these phases are at unit activity, the potentials of all of the three-phase subtriangles can be calculated with respect to each of the elements at the corners of the Gibbs triangle, from thermodynamic information, as discussed earlier. The values shown in that figure are voltages vs. pure lithium.

The potential range in which water is stable is bounded by the potentials of two triangles, the three-phase triangles that have H_2O at their corners. It can be seen that their potentials with respect to lithium are 2.23 and 3.46 V and are 1.23 V apart. The presence of the phases LiOH and $LiOH.H_2O$ caused their potentials to both shift in the positive direction relative to that of Li.

It was pointed out in Sect. 9B.3 that these calculations relate to a very basic aqueous electrolyte, with a pH value of 14. Conversion of the potential of the triangle that has both hydrogen and water present (2.23 V) to that at pH 0 can be done by adding the product of 14 and 0.059 V, the change in potential per pH unit. The result is 3.05 V, which is the value found in electrochemical tables for the potential of the standard hydrogen electrode (SHE).

13B.5.1 Lithium Cells in Aqueous Electrolytes

It was pointed out in Sect. 9B.3 that this relationship between these different potential scales means that if a material has a potential that is between 2.09 and 3.32 V positive of pure lithium and does not dissolve or otherwise react chemically, it will be stable vs. water containing LiOH. Thus, one can use electrodes that react with lithium in aqueous electrolytes if Li ions are present in the electrolyte. Lithium-based electrochemical cells can operate in aqueous electrolytes, so long as both electrodes react with lithium and their potentials are within this range. This has been demonstrated experimentally [3–6]. Figure 13B.3 shows cyclic voltammograms of $VO_2(B)$ in two different aqueous electrolytes, one containing Li ions, and the other not [6]. Since the only appreciable reaction occurs when the Li ions are present in the water, it is obvious that the electrode reacts with Li, rather than hydrogen or oxygen.

If the lithium activity is too high in such an electrode, i.e., it has a potential less than 2.23 V vs. pure lithium in water of pH 14, it will reduce water, forming H_2 and $LiOH \cdot H_2O$.

13B.6 Significance of Electrically Neutral Species

An important feature of the discussion of both nonaqueous and aqueous systems has been that the potentials and thermodynamics of electrically neutral species in

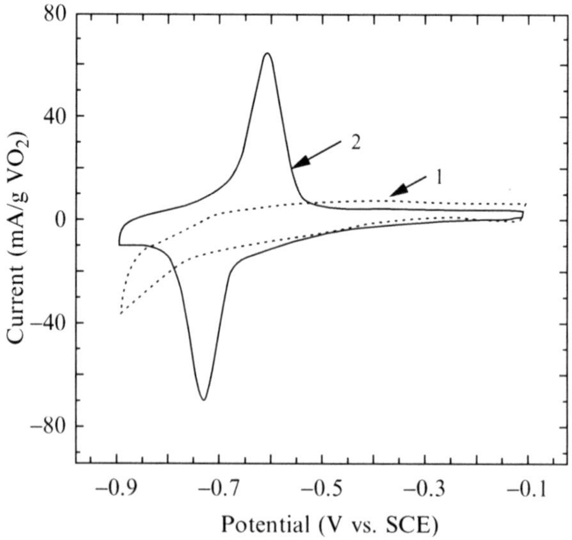

Fig. 13B.3 Cyclic voltammograms of VO_2 (B) in two different aqueous electrolytes. Scan 1 was made in one without lithium ions, whereas lithium ions were present in the electrolyte of scan 2

the electrodes are important. Potentials and voltages are independent of the identity, or even the character, of the electrolyte. They are directly related to the normal chemical thermodynamic properties of the electrode materials. Reference electrodes are typically elements or thermodynamically related polyphase mixtures that are electrically neutral.

13B.7 Reference Electrodes in Aqueous Electrochemical Systems

The examples discussed earlier indicate that the reference electrode situation is quite straightforward and is consistent with conventional thermodynamics in nonaqueous systems. However, this is quite different when dealing with aqueous systems, which are within the domain of traditional electrochemistry.

If one looks into the older electrochemical literature, he finds statements such as that reference electrodes were considered somewhat of a *black art* for many years, with information primarily passed on by word of mouth or in brief notes among workers in electrochemistry [7]. A significant step forward was the book entitled *Reference Electrodes* that was edited by Ives and Janz in 1961 [8]. Another source that is often cited is the review article by Butler in 1970 that dealt with reference electrodes in aprotic organic solvents [7].

This general approach to the reference electrode matter is quite different from that described earlier in this chapter. One major difference is that a property of the electrolyte, the hydrogen ion concentration, as expressed in the form of the pH, is generally considered important. This is different from the earlier examples, in which the electrolyte plays no role other than acting as an ion-pass and electronically impervious filter.

It is generally accepted in the electrochemical community that the primary reference electrode in aqueous systems should be the so-called *standard hydrogen electrode* (SHE). It is sometimes also called the *normal hydrogen electrode* (NHE). This electrode involves the use of H_2 gas at a pressure of 1 atm flowing over an inert metallic contact material (generally platinum) in an electrolyte in which the activity of hydrogen ions (not atoms or molecules) is unity. Pains are generally taken to obtain a large gas/metal contact area that is not blocked by the presence of intermediate products.

Pure water dissociates into its component ions H^+ (or more properly, H_3O^+, $H_5O_2^+$, or $H_9O_4^+$) and OH^- only to a small extent, with the degree of dissociation equal to about 1.4×10^{-9}. This means that there are more than 7×10^8 molecules of water for each H^+ or OH^- ion. As in the case of electrically charged defect pair equilibria in solids, the product of their concentrations is a constant. The ionic product of water, K_W, which is defined as

$$K_W = [H^+][OH^-] \qquad (13B.5)$$

has been found to be approximately 10^{-14}.

Table 13B.2 Examples of reference electrodes used in aqueous systems

Electrode	pH Voltage vs. SHE at pH = 0 (V)
Hg/HgO – 0.1 M NaOH	0.926
Hg/Hg_2Cl_2–0.5 M H_2SO_4	0.68
Hg/Hg_2SO_4–sat. K_2SO_4	0.64
Hg/Hg_2Cl_2–0.1 M KCl	0.3337
Hg/Hg_2Cl_2–1 M KCl	0.2801
Hg/Hg_2Cl_2	0.2681
Calomel – sat. KCl	0.2412
Calomel – sat. NaCl	0.2360
Ag/AgCl	0.2223

In solutions of acids or bases the relative concentrations of these two ionic species can vary over many orders of magnitude, mostly much less than unity. A logarithmic function, pH, was introduced as a measure of the concentration of one of them, the H^+ ions. Because of the ionic product equilibrium, the value of the other follows. The definition of pH is

$$pH = -\log[H^+]. \tag{13B.6}$$

Thus, in neutral water, the concentration of H^+ ions is equal to that of the OH^- ions, and both are $10^{-7}\,cm^{-3}$, so that the value of the pH is 7.

The assumption is generally made that the activity of H^+ ions is the same as their concentration, $[H^+]$. Thus, the value of the pH at the SHE reference electrode where the H^+ ion activity is unity must be 0. In experiments it is often not convenient to actually have an SHE, and the associated pH 0, in an experiment. A number of other types of electrodes are thus generally employed as secondary references. Some of these are listed in Table 13B.2.

13B.8 Historical Classification of Different Types of Electrodes in Aqueous Systems

In the electrochemical literature one often finds that electrodes used in aqueous systems have been historically classified into three main types: electrodes of the first kind, electrodes of the second kind, and redox, or oxidation–reduction, electrodes.

13B.8.1 Electrodes of the First Kind

Some of the common electrodes are sometimes called *cationic electrodes*, although there are also anionic examples. In addition to an inert electrical lead, they commonly consist of a single metal phase that is in contact with an electrolyte containing its cations. Common examples include metallic Ag, Bi, Cd, Hg, or Ni.

Another example is the *reversible hydrogen electrode* (RHE), in which bubbles of gaseous hydrogen at 1-atm pressure flow over a catalytic, but electrochemically inert, metallic surface that is in contact with the electrolyte. The general construction is the same as that of a SHE electrode, except that the pH of the electrolyte is not fixed at 0.

The potential of an electrode M of the first kind is generally given as

$$E = Constant + (RT/zF)\ln[a(M^+)/a(M)], \tag{13B.7}$$

where z is the number of electrons per cation in the electrolyte. The constant is called the *standard electrode potential, E°*. If M is an element, its activity is defined as unity, or simply:

$$E = E^0 + (RT/zF)\ln a(M^+). \tag{13B.8}$$

Thus, the electrode potential is a function of a property of the electrolyte, the activity, or concentration, of the M^+ ions.

In the case of the reversible hydrogen electrode, $a(H_2)$ is the pressure of the hydrogen gas. If this is 1 atm, this value is unity. Thus, the reversible hydrogen electrode potential can be approximately stated as

$$E = E^0 - 2.303(RT/F)(pH). \tag{13B.9}$$

If the pH is 0, this is equivalent to the SHE, so that the standard electrode potential E° is equal to 0. Thus, the difference between the RHE and the SHE is

$$E_{RHE} - E_{SHE} = -2.303(RT/F)(pH) \tag{13B.10}$$

The value of the first term on the right hand side is 0.059 V at 298 K. This difference (in volts) is then given as

$$E_{RHE} - E_{SHE} = -0.059(pH). \tag{13B.11}$$

There are also analogous *anionic electrodes* of the first kind that contain an elemental gaseous species, such as Cl_2, that enters the electrolyte as anions. In that case, n will have a negative value.

13B.8.2 Electrodes of the Second Kind

Electrodes of the second kind have two solid phases in contact with the (liquid) electrolyte, as well as an inert electrical lead. One of the phases is typically a metal, and the other is a *sparingly soluble salt*, or compound, of that metal. Examples of this type are (Ag,AgCl), (Hg,Hg_2Cl_2) (commonly called the *calomel electrode*), (Hg,Hg_2SO_4), and (Hg/HgO).

These are sometimes called *anionic electrodes*, and generally also are in contact with, or contain, a solution that has a salt of the same anion as that in the second solid phase. As an example, the *standard calomel electrode* (SCE) generally contains solid Hg and solid Hg_2Cl_2 in contact with a saturated solution of KCl. Under these conditions the potential of this electrode is 0.242 V positive of the primary reference SHE potential.

The complication is that experiments are often not performed under the same conditions as those required for the reference electrode. The main issue is the electrolyte. If the experimental and the reference electrode electrolytes are different, they can be connected by use of an additional intermediate electrolyte. Traditionally, this involved the use of a *salt bridge* containing a liquid electrolyte, and care was taken that it did not introduce a significant liquid junction potential. Many modern electrodes use solid electrolytes, such as special ionically conducting glasses, for this purpose.

If a reference electrode of this type is chemically isolated from the electrolyte its constitution cannot change, and it will have an electric potential that is independent of the composition of the electrolyte being used in the experiment.

An electrode of the second kind containing solid Hg and solid HgO is often used in alkaline electrolytes, where it is in direct contact with the experimental electrolyte. Under these conditions it behaves differently from the electrodes that contain chloride and sulfate species. In this case the electrode potential is given by

$$E = E^0 - (^{RT}/_{\mathrm{F}}) \ln a(OH^-), \tag{13B.12}$$

or

$$E = E^0 + 2.303(^{\mathrm{RT}}/_{\mathrm{F}}) \log[H^+], \tag{13B.13}$$

which is also

$$E = E^0 - 2.303(^{RT}/_{\mathrm{F}})(pH). \tag{13B.14}$$

This means that the potential varies with the pH of the electrolyte in the same way as does the reversible hydrogen electrode (RHE), although they have different values of E°.

Thus there is a difference between the electrolyte pH dependence of the potentials of these two types of electrodes and those discussed earlier with isolated chloride or sulfate species. This can be understood by use of *potential–pH plots*, often called *Pourbaix diagrams*, due to their development by M. Pourbaix [9]. These figures are very useful *thinking tools*, as they not only show how the potentials of various reactions vary with the pH of the electrolyte, but also indicate domains of stability of the different phases present. The general form of such a diagram is shown in Fig. 13B.4.

It can be seen that the potential of the SHE, with the requirement that the activity of the H^+ ions is unity, so that it must be physically isolated, is independent of the pH of the electrolyte, whereas the potential of the RHE varies with the pH. As shown earlier, the potential of the Hg/HgO electrode is pH-dependent as well, whereas the potentials of the Ag/AgCl, Hg,Hg_2Cl_2, and Hg,H_2SO_4 electrodes are not.

The Gibbs phase rule can be used to help understand these things, as well as the difference between the treatment of reference electrodes in nonaqueous and aqueous electrochemical systems.

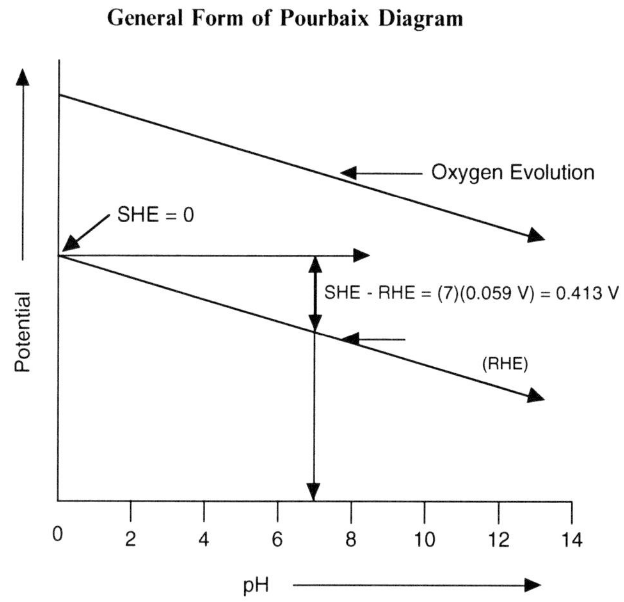

Fig. 13B.4 General form of an E vs. pH, or Pourbaix diagram. The influence of the pH on the potentials at which H_2 and O_2 have unit activity is shown. When the pH is 0, the potentials of the SHE and RHE electrodes are equal

13B.9 The Gibbs Phase Rule

The Gibbs Phase Rule [10, 11] plays an important role in the consideration of phase equilibria, and was discussed in Chap. 2 in connection with the variation of the electrical potential of electrodes with their composition. It will be briefly reviewed here, and its application to reference electrodes discussed.

The Gibbs phase rule can be written as:

$$F = C - P + 2, \tag{13B.15}$$

where C is the number of components (e.g., elements or electrically neutral stable entities), P is the number of phases present, and F is the number of degrees of freedom, or the number of thermodynamic parameters that must be specified in order to define the system and all of its associated electrical and chemical properties.

The eligible thermodynamic parameters are the temperature, the total pressure, and the chemical composition of each phase present. These are all intensive variables, so their values are independent of the amount of any of the materials present.

In binary systems $C = 2$, so that if the temperature and the overall pressure are held constant, and there are two phases present so that $P = 2$, the value of F is 0. This means that all of the intensive variables then have fixed values. These are thus *zero-degree-of freedom (ZDF) conditions,* and electrodes will have a fixed potential, regardless of the state of charge, and the amounts of the various phases present.

On the other hand, if there is only one phase present in a binary system, $C=2$, $P=1$, and thus $F=1$. This means that the properties are dependent upon the composition within that phase. An example of this is the variation of the potential of an insertion reaction electrode as its composition changes during charge or discharge in a battery.

Similar considerations apply to electrodes containing three components. In this case a useful thinking tool is the *isothermal Gibbs triangle*, or its approximation, the *ternary phase stability diagram*. The electrical potential is independent of the composition within subtriangles where three phases are in equilibrium. On the other hand, it is composition-dependent in two-phase and one-phase regions.

These conclusions have been thoroughly demonstrated experimentally in the case of binary alloy systems [12], and also ternary systems involving the reaction of lithium with binary metal oxides [13–15].

13B.10 Application of the Gibbs Phase Rule to Reference Electrodes

13B.10.1 Nonaqueous Systems

Application of these principles in nonaqueous systems is straightforward. If the temperature and total pressure are kept constant and there is only one component, e.g., a pure metal, the electrical potential must have a fixed value. On the other hand, if a binary phase, such as a metal oxide, is used as a reference, there must also be another phase in equilibrium with it, so that both C and P equal 2, in order to have a fixed potential. This second phase is often the metal component of the oxide, but it does not have to be. Instead, it could be a gas such as oxygen or air. The important thing is that this second phase should not introduce an additional component, for it would then be a ternary, instead of binary, system.

As mentioned earlier, if there are three components, i.e., a ternary system, there must be three phases in equilibrium with each other in order for $F=0$.

13B.10.2 Aqueous Systems

Aqueous systems introduce an additional feature. As it is expected that water will equilibrate with the electrode at their interface, the presence of water introduces an additional phase. In addition, it has two components, hydrogen and oxygen. All these have to be included in the consideration of the Gibbs phase rule.

In the case of a hydrogen gas electrode in contact with water, there are two components, hydrogen and oxygen, and two phases, water and hydrogen gas. Assuming constant temperature and pressure, $F=0$, and the system is thermodynamically fixed.

However, experiments show that the electrical potential of such an electrode depends upon the value of the pH. When the pH is 0, the electrode is equivalent to the standard hydrogen electrode, the SHE. However, at other values of electrolyte pH its potential varies, as it is then a reversible hydrogen electrode, an RHE. The electrical potential difference between these two situations was shown earlier to be

$$E_{RHE} - E_{SHE} = -0.059(pH) \tag{13B.16}$$

volts.

This hardly looks like a thermodynamically fixed situation. However, it must be recognized that normal chemical thermodynamics deals with the equilibria of electrically neutral species, and the pH is a measure of the concentration of a charged species, H^+ (or H_3O^+ or $H_5O_2^+$).

The chemical potential of a neutral species M, $\mu(\mathrm{M})$, can, in principle, be decomposed into two parts, the chemical potential of its ions, and the chemical potential of its electrons. This can be written as

$$\mu(M) = \mu(M^+) + \mu(e^-). \tag{13B.17}$$

Thus if the value of $\mu(\mathrm{M})$ is held constant, as is the case if the system is thermodynamically fixed, the values of the chemical potentials of the ions and the electrons can both vary, but their values will depend upon each other.

In the case of hydrogen:

$$\mu(H_2) = 2\mu(H^+) + 2\mu(e^-), \tag{13B.18}$$

which can be rearranged to give

$$\mu(e^-) = {}^1/_2\mu(H_2) - \mu(H^+). \tag{13B.19}$$

The chemical potential of the electrons, $\mu(e^-)$, is related to the electrically measured quantity E by

$$\mu(e^-) = zFE. \tag{13B.20}$$

Electrons carry a negative charge, so $z = -1$ in this case. Actually, as mentioned already, one always measures differences in E and thus of $\mu(\mathrm{e}^-)$, between electrodes, for absolute values of electrical potentials cannot be measured. The activities of individual ionic species also cannot be measured experimentally [16].

The chemical potential of the hydrogen ions can be written in terms of their concentration as

$$\mu(H^+) = \mu(H^+)^0 + RT\ln a(H^+), \tag{13B.21}$$

where $\mu(\mathrm{H}^+)^0$ is a constant. Substituting further,

$$\mu(H^+) = \mu(H)^0 + 2.303RT\log[H^+] \tag{13B.22}$$

or

$$\mu(H^+) = \mu(H^+)^0 - 2.303RT(pH). \tag{13B.23}$$

This can then be put back into the equation for the chemical potential of the electrons, giving

$$\mu(e^-) = 1/2\mu(H_2) - \mu(H^+)^0 + 2.303(RT)(pH) \tag{13B.24}$$

so that the electrical quantity E is related to the pH by

$$E = constant - 2.303(^{RT}/_{F})(pH), \tag{13B.25}$$

where the value of the constant depends upon the identity of the neutral chemical species.

The result is that the potentials of all neutral species with zero degrees of freedom will lie along parallel lines with a slope of −0.059 V per pH unit in a plot of potential vs. pH, i.e., in *Pourbaix diagrams*. Their vertical locations will be determined by their potentials relative to the reversible hydrogen electrode.

Thus there is a clear differentiation between reference electrodes with zero degrees of freedom (ZDF electrodes) and those where this is not true that is readily seen in their dependence upon the pH of an aqueous electrolyte. This is indicated schematically in Fig. 13B.5.

The result of this difference is that if one wants to compare electrodes in aqueous systems it is important to know whether they are ZDF electrodes or not, and if one

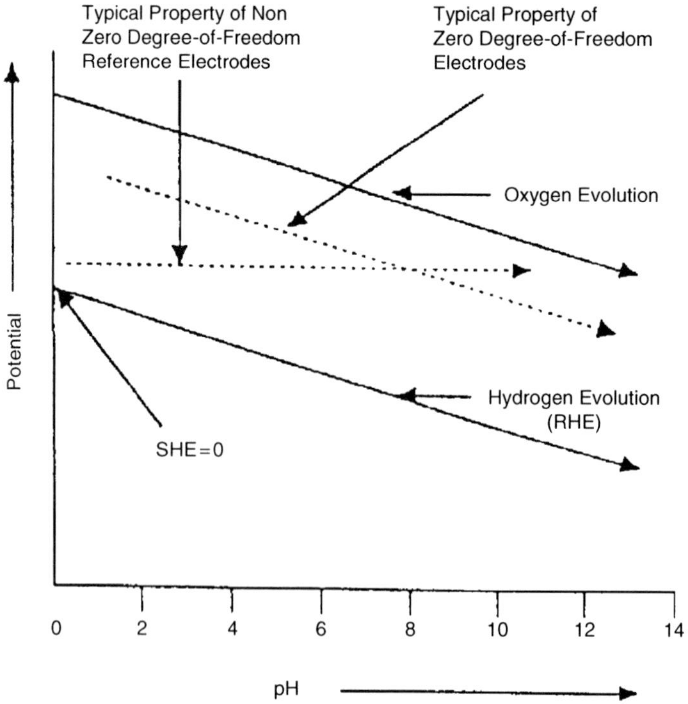

Fig. 13B.5 General E vs. pH diagram showing the difference between ZDF and non-ZDF electrodes

of them is not, additional information, generally the pH of the electrolyte, is needed in order to specify the thermodynamic state.

13B.11 Systems Used to Measure the pH of Aqueous Electrolytes

The difference in the potentials of ZDF electrodes and non-ZDF electrodes can be utilized to evaluate the pH of an electrolyte. An electrode to be used for this purpose will typically have a sealed inner compartment with a configuration that provides a fixed potential relative to the SHE. This is surrounded by a solid electrolyte, generally a glass with a relatively high ionic conductivity, whose exterior is exposed to the electrolyte whose pH is to be evaluated. A second, ZDF, electrode is also present in the electrolyte, and the voltage between the two is measured.

The construction of such a pH-measuring system is shown schematically in Fig. 13B.6.

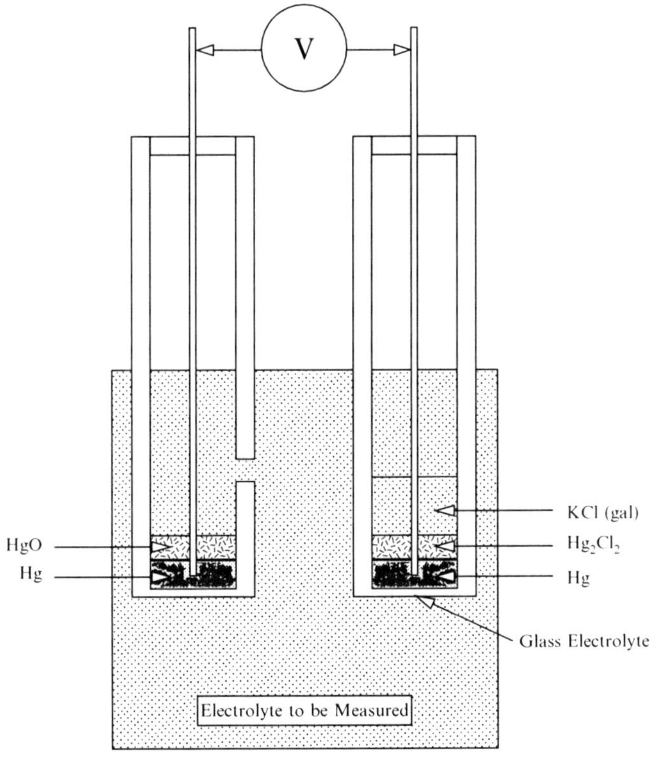

Fig. 13B.6 Schematic drawing of the construction of a system that can be used to measure the pH of a liquid electrolyte. A chemically isolated calomel electrode is in contact with the electrolyte through an ionically conducting glass membrane. A ZDF electrode (Hg/HgO) is in direct contact with the electrolyte

13B.12 Electrodes with Mixed-Conducting Matrices

As a final example, consider the use of a mixed-conducting matrix electrode containing a zero-degree-of-freedom (ZDF) reactant. Electrodes of this general class were first discussed some time ago [17,18]. The microstructure contains a phase that has a high chemical diffusion coefficient for the atoms of the electroactive species and is also an electronic conductor. Although it is not necessary, such a phase will generally have a relatively small compositional width, so that it does not have an appreciable electrochemical capacity. In addition, the microstructure contains phases that can undergo a reconstitution chemical reaction. If the number of such phases is equal to the number of components within them, this reaction will have zero degrees of freedom, and thus a composition-independent potential.

The result is that this type of electrode has a potential that is determined by the internal ZDF chemical reaction, even though the electrochemical reaction takes place elsewhere, at the electrode/electrolyte interface on the outside of the mixed-conductor material. This is illustrated schematically in Fig. 13B.7.

Whereas the initial example of this principle involved a Li–Si constant potential reaction inside a Li–Sn alloy mixed conductor at elevated temperatures, it has been demonstrated [19,20] that this concept can also be used at ambient temperature.

Electrodes of this type might be useful as secondary references in cases where one or more of the reactive phases is not stable in contact with the electrolyte. So far as the electrical potential is concerned, the identity of the electrolyte and the details of the interfacial reaction are not important.

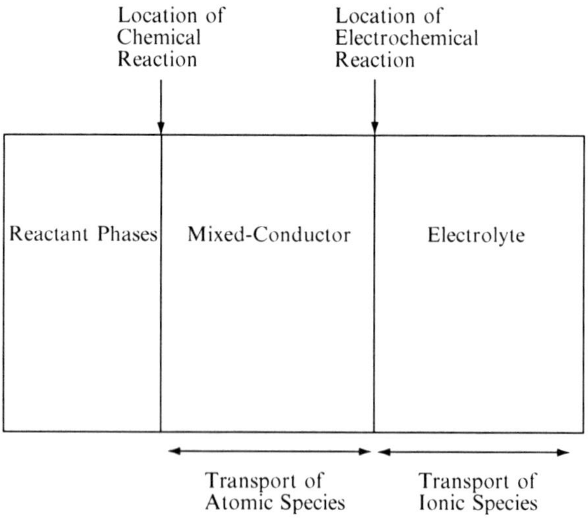

Fig. 13B.7 Schematic one-dimensional model of a mixed-conductor matrix electrode. The potential is determined by the electrically neutral chemical reaction in the interior, whereas the electrochemical reaction takes place at the interface between the mixed-conductor and the electrolyte

13B.13 Closing Comments

It is quite evident that the approach to reference electrodes is quite different in the nonaqueous and aqueous electrochemical communities. In the first case neutral chemical species are commonly used as references, and the details of the electrode/electrolyte reaction and the identity and concentrations of the species in the electrolyte are not considered. Electrical potential differences can be readily calculated from standard chemical thermodynamic data.

In the aqueous electrochemical community potentials are generally discussed in terms of the reactions at the electrode/electrolyte interface and the atomic and ionic species present there. The thermodynamic state of the bulk solid is not generally considered. In some cases the electrode potential depends upon the pH of the electrolyte, whereas in others it does not.

These different approaches can be rationalized by consideration of the Gibbs phase rule. When the electrode has zero degrees of freedom (ZDF) all the intensive variables, including the electrical potential, are fixed. On the other hand, if this is not true an additional variable must be specified, and this is commonly the pH in aqueous electrochemistry.

The characteristic dependence of the potential of ZDF electrodes upon the pH has been explained in terms of the two components of the chemical potential of neutral species.

The difference between ZDF electrodes and non-ZDF electrodes can be used to measure the pH, independent of the composition of the electrolyte. In addition, the potential of electrodes with a ZDF internal reaction and a mixed-conducting matrix has nothing to do with phenomena at the electrode/electrolyte interface where the electrochemical reaction takes place.

13C Potentials of Chemical Reactions

13C.1 Introduction

It has been known for some time that ions can be inserted or removed from insertion reaction materials by chemical, as well as electrochemical, means. This is one type of soft chemistry, or *Chemie douce*, as it was initially called in France. Much of the early attention to this possibility was focused on materials based upon graphite, and reviews of this work can be found in a number of places, e.g., in [22].

An important step in the development of the use of chemical methods to either modify or synthesize advanced battery electrode materials was the work of Armand, who used naphthalene complexes in a polar solvent to insert either sodium or potassium, and *n*-butyl lithium dissolved in hexane to introduce lithium, into insertion reaction materials [23]. He inserted these alkali metals into layer structures consisting of transition metal salts, such as CrO_3, between graphene planes. The presence of these very covalent species gives these graphite-related materials a very positive potential, so that they are interesting as potential positive electrode reactants.

This is actually quite different from much of the other work on graphite materials in lithium cells, in which the potential is much lower, so that they are interesting for use as negative electrode materials. Nevertheless, the principles are the same.

The use of *n*-butyl lithium, which is commercially available as a solution in hexane, to insert lithium into a material M, forming Li_xM and octane, C_8H_{18}, can be simply written as

$$xLiC_4H_9 + M = Li_xM + \frac{x}{2}C_8H_{18}. \quad (13C.1)$$

In the sodium naphthalene case the analogous reaction can be written as

$$xNaC_{10}H_8 + M = Na_xM + xC_{10}H_8. \quad (13C.2)$$

As discussed earlier in this text, the standard Gibbs free energy change in a reaction involving electrically neutral species can be readily converted to an electrical potential difference, or voltage. The Gibbs free energy of formation of *n*-butyl lithium is about $96.5\,\mathrm{kJ\,mol^{-1}}$, so that the potential that is attained upon its use to add lithium to electrode materials is about 1.0 V vs. elemental lithium. If the potential is greater than that value it will tend to decompose, providing lithium to react with the material M. That is, 13C.1 will tend to go to the right.

Thus it is possible to use such materials as reagents to chemically mimic electrochemical behavior, and thus screen or scan materials that might be considered as potential positive electrode reactants in lithium, or other alkali metal, cells. The amount of alkali metal uptake can be determined by assaying the resultant supernatant solution [24], and a rough indication of the kinetics of their reaction can be obtained by the observation of the temperature rise during the reaction [25]. In a number of cases an indication that a reaction has taken place is provided by a change in the color of the solution.

13C.2 Relation Between Chemical Redox Equilibria and the Potential and Composition of Insertion Reaction Materials

This situation can be represented schematically, as shown in Fig. 13C.1. If the potential of the solid is higher than the redox equilibrium potential, there will be a tendency for lithium to enter it from the adjacent liquid. On the other hand, if the potential of the solid is lower than the redox potential, there will be a tendency for the deletion of lithium, resulting in the potential increasing.

If two different reactants are used that have different redox potentials, the amount of lithium present in the solid can be changed between that characteristic of one of the redox potentials to that corresponding to the other. This is illustrated schematically in Fig. 13C.2.

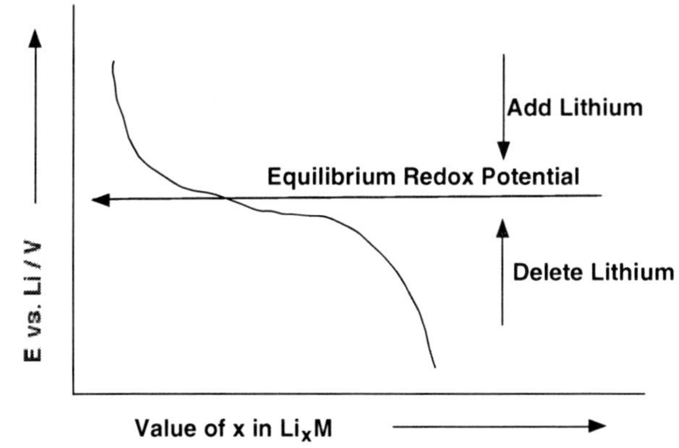

Fig. 13C.1 Illustration of the relationship between the potential and the amount of lithium in Li_xM

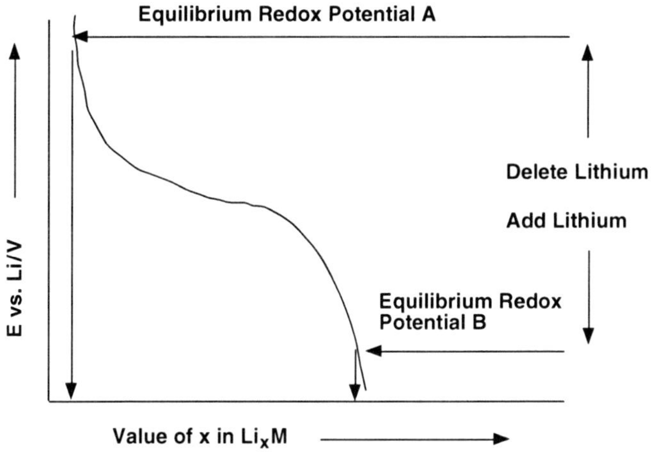

Fig. 13C.2 Illustration of the effect of using two different reagents, one at a high potential that causes a reduction in the amount of lithium in the solid, and the other, at a lower potential that increases the lithium content

13C.3 Other Examples

An example of a chemical reaction that can be used to delete lithium is the reaction of a material containing lithium with iodine to form LiI, which was discussed in Chap. 2. The formation and decomposition of LiI is potentially reversible. Its standard Gibbs free energy of formation is $-269.67\,\mathrm{kJ\,mol^{-1}}$ at 25°C, which converts to 2.8 V. This provides a good approximation for its equilibrium potential in solutions. However, as in all of the cases discussed here, the actual potential may vary somewhat, depending upon the solvent, reagent concentration, and the amounts and identities of other species present.

This means that if iodine is available and a material Li_xM is present that has a potential lower than about 2.8 V there will be a tendency for iodine to extract lithium from it, forming LiI, and raising its potential. This can be represented by the equation

$$\frac{x}{2}I_2 + Li_yM = xLiI + Li_{(y-x)}M. \qquad (13C.3)$$

As an example of the use of this method, solutions of iodine in acetonitrile, CH_3CN, were employed by Murphy et al. [26] to delete lithium from Li_xVS_2 and raise its potential. They also used *n*-butyl lithium to add lithium and reduce its potential [27].

In addition to *n*-butyl lithium and iodine, there are other oxidation or reducing agents that can be used. Bromine, in a solution of chloroform, $CHCl_3$, has been used to oxidize, and therefore reduce the lithium content of a number of materials. The standard Gibbs free energy of formation of lithium bromide is $-341.6\,kJ\,mol^{-1}$, so its equilibrium potential is quite high, about 3.54 V. Early examples of the use of bromine were the deletion of lithium from $LiVO_2$ [28], and from the more positive electrode material $LiCoO_2$ [29].

Other, more highly oxidizing, reagents were discussed by [30] and [31]. One example is the hexafluorophosphate salt NO_2PF_6, which can be dissolved in acetonitrile and has a potential about 4.45 V above that of lithium. Its reaction with a lithium transition metal dioxide can be written as

$$LiMO_2 + xNO_2PF_6 = Li_{1-x}MO_2 + xNO_2 + xLiPF_6. \qquad (13C.4)$$

A number of chemical reagents that have been used to modify mixed-conducting electrode materials in lithium systems are included in Table 13C.1. Some of these chemical reagents are reversible, whereas others are not. More information about organolithium materials can be found in [32].

Table 13C.1 Examples of lithium reaction materials and their approximate potentials vs. elemental lithium

Reagent	Solvent	*E* vs. Li volts	Color, higher *E*	Color, lower *E*
MoF_6	Acetonitrile	4.75	None	None
NO_2PF_6	Acetonitrile	4.45	None	None
Bromine	Acetonitrile	3.54	Brown	None
DDQ	Acetonitrile	3.5	None	Red
Iodine	Acetonitrile	2.8	Purple	None
Benzophenone	Tetrahydrofuran	1.5	None	Blue
n-Butyl lithium	Hexane	1.0	None	None
Benzophenone	Tetrahydrofuran	0.8	Blue	Purple
Naphthalene	Tetrahydrofuran	0.5	None	Green

DDQ is 2,3-dichloro-4,5-dicyanobenzoquinone

13C.4 Summary

There are a number of chemical equilibria that can act to influence the amount of inserted material; i.e., lithium, present in a mixed conductor in a manner analogous to the application of an electrochemical potential by the use of an electrochemical cell. Equilibria at low potentials are typically used to add lithium, and those with higher potentials are more commonly used to delete lithium from potential electrode materials.

13D Potential and Composition Distributions Within Components of Electrochemical Cells

13D.1 Introduction

The electrostatic, chemical, and electrochemical potentials inside condensed phases depend upon the nature and concentrations of the species that are present. These, in turn, vary with the local values of the relevant thermodynamic potentials, which are typically not uniform inside electrochemical cells.

A number of examples will be discussed in this section, including ionic conductors and mixed conductors between different types of nonblocking and selectively blocking electrodes under the imposition of either electrical or chemical potential differences.

Under charge transport conditions the *transference numbers* of individual species vary with position if a gradient in thermodynamic potentials is present. This can be readily understood by use of *defect equilibrium diagrams* as *thinking tools*.

These parameters can be experimentally evaluated by the use of proper sensors. One type measures the local value of the Fermi level of the electrons, and the other can be used to evaluate the local chemical potential or activity of neutral chemical species.

This is an important topic for several reasons. Local potentials, and their gradients, determine both the potentials and the kinetic behavior of electrodes in batteries. They also play critical roles in the properties of fuel cells.

13D.2 Relevant Energy Quantities

The energy of species inside solids is the sum of their chemical and electrical energies. The chemical energy is expressed as the chemical potential, and for species i,

$$E_{chem} = \mu_i. \tag{13D.1}$$

The electrical energy is the product of the charge and the local value of the inner potential.

$$E_{elect} = z_i q \phi \tag{13D.2}$$

The electrochemical potential η_i is the sum of the chemical and electrical energies

$$\eta_i = \mu_i + z_i q \phi. \tag{13D.3}$$

13D.3 What Is Different About the Interior of Solids?

Chemical potentials outside of solids always are referenced to electrically neutral chemical species. But chemical species inside solids are typically electrically charged ions. To achieve internal charge balance their charges must be balanced by the presence of either other charged ions or excess electrons or holes (a deficiency of electrons).

For equilibrium between an electrically neutral species M on the outside and the corresponding combination of an ionic species $M^+$and an electron on the inside:

$$\mu_M = \mu_{M^+} + \mu_{e^-}. \tag{13D.4}$$

Each of these species has a corresponding electrochemical potential, the combination of chemical and electrostatic potentials, or potential energies

$$\eta_{M^+} = \mu_{M^+} + z_{M^+} q \phi, \tag{13D.5}$$

$$\eta_{e^-} = \mu_{e^-} + z_{e^-} q \phi. \tag{13D.6}$$

Here z is the charge number, q the electronic charge, and ϕ the local inner electrical potential. Since $z_{M^+} = 1$ and $z_{e^-} = -1$, if we add these equations the terms containing the inner potential cancel each other, giving

$$\eta_{M^+} + \eta_{e^-} = \mu_{M^+} + \mu_{e^-}. \tag{13D.7}$$

This can be simply rearranged to become

$$\mu_M = \eta_{M^+} + \eta_{e^-}. \tag{13D.8}$$

The external chemical potential of a neutral species M is equal to the sum of the electrochemical potentials of its two related internal species, M^+ and e^-.

The sum of the gradients of the electrochemical potentials of the charged species inside a solid can be observed as an externally measurable gradient in the chemical potential of the neutral species. In the case of a one-dimensional physical system with a distance parameter x this can be written as

$$\frac{d\mu_M}{dx} = \frac{d\eta_{M^+}}{dx} + \frac{d\eta_{e^-}}{dx}. \tag{13D.9}$$

Gradients in their respective electrochemical potentials constitute the driving forces for the transport of species.

13D.4 Relations Between Inside and Outside Quantities

Information about potentials inside solids is typically obtained by use of external measurements. In the case of electronic species, this involves equilibration of the internal Fermi level with the Fermi level of an external metal probe. In the case of chemical species it is necessary to use an electrochemical cell and to balance chemical force with an equivalent electrical force

$$\Delta\mu_i = -z_i qE, \tag{13D.10}$$

where $\Delta\mu_i$ is the difference in chemical potential between neutral chemical species, z_i is the charge number of the ionic species under consideration, q is the elementary charge, and E is the voltage across the electrochemical cell.

13D.5 Basic Flux Relations Inside Phases

The particle flux density of any species i, J_i, is the number of particles of that type that cross a transverse area of $1\,\mathrm{cm}^2\ \mathrm{s}^{-1}$.

This can be expressed in terms of the concentration $[i]$ (particles per cm^3) and the macroscopic drift velocity $v_i(\mathrm{cm\,s}^{-1})$.

$$J_i = [i]v_i. \tag{13D.11}$$

The general mobility of species i, B_i, is defined as the ratio of the drift velocity and the negative gradient in the electrochemical potential, which is the force causing that drift.

$$B_i = -v_i \Big/ \frac{d\eta_i}{dx} \tag{13D.12}$$

Note that the general mobility B_i is different from the electrical mobility b_i, which is defined as the drift velocity of species i per unit internal electrical field.

$$b_i = -v_i \Big/ \frac{d\phi}{dx}. \tag{13D.13}$$

Thus the particle flux density of any species i can be written as

$$J_i = -[i]B_i \frac{d\eta_i}{dx}. \tag{13D.14}$$

Introducing the general definition of the electrochemical potential

$$J_i = -[i]B_i \left[\frac{d\mu_i}{dx} + z_i q \frac{d\phi}{dx} \right]. \tag{13D.15}$$

13D.6 Two Simple Limiting Cases

To understand these matters, two types of materials as simple limiting cases can be considered:

1. A metal, in which there is no internal electrical field, and therefore

$$J_i = -[i]B_i \left[\frac{d\mu_i}{dx} \right]. \tag{13D.16}$$

2. A chemically homogeneous material in which $d\mu_i/dx$ is 0, so that

$$J_i = -[i]B_i \left[\frac{d\phi}{dx} \right]. \tag{13D.17}$$

13D.7 Three Configurations

As examples, three simple configurations will be explored.

1. A solid electrolyte in which the transport of ionic species is blocked by the electrodes.
2. A mixed conductor in which the transport of electronic species is blocked by the electrodes.
3. A composite structure with a mixed conductor in series with a solid electrolyte.

13D.8 Variation of the Composition with Potential

As mentioned earlier, ionic and electronic species are present inside a solid, and their respective concentrations vary with the values of the relevant electrochemical potentials.

The chemical compositions of solids also depend upon the chemical potentials of the species present.

These features can be readily understood for simple cases, and as an example, the concentrations of both ionic and electronic defects will be calculated here as a function of the overall composition and the electrical potential for a simple binary phase MX. This is an approach that was pioneered by workers in Philips Laboratories [33–35]. It is useful to express the results in a *defect equilibrium diagram* (*DED*). It will be seen that such a graphical presentation of these matters can act as a very useful *thinking tool*.

13D.9 Calculation of the Concentrations of the Relevant Defects in a Binary Solid MX That Is Predominantly an Ionic Conductor

In a binary solid, four defects have to be considered, two electronic defects; electrons and holes, as well as two ionic defects, which might be interstitials and vacancies of one of the components.

Since there are four unknowns, there must be four relevant equations.

Two of these are mass action relations, one for the formation of electron-hole pairs

$$K_e = [e^-][h^o] \tag{13D.18}$$

in which the brackets indicate concentrations (number of particles per cm^3), and one for the formation of ionic defect pairs

$$K_i = [D_{M^o}][D_{X'}]. \tag{13D.19}$$

The notation that is used here is a modification of that generally referred to as *Kröger-Vink notation*. Instead of the common practice of using a dot to indicate a positive relative charge, a degree sign is used in this text, for that is a symbol that is readily available in typewriters and computers. In addition, the symbol D is used here as a general symbol for an ionic defect. D_{M° is an M-rich ionic defect, and $D_{X'}$ is a defect whose presence makes the material more X-rich.

In addition to (13D.18) and (13D.19), an expression is needed that relates to the overall chemical composition, and thus to the concentration of at least one of the ionic defects, in the phase MX. There are various ways in which this might be done. One is to assume that the MX is in equilibrium with its chemical environment, so that the chemical potentials of the species within it are the same as those in the environment. Consider the case in which it is assumed that the material is in equilibrium with a surrounding phase containing the species X.

The equilibrium between X species in the surrounding phase (assume that it is a gas and contains diatomic X_2 molecules) and singly charged X or M defect species inside the MX can be written as

$$X(g) = 1/2X_2(g) = D_{X'} + h^\circ \tag{13D.20}$$

in which the X-rich defect species $D_{X'}$ could be either an interstitial X ion or a vacancy on the M lattice. Either one of these has a negative effective charge, and as mentioned earlier, it makes no difference which one is actually present in the material in this calculation. In either case, a positively charged electron hole would also have to be present in order to achieve charge balance.

An additional equation can be the law of mass action relation that corresponds to the incorporation of one of the chemical species, and its charge-balancing electronic species, into the nominally MX phase. An example is the X-incorporation relation

$$K_X = \frac{[D_{X'}][h^o]}{(aX_2)^{1/2}}. \tag{13D.21}$$

It would have been possible to use a relation that involves M-rich defects instead. In that case, however, an electron would have to be present in order to maintain charge balance.

In addition to these law-of-mass-action type relations, there must also be an expression that reflects the requirement for overall electrostatic charge balance. This is sometimes called the *electroneutrality condition*. The number of negative charges introduced by the presence of the defects carrying negative effective charges must be balanced by the positive charge due to the presence of species carrying positive charges. This can be written as

$$[e'] + [D_{X'}] = [h^o] + D_{M^o}]. \tag{13D.22}$$

Simultaneous solution of these four equations can be used to obtain expressions to use to evaluate the four defect concentrations.

Because of the composition dependence in the incorporation equation (13D.20), the concentrations of all of the defects depend upon the composition of the phase, and thus upon the electrical potential.

To simplify matters it is useful to introduce a composition parameter F

$$F = K_X[a(X_2)]^{1/2}. \tag{13D.23}$$

By substitution into the electroneutrality condition, the concentrations of the various defects can then be calculated in terms of the values of the equilibrium constants and the composition parameter F.

$$[e'] = K_e\left[\frac{K_i + F}{F(K_e + F)}\right]^{1/2}, \tag{13D.24}$$

$$[h^o] = \left[\frac{F(K_e + F)}{K_i + F}\right]^{1/2}, \tag{13D.25}$$

$$[D_{X'}] = \left[\frac{F(K_i + F)}{K_e + F}\right]^{1/2}, \tag{13D.26}$$

$$[D_{M^o}] = K_i\left[\frac{K_e + F}{F(K_i + F)}\right]^{1/2}. \tag{13D.27}$$

These relations specify the concentrations of the four pertinent defects as functions of the overall composition of the MX phase, as expressed in terms of the value of the composition parameter F. They will each vary with temperature, as the values of the constants are temperature-dependent.

13D.10 Defect Equilibrium Diagrams

The form of these relations is illustrated in the *defect equilibrium diagram* shown in Fig. 13D.1. The concentrations of the four types of defects are plotted on a logarithmic scale against the logarithm of the parameter F. It has been assumed in this example that the material MX is primarily an ionic conductor. This will be the case if the Gibbs free energy necessary to form the ionic defect pair is less than that necessary to form the electronic defect pair. Thus, the value of K_i is significantly greater than the value of K_e. The constants used in this illustration are $K_e = 10^{10}$, $K_i = 10^{40}$, and $K_X = 10^{20}$. It can be seen that there are three general regions of behavior, labeled as Region I at low values of F when the material will be relatively M-rich, Region II at intermediate values of F, and Region III at high values of F, when the material will be relatively X-rich.

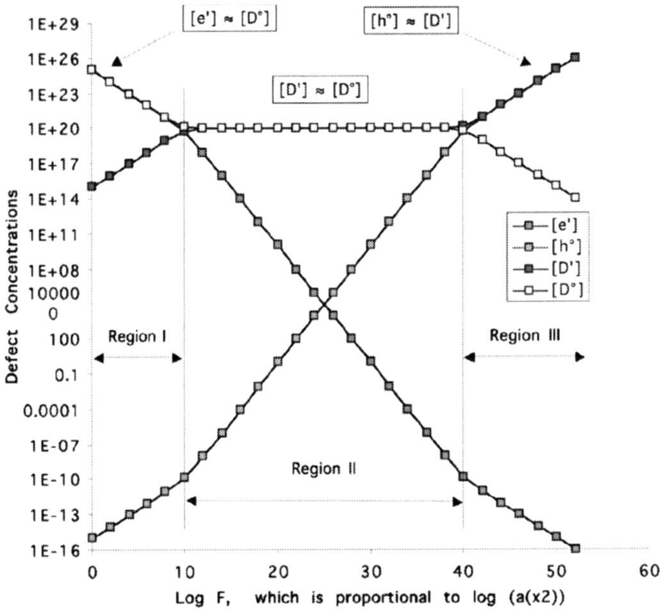

Fig. 13D.1 Defect equilibrium diagram showing concentrations of defect species inside solid MX as functions of the composition parameter F

13D.11 Approximations Relevant in Specific Ranges of Composition or Activity

It is often useful to work with approximations to the general relations that are applicable over these three different ranges of composition or activity. As mentioned earlier, the important criterion for the determination of useful approximations is the value of the composition parameter F. In Region I the X activity is very small and the M activity is correspondingly large, and F is very small, less than both K_e and K_i. In the central Region II, the value of F is between K_i and K_e. Likewise, when the value of the X activity is large and the M activity is small, F will be larger than both K_e and K_i.

At very low values of X activity F will be smaller than both K_e and K_i, giving the following approximations for the defect concentrations.

$$[e'] = K_e \left(\frac{K_i}{FK_e} \right)^{1/2}, \tag{13D.28}$$

$$[h^\circ] = \left(\frac{FK_e}{K_i} \right)^{1/2}, \tag{13D.29}$$

$$[D_{X'}] = \left(\frac{FK_i}{K_e} \right)^{1/2}, \tag{13D.30}$$

$$[D_{M^\circ}] = K_i \left(\frac{K_e}{FK_i} \right)^{1/2}. \tag{13D.31}$$

Likewise, at intermediate values of X activity $K_i > F > K_e$, and the defect concentrations can be approximated by:

$$[e'] = \frac{K_e K_i^{1/2}}{F}, \tag{13D.32}$$

$$[h^\circ] = FK_i^{-1/2}, \tag{13D.33}$$

$$[D_{X'}] = K_i^{1/2}, \tag{13D.34}$$

$$[D_{M^\circ}] = K_i^{1/2}. \tag{13D.35}$$

When the X activity is very large F becomes much greater than both K_e and K_i. The defect concentrations can be approximated by:

$$[e'] = K_e F^{-1/2}, \tag{13D.36}$$

$$[h^\circ] = F^{1/2}, \tag{13D.37}$$

$$[D_{X'}] = F^{1/2}, \tag{13D.38}$$

$$[D_{M^\circ}] = K_i F^{-1/2}. \tag{13D.39}$$

It can be seen from observation of the example defect equilibrium diagram shown in Fig. 13D.1 that the transition between Regions I and II occurs when $F = K_e$, and the transition between Regions II and III is where $F = K_i$. The slopes in Region II are ± 1, and in Regions I and III are $\pm 1/2$. Thus, it is very easy to draw the general form of such a figure, even without knowing the values of the relevant constants.

Figure 13D.1 shows that the defects that have the greatest concentrations, and which therefore play the dominant role in the electroneutrality relation and in determining the properties, are different in the three regions. The species with the highest concentrations in the central region II are both ionic defects. This indicates that the material may be primarily an ionic conductor in this range of composition. This is not true in the other regions, where the composition has more extreme values, either relatively M-rich in Region I, or relatively X-rich in Region III. In those cases, the dominant defect concentrations include one ionic defect and one electronic defect, and it is likely that there is important, if not dominant, electronic conductivity.

Since the mobilities of the electronic defects are generally much higher than those of ionic defects, this material will be an n-type conductor at higher values of F, and a p-type conductor at lower values of F than the transitions between the respective regions in Fig. 13D.1. Therefore it is obvious that if it is desired that a material acts as a solid electrolyte in an electrochemical cell it is important that the potentials of the two electrodes both be well within the central region of the *defect equilibrium diagram* for the material in question. This is shown schematically in Fig. 13D.2.

It will be shown in a later chapter that if this is not true, the measured voltage across the cell will be different from (less than) that expected from the difference in the chemical potentials at the electrodes.

13D.12 Situation in Which an Electrical Potential Difference Is Applied Across a Solid Electrolyte Using Electrodes That Block the Entry and Exit of Ionic Species

As discussed in a later chapter, electrical measurement methods are often used to determine the ionic conduction properties of materials that are being considered as solid electrolytes. There are two general strategies. One is to utilize electrodes that are essentially transparent to the ionic species, so that the overall impedance is dominated by what happens within the solid being measured. This often involves *DC measurements*. The other strategy is to use electrodes that are deliberately blocking to the ionic species, and to measure the system response to the application of alternating potential differences. Such AC measurement methods are often called *impedance spectroscopy*. The blocking-electrode case will be discussed here.

The flux of any species i is proportional to the gradient of its electrochemical potential.

$$J_i = -[i]B_i \frac{d\eta i}{dx}. \tag{13D.40}$$

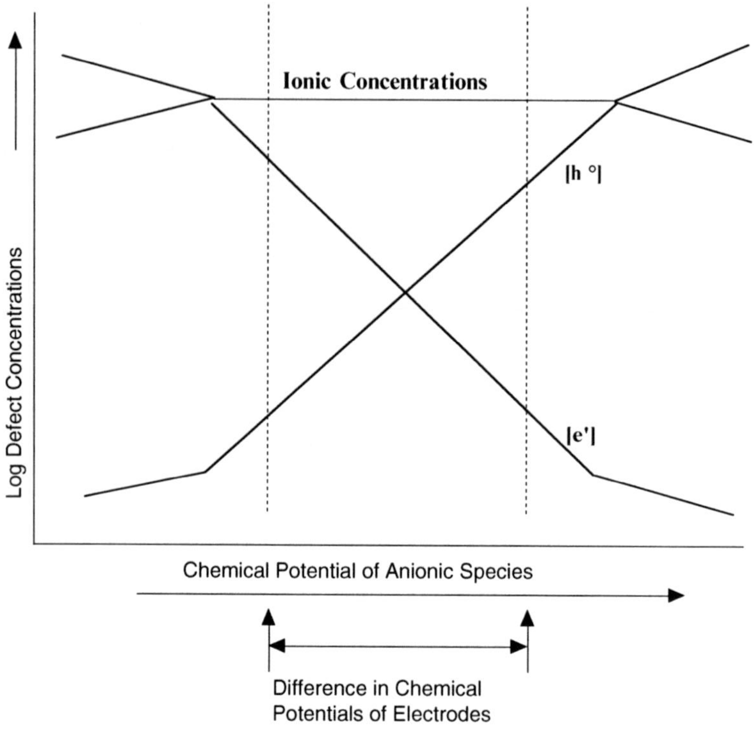

Fig. 13D.2 Schematic defect equilibrium diagram showing central region of solid electrolyte behavior

If one, or both, of the electrodes block the passage of the ionic species, there can be no ionic flux in the solid electrolyte material being investigated. Therefore, the gradient in the electrochemical potential of the ions inside must be zero.

$$\frac{d\eta_i}{dx} = \frac{d\mu_i}{dx} + z_i q \frac{d\phi}{dx} = 0. \quad (13D.41)$$

And if the potentials of both electrodes fall within the central region of the defect equilibrium diagram for the electrolyte, there is no gradient in the concentrations of the ionic species, so that

$$\frac{d\mu_i}{dx} = 0. \quad (13D.42)$$

And thus

$$\frac{d\phi}{dx} = 0. \quad (13D.43)$$

So there is no internal electrical field inside the solid, despite the imposition of an external electrical potential difference.

There will be gradients in the chemical potentials, and thus of the concentrations, of the electrons and holes, however. The result is that there is an electronic current across the cell, but it is due to the composition gradients of the holes and electrons in the interior of the electrolyte, not the presence of an electrical field.

Experimental observation of the magnitude and the voltage dependence of this current provides information about the separate contributions of the holes and electrons. This is known as the Hebb-Wagner experiment, and is discussed elsewhere in this text.

13D.13 The Use of External Sensors to Evaluate Internal Quantities in Solids

Electronic probes can be used to evaluate the electrochemical potential (Fermi level) of the electrons at specific locations. If there is no passage of current between the probe and the solid, their Fermi levels must be equal. Such measurements can be made as a function of position, and referenced to one of the electrodes of the cell, in order to provide information about the spatial variation of the electronic Fermi level along the material being investigated. This is shown schematically in Fig. 13D.3.

But information about the potential of the ionic species within the solid requires a different approach. As mentioned earlier, it is not possible to independently measure the properties of ions. Chemical potentials and forces within solids always relate to neutral species or combinations of species. The way to acquire this information is to use a probe that employs a suitable ionic conductor as electrolyte and an electronically conducting chemical reference electrode. By measurement of the voltage across this ionically conducting probe the difference in the chemical potential of the reference and the material being investigated can be obtained. If this is done as a function of position, information can be obtained about the chemical potential of the neutral chemical species present. This is shown schematically in Fig. 13D.4.

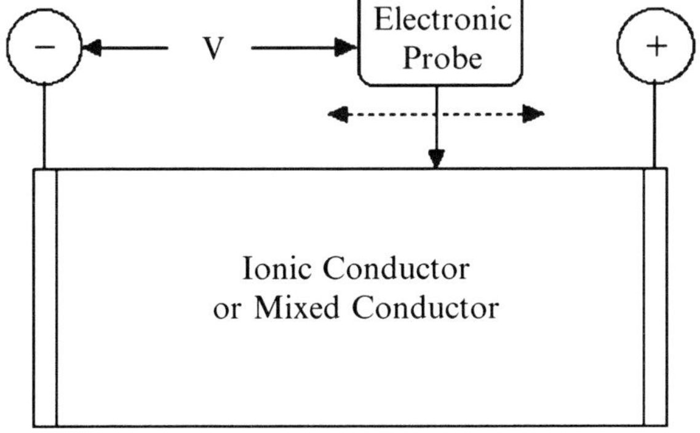

Fig. 13D.3 Use of an electronic probe (e.g. a metal wire) to measure the variation of the electrochemical potential of the electrons with position

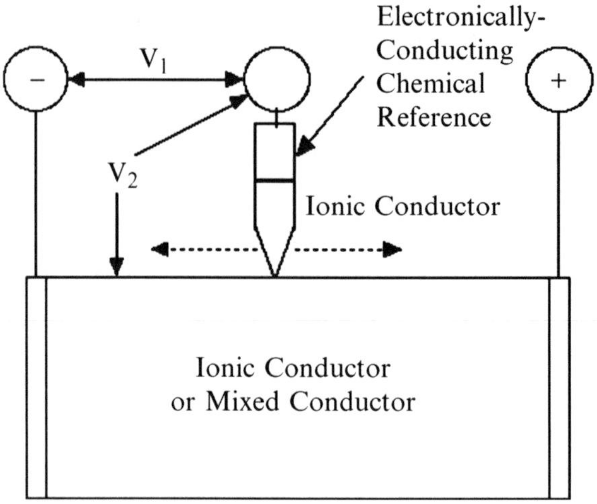

Fig. 13D.4 Use of an ionically conducting probe and reference electrode in order to obtain information about the variation of the chemical potential of the neutral chemical species present with position

13D.14 Another Case, a Mixed Conductor in Which the Transport of Electronic Species Is Blocked

Instead of the electrodes acting to block the transport of ionic species, it is possible to block the passage of electrons into and out of a mixed conductor or ionic conductor. This can be accomplished by putting an ionic conductor in the system, so that if current is passed between the electrodes there will be an ionic flux, but no electronic flux.

Even though there may be current flow through the system, external measurement of the Fermi Level with an electronic probe will show that there is no internal electrical field.

Instead of the relation

$$\frac{d\mu_M}{dx} = \frac{d\eta_{M^+}}{dx} + \frac{d\eta_{e^-}}{dx} \tag{13D.44}$$

if there is no gradient in the electrochemical potential of the electrons, this simplifies to

$$\frac{d\mu_M}{dx} = \frac{d\eta_{M^+}}{dx}. \tag{13D.45}$$

An experimental example is shown in Fig. 13D.5 [36]. In that case, an ionic probe was used to evaluate the electrochemical potential of silver ions within the mixed conductor Ag_2S, which was placed between two slabs of AgI, which is a pure silver ionic conductor. Thus, current flow resulted in the gradient within the Ag_2S when silver ions were transported from one silver metal electrode to the other.

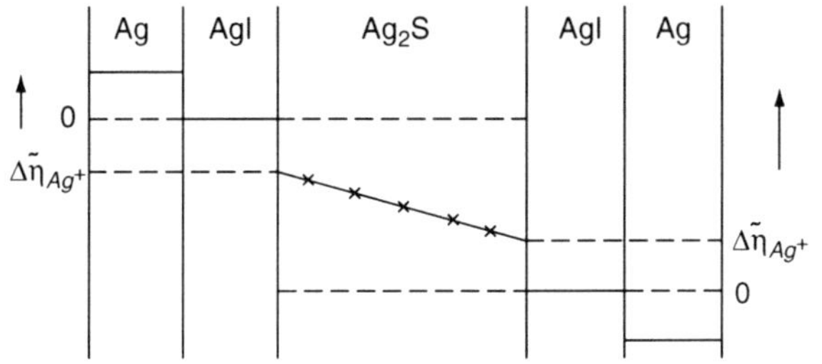

Fig. 13D.5 Variation of the chemical potential of silver with position within Ag_2S, a mixed conductor, during current passage, evaluated by the use of an ionic probe

13D.15 Further Comments on Composite Electrochemical Cells Containing a Mixed Conductor in Series with a Solid Electrolyte

One of the concepts that has been proposed for use in fuel cells involves the use of a *monolithic structure* that is a composite of an ionic conductor and a mixed conductor that both have the same crystal structure. The mixed conductor can then act as an electrode. The potential advantage of this approach is that it would minimize the generation of stresses due to local differences in thermal expansion.

Assuming that these two components have the same crystal structure and are of approximately the same composition, there are two general ways to do this.

One is by the use of localized doping of an ionic conductor to produce mixed conductor regions at the surface with increased electronic conductivity. These doped regions can then act as mixed conducting electrodes in series with the adjacent solid electrolyte region.

An alternative might be possible in some cases that would not require doping. This can be understood by considering the defect equilibrium diagrams discussed earlier. It can be seen that if the local electrical potential is sufficiently negative, there is the tendency for the presence of excess electrons, making the material an n-type mixed conductor. Alternatively, it may be possible to induce the local presence of holes at very positive potentials to make the material a p-type electronic conductor. The first of these is shown schematically in Fig. 13D.6.

When this is the case, the measured voltage is reduced below that which would be the case if the chemical potential difference were placed upon a purely ionic conductor. This is illustrated in Fig. 13D.7. Unfortunately, this means that the use of this monolithic concept in a solid electrolyte fuel cell necessarily means that the output voltage is reduced.

The spatial distribution of the chemical potential of the neutral chemical species M is shown in Fig. 13D.8.

The corresponding distributions of the chemical potentials of the M^+ ions and the electrochemical potential of the electrons are illustrated in Fig. 13D.9 for this situation.

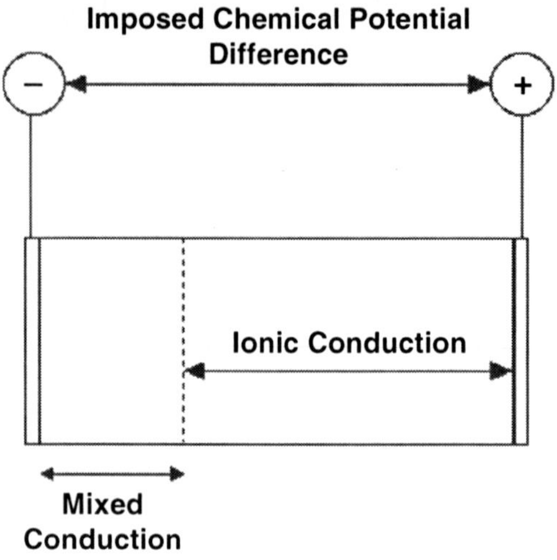

Fig. 13D.6 Schematic illustration of the case of an electrochemical cell in which an imposed chemical (or electrical) potential difference is such that the potential of the negative electrode is in the region of the DED in which the concentration of electrons is large enough to cause local mixed conduction

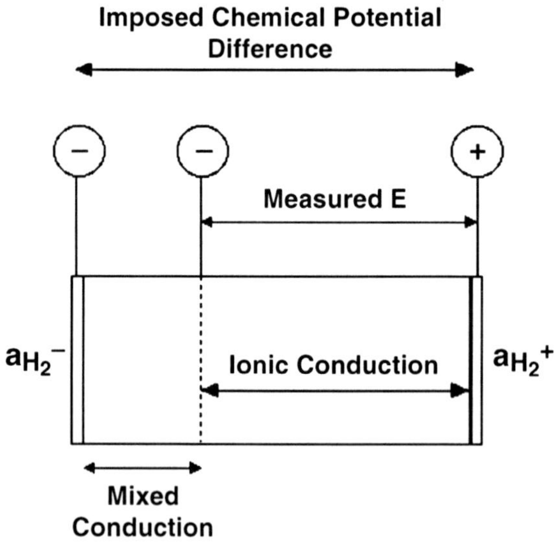

Fig. 13D.7 Schematic illustration showing that the electrical voltage is less than that corresponding to the imposed chemical potential difference if a portion of the material is mixed-conducting

It is seen that, although the chemical potential of the neutral chemical species, an externally measurable quantity, is a linear function of distance across the cell, the positional variations of the internal potentials of the two types of species are quite different.

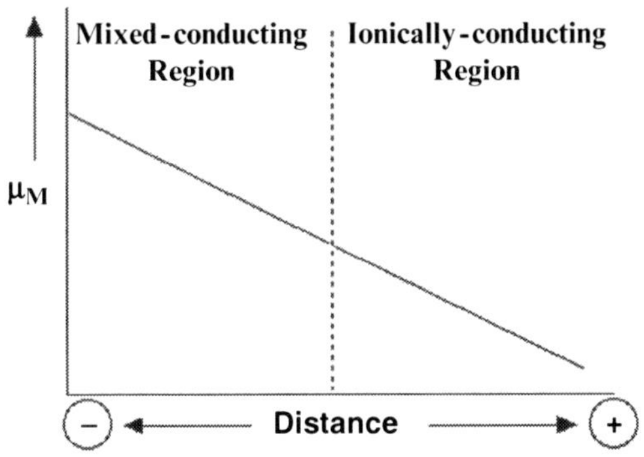

Fig. 13D.8 Spatial distribution of the chemical potential of the neutral chemical species *M* across the cell illustrated in Figs. 7 and 8

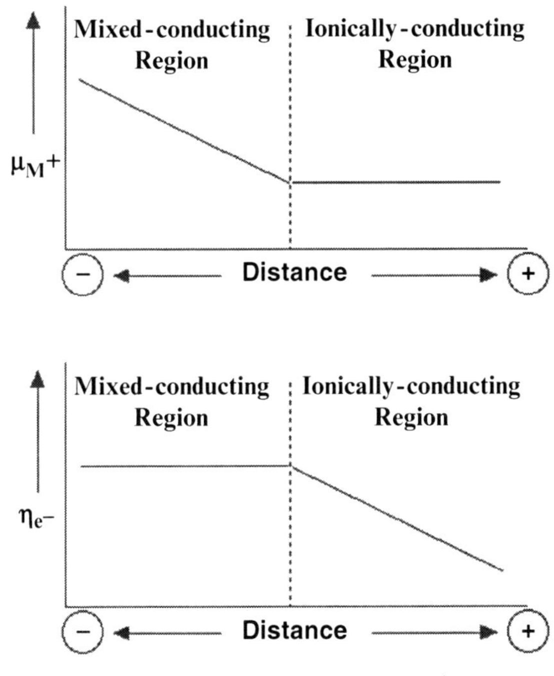

Fig. 13D.9 Spatial distribution of the chemical potential of the M^+ ions and the electrochemical potential of the electrons (Fermi level) are illustrated for the cell illustrated in Figs. 6 and 7

13D.16 Transference Numbers of Particular Species

The several situations that have been discussed here clearly indicate that the charge transport properties of a given material can vary significantly, depending upon the potentials imposed by its electrodes, as well as whether the electrodes – and other phases present – limit the passage of either ionic or electronic species.

A term used in this connection is the *Hittorf transference number* of a particular species. It indicates the fraction of the total charge current that is transported by that species.

Consideration of the defect equilibrium diagram shows that the concentrations of the various charged species present vary with the potentials applied by the electrodes, and the relation between chemical potentials and electrical potentials. Variation of these concentrations means that there is a variation in the transference numbers of the different species. Transference numbers can thus also vary with position within a solid.

But in addition, it is obvious that transference numbers also depend upon the experimental conditions - e.g. the properties of the electrodes used in a given experiment.

References

1. C.J. Wen, B.A. Boukamp, W. Weppner and R.A. Huggins, J. Electrochem. Soc. 126, 2258 (1979)
2. I. Barin, *Thermochemical Data of Pure Substances*, VCH, Weinheim, Germany (1989)
3. W. Li, J.R. Dahn and D.S. Wainwright, Science 264, 1115 (1994)
4. W. Li, W.R. McKinnon and J.R. Dahn, J. Electrochem. Soc. 141, 2310 (1994)
5. W. Li and J.R. Dahn, J. Electrochem. Soc. 142, 1742 (1995)
6. M. Zhang and J.R. Dahn, J. Electrochem. Soc. 143, 2730 (1996)
7. J.N. Butler, Adv. Electrochem. Electrochem. Eng. 7, 77 (1970)
8. D.J.G. Ives and G.J. Janz, *Reference Electrodes*, Academic Press, New York (1961)
9. M. Pourbaix, *Atlas of Electrochemical Equilibria*, Pergamon Press, Oxford (1966)
10. J.W. Gibbs, Trans. Conn. Acad. Sci. 108 (1875)
11. J.W. Gibbs, *The Scientific Papers of J. W. Gibbs, Vol. 1*, Dover, New York (1961)
12. W. Weppner and R.A. Huggins, J. Electrochem. Soc. 125, 7 (1978)
13. N.A. Godshall, I.D. Raistrick and R.A. Huggins, Mater. Res. Bull. 15, 561 (1980)
14. N.A. Godshall, I.D. Raistrick and R.A. Huggins, Proc. 16th IECEC, 769 (1981)
15. N.A. Godshall, I.D. Raistrick and R.A. Huggins, J. Electrochem. Soc. 131, 543 (1984)
16. J. Koryta, J. Dvorak and V. Bohackova, *Electrochemistry*, Methuen, London (1966)
17. B.A. Boukamp, G.C. Lesh and R.A. Huggins, J. Electrochem. Soc. 128, 725 (1981)
18. B.A. Boukamp, G.C. Lesh and R.A. Huggins, in *Proc. Symp. on Lithium Batteries*, ed. by H.V. Venkatasetty, Electrochem. Soc., Palo Alto, CA (1981), p. 467
19. A. Anani, S. Crouch-Baker and R.A. Huggins, in *Proc. of Symp. on Lithium Batteries*, ed. by A.N. Dey, Electrochem. Soc., Palo Alto, CA (1987), p. 382
20. A. Anani, S. Crouch-Baker and R.A. Huggins, J. Electrochem. Soc. 135, 2103 (1988)
21. I.D. Raistrick, A.J. Mark and R.A. Huggins, Solid State Ionics 5, 351 (1981)
22. L.B. Ebert, Intercalation Compounds of Graphite, in *Annual Review of Materials Science*, ed. by R.A. Huggins, Annual Reviews, Palo Alto, CA (1976), p. 181
23. M. Armand, New Electrode Material, in *Fast Ion Transport in Solids*, ed. by W. van Gool, North-Holland, Amsterdam (1973), p. 665
24. M.B. Dines, Mater. Res. Bull. 10, 287 (1975)
25. M.S. Whittingham and M.B. Dines, J. Electrochem. Soc. 124, 1387 (1977)
26. D.W. Murphy, C. Cros, F.J. DiSalvo and J.V. Waszczak, Inorg. Chem. 16, 3027 (1977)
27. D.W. Murphy, J.N. Carides, F.J. DiSalvo, C. Cros and J.V. Waszczak, Mater. Res. Bull. 12, 825 (1977)
28. K. Vidyasagar and J. Gopalakrishnan, J. Solid State Chem. 42, 217 (1982)

29. A. Mendiboure, C. Delmas and P. Hagenmuller, Mater. Res. Bull 19, 1383 (1984)
30. G.M. Anderson, J. Iqbal, D.W.A. Sharp, J.M. Winfield, J.H. Cameron and A.G. McLeod, J. Fluorine Chem. 24, 303 (1984)
31. A.R. Wizansky, P.E. Rauch and F.J. Disalvo, J. Solid State Chem. 81, 203 (1989)
32. B.J. Wakefield, *The Chemistry of Organolithium Compounds*, Pergamon Press, Oxford (1974)
33. F.A. Kröger, H.J. Vink and J. van den Boomgard, Z. Phys. Chem. 203, 1 (1954)
34. G. Brouwer, Philips Res. Rep. 9, 366 (1954)
35. F.A. Kröger, *The Chemistry of Imperfect Crystals*, 2nd ed., North Holland, Amsterdam (1974)
36. H. Schmalzried, M. Ullrich and H. Wysk, Solid State Ionics 51, 91 (1992)

Chapter 14
Liquid Electrolytes

14.1 Introduction

Most current battery systems have solid electrodes, separated by liquid electrolytes. Aside from considerations such as the magnitude of the ionic conductivity of liquids typically being considerably greater than those of solids, one of the major advantages of this arrangement is that the presence of the liquid reduces problems resulting from the volume changes that typically result from the changes in the composition of the electrode materials as they are charged and discharged.

A major consideration in connection with electrolytes has to do with the range of potentials over which they are stable. An obvious example of this is the fact that aqueous electrolytes cannot be used with negative electrodes that have high lithium activities. Organic solvent electrolytes must be used instead.

After a brief general discussion of the relative stability of electrolytes with different types of electrodes in this section, the following sections of this chapter will discuss several different types of electrolytes for use with lithium and hydrogen systems.

14.2 General Considerations Regarding the Stability of Electrolytes Vs. Alkali Metals

One can understand the relative stability of various materials classes by the use of the simple pseudo-ternary stability diagram in Fig. 14.1 [1].

Three general types of species are placed at the corners, very electropositive metals, e.g. the alkali metals, are at the lower left corner, and simple organic radicals such as methyl or ethyl groups are at the lower right corner. Highly oxidizing electronegative inorganic species, such as O_2, F_2, Cl_2, and S, are at the top corner.

A number of materials of interest with regard to the question of the stability of electrolytes can be located on this diagram.

R.A. Huggins, *Advanced Batteries*: Materials Science Aspects,

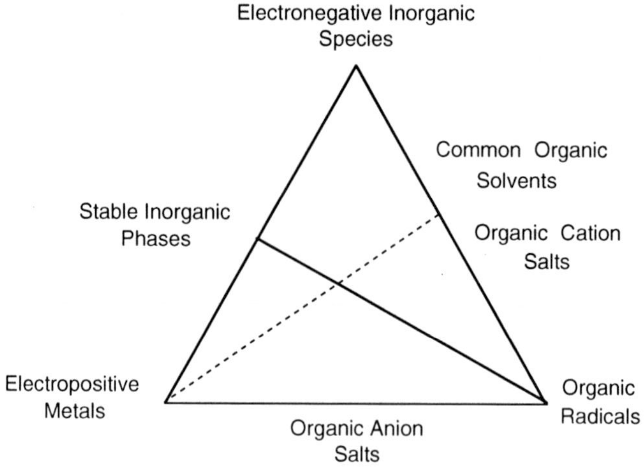

Fig. 14.1 Generalized pseudo-ternary stability diagram

Species at the corners will tend to react to produce materials of the type that are along their mutual edge. For example, electropositive metals react with electronegative inorganic species to form stable inorganic phases, such as Li_2O, Li_2S, LiF, NaCl, oxides and chalcogenides, as well as ternary salts such as Na_2SO_4 and Li_2CO_3.

A similar situation holds for the other two sides of the triangle.

Common organic solvents such as propylene and related carbonates, dioxolane, ethers, and also low temperature salts with organic cationic groups and inorganic anions can be thought of as formed by the reaction of organic groups with highly oxidizing inorganic species. Along the bottom of the triangle are compounds of electropositive metals and organic anionic groups.

Two possible tie lines are shown. According to the principles related to the use of ternary phase equilibrium diagrams discussed earlier, only one of these (i.e., the solid line) can be stable. Any reaction between the electropositive metals at the bottom left corner and materials above that line, i.e. electronegative inorganic species, organic solvents or organic cation salts, will result in the formation of a combination of phases at the ends of the stable tie line, i.e., stable inorganic phases and organic radicals.

This terminology may be a bit confusing. Remember that *electronegativity* relates to the affinity that a particular species has toward electrons when it is bonded to other species. When there is a covalent bond between two species with different *electronegativities*, the valence electrons tend to be displaced in the direction of the species with greater *electronegativity*, causing it to acquire a negative charge. The other species tends to acquire a positive charge. Thus, the electropositive elements such as the alkali metals behave as positively-charged species when combined with other elements, and their ions carry positive charges.

On the one hand, the *electric potentials* at which electropositive species are in equilibrium with their ions are very negative, and on the other hand, the *electric potentials* at which electronegative species are in equilibrium with their ions are relatively positive.

One feature of such isothermal ternary phase stability diagrams is that they can be understood and used in the same way as the other simple ternary phase stability diagrams that were discussed earlier.

For example, one can determine in which direction reactions tend to proceed, and thus which combination of phases are stable relative to other possibilities by the use of tie lines.

One can represent a possible displacement reaction as a pair of intersecting tie lines, one connecting the compositions of the pair of potential reactant phases and the other the compositions of the two potential product phases. Tie lines in ternary phase diagrams cannot cross, and it is possible to determine which one is more stable from information about the relative values of the standard Gibbs free energies of formation of the reactant and product phases.

In general, compounds on the upper left side of this diagram have much greater negative Gibbs free energies of formation than any of the other types of phases present. Therefore, tie lines will tend to run between such phases and the organic group corner of the diagram. Consequently, any phase further away from the electropositive metal corner of the diagram than that tie line will not be stable vs. the electropositive species.

Some of these product phases that form as reaction product layers at the interface are electronically insulating, but may have some ionic conductivity. The term *solid electrolyte interphase* (SEI) is often used in connection with organic solvent electrolytes in lithium cells, in which a reaction product is formed on the negative electrode along with the generation of an organic radical gas, such as propylene. This topic will be discussed in more detail in Sect. 16A.6.

14A Elevated Temperature Electrolytes for Alkali Metals

14A.1 Introduction

Some years ago, there was a great deal of interest in the development of relatively large scale batteries for use in vehicle propulsion. One of the early approaches involved the use of sodium beta alumina as a sodium-conducting solid electrolyte. The sodium battery systems that evolved, primarily the Na/Na_xS and the Zebra systems, are discussed elsewhere in this text.

Although it was recognized that lithium-based systems might have advantages such as a lower weight and higher voltages, no good lithium-conducting solid electrolytes were known. An alternative initiative appeared, involving the use of lithium-conducting molten salts as electrolytes and their employment in lithium-based systems.

14A.2 Lithium-Conducting Inorganic Molten Salts

A large amount of work was done using a LiCl–KCl molten salt with the eutectic composition, 42 mol% KCl and 56 mol% LiCl. The eutectic melting point is 356°C, but it was generally operated above about 380°C. At high currents, when the flux of lithium into or out of the electrode materials is large, there can be significant transient changes in the local salt concentration, resulting in changes in the melting point within the cell electrodes.

However, it was found that this electrolyte is not sufficiently stable in contact with unit activity lithium electrodes. There is partial decomposition resulting in the escape of potassium vapor, because of displacement by lithium. This also results in some electronic leakage, and the appearance of a black color [2, 3].

This problem is not present if the lithium activity in the negative electrode is reduced. This was commonly done by the use of lithium–aluminum or lithium–silicon alloys that have much higher melting points, which were desirable anyway, as the low melting point of lithium constitutes a potential danger of thermal runaway.

14A.3 Lower Temperature Alkali Halide Molten Salts

There have been some investigations of the possibility of reducing the melting point of alkali halide molten salts by the addition of other components. The LiF–LiCl–LiI eutectic temperature is 334°C, that of the LiF–LiCl–LiBr–LiI system is 325°C, and that of the LiF–LiBr–KBr system is 280°C [4]. However, there have been no very surprising or exciting results from this simple alkali halide approach. As will be seen in the next section, other combinations are more promising.

14A.4 Other Modest Temperature Molten (and Solid) Salts

In addition to the alkali halide systems, there have been investigations of electrolytes based upon the ternary chloride phase $LiAlCl_4$ [5–7]. This phase melts at 145°C, and is an interesting electrolyte for lithium both below and above that temperature.

The investigation of the thermodynamic properties of this system [6] resulted in the determination of the ternary phase stability diagram, which is shown in Fig. 14A.1 for the temperature 135°C, where $LiAlCl_4$ is still solid.

It is seen that the high lithium activity, or low voltage, limit is determined by the formation of LiCl and aluminum. At the low lithium activity, or high voltage, limit $LiAlCl_4$ decomposes to form $AlCl_3$ and chlorine gas.

It was found that the stability window of $LiAlCl_4$ varies with temperature, extending between 1.68 and 4.36 V vs. lithium at 25°C, and between 1.7 and 4.2 V vs. lithium at 135°C.

Because chlorine gas is present at the high potential limit of $LiAlCl_4$, it is possible to use measurements of the potential on the three-phase equilibrium at the

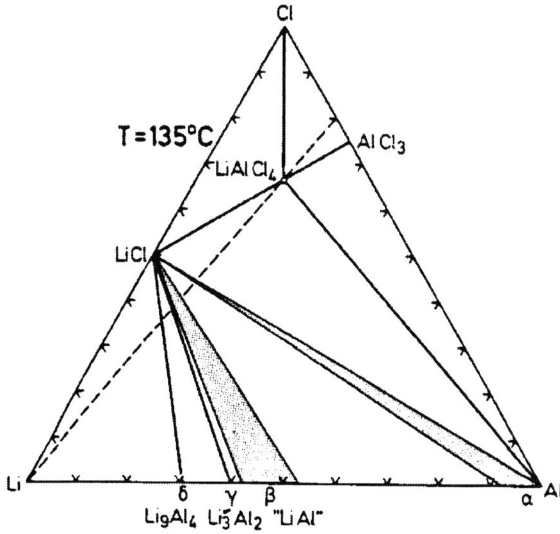

Fig. 14A.1 Phase stability diagram for the system Li–Al–Cl at 135°C

high potentials side of this phase to give information about the chlorine gas pressure present, i.e., to use this system as a chlorine sensor [6].

Furthermore, since the phase stability diagram indicates that $LiAlCl_4$ is stable vs. pure aluminum at relatively high lithium activities, it should not be surprising that it was shown [7] that it is possible to use it as a molten electrolyte to electro-deposit aluminum.

Subsequently, there has been further exploration of other low-melting salt combinations that are stable in the presence of lithium. For example, the eutectic in the LiCl–NaCl–KCl–$AlCl_3$ system has a melting point of 61°C, and the LiCl–KCl–$AlCl_3$ eutectic has a melting point of 86°C [8].

14A.5 Relation Between the Potential and the Oxygen Pressure in Lithium Systems

Another matter that must be kept in mind in the consideration of the use of lithium electrolytes is the limitation of the potential due to the formation of Li_2O, if there is any oxygen present, either in the environment or in the electrodes.

By considering the reaction

$$2Li + {}^{1}/_{2}O_2 = Li_2O \tag{14A.1}$$

and the general relation between the voltage and the change in Gibbs free energy,

$$E = -\Delta G_r{}^{\circ}/zF \tag{14A.2}$$

that can be converted to the Nernst equation, written in terms of the partial pressure of molecular oxygen gas, and with one atmosphere pressure as the standard state

$$E = -\left({}^{RT}/_{4F}\right)\ln p(O_2). \tag{14A.3}$$

Thus the voltage vs. pure lithium for any value of oxygen activity or partial pressure can be calculated. Conversely, this means that for every value of electrical potential in a lithium electrolyte system that has access to oxygen there is a corresponding value of oxygen activity or partial pressure.

At a temperature of 25°C, $\Delta G_r{}^\circ$ for the formation of Li_2O is -562 kJ mol^{-1}, so that the value of the voltage vs. pure lithium at which the oxygen pressure is one atmosphere is 2.91 V. At 400°C, where a number of measurements were made on lithium-transition metal oxides [4], the value of the standard Gibbs free energy of formation of Li_2O is -511.3 kJ mol^{-1}, so that the potential at which the oxygen pressure is one atmosphere is 2.65 V vs. lithium at that temperature.

As discussed in Sect. 9.4, it has been shown that measured values of electrical potential relative to lithium correlate well with calculated values of the oxygen activity in a number of ternary lithium-transition metal oxides [9].

It was found, however, that the thermodynamic properties of the LiCl–KCl molten salt electrolyte in experiments performed in a glove box at 400°C limited the maximum oxygen partial pressure attainable to about 10^{-25} atm. This was determined by experiments in which powders of a series of binary and ternary transition metal oxides were immersed in the LiCl/KCl salt at 400°C for 18 h. Subsequent X-ray analysis showed that all of the ones with calculated oxygen partial pressures over 10^{-25} atm reacted to form either lower oxide phases or amorphous products. The potential at which the oxygen partial pressure is 10^{-25} atm is 1.82 V vs. lithium at that temperature. All those whose electrical potentials were below that voltage remained stable but oxides that react with lithium at higher potentials could not be investigated by the use of the LiCl–KCl molten salt electrolyte.

Within the range of stability of the LiCl–KCl electrolyte, it was shown that the relation between the equilibrium voltage vs. lithium and the oxygen partial pressure of the electrode materials could be represented by

$$E = (RT\,/\,4F)\ln p(O_2) - \left(\Delta G_f{}^\circ(Li_2O)\,/\,2F\right), \tag{14A.4}$$

where $\Delta G_f{}^\circ(Li_2O)$ is the standard Gibbs free energy of formation of Li_2O at the temperature in question.

14A.6 Implications for the Safety of Lithium Cells

The relationship in (14A.4) is not just relevant to experiments in this particular electrolyte, but has important ramifications for the stability of other oxide reactants in any electrolyte, as well.

Electrode reactants that have high voltages vs. lithium have low lithium activities, and because of these relationships, they also have high oxygen activities, or pressures.

In the case of small closed cells with organic solvent electrolytes instead of chloride molten salts, the maximum voltage can reach much higher values, as discussed in Chap. 10. Thus the positive electrode reactants can have high oxygen pressures. This is surely one of the major factors in the thermal runaway safety problem that is known to be related to the evolution of oxygen from the positive electrodes.

14B Ambient Temperature Electrolytes for Lithium

14B.1 Introduction

As mentioned already, the main function of an electrolyte is to allow the transport of the electro-active ions, while blocking the passage of electronic species. They must also be stable at the potentials and in the chemical environments imposed by the two electrodes. In the case of lithium cells, this latter requirement can be quite demanding. Lithium reacts vigorously with water, forming LiOH and hydrogen, so that aqueous electrolytes are not suitable in the presence of high lithium activities. Instead, organic solvent electrolytes containing lithium salts are generally used in lithium cells at ambient temperatures at the present time.

14B.2 Organic Solvent Liquid Electrolytes

The most common electrolytes in lithium batteries are organic solvent lithium electrolytes. They can be thought of as having organic cations and carbonate anions. That is, they fall along the right hand side of the generalized ternary stability diagram discussed in Chap. 14. These materials are not stable in the presence of high activities of electropositive metals, such as lithium. Thus, they can only be used in lithium battery systems if there is an intermediate phase separating them from the high lithium activity negative electrode. This has been recognized for some time, and this interfacial region is generally called the "solid electrolyte interphase", or SEI [10]. This layer must serve to block any chemical reaction of the electrode material and the electrolyte, as well as allowing the transport of ionic species, but not electrons. This layer thus acts as a second electrolyte in series with the primary, organic solvent, electrolyte. It must also maintain its mechanical integrity in the face of any local configurational changes. The SEI will be discussed further in Sect. 16A.6.

A variety of different organic cation solvents have been explored for use in lithium cells, and a number of criteria have to be considered. For reasons that will

Table 14B.1 Melting points and dielectric constants of various organic solvents

Solvent	Abbreviation	Melting point/°C	Dielectric constant
Diethyl carbonate	DEC	−74.3	2.8
Ethyl methyl carbonate	EMC	−53	64.9
Propylene carbonate	PC	−48.8	39
Butyrolactone	GBL	−43.5	34
Acetonitrile	AN	−48.9	36
Ethylene carbonate	EC	36.4	53
Tetrahydrofuran	THF	−108.5	7.4
Dimethyl carbonate	DMC	4.6	3.1

be discussed shortly, these generally consist of two or more organic solutions, each have components selected for a specific purpose. Among the important criteria are the melting point and the dielectric constant, and values of these for a number of the common solvents are given in Table 14B.1.

Although they were of interest earlier, ethers have been largely supplanted, for they are not stable at the higher potentials of positive electrode components that have evolved in recent times.

For reasons that will be discussed in Sect. 16A.6, ethylene carbonate is very attractive. However, its melting point is relatively high, so that it is generally combined with another solvent. Propylene carbonate was an early choice as a cosolvent, but it was recognized that this combination caused defoliation of graphitic negative electrodes [11], as had been found in the early experiments on the use of graphite in the negative electrode of lithium cells [12].

In addition to compatibility with the negative electrode, stability at the potentials imposed by the positive electrode is required. The positive stability limits of a number of electrolytes were evaluated [13, 14]. It was found that solutions of ethylene carbonate with either dimethyl carbonate or diethyl carbonate are stable up into the 5 V range. This is interesting as neither of these two additives is stable at such positive potentials by itself. However, they remain the materials of choice at the present time.

14B.3 Lithium Salts

To operate as an electrolyte in lithium cells, these solvents have to contain lithium salts. There are several important criteria. One is that the salt must have a high solubility and should dissociate to provide the required charged mobile species. They must, of course, be stable at the potentials of the two electrodes, and should be nontoxic and inert to the other components of the electrochemical cell. Especially important is the question of the formation of the SEI layer on the negative electrode. Several such salts have been explored, including $LiClO_4$, $LiAsF_6$, $LiBF_4$, and

$LiPF_6$. At the present time, $LiPF_6$ is most widely used in commercial cells. The characteristics of these various salts are discussed in more detail elsewhere [15, 16].

Another salt that has become an interesting alternative recently is called LiBOB [17]. It is based upon a chelated borate anion, BOB [bis(oxalato)borate]. This salt is evidently stable in contact with graphite negative electrodes, as well as positive electrode reactants. However, it seems to be more reactive in contact with charged positive electrodes than electrolytes containing $LiPF_6$ [18].

14B.4 Ionic Liquids

Although the organic solvent electrolytes discussed earlier are being employed in lithium battery technology at the present time, there is increasing interest in an alternative group of materials that are generally called *ionic liquids*. One attractive feature for some applications is that they generally have much lower vapor pressures than the organic cation carbonates in which the organic species are relatively small, flammable, and sometimes toxic.

These materials conduct charge by the transport of one or both of their ions, as is also the case with a large number of other materials that become molten salts at elevated temperatures mentioned in Sect. 14.2.

The primary difference between the ionic liquids discussed here and the inorganic molten salts is that they have significantly lower melting points. The melting point is related to the energy necessary to break the bonds holding the component species together. This is sometimes called their lattice energy. In general, the larger the structural groups that act as ions, and the lower their electrical charges, the easier it is for thermal energy to break them apart from each other.

Thus organic cations with quaternary ammonium groups, designated as (R_4N^+), have low lattice energies, and have much lower melting points than their alkali metal cation analogs. As an example, NaBr melts at 747°C, whereas tetrabutylammonium bromide melts at 104°C.

Another factor is the symmetry of the component ions. Materials with ions that are less symmetrical and have less spherical shapes tend to have lower melting points as they are more difficult to fit into crystal lattices.

Thus low melting (below 100°C) ionic liquids tend to use bulky anions such as $(F_3CSO_2)_2N^-$, BF_4^-, $AlCl_4^-$, PF_6^-, and $F_3CSO_3^-$, and cations such as butyl pyridinium and butylmethylimidazolium.

Some low temperature ionic liquids of this type can dissolve a wide range of simple metal oxides, such as CoO, NiO, CuO, ZnO, MnO_2, V_2O_5, and Fe_3O_4, as well as organic materials such as sugar and cellulose, because of their high chlorine content activity.

The ionic conductivity of ionic liquids of this type is generally in the range of $mS\ cm^{-1}$, depending upon the chain length of the alkyl cation component.

Although there has been some concern about the stability of materials of this type under reducing conditions, such as in contact with high lithium activity negative electrodes, it has been shown [19] that some such materials can behave very well in lithium cells. The particular ionic liquid electrolyte that was used in this case was a ternary mixture: $PYR_{14}TFSI + x\ LiTFSI + y\ PEGDME$ (or TEGDME). $PYR_{14}TFSI$ is *N*-methyl-*N*-butylpyrrolidinium bis(trifluoromethane sulfonyl)imide, TEGDME is tetra(ethylene glycol)dimethyl ether, and PEGDME is poly(ethylene glycol)dimethyl ether. The thermal stability is best with increasing amounts of $PYR_{14}TFSI$, whereas increasing the amount of PEGDME or TEGDME reduces the viscosity and increases the ionic conductivity. Elemental lithium electrodes cycled well in this electrolyte, with no sign of filamentary growth or dendrites.

This ionic liquid electrolyte was also found to be stable in contact with some positive electrode reactants, sulfur and Li_xFePO_4. The measured values of capacity and cycle life were also favorable.

It is reasonable to expect to see more interest in the use of these electrolytes in lithium cells in the future.

14C Aqueous Electrolytes for Hydrogen

14C.1 Introduction

Aqueous electrolytes have been used for batteries for many years, and their major function is to transport hydrogen without electronic leakage. There are two general mechanisms, depending upon the pH of the electrolyte. In acid solutions protons are hydrated, so that the hydrogen transport mechanism involves the hopping of protons between adjacent species. This is called the *Grotthus mechanism*. Although it had generally been assumed that the hydrated species is H_3O^+, it now seems that two other, larger, groups are actually involved, the *Eigen cation*, $H_9O_4{}^+$, and the *Zundel cation*, $H_5O_2{}^+$, and that interconversion between these two structures accounts for proton motion [20].

In the case of alkaline solutions, OH^- ions are present and hydrogen is transported by hopping between water molecules and adjacent OH^- ions. As would be expected, the conductivity varies with the concentrations of these defect species. Some data are shown in Table 14C.1.

When discussing aqueous electrolytes that transport hydrogen, the obvious tendency is to think in terms of liquid water containing various simple salts that cause it to have a range of values of pH. There is, however, a very different topic that should be considered. This involves solid materials that are used as electrolytes in some battery and fuel cell systems that essentially act as nanoporous containers for aqueous solutions. Although they have the mechanical properties of solid polymers, their chemical and electrochemical properties are very close to those of water. The most well known of this class of materials is *Nafion*, a proprietary product of E.I. du Pont de Nemours and Company (DuPont). This, and some of the other alternatives, will be discussed in this chapter.

Table 14C.1 Dependence of the conductivity upon the composition for some aqueous electrolytes at 25°C

Salt	Concentration (M)	Conductivity (ohm^{-1} cm^{-1})
H_2SO_4	0.1	0.048
H_2SO_4	0.53	0.21
HCl	0.1	0.039
HCl	1.0	0.33
HCl	6.0	0.84
KCl	0.1	0.013
KCl	1.0	0.11
$LiClO_3$	0.1	0.089
KOH	0.1	0.31
KOH	0.3	0.50
KOH	0.5	0.29

14C.2 Nafion

At the present time Nafion is the primary choice for use as an electrolyte in low-temperature hydrogen/oxygen fuel cells because of its high proton conductivity and excellent chemical and thermal stability.

Its development was initiated by W.G. Grot at DuPont in the early 1960s [21], and its major large-scale application has been as a membrane in the chlor-alkali process. The focus on its use as an electrolyte in hydrogen-transporting fuel cells has appeared more recently. Thorough reviews of the state of knowledge relating to Nafion can be found in [22–24].

Nafion's unique ionic properties are the result of incorporating perfluorovinyl ether groups, terminated with sulfonate groups, onto a tetrafluoroethylene (teflon) backbone. It is produced from a precursor containing $-SO_2F$ functional groups that can be melt-fabricated. After reaching the desired shape, it is then given a treatment in hot aqueous NaOH that converts the sulfonyl fluoride ($-SO_2F$) groups into sulfonate groups ($-SO_3^-Na^+$). This form of Nafion, referred to as the neutral or salt form, is finally converted to the acid form containing the sulfonic acid ($-SO_3H^+$) groups.

Nafion was the first synthetic polymer developed that has ionic conduction properties. Materials in this class are now called *ionomers*. The presence of water in its structure plays a critical role in its conduction properties. Because of its Teflon backbone, Nafion is quite resistant to chemical attack. It is mechanically quite strong at ambient temperatures, but becomes thermoplastic and mechanically weak at elevated temperatures. A number of different types of Nafion membranes have been produced with different pore dimensions and properties for various applications.

Such hydrated acidic polymers combine, in one macromolecule, the high hydrophobicity of the backbone with the high hydrophilicity of the sulfonic acid functional group, which results in a constrained hydrophobic/hydrophilic nanoseparation.

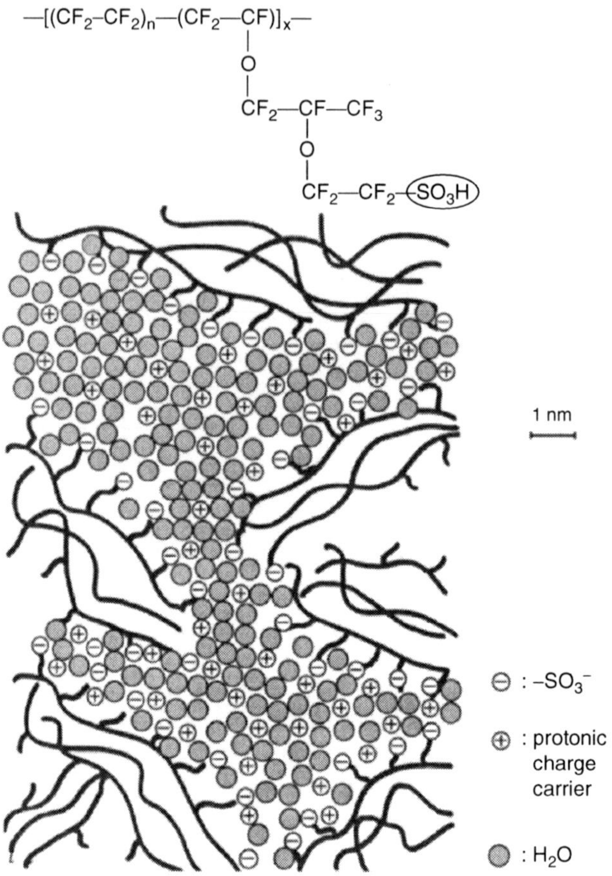

Fig. 14C.1 Schematic of the microstructural features of Nafion containing an intermediate amount of water

The sulfonic acid functional groups aggregate to form a continuous hydrophilic region in the structure that is hydrated upon absorption of water. It is within this continuous domain that ionic conductivity occurs. Protons dissociate from their anionic counter ion ($-SO_3^-$) and become solvated and mobilized by the hydration water.

The microstructure is illustrated schematically in Fig. 14C.1 [24].

High proton conductivity is only obtained at high levels of hydration, and water typically must be supplied to the electrolyte through humidification of the feed gases. It is also produced by the electrochemical reduction of oxygen at the cathode in fuel cells. Because of the requirement that water be present in the structure, the maximum operation temperature is limited to about the condensation point of water. This is, of course, 100°C for a water pressure of 1 atm (10^5Pa), but the cells are typically operated at several atmospheres pressure, which raises the temperature limit somewhat. Since the proton-containing species also contain water, protonic current leads to transport of water through the membrane. This must also be taken into account in the overall water balance.

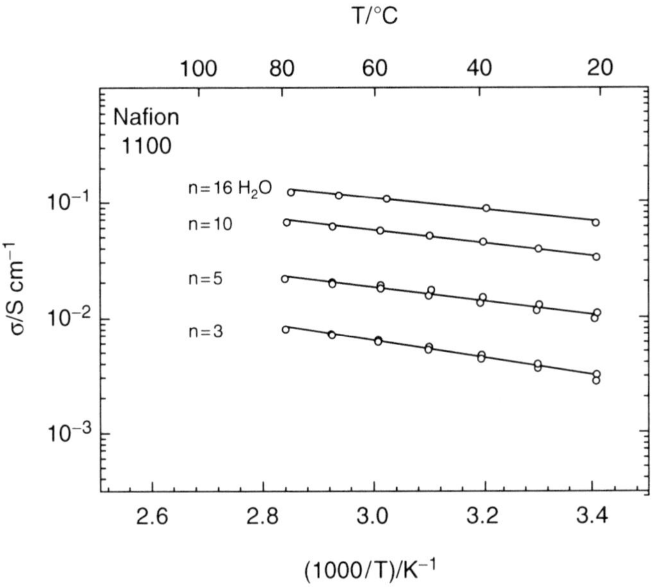

Fig. 14C.2 The dependence of the proton conductivity of Nafion 1,100 upon temperature and degree of hydration

The significant variation of the ionic conductivity of Nafion 1,100 with temperature and degree of hydration is shown in Fig. 14C.2 [24].

14C.3 Other Considerations Relating to Nafion

In addition to the requirement that care must be taken to continually control the water balance in electrochemical cells that have Nafion electrolytes, there are additional matters that must be considered.

The limited operating temperature and the acidity of the electrolyte make it necessary to use expensive platinum or platinum alloys as electrocatalysts to promote the electrochemical reactions in the anode and cathode structures. However, even with platinum, only rather pure hydrogen can be oxidized at sufficient rates, for even trace amounts of CO in the feed gas can adsorb and block the reactive sites on the catalyst surface. This is not such a problem with fuel cells that operate at higher temperatures, because of the increased desorption of CO as the temperature is raised.

There is a continued interest in the direct oxidation of methanol in fuel cells. In addition to the problem of the presence of CO as a result of the methanol production process, and the resultant catalyst contamination problem, the *crossover* of methanol to the positive electrode side of the cell is deleterious. This is a type of *chemical short-circuiting*, and results in a significant decrease in the cell voltage, as well as the efficiency of the use of the fuel.

14C.4 Alternatives to Nafion

Other polymer systems are continually being sought, and one approach involves the use of *adducts* (complexes) of basic polymers with *oxo-acids*. Examples are *polybenzimidazole (PBI)* and *phosphoric acid* [25].

Phosphoric acid is an almost ideal proton conductor. In contrast to water, which exhibits a high mobility for protonic defects but a very low intrinsic concentration of protonic charge carriers, liquid phosphoric acid shows both a high mobility and a high concentration of intrinsic protonic defects.

The total conductivity of phosphoric acid at its melting point (42°C) is 7.7×10^{-2}S cm^{-1}, with an estimated proton mobility of 2×10^{-5}cm^2 s^{-1}.This extremely high proton mobility has been explained by the correlated motion of the oppositely charged defects in the phosphate lattice ($H_2PO_4^-$, $H_4PO_4^+$) when they are close to one another. Such correlation effects are quite common in other proton conductors with high concentrations of charge carriers. The phosphate species have a low diffusion coefficient, and thus are essentially immobile in the structure.

The imidazole unit, $C_3H_4N_2$, has a cyclic structure with a five-member ring containing two nitrogens. Benzimidazole is bicyclic, with a combination of a six-member benzene and a five-member imidizole ring, and has the formula $C_7H_6N_2$.

A strong acid/base reaction occurs between the nonprotonated basic nitrogen of the polybenzimidazole (PBI) repeat unit and the absorbed phosphoric acid. The transfer of one proton leads to the formation of a benzimidazolium cation and a dihydrogen phosphate anion, forming a stable hydrogen-bonded complex. The very good proton conductivity of phosphoric acid is hardly affected by the interaction with PBI.

Details concerning the structure and the fundamental features of the transport phenomena in the adducts of polybenzimidazole and phosphoric acid are not as advanced as they are in the case of Nafion. Although the proton conductivity is more than a factor of 10 lower than that of Nafion, PBI-H_3PO_4 adducts have an important advantage when used with direct methanol fuel cells in that there is very little *methanol crossover.*

14D Nonaqueous Electrolytes for Hydrogen

14D.1 Introduction

There is continuous interest in materials that have high specific capacities for the storage of hydrogen, and this was discussed in Sect. 10B.3. There are two general types of applications that are of interest. One involves the reversible absorption of hydrogen from the gas phase, and is of particular interest for applications such as fuel cells. The other is related to the use of hydrogen in rechargeable batteries. Relatively small metal hydride/nickel batteries are manufactured and sold in very large numbers at the present time, and the materials that are currently used in such cells typically have satisfactory capacity densities, i.e., capacities per unit volume, for those applications.

In the case of larger scale applications, such as for use in electrically powered vehicles, weight is an especially important parameter. The current US government target is to find materials that can reversibly store 6% hydrogen by weight.

Cost can also be a significant problem, as well. Thus there is a continuous search for alternative hydrogen storage materials with more attractive properties.

Current metal hydride cells use alkaline aqueous electrolytes, as discussed already. Some of the materials that might be more favorable than the current choices cannot be used in such electrolytes. As an example, it would be attractive to be able to use metal hydride systems containing magnesium, for this element is relatively inexpensive and the related hydrides can have low weights. Some work on magnesium alloys is discussed later in this chapter.

However, the affinity of magnesium for oxygen and water is very high. This means that the surfaces of magnesium-containing materials can be easily poisoned by even minor amounts of such species, and this results in practical difficulties when impure hydrogen is present in the gas phase. Direct contact with aqueous electrolytes in electrochemical systems is also obviously out of the question with magnesium-based materials.

There are two ways to get around such problems in electrochemical systems. One is to use an electrolyte that is in true thermodynamic equilibrium with the species in the electrode, so that there is no driving force for the formation of a reaction product on the surface. The other is to arrange the thermodynamic conditions at the interface such that a reaction product is formed that is stable in contact with the electrolyte, and is also a good solid electrolyte for the electroactive species.

14D.2 Methods Typically Used to Study Materials for Hydrogen Storage

As discussed in Sect. 10B.3, much of the research on materials for hydrogen storage use in nonelectrochemical applications has involved the use of P-C-T methods, in which the relevant parameters are the hydrogen activity (pressure), the composition (amount of hydrogen in the hydrogen storage material), and the temperature. Pressure-composition isotherms are often measured at various values of the temperature.

Alternatively, electrochemical methods can be employed if a suitable electrolyte is used that conducts hydrogen ions, either protons (H^+) or hydride (H^-) ions. In this case, it is typical to measure the voltage across the electrochemical cell (related to the hydrogen activity) as a function of composition and as a function of the temperature.

These two methods are related by the dependence of the voltage across an electrochemical cell upon the difference in the chemical potential of the electroactive species i, whose charge number in the electrolyte is z, as discussed earlier

$$\Delta E = -(\Delta\mu_i/zF). \tag{14D.1}$$

The chemical potential of any species i is related to its activity or pressure by

$$\mu_i = \mu_i^0 + RT\ln(p_i/p_i^0), \tag{14D.2}$$

where μ_i^0 is its chemical potential in a standard state, and p_i^0 is the pressure at that standard state. These can be combined to give the well-known Nernst equation

$$\Delta E = -(RT/zF)\ln(p_i/p_i^0). \tag{14D.3}$$

Thus voltage measurements across an electrochemical cell are effectively equivalent to measurement of the logarithm of the pressure. On the one hand, in H^+ (proton) electrolytes, the charge number is $+1$, and the hydrogen activity (pressure) increases as the electrical potential becomes more negative. On the other hand, the charge number in H^- (hydride ion) electrolytes is -1, so the hydrogen activity (pressure) increases as the electrical potential becomes more positive.

14D.3 Potential Advantages of Electrochemical Methods

Electrochemical methods have several potential advantages over P-C-T methods. They generally involve measurements at relatively low hydrogen activities, and thus do not require the use of pressure vessels. No temperature change is required for absorption or desorption. Therefore, one can make isothermal experiments over a wide range of pressures. Large variations in hydrogen activity can be obtained by relatively small differences in the cell voltage, and the effective pressure can be changed easily and rapidly over several orders of magnitude. Coulometric titration can provide a simple and accurate method to change the hydrogen concentration in the material being studied, and quantitative evaluation of the reaction kinetics is also relatively easy.

14D.4 The Amphoteric Behavior of Hydrogen

As mentioned earlier, aqueous electrochemical systems cannot be used for studies of oxygen-sensitive or water-sensitive materials, such as magnesium, vanadium, or titanium.

There is, however, an alternative, due to the amphoteric nature of hydrogen, which has an intermediate value of electronegativity, 2.1 eV, according to

Pauling [26]. This can be readily seen, for if HCl dissolves in water or another solvent, the result is the presence of H^+ and Cl^- ions, but if LiH dissolves in a solvent, the result is the presence of Li^+ and H^- ions.

To have useful electrochemical cells that transport hydrogen in the form of hydride ions, it is necessary to have a solvent that dissolves a hydride salt, to have a counter electrode that can act to supply and absorb neutral hydrogen, and to have a suitable reference electrode. As will be seen below, the reference material may be a primary reference potential for some other species, and also act as a secondary reference for hydrogen.

Two examples will be discussed here, the use of alkali metal molten salts at temperatures of about 400°C to study hydrogen in vanadium and titanium, and the use of organic-anion molten salts cells at 140–170°C to study hydrogen in binary magnesium alloys, magnesium hydrides.

14D.5 Relationships Between the Potential and the Stability of Phases in Molten Salts

Some molten salts, such as the alkali halides discussed earlier, are stable over wide ranges of potential. This can lead to conditions in which a number of metals are immune against the formation of deleterious reaction product phases upon their surfaces. This can be readily seen from Fig. 14D.1, which shows the potential ranges of several metal chlorides, as well as LiH, at 425°C [27].

It is seen that KCl is stable over a greater potential range than LiH. In a mutually saturated LiCl–KCl molten salt at that temperature the potential of elemental Li is 3.64 V negative of that of Cl_2 gas at one atmosphere. This is a wider potential range than the stability ranges of $TiCl_2$ and VCl_2, which are 2.50 V and 2.25 V, respectively. This means that titanium will be immune to the formation of chloride

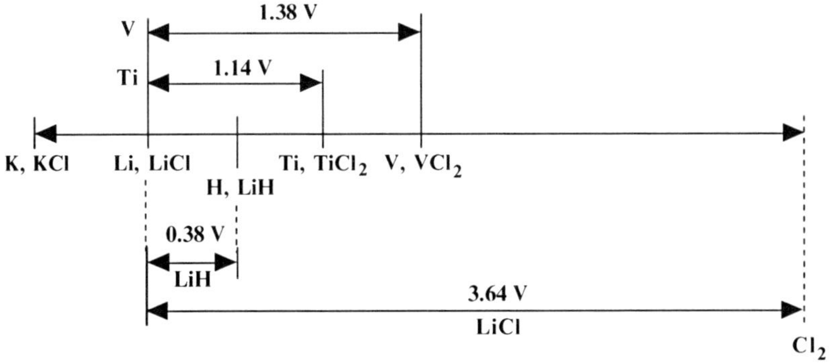

Fig. 14D.1 Potential ranges in several metal chloride systems at 700 K. The range of stability of LiH is also shown

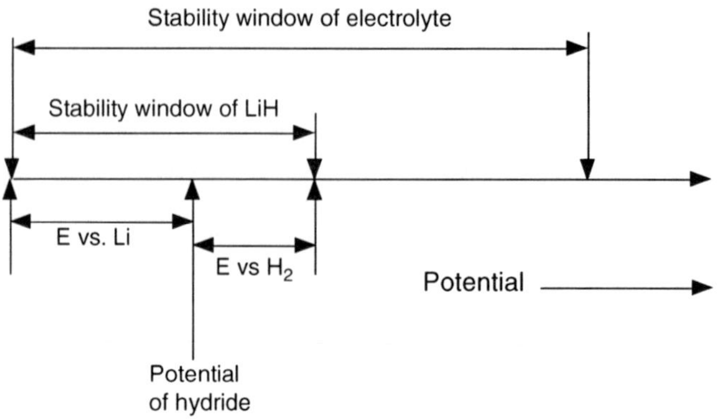

Fig. 14D.2 Potential relations when using a hydride-conducting electrolyte to investigate hydrogen storage materials

on its surface at all potentials more than 2.50 V negative of Cl_2, or less than 1.14 V positive of the potential of pure Li. The corresponding values for vanadium are 2.25 and 1.38 V.

LiH is stable over a range of 380 mV at that temperature. Thus the hydrogen pressure becomes one atmosphere when the potential reaches 380 mV positive of the potential of pure lithium. The potential relationships when using a hydride-conducting electrolyte to investigate a hydrogen storage material are illustrated in Fig. 14D.2. The potential of a hydride material can be evaluated against either the lithium or the hydrogen potentials. It is positive versus lithium, and negative vs. one atmosphere of hydrogen.

14D.6 Alkali Halide Molten Salts Containing Hydride Ions

As mentioned earlier, alkali halide molten salts have been of interest for battery applications for a number of years. The eutectic composition in the LiCl–KCl system, which melts at 356°C, was employed in a large development effort some years ago by the Department of Energy in the United States that was aimed toward the use of large Li,LiAl/FeS and Li,LiAl/FeS_2 cells for vehicular propulsion. It has been found that this salt can dissolve a significant amount of LiH.

The eutectic temperatures of this, and several other alkali halide salts that also dissolve LiH, and thus can contain hydride ions, are listed in Table 14D.1.

The all-lithium systems have been of interest because they do not generate lithium ion concentration gradients during high current operation. Deviations from the eutectic composition in salts containing two different cations can lead to localized solidification and a reduction in the conductivity. The salt with a composition in the LiF–LiBr–KBr system has a substantially lower melting point than the others [28], but it has two different cations, and so would not be useful in some applications.

Table 14D.1 Eutectic temperatures of some alkali halide salts

System	Eutectic temperature (°C)
LiF–LiCl–LiBr	436.0
LiCl–KCl	352.0
LiF–LiCl–LiI	334.3
LiF–LiCl–LiBr–LiI	325.4
LiF–LiBr–KBr	280.0

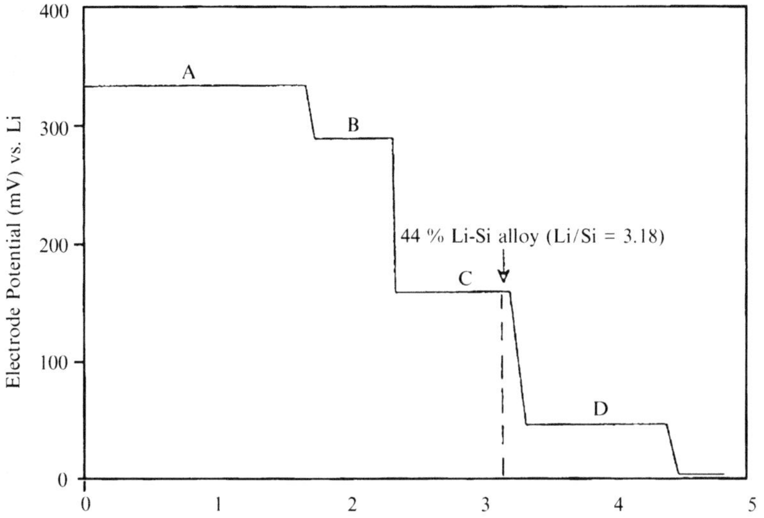

Fig. 14D.3 Electrochemical titration curve for the Li–Si system at 415°C

When lithium hydride is dissolved in such halide salts, charge transport can occur by the motion of hydride ions. If lithium hydride is dissolved up to its solubility limit the relation between the hydrogen activity and the lithium activity can be simply determined from the Gibbs free energy of formation of LiH, for

$$\mu_{LiH} = \mu_{Li} + {}^1\!/_2\mu_{H_2} = 1. \tag{14D.4}$$

Therefore, measurements of the lithium activity can be readily converted into values of the hydrogen activity. This means that a reference electrode that establishes a fixed value of the lithium activity can serve as a convenient secondary reference for the hydrogen activity.

A number of two-phase plateaus are known in binary lithium alloy systems that can be used for such purposes [29–32]. As an example, the lithium–silicon phase diagram contains four intermediate phases in the temperature range of the alkali halide molten salts. Their Li/Si stoichiometric ratios are about 1.714, 2.333, 3.25, and 4.4. The equilibrium titration curve, measured at 415°C [33], is shown in Fig. 14D.3.

The alloy with the Li/Si ratio of 3.18 has been used in commercial thermal batteries, and has a potential 158 mV positive of pure lithium at that temperature. It can be used as an indirect reference for hydrogen if the electrolyte is saturated with LiH.

14D.7 Solution of Hydrogen in Vanadium

An alkali halide electrolyte saturated with LiH has been used to investigate the solubility of hydrogen in vanadium at 425°C [34]. The results are illustrated in Fig. 14D.4. As mentioned earlier, the potential of 1 atm H_2 is 380 mV vs. Li/LiH at that temperature, and it is seen that this system remains as a single phase over the full range of those measurements. The hydrogen solubility is quite large, about 30 atomic percent, in 1 atm of hydrogen. It is also seen that there was very little hysteresis, even with measurements at the relatively high current density of 5 mA cm^{-2}.

14D.8 The Titanium–Hydrogen System

Another example of the use of an hydride-containing alkali metal molten salt were investigations of the reaction of hydrogen with titanium [35, 36]. In this case, several different hydrogen-titanium phases are stable within the range of the accessible experimental parameters.

The titanium–hydrogen phase diagram is shown in Fig. 14D.5. It is seen that three phases are present in different compositional ranges above a eutectoid reaction at about 300°C. The hexagonal close-packed alpha phase is stable at low hydrogen concentrations, a body-centered cubic beta phase is stable at intermediate compositions, and the delta phase forms at high hydrogen concentrations. There are

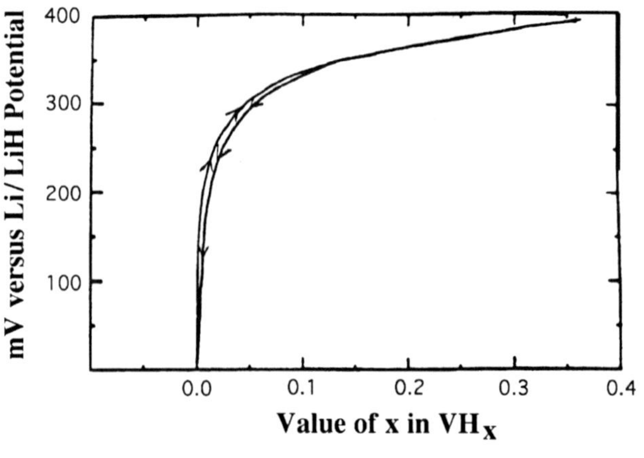

Fig. 14D.4 Potential of hydrogen in vanadium during a galvanic cycle at 700 K

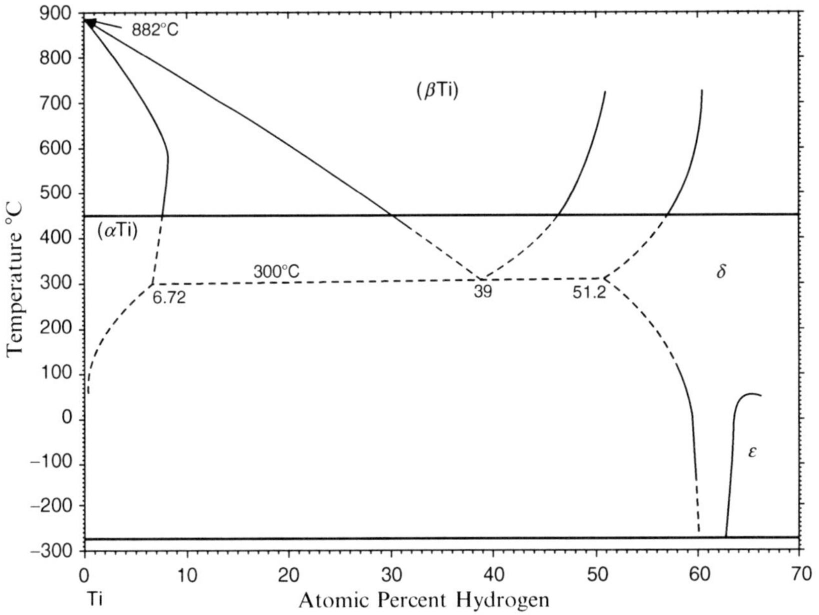

Fig. 14D.5 Titanium–hydrogen phase diagram

two-phase composition ranges in between each of these single-phase regions. The fact that some of the lines are dashed, rather than full lines, indicates that the temperature of the eutectoid reaction was not firmly known at the time of the diagram's publication.

The results of coulometric titration experiments at several temperatures are shown in Fig. 14D.6. The potential was measured relative to the two-phase Li–Si reference electrode, and converted to the standard hydrogen 1-atmosphere electrode potential at that temperature. As expected from the Gibbs phase rule and the phase diagram, the single-phase regions, in which the potential varies with the hydrogen content, and the constant-potential two-phase regions between them are clearly seen.

Only one long two-phase plateau was observed, between the α and δ phases, at 300°C. At the higher temperatures two plateaus were found, related to the α–β and β–δ equilibria in different ranges of composition.

14D.9 Use of Low Temperature Organic-Anion Molten Salt to Study Hydrogen in Binary Magnesium Alloys

It has been demonstrated experimentally that the organic-anion molten salt $NaAlEt_4$, where Et is ethylene, is stable in the presence of elementary sodium, and up to potentials about 1 V positive of that of sodium [37]. This salt can also dissolve appreciable quantities of NaH, and can be synthesized by direct reaction of sodium with $AlEt_3$.

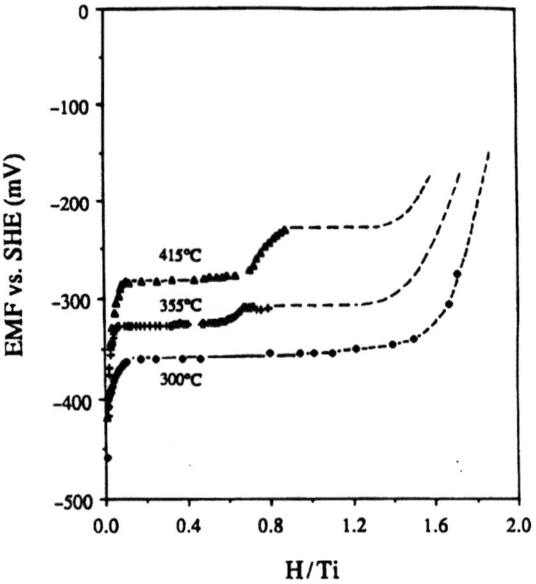

Fig. 14D.6 Electrochemical titration of hydrogen into titanium at several temperatures

This salt has been used to study the hydrogen absorption properties of three binary magnesium alloy families; the Mg–Al, Mg–Cu, and Mg–Ni systems, at 142 and 170°C [38, 39], as discussed in Chap. 4.

14D.10 Summary

The investigation of the properties of alternate materials for the storage of hydrogen in solids requires information about several important parameters. The method that is most commonly used involves measurements of pressure–composition–temperature relations. This requires equipment that can handle a range of gas pressures.

An alternative is the use of electrochemical methods. This can be particularly advantageous for the investigation of materials that are either oxygen or water-sensitive. One way to do this is to use molten salt electrolytes that contain hydride ions. Some such salts can be very reducing and remove surface oxides that block the passage of hydrogen.

These methods have been demonstrated by several examples, using a high temperature alkali halide molten salt and a lower temperature organic-anion molten salt. The solubility of hydrogen in single-phase vanadium has been evaluated as a function of potential, and the compositions of the different single and two-phase regions in the titanium–hydrogen system have been measured. In addition, the reaction of a number of binary alloys in the Mg–Al, Mg–Cu, and Mg–Ni systems with hydrogen has been studied.

References

1. G. Deublein and R.A. Huggins, Solid State Ionics 18/19, 1110 (1986)
2. R.J. Heus and J.J. Egan, J. Phys. Chem. 77, 1989 (1973)
3. R.N. Seefurth and R.A. Sharma, J. Electrochem. Soc. 122, 1049 (1975)
4. G. Deublein and R.A. Huggins, unpublished results (1986)
5. W. Weppner and R.A. Huggins, J. Electrochem. Soc. 124, 35 (1977)
6. W. Weppner and R.A. Huggins, Thermodynamic Stability of the Solid and Molten electrolyte $LiAlCl_4$, in *Fast Ion Transport in Solids*, ed. by P. Vashishta, J.N. Mundy and G.K. Shenoy, North-Holland, New York (1979), p. 475
7. I.D. Raistrick and R.A. Huggins, Use of Lithium Aluminum Chloride Molten Salt as an Electrolyte in Lithium Cells, in *Proceedings of the Fourth International Symposium on Molten Salts*, ed. by M. Blander, D.S. Newman, G. Mamantov, M.L. Saboungi and K. Johnson, Electrochemical Society, Pennington, NJ (1984), p. 82
8. G. Deublein, Personal communication (2007)
9. N.A. Godshall, I.D. Raistrick and R.A. Huggins, J. Electrochem. Soc. 131, 543 (1984)
10. E. Peled, J. Electrochem. Soc. 126, 2047 (1979)
11. R. Fong, U. von Sacken and J.R. Dahn, J. Electrochem. Soc. 137, 2009 (1990)
12. J.O. Besenhard and H.P. Fritz, J. Electroanal. Chem. 53, 329 (1974)
13. J.-M. Tarascon and D. Guyomard, J. Electrochem. Soc. 140, 3071 (1993)
14. J.-M. Tarascon and D. Guyomard, Solid State Ionics 69, 293 (1994)
15. K. Xu, Chem. Rev. 104, 4303 (2004)
16. J. Barthel and H.J. Gores, in *Handbook of Battery Materials*, ed. by J.O. Besenhard, Wiley-VCH, New York (1999), p. 457
17. K. Xu, S. Zhang, T.R. Jow, W. Xu and C.A. Angell, Electrochem. Solid State Lett. 5, A26 (2002)
18. J. Jiang, H. Fortier, J.N. Reimers and J.R. Dahn, J. Electrochem. Soc. 151, A609 (2004)
19. J.H. Shin ad E.J. Cairns, *Rechargeable Li Metal Cells Using N-Methyl-N-butyl pyrrolidinium Bis(trifluoromethane sulfonyl)imide Electrolyte Incorporating Polymer Additives*, Presented at Focussed Battery Technology Workshop III, Pasadena (2008)
20. N. Agmon, Chem. Phys. Lett. 244, 456 (1995)
21. W.G. Grot, US Patent 3,770,567 (1971)
22. K.-D. Kreuer, Chem. Mater. 8, 610 (1996)
23. K.A. Mauritz and R.B. Moore, Chem. Rev. 104, 4535 (2004)
24. K.-D. Kreuer, S.J. Paddison, E. Spohr and M. Schuster, Chem. Rev. 104, 4637 (2004)
25. J.S. Wainright, J.T. Wang, D. Weng, R.F. Savinel and M. Litt, J. Electrochem. Soc. 142, L121 (1995)
26. L. Pauling, The Nature of the Chemical Bond, Cornell Univ. Press, Ithaca, NY (1939), p. 60
27. G. Deublein, B.Y. Liaw and R.A. Huggins, Solid State Ionics 28–30, 1078 (1988)
28. G. Deublein and R.A. Huggins, unpublished results
29. R.A. Huggins, J. Power Sources 22, 341 (1988)
30. R.A. Huggins, in *Fast Ion Transport in Solids*, ed. by B. Scrosati, et al., Kluwer, Amsterdam (1993), p. 143
31. R.A. Huggins, in *Handbook of Battery Materials*, ed. by J.O. Besenhard, Wiley-VCH, New York (1999), p. 359.
32. R.A. Huggins, J. Power Sources, 81–82, 13 (1999)
33. C.J. Wen and R.A. Huggins, J. Solid State Chem. 37, 271 (1981)
34. G. Deublein and R.A. Huggins, J. Electrochem. Soc. 136, 2234 (1989)
35. G. Deublein, B.Y. Liaw and R.A. Huggins, Solid State Ionics 28–30, 1660 (1988)
36. B.Y. Liaw, G. Deublein and R.A. Huggins, J. Alloys Compounds 189, 175 (1992)
37. G. Deublein and R.A. Huggins, Solid State Ionics 18/19, 1110 (1986)
38. C.M. Luedecke, G. Deublein and R.A. Huggins, in *Hydrogen Energy Progress V*, ed. by T.N. Veziroglu and J.B. Taylor, Pergamon Press, New York (1984), p. 1421
39. C.M. Luedecke, G. Deublein and R.A. Huggins, J. Electrochem. Soc. 132, 52 (1985)

Chapter 15
Solid Electrolytes

15.1 Introduction

The surprising discovery of electronically insulating solids in which there can be very rapid long-range motion of charged ionic species, such that they can act as solid electrolytes, really marked the beginning of the era of modern batteries. The subsequent development of solids with insertion reactions has received more attention in recent years, however, as discussed elsewhere in this text.

Nevertheless, it is important to give some attention to the topic of solid electrolytes, and the structural and mechanistic features that make them different from most other, more common, materials.

One reason that the application of solid electrolytes in batteries is currently limited is that most, but certainly not all, of the important electrode materials undergo much substantial changes in their volume as they are charged and discharged. This can result in the imposition of large mechanical strains upon other components in electrochemical cells. Thus it makes sense to have at least one of the components as a liquid to accommodate this mechanical challenge.

Section 15.1 deals briefly with the history of the discovery of solid electrolytes, and the types of crystallographic defects that are involved in their unusual behavior.

Section 15A.4 discusses the influence of the details of crystal structures that make fast ionic motion possible, so that these materials can act as solid electrolytes.

Section 15B.12 deals with the specific case of lithium ion conductors, for that is where most of the current interest in solid electrolytes resides. One of the potential applications of solid-electrolyte cells that is currently of interest involves batteries that can operate at modestly elevated temperatures. They can be useful in connection with so-called downhole instruments used in the petroleum mining industry. The operating temperatures in that application are often too high to consider the use of liquid electrolytes.

R.A. Huggins, *Advanced Batteries*: Materials Science Aspects,

15A Solid Electrolytes: Introduction

15A.1 Introduction

The discussion thus far has been focused upon solids that function as electrodes, for they are the primary determinants of the voltages and capacities of electrochemical cells. When electrolytes have been mentioned, the implicit assumption has been that they are liquids. However, there are also a number of solids that can perform the *primary functions of an electrolyte*, i.e., allowing the transport of charged ionic species, and selectively preventing the transport of electronic species. Although there are many materials that are fine electronic insulators, the major feature that distinguishes materials that can act as *solid electrolytes* is the high mobility of ionic species within their crystal structures.

The phenomenon of electrical charge transport by the motion of ions in solids has actually been the subject of a large number of research investigations over many years. One of the first investigators to observe ionic conduction in solids experimentally was Michael Faraday. Actually, this phenomenon was employed in a practical broadband light source over 100 years ago [1]. This device, known as the *Nernst glower*, consisted of doped ZrO_2, which conducts electrical charge by the motion of defects in the oxide ion lattice at high temperatures [2]. Large-scale commercial success was prevented by the concurrent and more successful development of the tungsten filament electric light bulb.

This area of science has received greatly renewed emphasis in recent years because of the recognition of the *practical utility* that can be derived by the employment of solid ionic conductors for a wide variety of both scientific and technological applications. Recognition of the potential utility of these materials for a wide range of purposes arose from a group of important papers by Wagner and coworkers [3–6]. Somewhat later, this interest was accelerated by the discovery of two important families of materials that have unusually high values of ionic conductivity at surprisingly low temperatures. These were the silver-conducting *ternary silver iodides*, which were simultaneously discovered in England [7–9] and the USA [10, 11], and the *alkali metal-conducting β-alumina (ternary sodium-aluminum oxide) family*, which was discovered by workers at the Ford Scientific Laboratory [12, 13]. The latter was of special importance because it led to the recognition that a wide variety of ions can exhibit rapid transport in solids. Up to that time it had been generally assumed that this property was confined, for some unknown reason, to silver and copper compounds.

At about the same time the Ford group showed [14] that solid electrolytes such as sodium β-alumina could be employed in a *radical new design* of a secondary (reversible) battery. They proposed a *Na/Na_xS cell*, employing liquid electrode constituents and a solid electrolyte that could potentially have unusually large values of specific energy and power. This liquid electrode/solid electrolyte/liquid electrode $(L/S/L)$ configuration is the opposite of the conventional solid electrode/liquid electrolyte/solid electrode $(S/L/S)$ battery design.

As a result of the attention drawn by these discoveries, a large number of additional scientists and engineers were attracted into this area and activities began to increase rapidly. In addition to work on the development of practical devices employing solid ionic conductors as solid electrolytes and various types of electrochemical transducers, growing attention was given to the many scientific questions relating to the unusual behavior of these materials.

It has been known for a long time that solids in which electrical charge transport occurs by the motion of *vacancies* in the crystal structure, such as the alkali halides, have relatively low ionic conductivities and high activation enthalpies (and thus a strong dependence upon the temperature). On the other hand, materials such as some silver and copper halides that typically transport electrical charge by the motion of charged *interstitial species* generally have considerably greater conductivities and lower activation enthalpies.

The discovery of these materials showed that there is a third group of materials, sometimes called "fast ionic conductors", or even "superionic conductors" in some circles, that exhibit more extreme behavior, with surprisingly high values of ionic conductivity and unusually low values of activation enthalpy. This area of activity is part of what is now generally called "solid state ionics".

It is now clear that this unusual behavior is related to characteristics of specific ions and crystal structures. This will be discussed in the following Sect. 16A.6, where it will be seen that there are also intermediate cases, as well as situations in which a given material shifts from one type of behavior to another.

But first, it will be useful to discuss terminology and models that relate to the transport of electrical charge in nonmetallic solids.

15A.2 Structural Defects in Nonmetallic Solids

Analogous to the mechanical behavior of solids, in which the critical phenomena involve the properties and motion of two-dimensional crystallographic defects called *dislocations*, it is now recognized that the transport of electrical charge in nonmetallic solids occurs by the motion of atomic-sized deviations from the perfectly ordered crystal structure, called *point defects*.

Although there is no problem in visualizing the *configurational aspects* of various structural defects in solids, a great deal of confusion is caused by the fact that there are several different ways of describing the individual species present in a solid. This problem becomes particularly acute when dealing with nonmetallic solids for there are two different methods of notation used to describe the basic crystalline constituents themselves. Since these two different approaches to the description of the normal constituents of the structure also give rise to a different philosophy with respect to the description of defects, they will be briefly reviewed at this point.

These two methods of description are the *ionic* or *chemical* model on the one hand, and the *defect* model on the other. Perhaps the best way to illustrate the

M^+	X^-	M^+	X^-
X^-	M^+	X^-	M^+
M^+	X^-	M^+	X^-
X^-	M^+	X^-	M^+

Fig. 15A.1 Ionic model of simple crystal MX

differences between these two types of notation is by means of an example or two. Consider a simple ionic crystal such as sodium chloride, which we shall call by the generic label MX. The ionic model is based upon the fact that a free atom must be electrically neutral; that is, the positive charge on the nucleus is balanced by the negative charge of its electron cloud. When two such neutral atoms come together to form an ionic crystal the *bonding* that occurs results from the transfer of one or more electrons from the more electropositive element to the more electronegative element. In the case of sodium chloride this means that an electron leaves the environment of the sodium nucleus and adopts a location in the environment of the chlorine nucleus. After such an electron transfer it is reasonable to look upon the sodium atom as a positively charged ion M^+ and the chlorine atom new as a negatively charged ion X^-. One can then represent the structure of such a crystal schematically as shown in Fig. 15A.1. In this case M^+ and X^- represent the positive sodium and negative chlorine ions, respectively.

If the crystal was made up of divalent cations and divalent anions one would write M^{2+} and X^{2-} instead, representing the transfer of two electrons from each cation and the acceptance of two electrons by each anion.

Consider how this situation is described by means of the *defect* model. In this case, the ground state is not taken as a collection of neutral free atoms, but instead, as the neutral crystal after bonding has taken place, so if a cation is expected to give an electron to the crystal and an anion is expected to acquire an electron by the bonding process that takes place when the crystal is formed, the notation used by the defect model refers to the *as-bonded* state. The charges that are attributed to each constituent are *relative* charges compared with the charges that a constituent would normally have *after* the crystal has been formed. What this means is illustrated in Fig. 15A.2, which shows the same crystal as depicted in Fig. 15A.1, but uses the notation of the defect model.

In this case the superscript x indicates the normal charge of that species within the crystal after bonding has taken place. This symbol indicates that the *effective* or *relative* charge for that species is zero. That is, it is the same for all species, and it is the normally expected charge for that particular constituent in its normal position in the lattice. In the case of a perfect crystal in which all of the constituents, both M and X, are represented as having their normal charges in accordance with the defect notation convention the representation can be simplified by omitting the superscript x; this simplified representation would then appear as shown in Fig. 15A.3.

M^x	X^x	M^x	X^x
X^x	M^x	X^x	M^x
M^x	X^x	M^x	X^x
X^x	M^x	X^x	M^x

Fig. 15A.2 Defect model of simple crystal MX

M	X	M	X
X	M	X	M
M	X	M	X
X	M	X	M

Fig. 15A.3 Schematic representation of a perfect crystal MX according to the simplified defect notation convention

One advantage of the defect notation convention is now apparent, for it places emphasis upon any defects or irregularities present in the solid, since they will be the only constituents that carry *effective* or *relative* charges in this scheme. This emphasis upon defects, rather than upon the background crystal, is both appropriate and handy.

15A.3 Various Types of Notation That May Be Used to Describe Imperfections

In addition to the two basically different philosophies that can be used to describe the charge distribution within a perfect crystal, there are a number of different schemes used to represent imperfections or deviations from the normal distribution of species within the perfect crystal. These fall into two general classes: those in which the electrical charge on a defect is described on the basis of the *ionic model* of the perfect crystal on the one hand, and those in which the electrical charge carried by a defect is based upon the *defect model* of the perfect crystal. To be useful, any notation scheme must include information concerning the following three features of any defect: (a) the chemical or atomic species being described, (b) its location or site in the lattice, and (c) its electrical charge. As mentioned earlier, the value of the electrical charge carried by a defect species depends upon which of the two basic philosophies, the *ionic* or the *defect model* is followed.To see how this works, consider the simple case of a crystal MX containing two defects, a divalent cation that sits upon a normal (monovalent) cation site, and a vacancy at a different cation site. This latter defect can, of course, be described by saying that one of the normal monovalent

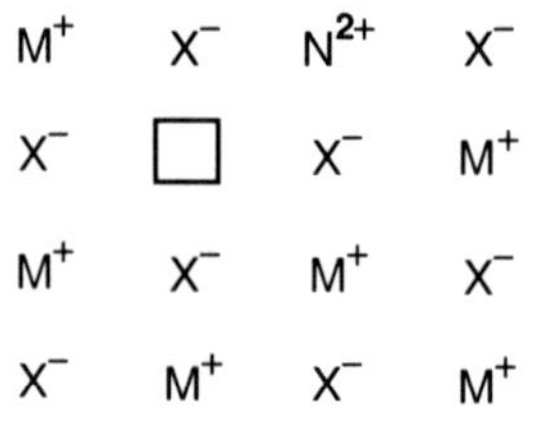

Fig. 15A.4 Representation of a portion of a crystal MX containing two defects, a divalent cation N and a vacancy upon a cation site

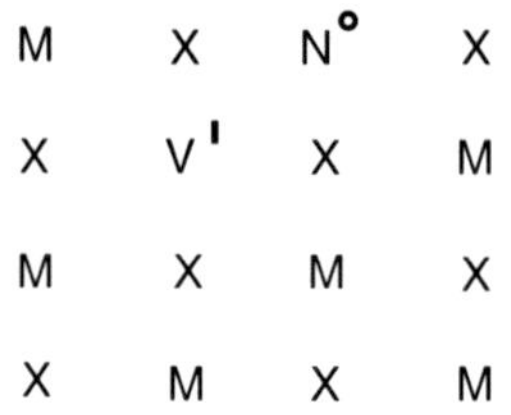

Fig. 15A.5 Defect representation of a portion of a crystal MX containing two defects, a divalent cation N^{2+} and a vacancy, both on normal cation sites

cations is missing. A portion of a crystal containing these two defects can be represented schematically according to the *ionic model* as shown in Fig. 15A.4. In this case, the square is used to represent the site in the crystal from which the cation is missing, or at which the vacancy exists.

The electrical charge balance for this scheme is expressed as

$$2M^{+} = N^{2+}\square \tag{15A.1}$$

A portion of crystal MX containing the same two defects is illustrated in Fig. 15A.5 according to the *defect* convention. In this case two new symbols are introduced. The degree sign over the N indicates that the N species carries a charge that is one unit more positive than that of the species that normally occupies that site. That is, the *relative* charge of the cation N is $+1$. Likewise, the apostrophe is used to indicate that the *relative* charge of the vacancy is -1. That is, the vacancy carries one unit less positive charge than the monovalent cation species that normally occupies this same site. This method of representing the relative charge of species was introduced by Kröger and Vink [15], and is generally called *Kröger–Vink notation.*

The charge balance in this crystal is represented according to the *defect model* by

$$\mathrm{N_M}^{\circ} = \mathrm{V_{M'}}, \tag{15A.2}$$

where the subscript M indicates the location of the defect within the crystal; that is, both defects in this example reside upon normal cation (M) sites. In this text the simplified defect notation convention will be used. Table 15A.1 gives examples

Table 15A.1 Symbols used to denote various isolated defects

Type of defect	Symbol and expected relative charge	
	Where MX is $M^{+}X^{-}$	Where MX is $M^{2+}X^{2-}$
Native (intrinsic) defects		
Vacancy on cation site	$V_M{}'$	$V_M{}''$
Vacancy on anion site	$V_X{}^{\circ}$	$V_X{}^{\circ\circ}$
Interstitial cation	$M_i{}^{\circ}$	$M_i{}^{\circ\circ}$
Interstitial anion	$X_i{}'$	$X_i{}''$
Cation on anion site	$M_X{}^{\circ\circ}$	$M_X{}^{\circ\circ\circ\circ}$
Anion on cation site	$X_M{}''$	$X_M{}''''$
Quasi free electron	e'	e'
Quasi free hole	h°	h°
Foreign (extrinsic) defects		
Monovalent foreign on cation site	$N_M{}^{x}$	$N_M{}'$
Divalent foreign cation on cation site	$N_M{}^{\circ}$	$N_M{}^{x}$
Trivalent foreign cation on cation site	$N_M{}^{\circ\circ}$	$N_M{}^{\circ}$
Interstitial foreign monovalent cation	$N_i{}^{\circ}$	$N_i{}^{\circ}$
Interstitial foreign divalent cation	$N_i{}^{\circ\circ}$	$N_i{}^{\circ\circ}$
Monovalent foreign anion on anion site	$Y_X{}^{x}$	$Y_X{}^{\circ}$
Divalent foreign anion on anion site	$Y_X{}'$	$Y_X{}^{x}$
Interstitial foreign monovalent anion	$Y_i{}'$	$Y_i{}'$

of various defects and their representation by this scheme. In the simple examples included in that table, it is assumed that there are only two types of sites for cations and two types for anions, *normal sites* and *interstitial sites*. Contrary to this, there may be more than one type of interstitial sites. In such cases it may be desirable to introduce an additional variation.

An important consideration is the local concentration of any particular defect. Concentrations can be specified in any of several ways, by weight fraction, by site fraction, by mole fraction, or by specific concentration (number per cm^3). It is important to specify which of these is being used in any given case.

15A.4 Types of Disorder

It has long been known that *compounds* do not have fixed compositions, but instead, can exist over a range of composition. In the case of a phase MX, this means a range of values of M/X.

There are other ways by which the composition of a solid can be changed, but they will not be discussed here. It is important now, however, only to recognize that such compositional variations may occur, either between different samples of the same nominal material MX, or between different locations within a given sample. Such compositional variations necessarily involve the presence of defects within the crystal structure.

It is a common practice to classify solids with respect to their predominant defect types. Another way of saying this is to indicate which pairs of defects tend to be present in a given material, and can accompany variations in its overall composition by changes in their relative concentrations.

There are three pairs of defects that can be important in this connection, and whose relative concentrations determine the composition of a particular material. Which of these pairs is characteristic of a given material is designated by its *disorder type*.

The four primary disorder types are as follows:

(a) *Schottky disorder*, in which the primary native defects are cation vacancies and anion vacancies, $V_{M'}$ and V_{X°. The relative charges of these vacancies depend upon the material under consideration. Examples of materials having this type of disorder are NaCl, LiF, and MgO.
(b) *Frenkel disorder*, in which the predominant native defects are cation vacancies and interstitial cations, $V_{M'}$ and M_{i°. The relative charges carried by these defects are, as mentioned earlier, dependent upon the particular material under consideration. Examples of materials having this type of disorder are AgI and FeO.
(c) *Redox Frenkel disorder*, in which foreign interstitial cations, N_{i°, are present, and their charges are balanced by a reduction in the charge of one of the native species, typically a cation, $M_{M'}$. The native species whose charge can be thus changed are said to undergo redox reactions. This type of disorder is very common in materials that are used as electrode reactants in battery systems.
(d) *Anti-Frenkel disorder*, in which the predominant native defects are anion vacancies and interstitial anions, V_{X° and $X_{i'}$. Again, the values of the relative charges of these defects depend upon the particular material. Examples of materials having this type of disorder are CaF_2, ZrO_2, and UO_2.

One may also expect to find a so-called *antistructure disorder* in some cases in which the difference in electronegativity between the cations and the anions is relatively small. In this kind of disorder the important defects would be cations on anion sites and anions on cation sites, M_X and X_M. The relative charges of these defects are normally negative and positive, respectively.

One might also expect to find examples of *anti-Schottky disorder* in which the predominant defect types are the two interstitials. However, no materials have yet been definitely shown to have this type of disorder.

The type of disorder present in a given material relates to the question of the free energy necessary to form particular types of defects. As a general rule, the defects that determine the type of disorder that a particular material will have are those that have the lowest values of free energy of formation.

Although one often sees references in the literature to *Frenkel* or *Schottky defects*, such use of these terms is improper; instead, one should speak of *Schottky* or *Frenkel disorder*. For the very special case in which a compound has an exactly stoichiometric composition, the defect concentrations are not zero, but instead, the concentrations of the two predominant defects must be equal to each other.

15B Mechanism and Structural Dependence of Ionic Conduction in Solid Electrolytes

15B.1 Introduction

As mentioned in Sect. 15.1, it is now quite apparent that materials that exhibit unusually fast ionic conduction do so because of special characteristics related to their crystal structures. It is the purpose of the present chapter to discuss this crystal structure dependence and the reasons for the crystallographic influence on ionic transport kinetics.

There have been a number of reviews and compilations published over many years that have discussed the considerable number of solids that have been recognized to have large values of ionic conductivity. Some of the early ones are listed in the references of this chapter [22–28]. There is no intention to repeat such information in detail here. Instead attention shall be focused upon the characteristics of specific crystal structures that cause them to exhibit rapid ionic motion. In addition, several new groups of cationic conductors that have become recognized more recently will be discussed in this context. Some attention will also be given to mixed-conducting materials, in which both the ionic and electronic conductivities are relatively high.

15B.2 Characteristic Properties

The highest value of ionic conductivity yet reported for a solid at room temperature is found in rubidium silver iodide ($RbAg_4I_5$). Measurements indicated a value of about 0.3 $(\text{ohm-cm})^{-1}$ at 25°C [10]. This is not only within an order of magnitude of typical values for liquid electrolytes commonly used in battery systems, but is about 14 orders of magnitude greater than the ionic conductivity of sodium chloride, a typical alkali halide, at the same temperature. While the difference between the ionic conductivities of these two solids is considerably reduced at elevated temperatures because their temperature dependences are vastly different, the great disparity between the conductivity values of materials such as $RbAg_4I_5$ and the alkali halides, traditionally considered model ionic conductors, indicates that there is an appreciable substantive difference in the transport phenomena in these two groups of materials.

To illustrate how extremely rapid ionic transport in some fast ionic conductors can be, it is interesting to utilize a simple model [29] to calculate the maximum value that the diffusion coefficient might have in any solid with fixed lattice positions. If we make the assumption that the ions move at thermal velocity v directly from one lattice site to the next at distance d without oscillating at each site, we can calculate the jump frequency υ from $\upsilon = v/d$. Assuming a jump distance of 1 Å, and picking a temperature of 300°C, υ is found to be $3.4 \times 10^{12}\,\mathrm{s}^{-1}$. Inserting this value into the relation $D = \alpha d^2 \upsilon$ and setting $\alpha = 1/6$ for a cubic structure, it is found that $D = 5.6 \times 10^{-5}\,\mathrm{cm}^2\,\mathrm{s}^{-1}$.

Experimental conductivity data for AgI, a fast ion conductor, can be compared with this limiting value of the diffusion coefficient by use of the Nernst-Einstein relation, as explained later. The maximum theoretical value of the conductivity thus obtained is 2.8 $(\mathrm{ohm\,cm})^{-1}$ at 300°C.

Thus it can be seen that the magnitude of the ionic conductivity in this material is very close to the value that would be predicted by this simple limiting model that assumes that all of the silver ions are free to move and that every jump attempt is successful. These experimental conductivity data also lead to a value of diffusion coefficient for the silver ions inside solid AgI that is comparable to those typically found in liquids.

As an even more extreme example of the unusually rapid mass transport that can sometimes be found in fast ionic conductors, measurements [30] have shown that the chemical diffusion coefficient for silver in $\alpha - Ag_2S$ is $0.47\,\mathrm{cm}^2\,\mathrm{s}^{-1}$ at 200°C. This is a number of orders of magnitude greater than even the largest values found for diffusion coefficients in typical metals and close-packed ionic materials. In this case, there are two factors of importance. Not only are the silver ions relatively mobile in this phase of silver sulfide, but this material is a mixed conductor, that is, both the electrons and ions contribute to the electric charge transport. The interaction of the fluxes of these two species in the diffusion process results in a large internal electrical field and great enhancement of the rate of ionic motion and thus of the chemical diffusion coefficient. This behavior, sometimes expressed in terms of a *thermodynamic enhancement factor* (see Sect. 17B), can lead to values of diffusion coefficient several orders of magnitude greater than the limit derived from the simple model discussed earlier. Therefore, it becomes quite obvious that a simple single-jump model cannot be realistic for such cases. There must be some sort of cooperative motion or multiple-site jumps taking place.

Other important differences between fast ionic conductors and more conventional materials can be seen in Fig. 15B.1, where the variation with temperature of the ionic conductivity of four materials is shown: KCl (an alkali halide with Schottky (vacancy) disorder), AgCl (which has Frenkel (interstitial) disorder), AgI, and sodium beta alumina (with the nominal formula $MAl_{11}O_{17}$), the last two being examples of fast ionic conductors. The AgI undergoes a phase transformation at 146°C, and the low temperature β phase is not in the fast ionic conductor class. On the other hand, beta alumina appears to be stable in the high conductivity form to well below room temperature.

Examination of the temperature dependence of the ionic conductivity of many materials has shown that their behavior falls into three general categories, depending

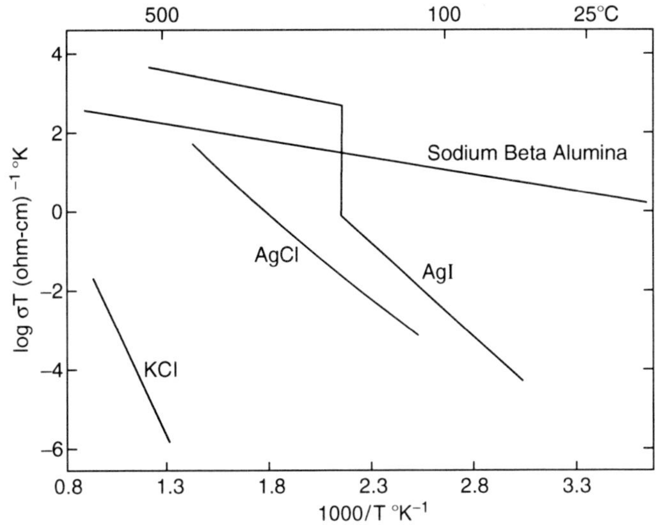

Fig. 15B.1 Temperature dependence of the ionic conductivity of several materials, illustrating the large variation in behavior that is found experimentally

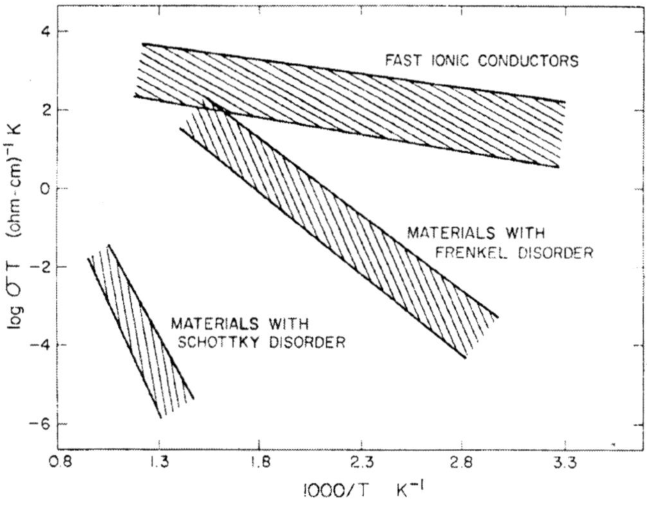

Fig. 15B.2 Bands of behavior found for materials with different ionic conduction mechanisms

upon the types of ionic disorder, and thus the ionic conduction mechanism. These are illustrated in Fig. 15B.2.

From Figs. 15B.1 and 15B.2 it can be seen that there are three principal features of the experimental data relating to fast ionic conductors that differentiate them from normal ionic conductors. The first, that has already been mentioned, is that the absolute magnitude of the ionic conductivity can be unusually high.

The second unusual experimental characteristic of fast ionic conductors is the small temperature dependence of the conductivity. In some cases, the activation energy can be as low as 2–5 kcal mol^{-1}. Since the mobile carrier concentration is very large as well as temperature-independent, this is interpreted as a low enthalpy of motion.

The third unique feature is the very small value of σ_0 found in these materials, of the order of 100 (ohm-cm)$^{-1}$ K. (σ_0/T is the preexponential of the conductivity.) As will be shown shortly, this implies either an especially low value of the attempt frequency or a very small value of the entropy of migration.

These novel characteristics are, of course, interrelated, rather than independent. However, the important point is that they clearly indicate that ionic motion in fast ionic conductors is vastly different from that found in the groups of materials with which the materials science community has been traditionally concerned.

15B.3 Simple Hopping Model of Defect Transport

The traditional atomistic model of defect transport involves the random walk of either vacancies or interstitial ions by isolated jumps between lattice sites in the crystal structure. In this formulation, the diffusion coefficient of a particle species is expressed as

$$D = \alpha d^2 \upsilon, \tag{15B.1}$$

where α is a geometric factor, which is the reciprocal of the number of possible jump directions from a given site. The value of α is 1/6 for a close packed cubic lattice, and ½ if jumps are constrained to only one dimension. In this expression, d is the distance moved during a single jump and υ the average jump frequency. It should be mentioned that this equation would have to include the correlation factor if one were dealing with the diffusion of radiotracers.

The average jump frequency can be written in terms of the fraction of the species that is free to move, β, the attempt frequency υ_0, and the probability of success, $\exp(-\Delta G/RT)$, where ΔG is the free energy of migration,

$$\upsilon = \beta \upsilon_0 \exp(-\Delta G/RT). \tag{15B.2}$$

If the free energy is divided into its entropy and enthalpy components, the following expression for the diffusion coefficient results:

$$D = \alpha d^2 \beta \upsilon_0 \exp(\Delta S/R) \exp(-\Delta H/RT). \tag{15B.3}$$

One can go one step further by consideration of the factor β. When transport occurs by the motion of defects, the value of β is obviously the defect concentration. In conventional high-purity ionic conductors where the defect concentration is dominated primarily by the probability of the formation of intrinsic defect pairs, a further temperature dependence is introduced, for

$$\beta = \exp(-\Delta G_f/RT), \tag{15B.4}$$

where ΔG_f is the free energy of formation of the intrinsic defect pair. In such a case, β can be quite small at lower temperatures, severely constraining the magnitude of the diffusion coefficient. The slope of a plot of ln D vs. T^{-1} is then given by $-(\Delta H + \Delta H_F)/R$, where ΔH_F is the enthalpy of defect pair formation. In fast ionic conductors, however, essentially all of the ions on one of the sublattices can be considered mobile. As a result, β is both very large and effectively independent of temperature. When this is the case, the slope of the ln D vs. T^{-1} plot becomes only $-\Delta H/R$.

If one can assume that charge and mass are transported by the same mobile species, the ionic conductivity is related to the particle diffusion coefficient by the *Nernst-Einstein relation*

$$\sigma = DCz^2F^2/RT, \tag{15B.5}$$

where C is the concentration of the mobile species, z their charge number, and F the Faraday constant.

In accordance with the Nernst-Einstein relation and the temperature dependence of the diffusion coefficient in (15B.3), one expects a linear relationship between the logarithm of the product σT and T^{-1}, as shown in Fig. 15B.1.

This relationship can be expressed as

$$\sigma = (\sigma_0/T)\exp(-\Delta H/RT), \tag{15B.6}$$

where σ_0 is the value of the extrapolated intercept at $T^{-1} = 0$ on such a plot. By combining (15B.3) and (15B.5), it follows that

$$\sigma_0 = C\beta d^2 \upsilon_0 \alpha (z^2F^2/R)\exp(\Delta S/R). \tag{15B.7}$$

The interesting thing to note is that the fast ion conductors clearly have a much lower value of σ_0 than do the normal ionic conductors. This difference can be as large as 5 orders of magnitude. Which terms in (15B.7) could be responsible for this? The value of C, d, α, and z do not vary appreciably from one material to the next, so the only possibilities are β, υ_0, and ΔS. However, β is expected to be both temperature-independent and much larger in fast ion conductors than those in which the defect concentration is determined by intrinsic thermally activated defect-pair formation, so this is obviously not the dominant effect.

Values of ΔS for the migration of species in fast ion conductors are not presently known, but it is reasonable to assume that they may be unusually small, as motion of the fast ions may entail relatively insignificant changes in the lattice frequencies of the surrounding crystal structure. The attempt frequency υ_0 may also be unusually low in such materials if the mobile ions sit in relatively shallow potential wells.

From these simple considerations, it seems clear that any useful theoretical approaches to the understanding of the transport process in fast ionic conductors must address themselves not just to the observations of very low activation energy, but also to the existence of unusually low values of either ΔS or υ_0.

Another important question is, of course, why the phenomenon of unusually fast ionic conduction occurs in some crystal structures, but not in others. This will be addressed in the following sections.

15B.4 Interstitial Motion in Body-Centered Cubic Structures

One of the materials first recognized to have unusually high values of ionic conductivity was the high-temperature (above 146°C) α form of AgI [31]. It was found to have such an extremely high value of ionic conductivity in the solid state that there is a drop of more than 20% upon melting. This is contrary to the general expectation that ionic motion is faster in liquids than in solids. These results were confirmed by Qvist and Josefson [32], and are illustrated in Fig. 15B.3.

The crystal structure of this phase can be described simply as a body-centered cubic arrangement of iodine ions, within which the silver ions move interstitially [33,34]. The silver ions are so mobile in the crystal structure that their positions are hard to determine by X-ray diffraction methods. They are thought to move rapidly among sites with three-fold and four-fold coordination within the relatively static iodine structure. A unit cell of the α-AgI structure is shown in Fig. 15B.4. A three-dimensional view of the iodine sublattice is shown in Fig. 15B.5. The open tunnel space in the cube-edge direction is clearly visible.

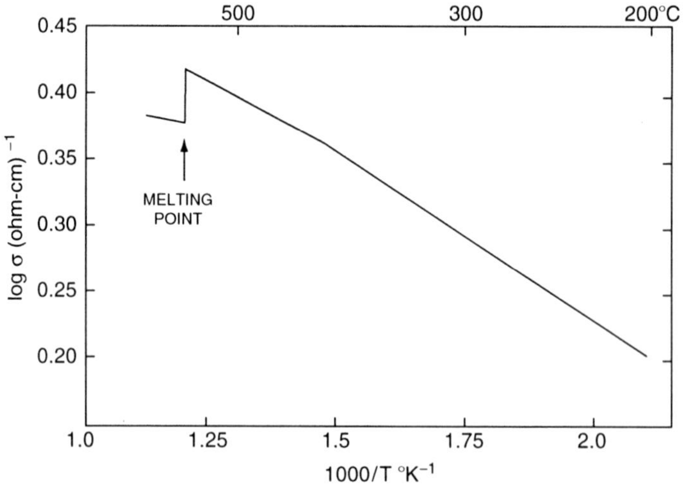

Fig. 15B.3 Temperature dependence of the ionic conductivity of a-AgI showing the drop at the melting point

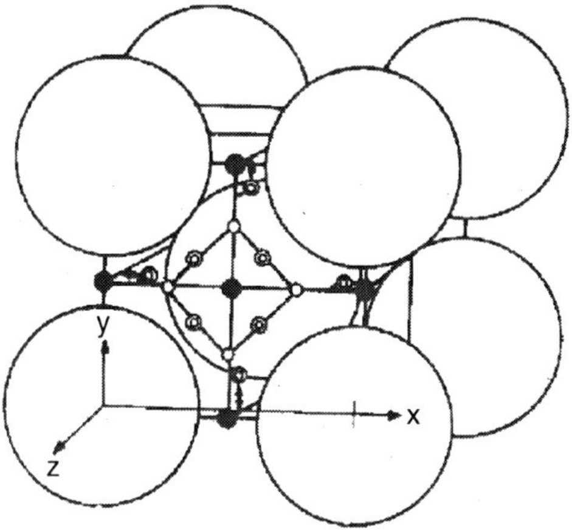

Fig. 15B.4 Unit cell of the α-AgI structure, showing three-fold and four-fold interstitial sites

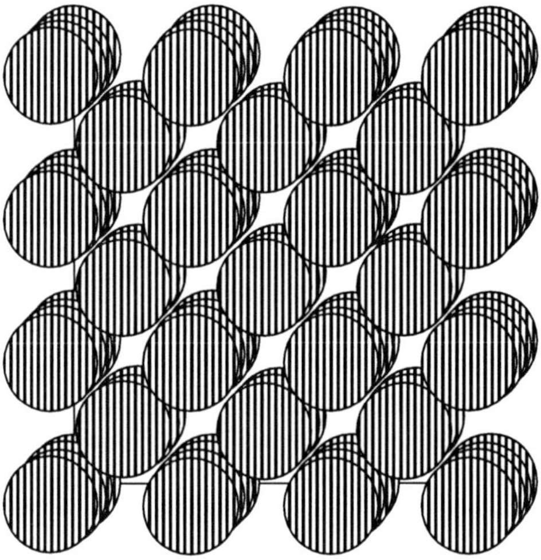

Fig. 15B.5 Three-dimensional view of the iodine sublattice in the α-AgI structure

Another way of describing such structures is in terms of a *molten sublattice* permeating a framework defined by the other ions present. This molten sublattice model is consistent with thermodynamic data, for it was shown [35] that the entropy change upon melting is considerably lower for such materials than it is for normal materials. Thus these materials behave as though part of the structure behaves thermally as though it is already molten before the melting point of the structure as a whole is reached.

The properties of several other materials that have this same structure with rapid cation motion through a relatively immobile bcc anion lattice are shown in Fig. 15B.6.

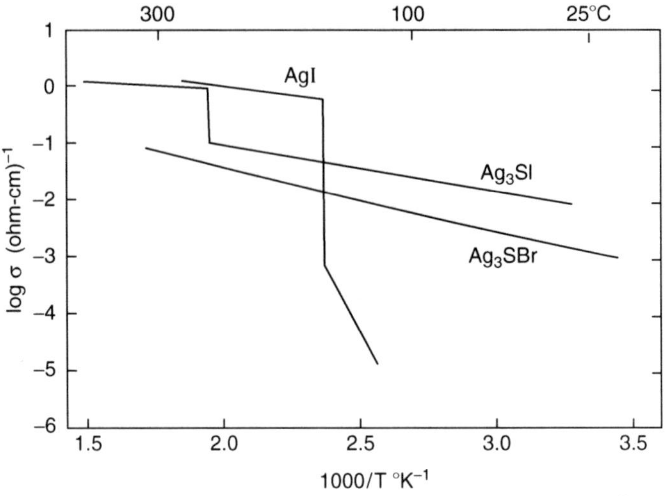

Fig. 15B.6 Ionic conductivity of several materials with mobile cations in a bcc anion lattice

15B.5 Rapid Ionic Motion in Other Crystal Structures

This phenomenon can occur in materials with other anion arrangements. Figure 15B.7 shows some examples of materials with mobile cations in face-centered cubic anion lattices.

This phenomenon is not limited to mobile cations in a stable anion arrangement. The opposite can also be true, and three examples in which fluorine anions have very high values of mobility are shown in Fig. 15B.8.

To understand this phenomenon, it is useful to consider the location and transport paths of the mobile cations within the anion environment of the crystal structure. One possible method is to use X-ray diffraction. This can provide information about the time average location of species in solids by careful measurement of diffraction intensities. The result can be expressed as an electron density map. The result of such an experiment [36] for the case of high chalcocite (Cu_2S) is shown in Fig. 15B.9. It is seen that the copper ions do not occupy fixed places, but are found along paths that extend between what would normally be considered crystallographic sites.

The α-AgI phase is often considered to be a prototype of the simple fast ionic conductors, and deserves special attention, as it illustrates several principles involved in such materials. It has been the subject of a large number of investigations, and a thorough review was presented by Funke [37].

The high-temperature phase of Ag_2S has a structure that is similar to that of AgI. In this case, however, Ag ions move interstitially through tunnels in a body-centered cubic array of sulfur ions. This phase has been known for some time [3]

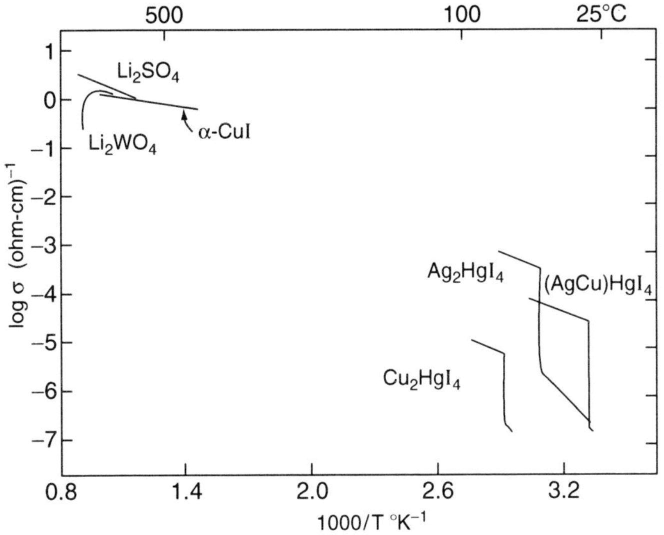

Fig. 15B.7 Ionic conductivity of some materials with mobile cations in an fcc arrangement of anions

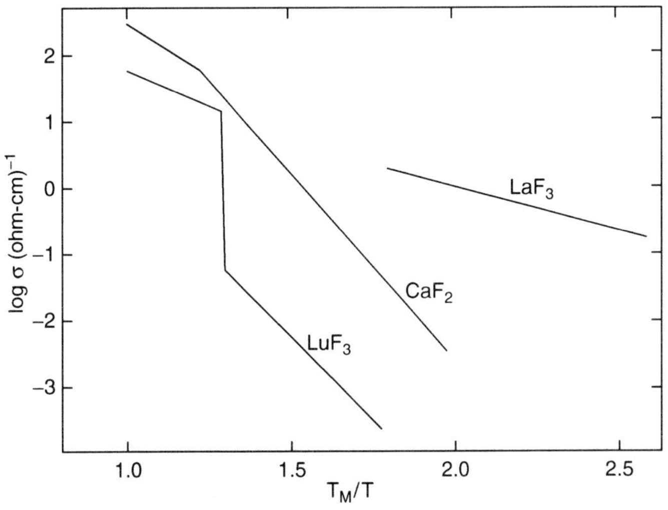

Fig. 15B.8 Examples of materials with highly mobile anionic species

to be a mixed conductor in which the Ag ions are very mobile, although the charge transport is dominated by electronic conduction. It was mentioned in Sect. 15B.2 that measurements of the chemical diffusion coefficient [30] have shown extremely high values ($0.47\,cm^2\,s^{-1}$ at 200°C). This is due in part to a large thermodynamic enhancement factor, which makes the chemical diffusion coefficient much greater

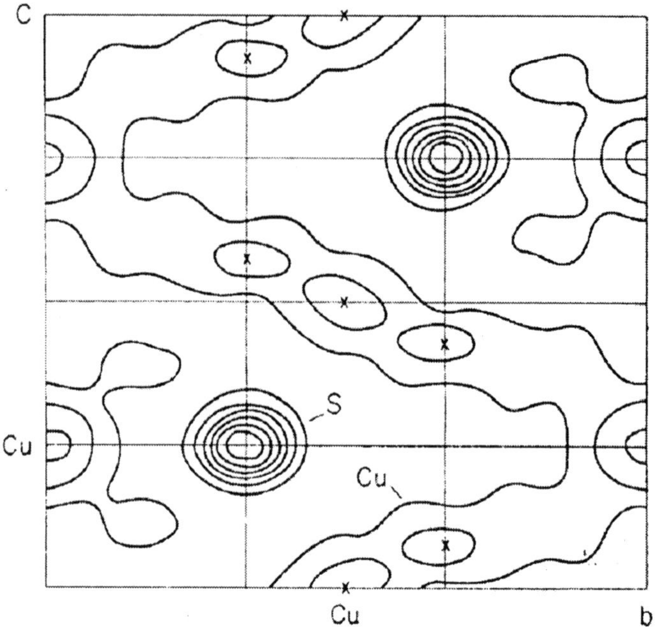

Fig. 15B.9 Electron density section (0 *x y*) through the orthohexagonal cell of high chalcocite

than the self-diffusion coefficient and causes accelerated ionic transport when a concentration gradient is present.

It is also interesting that hydrogen diffusion in body-centered cubic metals has a number of characteristics that are similar to those found in materials with the α-AgI structure that exhibit fast ionic conduction [26, 38].

15B.6 Simple Structure-Dependent Model for the Rapid Transport of Mobile Ions

To provide some understanding of the basis for the dependence of the ionic conductivity of different materials upon the features of their crystal structures, a simple *structure-dependent model* for ionic transport was developed a number of years ago [39–41]. Those calculations involved the determination of the *potential profiles* of the mobile species within the specific crystallographic arrangements of the other ions, with the basic assumption that the other constituents in the lattice remain fixed in position on the time scale of the motion of the mobile species.

Although much more sophisticated methods have been developed more recently, bolstered by the availability of much greater computer power, this simple approach served to provide insight into the mechanism of ionic transport and the major factors that lead to its dependence upon crystallographic features.

This approach, first applied to the body-centered cubic α-AgI structure [39], followed the general method initiated by Born and Mayer [42], in which the total energy, E_T, is assumed to be the sum of two-body interaction energies between the mobile ion i and the surrounding lattice ions j. This interaction is expressed as the sum of three types of terms, the electrostatic Coulombic interaction E_C, dipolar polarization (van der Waal) interactions, E_P, and overlap repulsion between the closed shell ions, E_R. That is,

$$E_T = E_C + E_P + E_R, \tag{15B.8}$$

where

$$E_C = -e^2 \sum_j (q_i q_j)/r, \tag{15B.9}$$

$$E_P = -(e^2/2) \sum_j (\alpha_j q_i)/r^4, \tag{15B.10}$$

$$E_R = b \sum_i \exp[r_i + r_j + r_{ij}/\rho], \tag{15B.11}$$

and q_i and q_j are the charges, a_j the polarization, r_i and r_j repulsion radii, b and r constants, and r_{ij} the distance between the mobile cation and the jth static lattice ion.

With the use of a small computer, the *total interaction energy* between a single mobile ion arbitrarily placed at any position within the crystal structure and the other atoms in the lattice could be calculated, with the simplifying assumption that all of the others remain fixed in position, rather than relaxing to accommodate the position assumed for the mobile ion. It was also assumed that there is no interaction between nearby mobile species, but instead all of the energy resides in the mobile ion-static lattice interaction.

By this method the variation of the total energy with the assumed position of the mobile ion within the tunnels that run in the cube-edge direction through the body-centered cubic anion lattice was found for a series of cases assuming different values of the cation repulsion radius. In addition to a series of points along the centerline of the interstitial tunnel, a three-dimensional array of off-center positions was also investigated.

One of the important results from this rather simple early theoretical approach was the conclusion that the potential profile is such that the *minimum energy path* through the tunnel that runs between the anions typically does not follow the centerline, but instead, its location is strongly influenced by the value of the cation radius. In the case of small mobile cations there is a symmetrical pair of preferred paths that deviate toward the nearby anions along the tunnel wall, due to the relatively large influence of the attractive polarization energy term compared with the repulsive term. In the case of larger cations the opposite is true, and a pair of equivalent minimum energy paths were found, which deviate from the centerline in the opposite sense, away from the nearby anions, as a result of the predominance of the repulsion term. The positional variation of both these short-range interactions is greater than that due to the Coulombic term in this structure. The result is illustrated

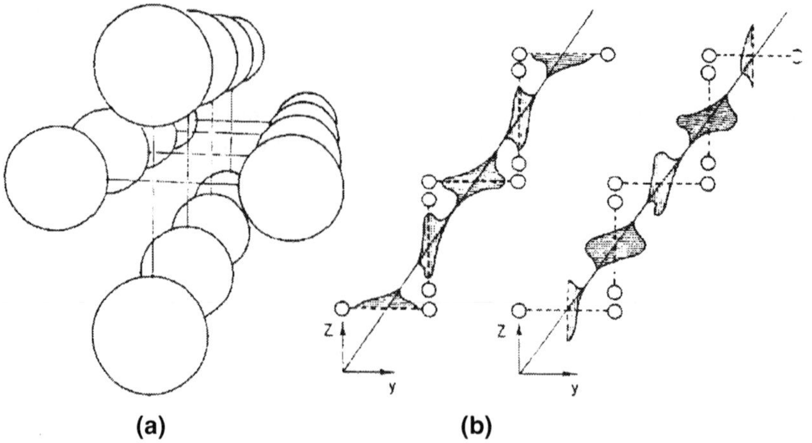

Fig. 15B.10 (**a**) Perspective drawing showing the anion arrangement along a tunnel in the cube-edge direction of the a-AgI crystal structure. (**b**) Schematic illustration of the minimum energy paths for small (left) and large (right) ions within the tunnel

in Fig. 15B.10, which also includes a perspective drawing showing the tunnel structure in the AgI lattice.

As an ion progresses along the tunnel it moves through an array of anions, which may be described as a series of alternating north–south and east–west dumbbell pairs that define a series of two-dimensional apertures through which the mobile ions move. This type of aperture can be seen to be quite fortuitous, for it permits a wide variety of paths to accommodate the relative magnitudes of attractive and repulsive forces as the mobile ion progresses through the tunnel.

After determining the potential profile within the tunnel and the minimum energy path, the variation of the (minimum) energy with position along the tunnel can be calculated for ions of any radius. The result is shown in Fig. 15B.11 for a few different ionic sizes. It is seen that the competing effects of the polarization and repulsive terms, which are out of phase, result in a flatter potential profile for ions of intermediate size. If one assumes that the peak-to-valley distance can be treated as the activation enthalpy for motion, the variation of that quantity with ionic size can be obtained, as shown in Fig. 15B.12. This result predicts that ions of intermediate size should be more mobile in this structure than either smaller or larger ones.

15B.7 Interstitial Motion in the Rutile Structure

The rutile (TiO_2) structure has tetragonal symmetry in which the cations are octahedrally coordinated by anions, and these octahedra are joined into a three-dimensional framework by sharing corners such that each anion is adjacent to three cations. This arrangement produces a relatively open tunnel in the *c direction* made up of a string of distorted edge-sharing tetrahedral sites that are all empty in the

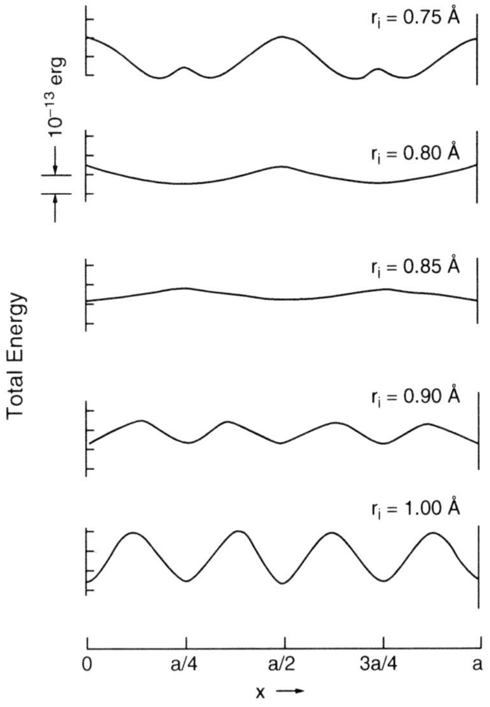

Fig. 15B.11 Variation of the calculated energy along the minimum energy path for interstitial ions of different sizes

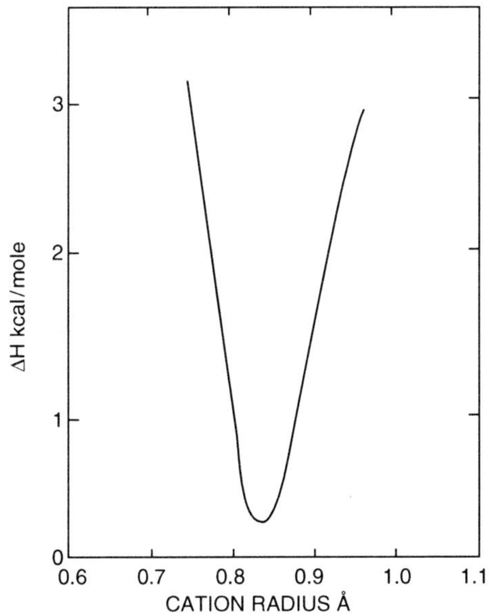

Fig. 15B.12 Variation of the calculated activation energy for motion along the minimum energy path for cations of different sizes

ideal structure. If an ion were to reside in these sites, motion between them would involve passing through a two-coordinated aperture (the shared edge) similar to that in the α-AgI structure.

Diffusion experiments have shown that Li ions are quite mobile in TiO_2 [43,44], and that ionic transport is highly directional, indicating a strong preference for solute motion through these unidirectional tunnels. Electrochemical measurements utilizing a molten salt electrolyte [45] confirmed these results, and showed the large effect of the thermodynamic enhancement factor upon the chemical diffusion coefficient in this case, since the introduction of the lithium solute makes this material a mixed conductor. Related observations of fast and very anisotropic diffusion have also been reported for H^+ and D^+ ions in this material [46].

Calculations were also made on this structure [41] using the minimum energy path approach described earlier. However, the method must be modified in this case because the anions are not in positions of cubic symmetry in this structure and thus have a permanent dipole moment. This involves the introduction of an additional monopole–permanent dipole interaction energy E_M, where

$$E_M = (\mu_j r_{ij})/r_{ij}^2 \tag{15B.12}$$

and μ_j is the permanent dipole moment of the polarizable ion in the structure, and r_{ij} is the unit vector from the ith to the jth ion. The value of μ_j must be calculated from the polarizability of the ion and the geometry of the structure.

An additional feature that was included in the calculations made on the rutile structure involved the repulsion expression. It is often found in such calculations that the attractive polarization force is greater than the repulsion force at all distances, leading to the so-called *polarization catastrophe*. It was found that this problem could be avoided in this case by a separate calculation of the preexponential term in the repulsion expression. The resulting values were found to vary substantially from Pauling's calculated values [47] close to unity, which fit quite well for alkali halides, in the case of many oxygen, sulfur, and fluorine compounds. Use of these larger values eliminated the polarization catastrophe problem in this structure.

This structure is also different from α-AgI in that it also contains highly charged static cations, which provide an important additional Coulombic repulsive force upon the mobile ions. Primarily because of this fact, it was found that the minimum energy path does not deviate from the centerline of the tunnel in this crystal structure.

The variation of the energy with distance along the c axis in TiO_2 is plotted for several assumed values of the mobile ion radius in Fig. 15B.13. Similar calculations were made for the case of interstitial motion in MgF_2, and the variation in calculated values of motional activation enthalpy with the radius of the mobile cation in the tunnels is shown in Fig. 15B.14. It was found that the primary reason for the lower activation enthalpy in the fluoride structure was the decrease in the overlap repulsion term caused by the dependence of the preexponential factor upon the charges and polarizabilities of the ions present.

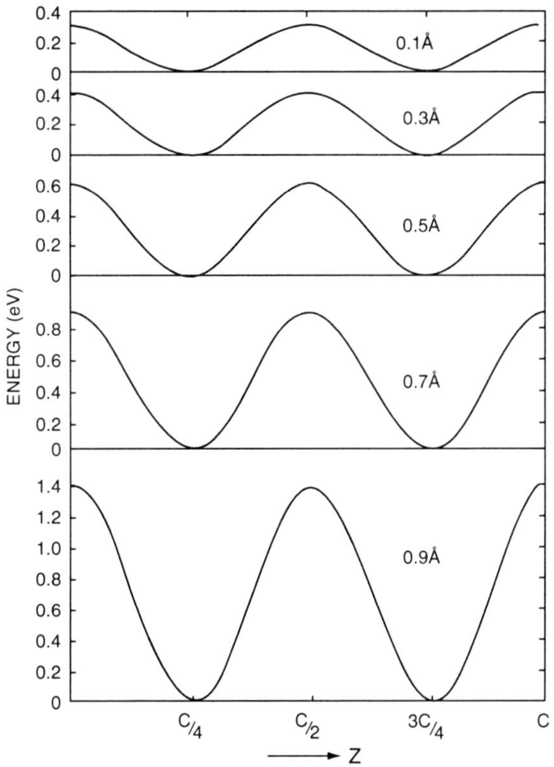

Fig. 15B.13 Influence of ionic size on the energy profile along the minimum energy path in the TiO_2 (rutile) structure

Further calculations showed that cationic motion from an interstitial tetrahedral site in the *c*-direction tunnel into a second type of tetrahedral site outside the tunnel involves a much greater activation enthalpy. Thus this model is in accordance with experimental evidence, which has shown that diffusion perpendicular to the *c* axis is much less rapid than that parallel to it.

The actual values of motional activation enthalpy calculated by this method were somewhat greater than those obtained experimentally, as might be expected from the assumptions employed in this simplified approach.

15B.8 Other Materials with Unidirectional Tunnels

There are also a number of other materials containing unidirectional crystallographic tunnels within which ionic species can be quite mobile. While some of these are primarily ionic conductors, others show mixed ionic–electronic conduction. Some of those that received the greatest attention in early work in this area were the

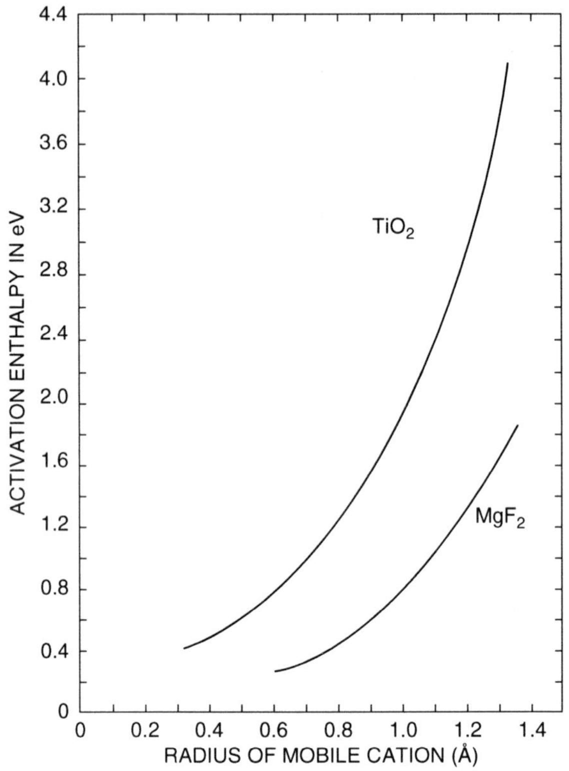

Fig. 15B.14 Calculation variation of the activation enthalpy with size of the mobile cation in the TiO_2 and MgF_2 structures

alkali metal vanadium oxide bronzes, the quaternary titanium oxide hollandites, the alkali aluminosilicate β-eucryptite, and the lithium titanium oxide phase ramsdellite.

15B.9 Materials with the Fluorite and Antifluorite Structures

A number of materials with the fluorite (CaF_2) structure have been recognized for some time as having quite high values of anionic conductivity at high temperatures. The most common of these is the ZrO_2 family, in which the mobile species is the oxide ion vacancy.

By doping with aliovalent cations the conductivity of ZrO_2 can be enhanced appreciably, and these materials, along with other oxide ion conductors, will be discussed further in Sect. 16B.3.

There are also a number of fluorides and chlorides with this same structure that also show large values of anionic conductivity at elevated temperatures. Of special interest because of its very large conductivity (for fluoride ions) is PbF_2, which has been a subject of a number of investigations.

Experiments have also been reported on several materials with the antifluorite structure that have been found to be interesting cationic, rather than anionic, conductors, especially for lithium ions. It has been found possible to produce structures, e.g., Li_5AlO_4, Li_5GaO_4, and Li_6ZnO_4, that have large concentrations of built-in vacancies [27,48], and high values of lithium ionic conductivity at moderate temperatures.

Rapid ionic diffusion is also present in the mixed-conducting phases Li_3Sb and Li_3Bi, which have cubic structures similar to the antifluorites [49,50]. In these cases, however, both the tetrahedral and octahedral interstices in the face-centered cubic Sb (or Bi) lattice are occupied by Li ions at the stoichiometric composition. Ionic transport occurs by the motion of Li ions, and it was found that the interaction of ionic and electronic fluxes produces large values (up to 70,000 at 360°C in Li_3Sb) of the *thermodynamic enhancement factor*, and thus high *chemical diffusion* coefficients in the presence of compositional gradients.

Because one finds rather high values of both anionic (in the fluorite structure) and cationic (in the antifluorite structure) transport, it is worth giving consideration to the unique or special features of this type of crystal structure.

Calculations were also made on this structure using the simple *minimum energy path model* discussed earlier [40]. The mobile ions reside in tetrahedrally coordinated sites within the face-centered cubic sublattice of the other (static) species. While it has often been assumed in the literature that the mobile ions jump directly from tetrahedral to tetrahedral sites, an alternate path is possible, involving motion through an intermediate normally empty octahedral site.

Using the procedures described earlier, it was found that the *minimum energy path*, as defined by the interaction with the static lattice alone, involves motion from one tetrahedral site to the next by following a path that passes close to, but not through the center of, the intermediate octahedral site. This path is shown schematically in Fig. 15B.15.

After determining the preferred minimum energy path, the energy profile along the path was determined for several materials with the fluorite structure. As examples, the results of BaF_2 and CaF_2 are shown in Fig. 15B.16. It is seen that there are

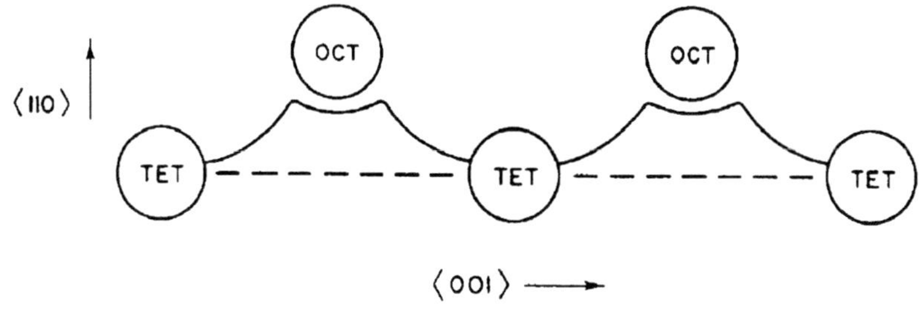

Fig. 15B.15 Schematic representation of the minimum energy path in the fluorite structure

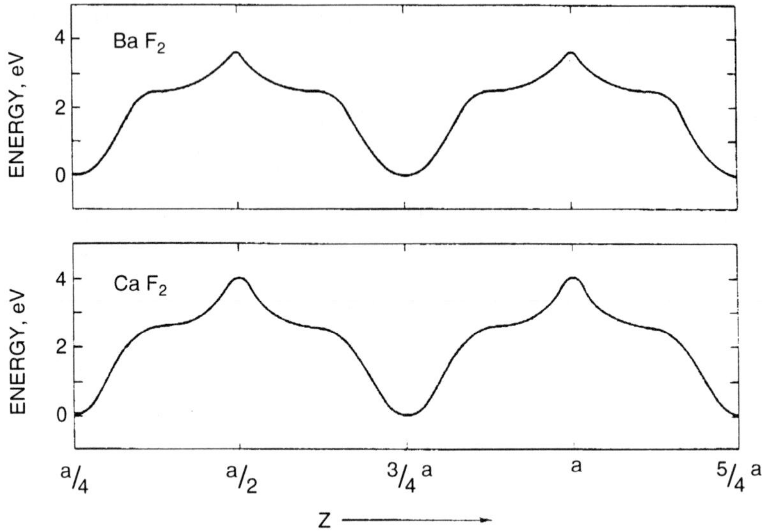

Fig. 15B.16 Calculated variation of potential energy along the minimum energy path for BaF_2 and CaF_2

shoulders on these curves, indicating a set of eight metastable (or almost metastable) positions near, but slightly displaced from, the centers of the octahedral sites. It is interesting that the off-center displacement found from this model is about the location of the interstitial position reported by Willis [51] and Cheetham et al. [52] from neutron diffraction experiments.

The energy profile along a direct tetrahedral–tetrahedral path was found to produce a considerably greater activation enthalpy than for the tetrahedral–octahedral–tetrahedral path in all cases studied.

Because these calculations on the CaF_2 structure related to the motion of normal lattice species, rather than interstitials, the range of ionic sizes, relative to the lattice parameter, that were explored was not very large. As a result, a large dependence upon ionic size was not found in this case. It should also be noted that in this structure the *Coulombic term* plays a relatively important role, whereas in the other two cases mentioned earlier, shorter range interactions were more important.

15B.10 Materials with Layer Structures

The β-alumina family comprises the presently most important and visible group of solid electrolyte materials with layer-type crystal structures. A number of reviews are available covering the extensive early work [25, 26, 53, 54], thus the comments here will be rather brief.

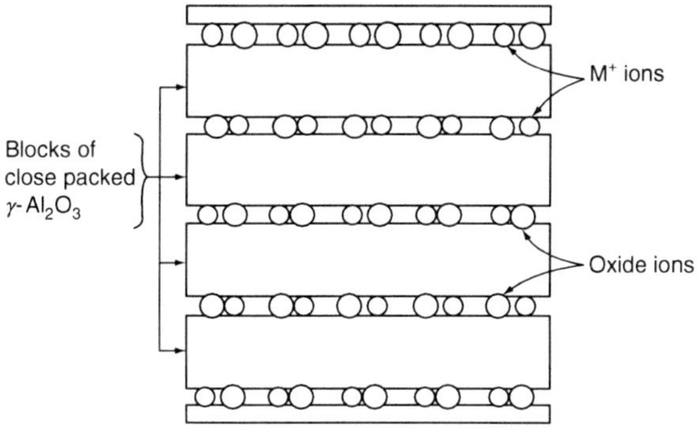

Fig. 15B.17 Schematic representation of the structure of materials of the beta alumina family showing spinel blocks separated by bridging layers containing M^+ and oxide ions

The structure of this family of materials can be visualized in terms of blocks of close-packed oxide ions containing cations in the arrangement typical of the *spinel structure*, separated by galleries in which only one-third of the oxide ion sites are occupied by oxide ions. One-third are occupied by M^+ cations such as Na^+, and the other third are empty. The oxide ions in the *gallery layer* are relatively immobile and act to bridge between the adjacent spinel blocks, but the cations can move readily through the half-occupied balance of the structure. This is shown schematically in Fig. 15B.17.

The arrangement of ions in the bridging layer of beta alumina, assuming the ideal stoichiometry is shown in Fig. 15B.18. Actual materials always have excess cations, for which there are two different types of sites. In this figure, oxide ions in the adjacent close-packed (111) plane of the spinel structure are shown by dashed circles, the oxide ions in the bridging layer by heavy circles, and the mobile cations by smaller cross-hatched circles.

The lithium nitride structure is also of the layer type. It is a lithium conductor, and there are two types of sites for the lithium ions, one in the hexagonal Li_2N layers, and the other in the relatively open intermediate layers where they form N–Li–N bridges [55, 56]. Ionic conductivity results on polycrystalline [57] and single crystalline [58] samples have indicated that lithium ion motion is very fast in this structure. It is also very anisotropic, as would be expected from the structure. The structure of lithium nitride is shown in Fig. 15B.19.

Rapid motion of cationic solute ions can also be found in a number of mixed-conducting materials with layer structures, such as those with the CdI_2 structure, in which the layers of the binary solvent composition are bound together primarily by *van der Waals forces*, rather than being ionically bridged, as in the β-aluminas. Examples of such materials are the transition metal disulfides, such as TiS_2.

Another structure that has been found to exhibit rapid anionic, rather than cationic, diffusion is the tysonite (LaF_3) type, LaF_3 itself is used in fluoride ion-selective electrodes [59], and there have been several investigations of fluoride ion

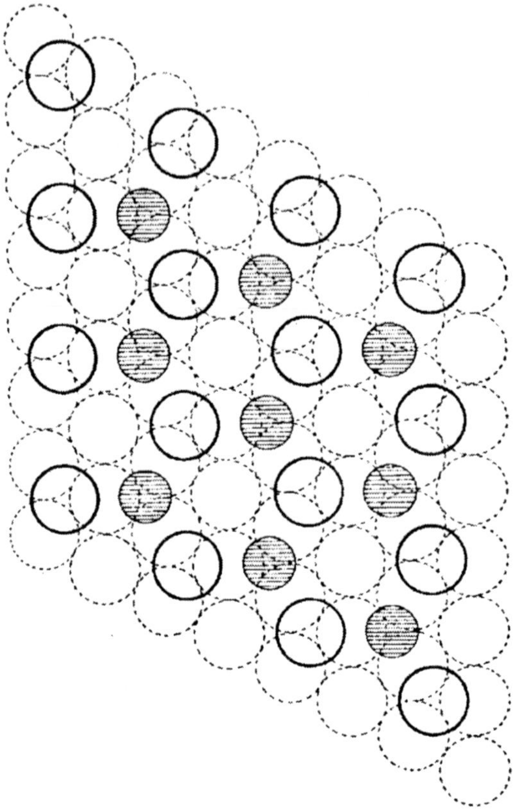

Fig. 15B.18 Arrangement of ions in the bridging layer of the ideal structure of beta alumina

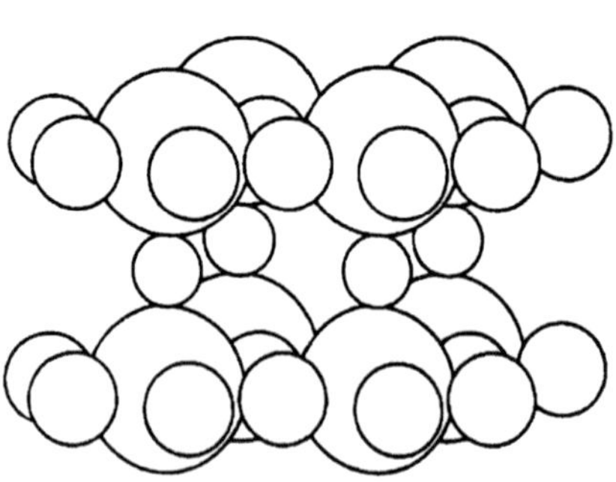

Fig. 15B.19 Structure of Li_3N. The large spheres are nitrogen ions, and the small spheres lithium ions

transport in LaF_3-based materials [15, 60]. Materials with this structure, but based upon CeF_3 instead of LaF_3, have also been investigated and found to have high values of fluoride ionic conductivity [61].

The mobile anions reside in three different types of sites in this structure, and according to NMR studies [62] it appears that transport of ions in and among these sites involves different values of activation enthalpy, which become important in the overall transport process in different temperature ranges.

15B.11 Materials with Three-Dimensional Arrays of Tunnels

Of special interest has been the effort to discover and design materials with crystal structures containing atomic-sized tunnels that are oriented in all three directions, so as to produce relatively isotropic ionic transport.

The structures of the ternary silver iodides of the $RbAg_4I_5$ family are of this type and have very high Ag^+ conductivity at ambient temperatures [9, 63].

Several groups of materials have been found, which have skeleton or network structures composed of various arrays of corner- and edge-shared tetrahedra and octahedra, which are permeated by tunnels that are dilutely populated by mobile monovalent ions [64–68].

There are several families of these, including those called *Nasicon* and *Lisicon*. More recently, high ionic conductivity has subsequently been found in doped *ternary titanate* materials, as well as some based upon the cubic *garnet structure*. More is said about these materials elsewhere in this text.

15B.12 Structures with Isolated Tetrahedra

Several materials have been found that are relatively good conductors for lithium ions whose structures are characterized by isolated tetrahedral anionic groups between which the cations percolate.

One example of this group of materials is Li_4SiO_4, lithium orthosilicate, in which 8 Li ions have been reported [69] to be distributed among the 18 interstitial sites between the tetrahedral SiO_4 anionic groups. Several studies [48,57,67,70,71] showed that the conductivity can be considerably enhanced in Li_4SiO_4–Li_3PO_4 solid solutions. An analogous orthogermanate Li_4GeO_4 has a similar structure, but the Li ion conductivity is lower in this case [72].

Several alkali metal chloroaluminates have structures with alkali metal ions in between isolated $AlCl_4$ groups and relatively low melting points. Conductivity measurements on $LiAlCl_4$, $NaAlCl_4$, and $KAlCl_4$ have shown relatively rapid alkali metal ion motion in these materials in the solid state [73, 74].

The amount of work that has been done to analyze the structures of this class of materials is not sufficient, however, to provide a simple basis for understanding the mechanism of ionic transport through the space within the various arrays of tetrahedral anions. This type of structure is shown schematically in Fig. 15B.20.

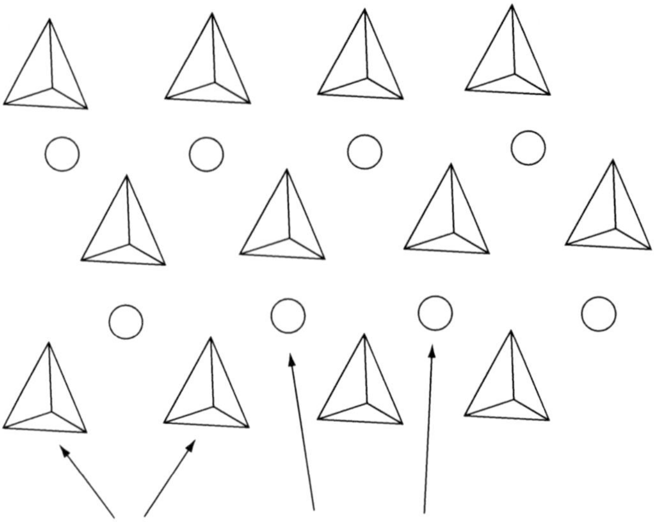

Fig. 15B.20 Schematic drawing of a structure with mobile Li ions between isolated tetrahedra

15C Lithium Ion Conductors

15C.1 Introduction

A number of materials that conduct lithium ions were mentioned in earlier chapters. This chapter will discuss only some of the more recent developments.

There are two major reasons why lithium-conducting solid electrolytes are not being used in common batteries at the present time. One is that they have conductivities that are competitive with the organic liquid solvent materials that are currently used. The other is that the solid electrode materials that are now used in lithium systems undergo volume changes as lithium is added or deleted. Because the electrodes *breathe* as they are cycled, there are significant mechanical stresses placed upon the electrolyte. This causes problems with thin solid electrolyte layers.

Lithium solid electrolytes might be useful in cells that operate at elevated temperatures, where the common liquid electrolytes are no longer stable, however. It was mentioned earlier that one such application is supplying electrical power to the instruments used for *down-hole* measurements, where the temperature can reach several hundreds of degrees.

15C.2 Materials with the Perovskite Structure

A considerable amount of excitement was generated by the introduction of the concept of replacing some of the Li^+ ions in the A-site of materials with the perovskite

ABO_3 structure by other small cations with greater positive charges [77–79]. This results in the introduction of vacancies in the lithium part of the crystal lattice. As an example, one La^{3+} ion can replace three Li^+ ions in such structures, introducing two vacancies.

The perovskite structure can be simply thought of as having a cubic unit cell with B ions at the corners, each bonded to six oxide ions. The A ions reside in the middle of the cube cell, and can move from cell to cell by hopping through a square set of four oxide ions on the cube surface if there is a vacancy in the adjacent cell. High ionic conductivity values were reported for some (La,Li) titanates, up to about $10^{-3}\,S\,cm^{-1}$ at ambient temperatures.

The question of the use of the interesting solid electrolytes in lithium batteries arose when it was observed that they showed rapid darkening when placed in contact with lithium [80]. This indicated that lithium reacts with this material at high lithium activities.

Coulometric titration experiments, illustrated in Fig. 15C.1, showed that lithium becomes inserted into this material at potentials lower than 1.7 V vs. pure lithium.

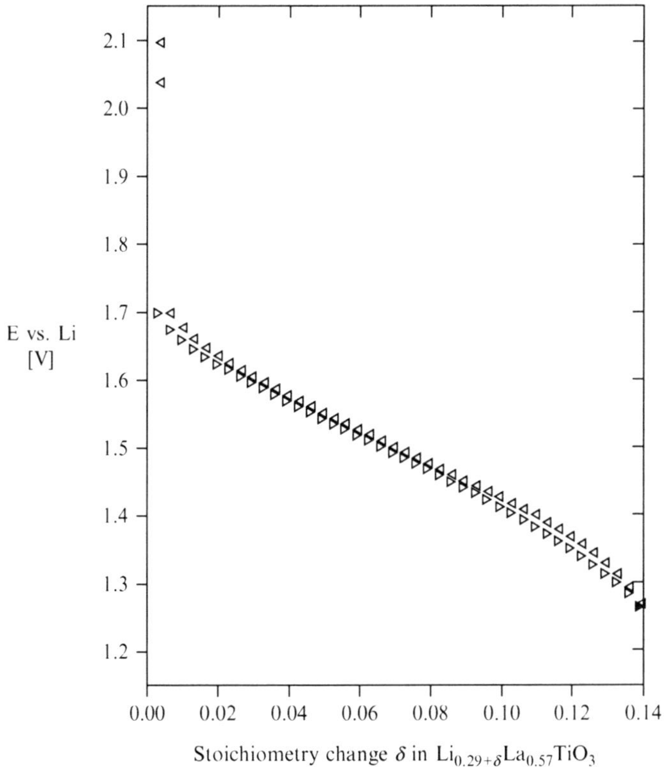

Fig. 15C.1 Coulometric titration curve showing lithium insertion into a material that was initially $Li_{0.29}La_{0.57}TiO_3$

Measurements of the chemical diffusion coefficient of lithium into this material were made using the GITT method [49] discussed in Sect. 17B. The values varied somewhat with composition, but were all in the range 10^{-7} to $10^{-6.5}\,cm^2\,s^{-1}$. These are very high, and indicate that this material would act as a rapid insertion reaction electrode in lithium systems. It will only be of limited use as an electrolyte, due to its limited potential range within which it is an ionic conductor.

15C.3 Materials with the Garnet Structure

It has been demonstrated that the strategy of replacing Li^+ ions with other ions with more positive charges so as to generate vacancies through which lithium ions can move is also applicable to materials with crystal structures related to the mineral garnet [81–84]. The prototype material of this family is $Ca_3Al_2Si_3O_{12}$. It is one of the group of materials that are called orthosilicates, which can be viewed as having isolated tetrahedral SiO_4 groups. The oxide ions are in layers, in a slightly distorted hexagonal close packing. Several materials with this structure were discussed briefly in Sect. 15A.4. Another example that has the related olivine structure is the important positive electrode material $LiFePO_4$ that was discussed in Sect. 9.4. In garnet the Ca^{2+}, Al^{3+}, and Si^{4+} ions have 8-, 6-, and 4- coordination to nearby oxide ions. Substitution of other similarly charged cations in these locations is often found in this family of minerals.

One example of this family of lithium-conducting solid electrolytes is $Li_5La_3M_2O_{12}$, in which the M ion could be Nb or Ta. It was reported that these materials have lithium ion conductivities of about $10^{-6}\,S\,cm^{-1}$ at ambient temperatures, which is about three orders of magnitude lower than data reported for the $(La,Li)TiO_3$ materials mentioned earlier [78].

However, at least some of these materials can have an important advantage compared with the titanium-containing materials, for it has been reported that the version containing Ta retains its white color after being in contact with molten lithium. This apparent stability against high-activity lithium means that it might be used in cells with a higher output voltage than those that are limited to operation at higher potentials.

The temperature dependence of the ionic conductivity of several lithium-conducting materials is shown in Fig. 15C.2 [81].

Subsequently, the influence of replacement of some of the La^{3+} ions by divalent alkaline earth ions has also been investigated [82], and it was found that the size of the substituted ion had an influence upon the measured ionic conductivity, the case of Ba^{2+} ion substitution giving the best conductivity values. Although the conductivity of this material, $Li_6BaLa_2M_2O_{12}$, was found to be slightly lower than that of the original material, it was found to have less problem with an additional grain boundary impedance.

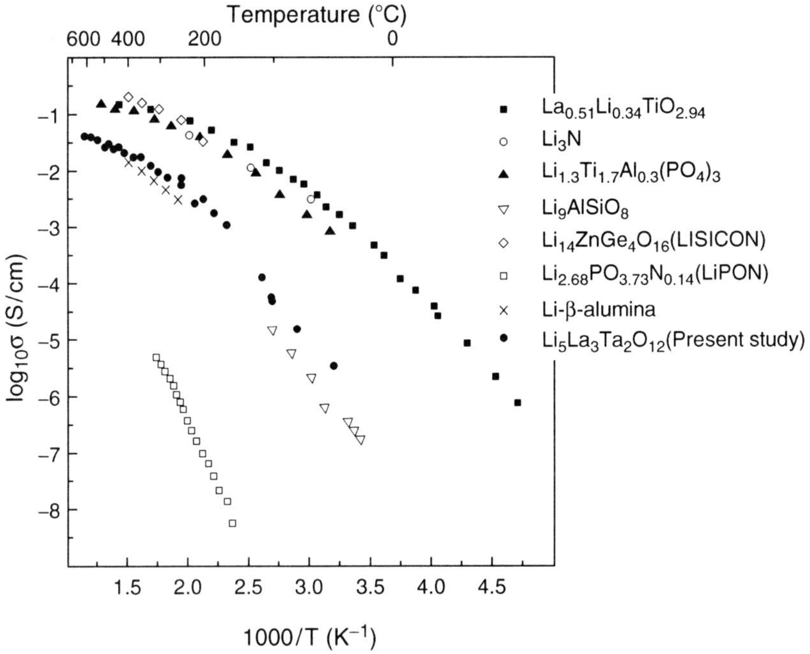

Fig. 15C.2 Temperature dependence of the ionic conductivity of several materials containing mobile lithium ions

In addition to being stable in contact with elemental lithium, it has been found that these garnet materials do not seem to react with a number of the commonly used positive electrode materials [83].

References

1. W. Nernst, Z. Elektrochem. 6, 41 (1900)
2. C. Wagner, Naturwissenschaften 31, 265 (1943)
3. C. Wagner, J. Chem. Phys. 21, 1819 (1953)
4. C. Wagner, Proc. Int. Committee Electrochem. Thermo. Kinet. (CITCE) 7, 361 (1957)
5. K. Kiukkola and C. Wagner, J. Electrochem. Soc. 104, 308 (1957)
6. K. Kiukkola and C. Wagner, J. Electrochem. Soc. 104, 379 (1957)
7. J.N. Bradley and P.D. Greene, Trans. Faraday Soc. 62, 2069 (1966)
8. J.N. Bradley and P.D. Greene, Trans. Faraday Soc. 63, 424 (1967)
9. J.N. Bradley and P.D. Greene, Trans. Faraday Soc. 63, 2516 (1967)
10. B.B. Owens and G.R. Argue, Science 157, 308 (1967)
11. B.B. Owens and G.R. Argue, J. Electrochem. Soc. 117, 898 (1970)
12. Y.F.Y. Yao and J.T. Kummer, J. Inorg. Nucl. Chem. 29, 2453 (1967)
13. R.H. Radzilowski, Y.F. Yao and J.T. Kummer, J. Appl. Phys. 40, 4716 (1969)
14. N. Weber and J.T. Kummer, Proc. Ann. Power Sources Conf. 21, 37 (1967)
15. F.A. Kröger and H.J. Vink, Solid State Phys. 3, 307 (1956)

16. M.L. Huggins, J. Phys. Chem. 58, 1141 (1954)
17. M.L. Huggins, J. Am. Ceramic Soc. 38, 172 (1955)
18. M.L. Huggins, Bull. Chem. Soc. Jpn. 28, 606 (1955)
19. M.L. Huggins, J. Am. Chem. Soc. 77, 3928 (1955)
20. R.A. Huggins and M.L. Huggins, "Structural Defect Equilibria in Vitreous Silica and Dilute Silicates", J. Solid State Chem. 2, 385 (1970)
21. R.A. Huggins, Structural Defect Equilibria in Vitreous Oxides Based upon the Structon Model, in *Reactivity of Solids*, ed. by J.S. Anderson, M.W. Roberts and F.S. Stone, Chapman and Hall, London (1972), p. 186
22. D.O. Raleigh, Prog. Solid State Chem. 3, 83 (1967)
23. W. Van Gool, ed. *Fast Ionic Conduction in Solids*, North-Holland, Amsterdam (1973)
24. W. Van Gool, *Ann. Rev. Mater.* Sci. 4, 311 (1974)
25. R.A. Huggins, "Very Rapid Transport in Solids," in *Diffusion in Solids: Recent Developments*, ed. by A.S. Nowick and J.J. Burton, Academic Press, New York (1975), p. 445
26. R.A. Huggins, *Adv. Electrochem. Electrochem. Eng.* 10, 323 (1977)
27. R.A. Huggins, Electrochim. Acta 22, 773 (1977)
28. P. Vashishta, J.N. Mundy and G.K. Shenoy, eds., *Fast Ion Transport in Solids*, Elsevier/North-Holland, Amsterdam (1979)
29. H. Rickert, in *Fast Ionic Conduction in Solids*, ed. by W. Van Gool, North-Holland, Amsterdam (1973), p. 3
30. W.F. Chu, H. Rickert and W. Weppner, in *Fast Ionic Conduction in Solids*, ed. by W. Van Gool, North-Holland, Amsterdam (1973), p. 181
31. C. Tubandt and E. Lorenz, Z. Phys. Chem. 87, 513 (1914)
32. A. Kvist, in *Physics of Electrolytes*, Vol. 1, ed. by J. Hladik, Academic Press, New York (1972), p. 319
33. L.W. Strock, Z. Phys. Chem. B 25, 441 (1934)
34. L.W. Strock, Z. Phys. Chem. B 31, 132 (1936)
35. M. O'Keefe, Science 180, 1276 (1973)
36. M.J. Buerger and B.J. Wuensch, Science 141, 276 (1963)
37. K. Funke, Prog. Solid State Chem. 11, 345 (1976)
38. J. Volkl and G. Alefield, in *Diffusion in Solids: Recent Developments*, ed. by A.S. Nowick and J.J. Burton, Academic Press, New York (1975), p. 231
39. W.F. Flygare and R.A. Huggins, J. Phys. Chem. Solids 34, 1199 (1973)
40. O.B. Ajayi, Ph.D. Dissertation, Stanford University, Palo Alto, CA (1975)
41. O.B. Ajayi, L.E. Nagel, I.D. Raistrick and R.A. Huggins, J. Phys. Chem. Solids 37, 167 (1976)
42. M. Born and J.E. Mayer, Z. Phys. 75, 1 (1932)
43. J.P. Hardy and J.W. Flocken, CRC Crit. Rev. Solid State Sci. 1, 606 (1970)
44. O.W. Johnson, Phys. Rev. 136, A284 (1964)
45. B. E. Liebert, PhD Dissertation, Stanford University, Palo Alto, CA (1977)
46. O.W. Johnson, S.-H. Paek and J.W. DeFord, J. Appl. Phys. 46, 1026 (1975)
47. L. Pauling, Z. Kristallogr. 67, 377 (1928)
48. I.D. Raistrick, C. Ho and R.A. Huggins, Mater. Res. Bull. 11, 953 (1976)
49. W. Weppner and R.A. Huggins, J. Electrochem. Soc. 124, 1569 (1977)
50. W. Weppner and R.A. Huggins, J. Solid State Chem. 22, 297 (1977)
51. B.T.M. Willis, Proc. Brit. Ceram. Soc. 1, 9 (1964)
52. A.K. Cheetham, B.E.F. Fender, and M.J. Cooper, J. Phys. C 4, 3107 (1971)
53. J.T. Kummer, Prog. Solid State Chem. 7, 141 (1972)
54. M.S. Whittingham and R.A. Huggins, in *Solid State Chemistry*, ed. by R.A. Roth and S.J. Schneider, Nat. Bur. Std. Spec. Publ 364, Washington, DC (1972), p. 139
55. E. Zintl and G. Brauer, Z. Elektrochem. 41, 102 (1935)
56. A. Rabenau and H. Schulz, J. Less Common Metals 50, 155 (1976)
57. B.A. Boukamp and R.A. Huggins, Phys. Lett. A 58, 231 (1976)
58. U. von Alpen, A. Rabenau and G.H. Talat, Appl. Phys. Lett. 30, 621 (1977)
59. M.S. Frant and J.W. Ross, Science 154, 1553 (1966)
60. A. Sher, R. Solomon, K. Lee, and M.W. Muller, Phys. Rev. 144, 593 (1966)

61. T. Takahashi, H. Iwahara and T. Ishikawa, J. Electrochem. Soc. 124, 280 (1977)
62. K. Lee, Solid State Commun. 7, 363 (1969)
63. S. Geller, Science 157, 310 (1967)
64. H.Y.-P. Hong, J.A. Kafalas and J.B. Goodenough, J. Solid State Chem. 9, 345 (1974)
65. H.Y.-P. Hong, Mater. Res. Bull. 11, 173 (1976)
66. J.B. Goodenough, H.Y.-P. Hong and J.A. Kafalas, Mater. Res. Bull. 11, 203 (1976)
67. R.D. Shannon, B.E. Taylor, A.D. English and T. Berzins, Electrochim. Acta 22, 783 (1977)
68. B.E. Taylor, A.D. English and T. Berzins, Mater. Res. Bull. 12, 171 (1977)
69. H. Völlenkle, A. Wittman and H. Nowotny, Mh. Chem. 99, 1360 (1968)
70. Y.-W. Hu, I.D. Raistrick and R.A. Huggins, Mater. Res. Bull. 11, 1227 (1976)
71. Y.-W. Hu, I.D. Raistrick and R.A. Huggins, J. Electrochem. Soc. 124, 1240 (1977)
72. B.E. Liebert and R.A. Huggins, Mater. Res. Bull. 11, 533 (1976)
73. W. Weppner and R.A. Huggins, Phys. Lett. A 58, 245 (1976)
74. W. Weppner and R.A. Huggins, J. Electrochem. Soc. 124, 35 (1977)
75. J. Gendell, R.M. Cotts and M.J. Sienko, J. Chem. Phys. 37, 220 (1962)
76. T.K. Halstead, W.U. Benesh, R.D. Gulliver II and R.A. Huggins, J. Chem. Phys. 58, 3530 (1973)
77. A.G. Belous, G.N. Novitsukaya, S.V. Polyanetkaya and Y.I Gornikov, Izv. Akad. Nauk SSSR Neorg. Mater. 23, 470 (1987)
78. Y. Inagumi, C. Liquan, M. Itoh, T. Nakamura, T. Uchida, H. Ikuta and W. Wakihara, Solid State Commun. 86, 689 (1993)
79. H. Kawai and J. Kuwano, J. Electrochem. Soc. 141, L78 (1994)
80. P. Birke, S. Scharner, R.A. Huggins and W. Weppner, J. Electrochem. Soc. 144, L167 (1997)
81. V. Thangadurai, H. Kaack and W. Weppner, J. Am. Ceram. Soc. 86, 437 (2003)
82. V. Thangadurai and W. Weppner, J. Am. Ceram. Soc. 88, 411 (2005)
83. V. Thangadurai and W. Weppner, J. Power sources 142, 339 (2005)
84. R. Murugan, V. Thangadurai and W. Weppner, J. Electrochem. Soc. 155, A90 (2008)

Chapter 16
Electrolyte Stability Windows and Their Extension

16.1 Introduction

While the potential utility of electrolytes surely depends primarily upon the magnitude and selectivity of the ionic conductivity, the practical utilization of such materials also requires that they meet the relevant stability requirements. As mentioned earlier, this means that they must be stable with respect to thermal decomposition, and reactions with other species in their environments. In addition, they must be utilized under conditions in which ionic, rather than electronic, species conduct most of the charge. Stated another way, such materials must be utilized within their appropriate stability ranges.

Thus, the practical utilization of materials as electrolytes is often limited to restricted ranges of temperature, pressure, and chemical potentials. This matter has received relatively little attention to date, despite its obvious practical importance.

The electrolyte must be stable in the presence of both the reducing conditions imposed by the negative electrode and the oxidizing conditions imposed by the positive electrode. The concept and the term *stability window* were introduced to describe the useful potential range of an electrolyte a number of years ago [1].

Although kinetic considerations may be important in some cases, this is primarily a question of thermodynamics.

It can be a particularly important problem on the negative potential side, for a number of systems considered today involve the use of very aggressive materials, such as the alkali metals or their alloys, as negative electrode constituents.

If one of the electrodes has a potential that is not within the stability window of the electrolyte, some sort of reaction will take place at the electrode/electrolyte interface. This can also happen at both electrodes. A simple example of this is what happens if two chemically inert electrodes, e.g., platinum wires, are put into a simple aqueous electrolyte cell and the voltage between them is raised until it is higher than the voltage equivalent to the standard Gibbs free energy of the formation of water, which is about 1.23 V at ambient temperature. The aqueous electrolyte is not stable,

R.A. Huggins, *Advanced Batteries*: Materials Science Aspects,

hydrogen gas is evolved at the negative electrode, and oxygen gas is evolved at the positive electrode. This process is, of course, called electrolysis.

16.2 Binary Electrolyte Phases

If the electrolyte is a binary phase containing an alkali metal, the question of stability is very simple if the phase diagram and some thermodynamic data are known. The terminal (most alkali-rich) phase in the binary system will be thermodynamically stable in contact with the alkali metal or its saturated solution. Any other will not.

The stability window can be readily calculated in a number of cases. Both LiI and Li_3N are particularly simple examples of solid electrolytes that are binary phases in which one of the components is the alkali metal lithium. They both exhibit ionic conduction, related to the motion of lithium within their crystal structures. In the Li_3N case, it was found that the ionic conductivity is very high at ambient temperatures [2–4].

One can calculate the thermodynamic stability limit ΔE in such cases directly, if the standard Gibbs free energy of formation of the electrolyte phase, $\Delta G_f{}^{\circ}$ is known.

$$\Delta E = -\left({}^{\Delta G_f{}^{\circ}}/_{zF}\right). \tag{16.1}$$

The result is that LiI will be stable over a range of 2.79 V at 25°C. If the negative electrode is pure lithium, the positive electrode potential may be up to 2.79 V more positive. In the case of Li_3N, the thermodynamic data indicate that it is only stable over a range of 0.44 V at that temperature. If positive electrodes are used with potentials exceeding these values, the solid electrolyte will decompose.

Soon after the discovery of the unusual ionic conductivity of Li_3N, efforts were undertaken to develop high-voltage lithium-conducting batteries in some quarters. Unfortunately, the limitations of the stability window were not considered, and those efforts were unsuccessful.

Both LiI and Li_3N are stable in contact with elemental lithium. However, additional complications can arise in other cases. Several other terminal binary phases dissolve some of the alkali metal at elevated temperatures. The excess alkali metal exists in the electrolyte as alkali metal ions and electrons. As a result, there can be an appreciable amount of electronic conductivity in such materials at high alkali metal activities. The alkali halides such as lithium and sodium chloride are examples of this, as they all dissolve their respective alkali metal components somewhat when molten – the lithium salts the least, those containing the heavier alkali metals more. This electronic conduction in the electrolyte leads to self-discharge in batteries containing such materials.

The stability windows of binary oxide phases can be similarly determined if the Gibbs free energy of the phase in question is known. In such cases, it is assumed that the oxide phase is stable in air, so that the issue is how far negative the potential of the negative electrode can be.

16.3 Ternary Electrolyte Phases

The stability of electrolytes that are composed of more than two components is more complicated. The stability windows of ternary (three component) phases will be considered briefly here. The relevant principles were developed some time ago [1].

In Chap. 4 it was shown that when sufficient information is available about the stable phases in a ternary system, one can use data on the standard Gibbs free energy of formation of the various phases to construct the relevant ternary phase stability diagram and identify the stable tie lines. This then provides information about which phases are thermodynamically stable in the presence of others. For example, only electrolyte phases that share two-phase tie lines or three-phase triangles with alkali metals will be stable in contact with them. Likewise, one can readily identify the phases that result from either reduction or oxidation reactions.

In order to show the principles and methodology involved, a hypothetical ternary system containing components Li, a second metal M, and oxygen will be briefly discussed. For simplification, it can be assumed that only two binary oxides of M, nominally MO and MO_2, are stable at the temperature of interest, that there are no phases between Li and M, and that there is only one ternary phase, $Li_2M_2O_3$. Li_2O must also be present. The locations of these various phases on a ternary phase stability diagram are shown in Fig. 16.1.

If the phase $Li_2M_2O_3$ is known to be a solid electrolyte that conducts lithium ions, what can be determined about the range of lithium activity (voltage vs. Li) over which it is stable?

In order to see how this question can be answered, it can be assumed that the values of the standard Gibbs free energy of formation $\Delta G_f{}^\circ$ of all the phases are

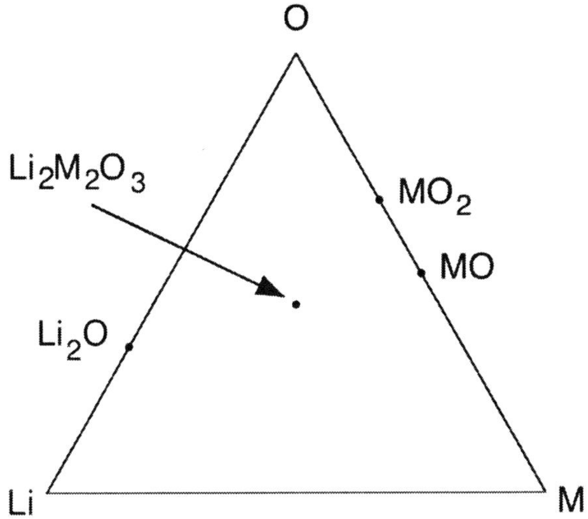

Fig. 16.1 Locations of phases assumed to be stable in the Li–M–O ternary phase stability diagram

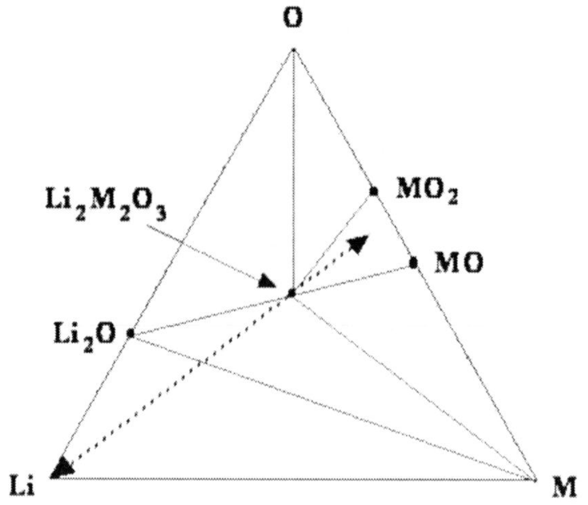

Fig. 16.2 Phase stability diagram showing the stable tie lines calculated from the thermodynamic data. Also shown is a line indicating the locus of compositions that would result if lithium were added to, or deleted from, the $Li_2M_2O_3$ phase

known. If they are not all known, there are several ways in which some of them may be deduced from targeted experiments [5].

Using the principles introduced in Chap. 4, the stable tie lines can be identified. This results in a phase stability diagram consisting of a number of ternary subtriangles, shown in Fig. 16.2. Superimposed upon this diagram is a line indicating how the overall composition would change if lithium would be either added to, or deleted from, the phase $Li_2M_2O_3$.

As discussed in Chap. 4, the microstructure of all compositions within a given subtriangle will contain the appropriate amounts of the three phases that are at its corners. Under equilibrium conditions, the Gibbs phase rule dictates that all the intensive variables of the system, at a particular temperature and pressure, are fixed when three phases are present in equilibrium in a ternary system. What this means in this context is that the activities, partial pressures, and chemical potentials of all of the components (ie., Li, M, and oxygen) are independent of composition within any given triangle. Their values can be calculated from the relevant thermodynamic data. Likewise, the voltage vs. each of the components will be constant for all compositions within each triangle, and can be calculated, assuming an appropriate electrochemical cell.

16.3.1 Stability Limits Relative to Lithium

In the case of the possible electrolyte phase, $Li_2M_2O_3$, the thermodynamic parameters of the triangle $Li_2M_2O_3 - Li_2O - M$ will determine its limit of stability in the low-potential lithium-rich direction, and those in the triangle $Li_2M_2O_3 - MO_2 - MO$ will determine its limit of stability in the high-potential lithium-poor direction.

In the low-potential case, the relevant reaction is

$$4Li + Li_2M_2O_3 = 3Li_2O + 2M, \tag{16.2}$$

and the minimum voltage vs. lithium is determined from

$$E = -\left({}^{\Delta G_r{}^\circ}/_{4F}\right), \tag{16.3}$$

where

$$\Delta G_r{}^\circ = 3\Delta G_f{}^\circ(Li_2O) - \Delta G_f{}^\circ(Li_2M_2O_3). \tag{16.4}$$

Likewise, the maximum potential at which this phase remains stable is determined by the relation

$$2Li + MO_2 + MO = 2Li_2M_2O_3. \tag{16.5}$$

This can be converted to a voltage vs. lithium from

$$E = -\left({}^{\Delta G_r{}^\circ}/_{2F}\right), \tag{16.6}$$

where

$$\Delta G_r{}^\circ = 2\Delta G_f{}^\circ(Li_2M_2O_3) - \Delta G_f{}^\circ(MO_2) - \Delta G_f{}^\circ(MO). \tag{16.7}$$

These voltage limits could also be converted to the equivalent limiting values of lithium activity on the two sides of the stability of the potential solid electrolyte phase, as discussed earlier. But this is not necessary at this point, for the important practical parameters are the voltages vs. some reference, such as pure lithium.

16.3.2 Stability Limits Relative to Oxygen

On the other hand, the range of oxygen pressure over which the phase $Li_2M_2O_3$ is stable can also be important. It can be seen from the phase stability diagram of Fig. 16.2 that it is stable in contact with oxygen at one atmosphere. What about the low oxygen partial pressure limit?

This can be determined from the ternary subtriangle on the opposite side of the phase, where the relevant oxygen reaction is

$$O_2 + Li_2O + 2M = Li_2M_2O_3. \tag{16.8}$$

The voltage vs. unit activity (1 atm) of oxygen is given in this triangle by

$$E = -\left({}^{\Delta G_r{}^\circ}/_{4F}\right). \tag{16.9}$$

This voltage can be converted into the related oxygen partial pressure by the use of the Nernst equation

$$E = -\left({}^{RT}/_{zF}\right) \ln \left({}^{p(O_2)^{II}}/_{p(O_2)^{I}}\right), \tag{16.10}$$

where z is the number of electrons involved in the transport of oxygen across the assumed electrochemical cell, and $p(O_2)^I$ is the partial pressure of oxygen in its standard state, which is 1 atm pressure in this case.

16.4 Summary

It has been shown that it is possible to calculate the stability windows of both binary and ternary phases that might be interesting as solid electrolytes by the use of thermodynamic information. Phase stability diagrams can be very useful in understanding what factors determine the ranges of stability of particular phases, and what adjacent phases are formed if the bounds of the stability window are exceeded.

16A Composite Structures That Combine Stability Regimes

16A.1 Introduction

The matter of the range of chemical stability of phases was discussed in Chap. 16, with particular reference to their use as electrolytes. One way in which the electrolyte part of an electrochemical cell can be extended is obviously by putting two electrolytes in series, such that the total voltage across them is distributed in such a way that each one operates over an electrical potential range within which it is predominantly an ionic conductor.

This can be done if both electrolytes are solids, and they are otherwise stable in contact with each other. It can also be done if one is solid, and the other is liquid. The use of two liquid electrolytes would require that provisions are taken so that they do not mix or get displaced, of course.

There is another issue regarding electrolytes that was discussed in Chap. 16. That is that there are a number of cases in which the chemical stability window of an electrolyte may be broader than its ionic conduction window. This can be represented schematically as shown in Fig. 16A.1.

Such a material can only be useful as an electrolyte within the range of electrical potential in which charge is carried primarily by the transport of ionic species.

It will be seen in the following brief discussion that there are a number of situations in which this problem can be mitigated by the use of two electrolytes in series.

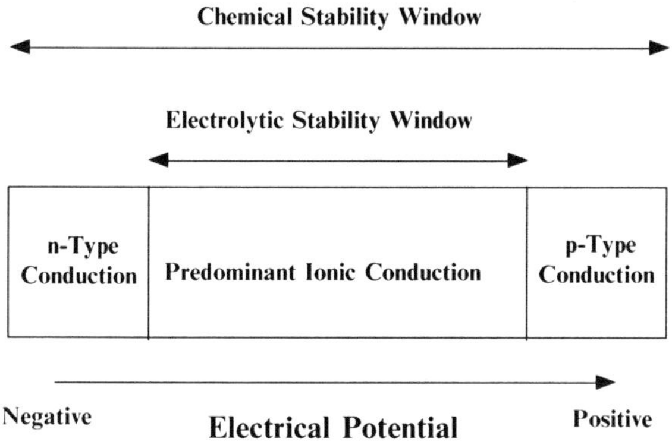

Fig. 16A.1 Schematic representation of the relationships between the chemical and electrolytic stability windows for the case in which the chemical stability window extends beyond the electrolytic window

16A.2 Two Solid Electrolytes in Series

One example of the concept of the use of two electrolytes in series is the use of a layer of a high-conductivity oxide that is only electrolytic over a limited range in series in contact with a layer of another, phase, maybe even of lower conductivity, that extends the stability range of the composite. An example is the use of ceria-based solid electrolyte materials in series with high-conductivity bismuth oxide phases that have limited stability [5, 6].

16A.3 Solid Electrolyte in Series with Aqueous Electrolyte

The voltage range of aqueous electrolyte electrochemical cells is limited by the decomposition voltage of water. But it is possible to obtain much higher voltages if there is an additional solid electrolyte between the water and one of the two electrodes. An example of this is the development of a composite system in which the negative electrode is lithium that is covered by a layer of lithium-transporting solid electrolyte [7].

In this case, the lithium solid electrolyte is a glassy material with a composition $LiM_2(PO_4)_3$, where M is a transition metal cation. This material is not sufficiently stable against lithium metal, so an additional thin layer of Li_3N is placed between them. This Li_3N layer is formed by depositing Cu_3N on the glass surface and reacting it with lithium. Thus, this configuration actually includes two solid electrolytes in series with an aqueous electrolyte.

The resulting configuration is shown schematically in Fig. 16A.2.

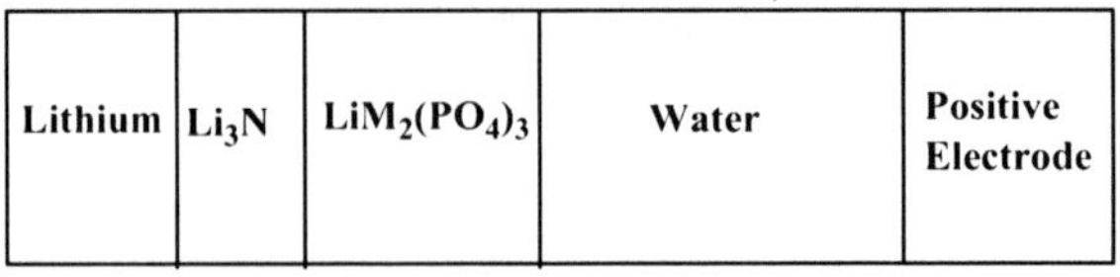

Fig. 16A.2 Schematic representation of a system with two solid electrolytes between lithium and water

16A.4 Solid Electrolyte in Series with Molten Salt

An example in which there is a solid electrolyte in series with a liquid molten salt will be discussed in Sect. 16B.3. In that case, the solid electrolyte is formed as a reaction product between lithium and a nitrate molten salt.

Another example is the Zebra battery system that was discussed in Chap. 4. In that case there are also two electrolytes in series, but this is not only to extend the electrolytic stability range.

One of them, the liquid electrolyte, $NaAlCl_4$, is not stable in contact with the negative electrode reactant, liquid sodium. So the sodium is contained in a second, solid electrolyte, sodium beta alumina. However, the liquid electrolyte serves another function, for it provides good electrochemical contact to the surfaces of the fine particle solid positive electrode. Such good contact is not possible with only a solid electrolyte.

16A.5 Formation of a Second Electrolyte by Topotactic Reaction Between a Liquid and a Solid Mixed Conductor Electrode

The "nickel" electrode was discussed in Sect. 11A.3. In that case, H_2NiO_2, or $Ni(OH)_2$, is topotactically formed on the surface of the mixed conductor $HNiO_2$ during the discharge of this electrode by the insertion of protons. The H_2NiO_2 phase is a solid electrolyte, and separates the aqueous electrolyte from the $HNiO_2$. Its thickness varies as the electrode is charged and discharged.

16A.6 Formation of a Protective Reaction Product Layer Between the Negative Electrode and the Organic Solvent Electrolyte in Lithium Cells

Most current ambient temperature lithium batteries, such as those that are used in small electronic devices, have electrolytes that are composed of organic solvents containing lithium salts. A common example is a solution of ethylene carbonate (EC) and diethylene carbonate (DEC) and $LiPF_6$.

These organic solvents are not stable at very negative potentials, and decompose to produce a reaction product layer, typically called a *solid electrolyte interphase* (SEI) on the surface of the negative electrode. An important feature of these layers is that they allow the transport of lithium ions and extend the total useful electrolytic potential range.

As will be discussed in Sect. 16A.6, these reaction product layers generally form at potentials not far from that of lithium. Their properties can be deliberately varied by the inclusion of extra species, such as vinyls, in the electrolyte. It will also be pointed out that other materials are sometimes deliberately introduced into organic solvent electrolytes in order to modify their behavior at high potentials.

If these product layers allow the transport of only ionic species, they will maintain a stable thickness. Their growth requires either the transport of neutral species, or of neutral combinations of species, from one side to the other.

16B The SEI in Organic Solvent Systems

16B.1 Introduction

It has already been mentioned in Sect. 14A.6 that organic solvent electrolytes are not stable at low potentials, where the chemical potentials of alkali metals, such as lithium, are high. They react with lithium and decompose to form a new phase that acts as a protective layer that blocks further reaction and is also an ionic conductor [8]. This has been called an *SEI*, or *solid electrolyte interphase* [6].

This SEI layer is a very important component of current lithium batteries that employ organic cation solvent liquid electrolytes, in which the anionic component is typically a carbonate. Understanding the structure, composition, and properties of such layers is difficult, for the layers are typically amorphous or contain amorphous phases.

Since the negative electrodes in current lithium-ion batteries are lithium carbons, the formation and stability of the SEI layer on the surface of carbons is of special interest, and will be discussed shortly. First, however, it is important to point out that SEI formation occurs because the stability range of the electrolyte is exceeded, and occurs on many types of negative electrodes, not just on carbons. But in addition, there may be some impurity species on the surface of the electrode material that also electrochemically react with the electrolyte as the potential becomes very negative.

16B.2 Interaction of Organic Solvent Electrolytes with Graphite

The history of the use of carbons in lithium cells was discussed briefly in Sect. 7A. One of the items mentioned was that the first attempts to electrochemically insert lithium into graphite as a lithium battery electrode were unsuccessful [10]. This involved the use of an organic solvent electrolyte containing propylene carbonate (PC). Lithium ions are solvated by polar solvent molecules, and the insertion of these groups can cause severe dimensional changes in the graphite structure. In some cases exfoliation of the graphite structure – either individual layers or groups of layers – can occur. These effects depend directly upon the identity of the species in the organic solvent, and it is now known that propylene carbonate is particularly bad in this regard.

It has subsequently been found that the solvent component that is most desirable is ethylene carbonate (EC) [11]. However, it has a high viscosity, and thus a relatively low conductivity. Thus, it is generally mixed with species with lower viscosities, such as diethyl carbonate (DEC), dimethyl carbonate (DMC), or ethyl methyl carbonate (EMC).

SEI layers form between the lithium graphite and the liquid electrolyte during the first charging cycle as the local potential is reduced in the direction of lithium deposition. The potential at which formation of these layers starts to occur, as well as the amount of charge that the layers consume depends upon the composition of the electrolyte, as well as the details of the nature of the lithium graphite. This process gets rather complex, and has been studied by many authors. Reviews of this topic can be found in several places [12–17].

The evidence of the irreversible formation of an SEI layer on a sample of natural graphite during the first charging cycle of a lithium cell is shown in Fig. 16B.1 [11]. Several features are evident. Charge begins to pass through the cell, indicating the deposition of lithium, at about 1.2 V vs. Li, with the amount increasing as the potential becomes lower. A second, and larger, phenomenon occurs below about 0.75 V, and it continues to grow down to a potential of about 0.2 V, when the reversible

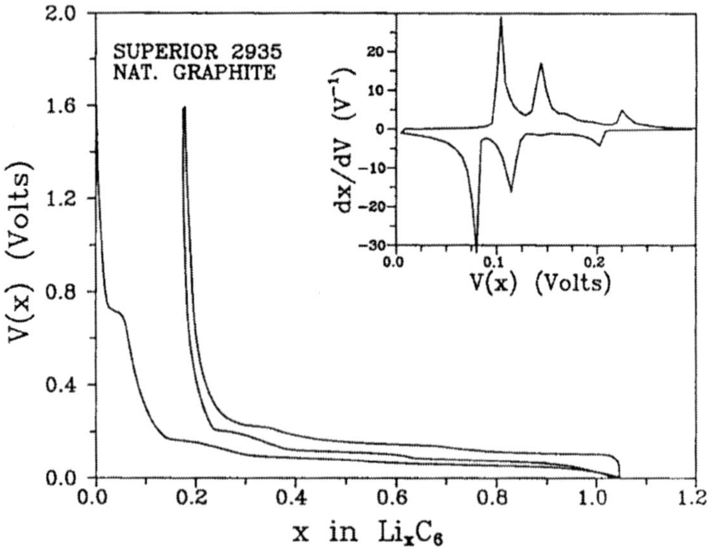

Fig. 16B.1 Relation between the potential of a natural graphite electrode and the amount of lithium reacted upon first charging. Also shown is the subsequent discharge/recharge cycle

insertion of lithium into the graphite begins. Both of these phenomena are obviously irreversible, for they do not appear during the second cycle.

It was found that the amount of charge consumed in the higher potential reaction was proportional to the surface area of the carbon, indicating that this involves the growth of a surface layer. The amount of this irreversible capacity was about $0.003\,\text{Ah}\,\text{m}^{-2}$, which indicated that the layer was about 20 Å thick. This is about the thickness that had been estimated to be necessary to prevent electron tunneling [9]. Electron transport is necessary for film growth.

The lower potential but larger reaction has been a quandary. The amount of charge consumed has been found to be independent of the magnitude of the surface area.

There has been some confusion and conflict concerning the composition of the SEI layer. It was first assumed that the layers formed by the decomposition of organic carbonates on elemental lithium were mainly composed of Li_2CO_3 [18]. However, later spectroscopic work showed evidence for the presence of lithium alkyl carbonate in the SEI [19], especially when ethylene carbonate is present. In the presence of trace amounts of water, LiF is found [20].

There are indications that there may be a composition gradient within the SEI, with lithium carbonate, and perhaps Li_2O, at the higher lithium activity region, and alkyl carbonate in the outer region nearer the liquid electrolyte. This would be consistent with the fact that there must be a potential, and thus a lithium activity, gradient across the solid electrolyte.

When the negative electrode is a lithium carbon, instead of elemental lithium, the possibility of the insertion of species from the electrolyte, or the SEI layer, into the graphite crystal structure, as mentioned earlier, must be also taken into account in discussions of the SEI layer. In the case of propylene carbonate, it is believed that solvated lithium species enter between the graphene sheets when the potential is below about 0.8 V vs. Li. The intercalation of solvated species between graphene sheets in graphitic materials can be considered as the formation of ternary compositions, $Li_x(\text{solv})_yC_n$. The difference in the structure of the graphite when such solvated species are present is illustrated schematically in Fig. 16B.2 [13].

It can be seen that the insertion of the larger species results in an extreme increase in the interlayer spacing. The insertion must, of course, start at the surface with the electrolyte (or the SEI layer). This causes severe bending of the graphene layers, with related mechanical work.

It has been pointed out that from purely geometrical arguments the thickness of graphite flakes will influence the extent of this behavior. This is shown schematically in Fig. 16B.3 [13].

It is also possible for ternary solvated species to enter the graphite structure, and then decompose, providing unsolvated lithium that can move further into the interior without being accompanied by organic species. This sequence is illustrated in Fig. 16B.4 [13].

There is also evidence that an irreversible reaction begins to appear on lithium carbons at about 1.2 V vs. Li, and is significantly more noticeable at about 0.7–0.8 V vs. Li in the case of other organic carbonates [11]. This is, again, evidence of

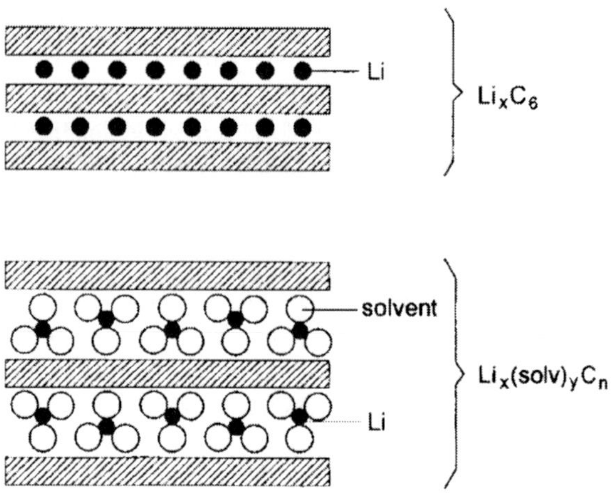

Fig. 16B.2 Schematic of the graphite structure intercalated by lithium ions (*top*), and by solvated lithium ions (*below*)

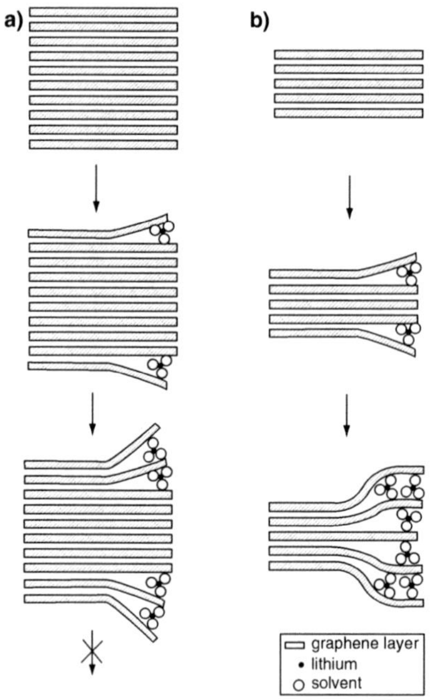

Fig. 16B.3 Schematic model showing the influence of the thickness of graphite flakes upon the extent of the insertion of solvated species

a two-step process in SEI formation, and these phenomena amount to irreversible capacity during the first charging cycle. They do not reappear in subsequent cycles, so long as the potential of the electrode does not go so far positive that the SEI layer becomes reoxidized.

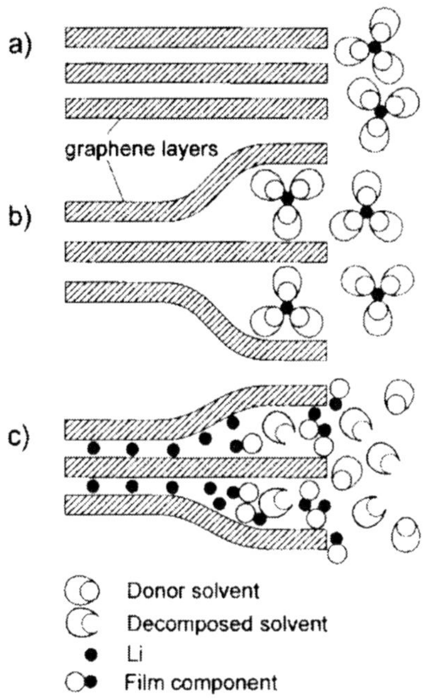

Fig. 16B.4 Schematic model of the transient insertion of solvated species into the surface region of graphite

Another type of experiment that is relevant when considering the formation of SEI layers has been the observation of gas that is evolved during SEI formation, and its relation to the composition of the electrolyte [21]. Propylene is observed from propylene carbonate, and ethylene from ethylene carbonate.

It is fair to say that it is not yet clear what actually happens in what appears to be a two-step process, and why propylene carbonate behaves so differently from ethylene carbonate with regard to exfoliation. One possibility is that ethylene carbonate forms a stable and protective SEI layer right away, preventing further reaction and intercalation, whereas the driving force for the formation of such a blocking SEI layer in the presence of propylene carbonate is not so strong, allowing it to penetrate into the graphite.

16B.3 Electrolyte Additives

It has been found that it can be very useful to add additional components to the liquid electrolyte, and there has been much work along this line. This generally involves changes in the composition, structure, and properties of the SEI layer. To have any effect, these materials must be reduced at potentials more positive than those at which the normal SEI layer would form. They are generally selected such that no solvent intercalation, and resultant exfoliation, will occur. Especially attractive are materials that make it possible to use propylene carbonate without its normal insertion and exfoliation properties. This is desirable because propylene carbonate, either alone or in mixtures, is intrinsically attractive, due to its high conductivity and wide temperature stability range. A number of materials that have been considered for this purpose are mentioned in [13] and [15]. Much of the work that has been done in this area is proprietary, and not readily available.

One of the materials that appear to be attractive for use with propylene-based electrolytes is ethylene sulfite (ES) [22]. The presence of only 5% ES completely eliminates the graphite exfoliation problem normally found when using propylene carbonate. However, the additional irreversible capacity resulting from the reduction of ES is a disadvantage.

Another additive that has attracted a lot of attention is vinylene carbonate (VC) [23, 24]. It apparently introduces very little additional irreversible capacitance, and the impedance of the SEI layer is reduced. Another apparent advantage is a reduction in capacity loss of lithium cells upon cycling at elevated temperature [25]. Catechole carbonate (CC) has also received some attention as a useful additive to graphite negative electrodes [26]. This material has a benzene ring that is fused with a structural unit similar to EC.

Thus it is obvious that this is an area of considerable current interest, both in the scientific and commercial communities. Minimization of first cycle irreversible capacitance, that requires the presence of otherwise useless lithium in the positive electrode, and improved cycle life are both very important in the practical utilization of lithium batteries.

16C Combination of a Solid Electrolyte and a Molten Salt Electrolyte

16C.1 Introduction

It is possible to use a molten salt electrolyte under conditions in which electrode potentials do not have to be within its stability range. This can be accomplished by use of electrodes upon which a second electrolyte is formed in situ. One such example is discussed here.

16C.2 The Lithium–Nitrogen–Oxygen System

It has been shown that it is possible to use lithium electrodes with a molten salt electrolyte composed of the 41%/59% eutectic composition in the $LiNO_3 - KNO_3$ system at temperatures at which it is liquid [27–29]. This composition has an eutectic melting point of 135°C and can act as a lithium-transporting electrolyte. Lithium melts at about 180°C, so there is a temperature range within which it is possible to have solid lithium metal in the presence of this molten salt.

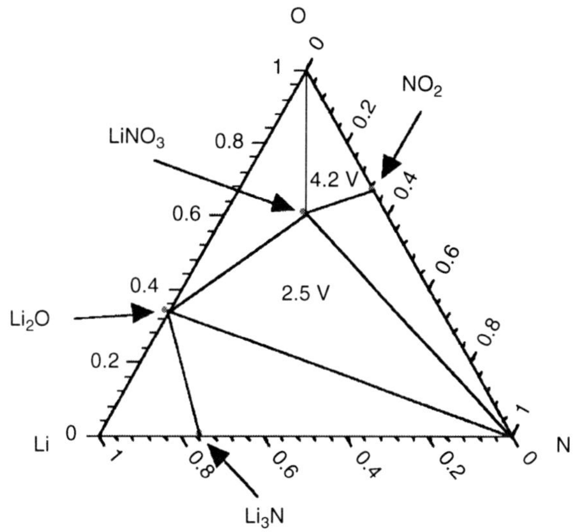

Fig. 16C.1 The calculated ternary phase stability diagram for the Li–N–O system at 150°C

The potassium in the molten salt is essentially inert, not reacting with the lithium phases present, so it is possible to think of this situation in terms of the ternary Li–N–O system.

The stable phases known to be present in this ternary system above 135°C are Li_2O, Li_3N, NO_2, and $LiNO_3$. From the thermodynamic data available, it is possible to determine the stable tie lines in the ternary stability diagram, and also calculate the potentials in the different subtriangles. The result is shown in Fig. 16C.1.

It is seen that the liquid electrolyte phase $LiNO_3$ is at the corners of two subtriangles that are at different distances from the lithium corner of the diagram. The potentials of these two triangles determine the stability window of this phase. As shown in the diagram, $LiNO_3$ is stable at potentials between 2.5 and 4.2 V vs. lithium at 150°C. Thus, it could act as the electrolyte in a cell whose electrodes both have potentials within that range.

At potentials above 4.2 V vs. lithium, the $LiNO_3$ will tend to decompose into NO_2 and oxygen. Below 2.5 V vs. lithium, the overall composition moves into a subtriangle in which Li_2O and nitrogen will tend to form. At even lower potentials Li_3N will also tend to form.

16C.3 Extension of the Effective Potential Range by the Formation of a Second Electrolyte In Situ

If lithium metal is present as the negative electrode, lithium oxide and lithium nitride will tend to form on its surface. If they are stable in the presence of the liquid nitrate, they will form a protective layer separating the lithium from the molten salt. They are both known to be good solid electrolytes for lithium. Measurements on lithium oxide were made many years ago [30], and information on Li_3N has been included in Chap. 16.

The result is that a solid electrolyte can form in situ on lithium. The negative electrode can thus have a potential of 0 V, instead of at least 2.5 V, vs. lithium. This means that the cell voltage could be up to 4.2 V if an appropriate positive electrode could be found.

16C.4 A Primary Lithium/Carbon Cell

An inert electrode material can be employed instead of a reversible reactant on the positive side of the cell. One such example is carbon. It can be seen from the ternary phase stability diagram that the carbon will have a potential of 2.5 V vs. lithium.

If current flows through such a lithium/$LiNO_3$/carbon cell a combination of lithium oxide and nitrogen will form on the positive side. This has been demonstrated experimentally, using relatively porous carbon foam [29]. The configuration of such a cell is shown schematically in Fig. 16C.2.

The maximum theoretical specific energy for such a $Li/LiNO_3/C$ cell is quite impressive. Using the general relation presented near the beginning of the text

$$MTSE = (26,805)(zE)/W, \tag{16C.1}$$

it can be found that the maximum theoretical specific energy is $3,880\,\mathrm{Wh\,kg^{-1}}$. This is a large number.

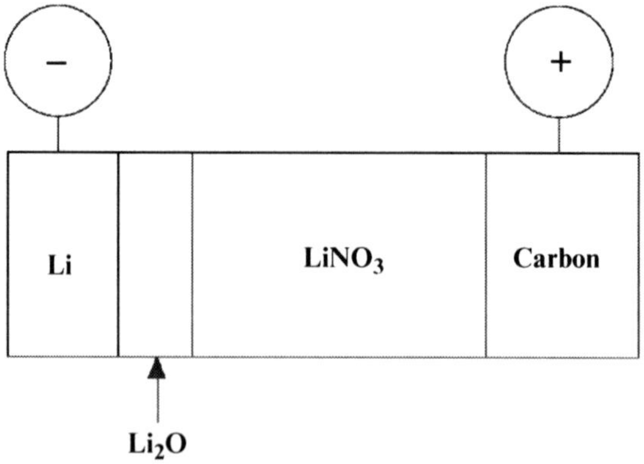

Fig. 16C.2 Schematic representation of the configuration of a $Li/LiNO_3/C$ cell

16C.5 Problems with This Concept

There are two problems with this approach, however. One is that the temperature must be maintained below the melting point of elemental lithium, for molten lithium can react with the electrolyte very rapidly, leading to thermal runaway. One could, of course, use lithium alloys that have higher melting points in order to avoid this problem.

The other problem is that there can be a small amount of self-discharge, due to the slow dissolution of lithium oxide into the molten nitrate electrolyte.

References

1. R.A. Huggins, "Evaluation of Properties Related to the Application of Fast Ionic Transport in Solid Electrolytes and Mixed Conductors," in *Fast Ion Transport in Solids*, ed. by P. Vashishta, J.N. Mundy and G.K. Shenoy, North-Holland, New York (1979), p. 53
2. B.A. Boukamp and R.A. Huggins, Phys. Lett. A58, 231 (1976)
3. U.V. Alpen, A. Rabenau and G.H. Talat, Appl. Phys. Lett. 30, 621 (1977)
4. B.A. Boukamp and R.A. Huggins, Mater. Res. Bull. 13, 23 (1978)
5. E.D. Wachsman, P. Jayaweera, N. Jiang, D.M. Lowe and B.G. Pound, J. Electrochem. Soc. 144, 233 (1997)
6. E.D. Wachsman, Solid State Ionics 152–153, 657 (2002)
7. S. Visco, Presentation at Meeting of the Materials Research Society, San Francisco, CA (2006)
8. A.N. Dey, Presentation at the Fall Meeting of the Electrochemical Society (1970), Abstract 62
9. E. Peled, J. Electrochem. Soc. 126, 2047 (1979)
10. J.O. Besenhard and H.P. Fritz, J. Electroanal. Chem. 53, 329 (1974)
11. R. Fong, U. von Sacken and J.R. Dahn, J. Electrochem. Soc. 137, 2009 (1990)
12. J.R. Dahn, A.K. Sleigh, H. Shi, B.M. Way, W.J. Weydanz, J.N. Reimers, Q. Zhong and U. von Sacken, "Carbons and Graphites as Substitutes for the Lithium Anode," in *Lithium Batteries*, ed. by G. Pistoia, Elsevier, Amsterdam (1994), p. 1
13. M. Winter and J.O. Besenhard, "Lithiated Carbons," in *Handbook of Battery Materials*, ed. by J.O. Besenhard, Wiley-VCH, Weinheim (1999), p. 383
14. E. Peled, D. Golodnitsky and J. Pencier, "The Anode/Electrolyte Iinterface," in *Handbook of Battery Materials*, ed. by J.O. Besenhard, Wiley-VCH, Weinheim (1999), p. 419
15. M. Winter, K.-C. Moeller and J.O. Besenhard, "Carbonaceous and Graphitic Anodes," in *Lithium Batteries*, ed. by G.-A. Nazri and G. Pistoia, Kluwer, Boston, MA (2004), p. 144
16. M. Nazri, B. Yebka and G.-A. Nazri, "Graphite-Electrolyte Interface in Lithium-Ion Batteries," in *Lithium Batteries*, ed. by G.-A. Nazri and G. Pistoia, Kluwer (2004), p. 195
17. K. Xu, Chem. Rev. 104, 4303 (2004)
18. A.N. Dey, Thin Solid Films 43, 131 (1977)
19. D. Aurbach, M.L. Daroux, P.W. Faguy and E. Yeager, J. Electrochem. Soc. 134, 1611 (1987)
20. D. Aurbach, A. Zaban, A. Schecheter, Y. Ein-Eli, E. Zinigrad and B. Markovsky, J. Electrochem. Soc. 142, 2873 (1995)
21. Z.X. Shu, R.S McMillan and J.J. Murray, J. Electrochem. Soc. 140, 922 (1993)
22. G.H. Wrodnigg, J.O. Besenhard and M. Winter, J. Electrochem. Soc. 146, 470 (1999)
23. B. Simon and J.-P. Boeuve, U.S. Patent 5,626,981 (1997)
24. J. Barker and F. Gao, U.S. Patent 5,712,059 (1998)
25. D. Aurbach, K. Gamolsky, B. Markovsky, Y. Gofer, M. Schmidt and U. Heider, Electrochim. Acta 47, 1423 (2002)
26. C. Wang, H. Nakamura, H. Komatsu, M. Yoshio and H. Yoshitake, J. Power Sources 74, 142 (1998)

27. I.D. Raistrick, J. Poris and R.A. Huggins, "Use of Alkali Nitrate Molten Salts as Electrolytes in Intermediate Temperature Lithium Batteries," in *Proceedings of the 16th Intersociety Energy Conversion Engineering Conference*, Atlanta, GA, American Society of Mechanical Engineers New York (1981), p. 774
28. I.D. Raistrick, J. Poris and R.A. Huggins, "Nitrate Molten Salt Electrolytes for Use in Intermediate Temperature Lithium Cells," in *Proceedings of the Symposium on Lithium Batteries*, ed. by H.V. Venkatasetty, Electrochemical Society, Pennington, NJ (1981), p. 477
29. J. Poris, I.D. Raistrick and R.A. Huggins, "Behavior of Lithium and Positive Electrode Materials in Molten Nitrate Electrolytes," in *Proceedings of the Symposium on Lithium Batteries*, ed. by H.V. Venkatasetty, Electrochemical Society, Pennington, NJ (1981), p. 459
30. R.M. Biefeld and R.T. Johnson, Jr., J. Electrochem. Soc. 126, 1 (1979)

Chapter 17
Experimental Methods to Evaluate the Critical Properties of Electrodes and Electrolytes

17.1 Introduction

Because of the desire to discover and optimize materials and configurations to be used in the components of advanced batteries, it is important to have a set of tools to evaluate their properties. A number of methods have been developed for this purpose, and the most useful and important ones are described and discussed in the following pages.

It is critical when using any such tool to design experiments that isolate the particular property that is to be evaluated. This is sometimes difficult to do, for battery materials and systems often consist of a number of components, with a significant degree of interaction among them.

The interpretation of the experimental results, and their meaning, is also very important, and sometimes difficult. It is often relatively easy to acquire electronic instruments to perform experiments, and software is also readily available to extract parameters from them. However, it is all too easy to improperly interpret what is actually being measured, and what the results mean. The professional literature is full of examples that illustrate this problem.

The material in the following pages is intended to help understand the important phenomena and measurement methods, and their interpretation. It is hoped that this will help to clear up some of the confusion and misunderstanding in this area.

17A Use of DC Methods to Determine the Electronic and Ionic Components of the Conductivity in Mixed Conductors

17A.1 Introduction

Electrical charge can be transported within solids by the net macroscopic motion of either electronic or ionic species. In materials that are of interest for use as solid electrolytes it is important that the charge transport is predominantly related to ionic

R.A. Huggins, *Advanced Batteries*: Materials Science Aspects,

motion. Minority electronic conduction is generally considered to be deleterious. On the other hand, ionic charge transport is generally not desired in materials such as electronic semiconductors, for it can cause changes in the properties of junctions as well as influence the concentrations of traps and other features of the microstructure. Between these two extremes are a number of materials that are mixed ionic and electronic conductors. Such mixed conductors play important roles in a number of different technologies, being employed as oxidation and hydrogenation catalysts, as chemical separation membranes, as reactants or mixed-conducting matrices in the microstructure of battery electrodes, and as fuel cell electrodes.

A number of methods have been developed to evaluate the separate electronic and ionic components of the total charge transport. Several direct current techniques are briefly described in this chapter. Transient and alternating current methods are discussed in Sects. 17A.6 and 17B.5.

17A.2 Transference Numbers of Individual Species

If more than one species can carry charge in a solid or liquid it is often of interest to know the relative conductivities or inversely, the impedances of the individual species. The parameter that is used to describe the contributions of individual species to the transport of charge is the transference number, sometimes called the *Hittorf transference number* [1]. This is defined as the fraction of the total electrical current that is carried by a particular species when an electrical potential difference is imposed upon the adjacent electrodes.

In the simple case wherein electrons and one type of ions can move through such an electrochemical cell, we can define the transference number of ions as t_i, and that of the electrons as t_e, where

$$t_i = i_i/(i_i + i_j) \tag{17A.1}$$

and

$$t_e = i_e/(i_i + i_j) \tag{17A.2}$$

and i_i and i_e are their respective partial currents. It can readily be seen that the sum of the transference numbers of all mobile charge-carrying species is unity. In this case

$$t_i + t_e = 1. \tag{17A.3}$$

Instead of expressing transference numbers in terms of currents, they can also be written in terms of the related impedances, Z_i and Z_e. For the case of a material containing these two mobile species, and the transport of charge by the motion of the ions under the influence of an applied voltage E_{appl},

$$t_i = (E_{appl}/Z_i)/[E_{appl}/Z_i + E_{appl}/Z_e] = (Z_e/Z_i + Z_e)) \tag{17A.4}$$

and likewise for electrons:

$$t_e = Z_i/(Z_i + Z_e). \tag{17A.5}$$

Whereas these parameters are often thought of as properties of the electrolyte or mixed conductor, in actual experiments they can also be influenced by what happens at the interfaces between the electrolyte, or mixed conductor, and the electrodes. They are only properties of the electrolyte or mixed conductor alone if there is no impedance to the transfer of either ions or electrons across the electrolyte/electrode interface or mixed-conductor/electrode interface, or within the electrodes.

The transference number is also not a constant for a given material, but instead is dependent upon the chemical potentials of the constituent components within it. Thus, it is dependent upon the local composition and will not be uniform throughout an electrolyte in a galvanic cell, or battery, in which the electrodes have different values of the chemical potentials.

17A.3 The Tubandt Method

Many years ago Tubandt [2] introduced a DC method to evaluate the ionic fraction of the total electrical conductivity in a solid. This involves the use of electrodes that are not blocking to either ions or electrons, and the imposition of a DC current of known time and magnitude. Changes in the weights of the two electrodes are measured, one getting lighter, and the other heavier. Assuming Faraday's law and knowledge of the atomic weight of the transporting ionic species, this can be used to determine the number of moles of material that were ionically transferred. Comparison with the number of Faradays of charge that passed through the system gives the value of the ionic transfer number, and thus also of the electronic transference number.

17A.4 The DC Assymetric Polarization Method

A different method was introduced by C. Wagner that can be especially useful for the evaluation of low levels of electronic conductivity in materials that are primarily ionic conductors.

The general philosophy behind this type of measurement that is generally called a Hebb–Wagner experiment [3–5] involves the independent measurement of the current carried by minority electronic species under conditions such that ionic transport is prevented. This requires the use of one electrode that is reversible to both ions and electrons, and another that is blocking to the mobile ionic species. It is accomplished by using an electrode that does not contain atoms of the mobile ionic species, and polarizing the cell such that they tend to move away from that electrode into the electrolyte. Since the electrode cannot supply those ions, the electrolyte becomes

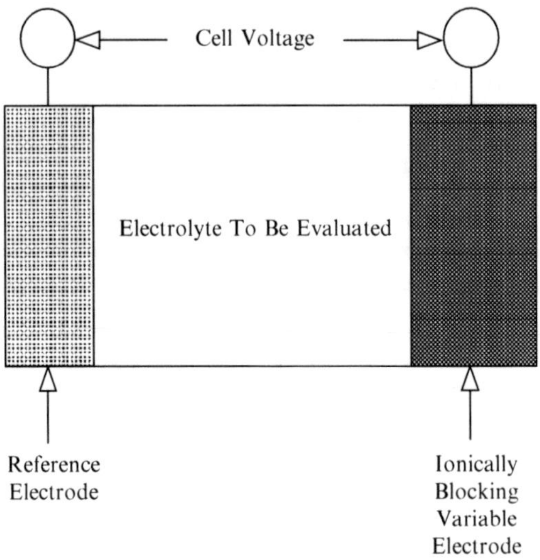

Fig. 17A.1 General configuration of an electrochemical cell used for Hebb–Wagner polarization measurements

locally starved, and ionic transport is prevented. The general configuration of an electrochemical cell used for this type of measurement is shown in Fig. 17A.1.

The polarity must be correct. If the mobile ionic defects are positively charged, this electrode must be made positive relative to the other electrode. If the mobile ionic defects carry negative charges, this ionic defect-starving electrode must be on the negative side of the experimental cell.

As the other electrode is reversible to the mobile ionic species, its composition and electrical potential define the local composition and chemical potentials of the material (electrolyte) that is being measured. The steady state current through the cell is evaluated as a function of the voltage between this reference electrode and the other variable ion-blocking electrode.

17A.5 Interpretation of Hebb–Wagner Asymmetric Polarization Measurements in Terms of a General Defect Equilibrium Diagram

The results obtained from a Hebb–Wagner experiment can be understood in terms of a simple defect equilibrium diagram (DED) for a binary material *MX*, a topic that was discussed in Sect. 13C.4. In this case, it is assumed that the material is predominantly an ionic conductor, so that K_i is much larger than K_e, and that the potential is in the central Region II of the DED. The parameters that are of interest relate to the minority electronic defect species in that compositional range.

As shown earlier, in the central Region II the electronic defect concentrations can be approximated as:

$$[e'] = K_e \left(\frac{K_i^{1/2}}{F} \right) \tag{17A.6}$$

and

$$[h^\circ] = \left(\frac{F}{K_i^{1/2}} \right), \tag{17A.7}$$

where

$$F = K_X a(X_2)^{1/2}. \tag{17A.8}$$

The reference electrode must have a fixed composition, and thus a fixed value of the parameter F. In the illustrative example shown in Fig. 17A.2, it is assumed to have a value of log F of about 17.

The potential of the other electrode, which is blocking to the mobile ionic species, can be varied by the imposition of a voltage across the cell. As an example, it is shown at a position where log F is about 22 in this figure.

As mentioned earlier, the voltage applied across the electrolyte in an electrochemical cell imparts an electrostatic force upon all charged species within it.

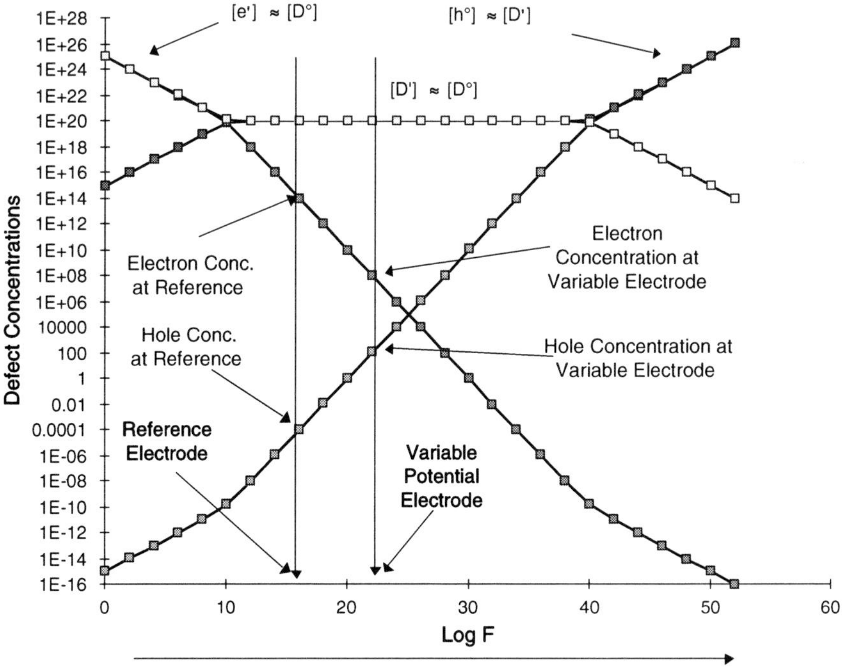

Fig. 17A.2 Illustration of the influence of the electrical potentials of the reference and variable electrodes upon minority defect concentrations using a defect equilibrium diagram

Negatively charged defects, such as excess electrons, e', are repulsed from the negative side, and attracted toward the positive side. The opposite is true for positively charged species.

However, if one of the electrodes blocks the passage of charged ionic species, they can have no net motion through the electrolyte. A gradient in their chemical potential is formed near the blocking electrode that provides an equal, but opposite, chemical force upon them. The result is that there is no electrostatic field in the interior of the electrolyte.

The difference in electric potential at the two electrodes, which has been shown to be proportional to a difference in the logarithm of the parameter F, establishes a gradient in the concentration of the electronic species, which are not blocked by the electrodes. This can readily be seen by observation of the DED.

As the result of their concentration gradients, which will be spatially uniform under steady-state conditions if their mobilities are not concentration-dependent, both negatively charged electrons and positively charged holes will move through the electrolyte by concentration-driven chemical diffusion. Their partial currents will be given by

$$I_{e'} = -\frac{\Delta[e']}{\Delta x} D_{e'} z_{e'} qA \tag{17A.9}$$

and

$$I_{h^\circ} = -\frac{\Delta[h^\circ]}{\Delta x} D_{h^\circ} z_h qA, \tag{17A.10}$$

where $D_{e'}$ and D_{h° are their respective chemical diffusion coefficients, Δx is the distance between the electrodes, and A is the cross-sectional area.

Now, it is possible to calculate the difference between the defect concentrations at the two electrodes. By combining (17A.6) and (17A.8) the relation between the concentration of electrons and the activity of the X_2 species is

$$[e'] = \frac{K_e K_i^{1/2}}{K_X} \left(\frac{1}{a(X_2)^{1/2}} \right). \tag{17A.11}$$

The Nernst equation can be used to relate the ratio of the activities of any species at the two electrodes to the voltage across an electrochemical cell. If the activity at one of the electrodes is assumed to be at a standard state, and thus defined to be equal to unity, this can be written for any species j at electrode 2 as

$$E_2 - E^\circ = -\frac{kT}{z_j q} \ln a(j)_2 \tag{17A.12}$$

because the logarithm of the activity at the standard state is 0. Since all potentials are relative, rather than absolute, E^{O} can be taken as 0. If (17A.12) is rearranged and the exponential taken of both sides,

$$a(j)_2 = \exp\left[-E_2 \left(\frac{z_j q}{kT} \right) \right]. \tag{17A.13}$$

Putting this into (17A.11) and introducing the parameter

$$W = \left(\frac{q}{kT}\right) \tag{17A.14}$$

and recognizing that z for X_2 is –2, the relation between the concentration of the electrons and the potential E_2 becomes

$$[e']_2 = \left(\frac{K_e K_i^{1/2}}{K_X}\right) \exp[-WE_2]. \tag{17A.15}$$

This can also be done to get the electron concentration at the other electrode at potential E_1

$$[e']_1 = \left(\frac{K_e K_i^{1/2}}{K_X}\right) \exp[-WE_1]. \tag{17A.16}$$

In both cases, the electron concentration decreases as the electrode potential becomes more positive. The difference in the electron concentrations at the two electrodes is simply:

$$[e']_2 - [e']_1 = \left(\frac{K_e K_i^{1/2}}{K_X}\right) (\exp[-WE_2] - \exp[-WE_1]). \tag{17A.17}$$

Because this is a difference, the standard state to which of the electrodes is compared is not important. For the situation in which electrode 1 is a reference electrode with unit activity of the species X_2, E_1 can be set equal to 0. In this special case, which is the one discussed by Wagner [5, 6], the difference in electron concentrations simplifies to become

$$[e']_2 - [e']_1 = \left(\frac{K_e K_i^{1/2}}{K_X}\right) (\exp[-WE_2] - 1). \tag{17A.18}$$

The form of this function is such that if the potential of the variable electrode becomes positive with respective to the reference potential, the difference in the electron concentration is negative, and approaches a plateau value

$$([e']_2 - [e']_1) \rightarrow -\left(\frac{K_e K_i^{1/2}}{K_X}\right). \tag{17A.19}$$

On the other hand, when the potential of the variable electrode is negative with respect to the reference potential the difference in electron concentrations is positive and increases exponentially as the cell voltage is increased.

The influence of the potential upon the concentration of holes in the central region of the defect equilibrium diagram can be calculated in a similar manner. In that case, introduction of (17A.8) into (17A.7) gives

$$[h^\circ] = \frac{K_X}{K_i^{1/2}} a(X_2)^{1/2}. \tag{17A.20}$$

Proceeding as before in the case of electrons,

$$[h^\circ]_2 = \left(\frac{K_X}{K_i^{1/2}}\right) \exp(WE_2). \tag{17A.21}$$

Thus,

$$[h^\circ]_2 - [h^\circ]_1 = \left(\frac{K_X}{K_i^{1/2}}\right) (\exp[WE_2] - \exp[WE_1]) \tag{17A.22}$$

and for the special case that electrode 1 is a reference electrode with unit activity of species X_2, E_1 can be set equal to 0. Then

$$[h^\circ]_2 - [h^\circ]_1 = \left(\frac{K_X}{K_i^{1/2}}\right) (\exp[WE_2] - 1. \tag{17A.23}$$

The form of this equation is such that the hole concentration increases exponentially as the potential of the variable electrode becomes more positive, and approaches a negative plateau as it becomes more negative.

$$([h^\circ]_2 - [h^\circ]_1) \rightarrow -\left(\frac{K_X}{K_i^{1/2}}\right) \tag{17A.24}$$

This is just the opposite of the influence of the potential upon the concentration of electrons.

The total minority electronic species current I_t that would flow through a material that is predominantly an ionic conductor in the potential range characteristic of Region II of its defect equilibrium diagram can now be calculated.

From (17A.9) and (17A.10) and introducing the charge numbers of electrons and holes,

$$I_t = \left(\frac{qA}{\Delta x}\right) (D_{e'}\Delta[e'] - D_{h^\circ}\Delta[h^\circ]). \tag{17A.25}$$

Using (17A.17) and (17A.22), the general relation can be written as

$$I_t = \left(\frac{qA}{\Delta x}\right) \left[D_{e'}\left(\left(\frac{K_e K_i^{1/2}}{K_X}\right) (\exp[-WE_2] - \exp[-WE_1])\right)\right.$$
$$\left. -D_{h^\circ}\left(\left(\frac{K_X}{K_i^{1/2}}\right) (\exp[WE_2] - \exp[WE_1])\right)\right] \tag{17A.26}$$

and for the special case discussed by Wagner, where E_1 is defined as 0, this becomes

$$I_t = \left(\frac{qA}{\Delta x}\right)[A(\exp[-WE_2]-1)+B(1-\exp[WE_2])], \tag{17A.27}$$

where

$$A = D_{e'}\left(\frac{K_e K_i^{1/2}}{K_X}\right) \tag{17A.28}$$

and

$$B = D_{h^\circ}\left(\frac{K_X}{K_i^{1/2}}\right). \tag{17A.29}$$

An alternate standard state can be considered; the situation when the current due to the transport of electrons is equal to that due to the transport of holes. Then, from (17A.27)

$$A(\exp[-WE_2]-1 = B(1-\exp[WE_2]). \tag{17A.30}$$

If it can be assumed that $\exp(WE_2) \gg 1$, this simplifies to

$$A/B = \exp(WE_2), \tag{17A.31}$$

which can be rewritten to give the electrical potential for this reference standard state as

$$E^0 = 2.3\log\left[\left(\frac{D_{e'}}{D_{h^\circ}}\right)\left(\frac{K_e K_i}{K_{X^2}}\right)\right]. \tag{17A.32}$$

Often, the contribution of either electronic conduction or hole conduction will dominate, and the dominant species can be identified from the shape of the I/E curve. However, it has been shown that the relative magnitudes of the electron and hole currents, and thus the shape of the I/E curve, which is the result of the transport of both electrons and holes, are all dependent upon the potential of the reference electrode [6]. This can make the separate evaluation of these factors difficult in some cases.

This feature is illustrated in Figs. 17A.3–17A.5, in which the potential of the reference electrode is varied. It is readily evident that the shape of the data is highly dependent upon both the polarity of the electrodes and the potential of the reference electrode. These factors are often neglected when such experiments are undertaken.

One could use a variable-composition mixed conductor as the reference electrode, and control its potential by the use of a simple double electrochemical cell configuration such as that shown schematically in Fig. 17A.6. In this way the partial electronic and hole conductivities can be varied over a relatively wide range in order to clearly separate them.

Another way to separate these two factors was presented by Patterson [7], who showed that (17A.6) can be rearranged to give

$$I_{total}/[\exp(u)-1] = [(RTA)/(LF)][\sigma_n\exp(-u)+\sigma_p]. \tag{17A.33}$$

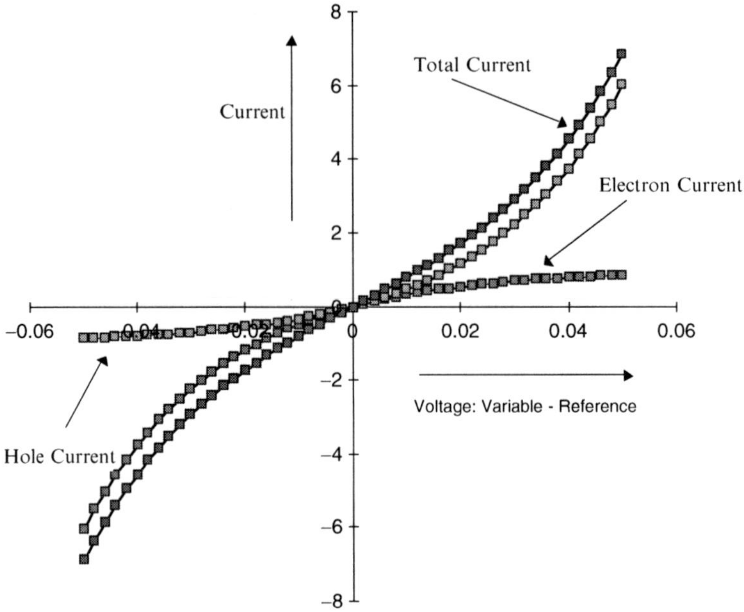

Fig. 17A.3 Dependence of the electron, hole, and total currents in a Hebb–Wagner experiment upon the relative potentials of the reference and variable electrodes when the reference electrode potential is where the electron and hole concentrations are equal

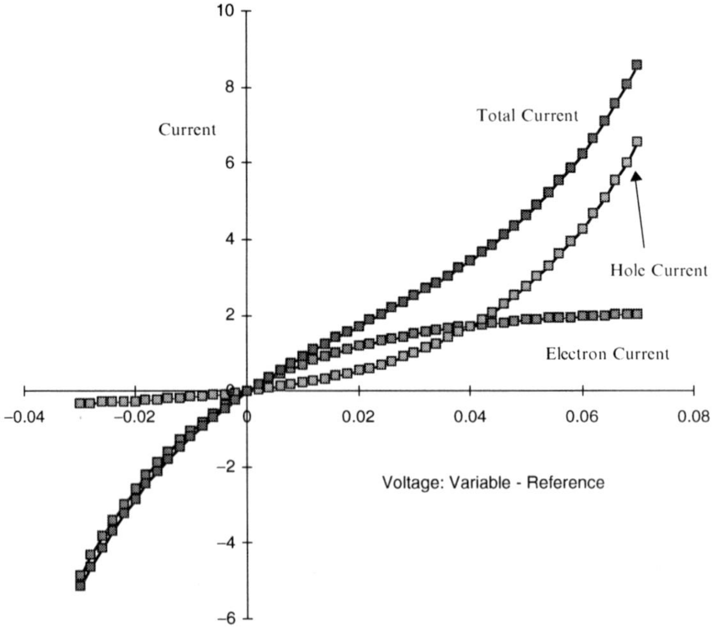

Fig. 17A.4 Example of the dependence of the electron, hole, and total currents upon the relative potentials of the reference and variable electrodes when the reference electrode potential is 20 mV negative of where the electron and hole concentrations are equal

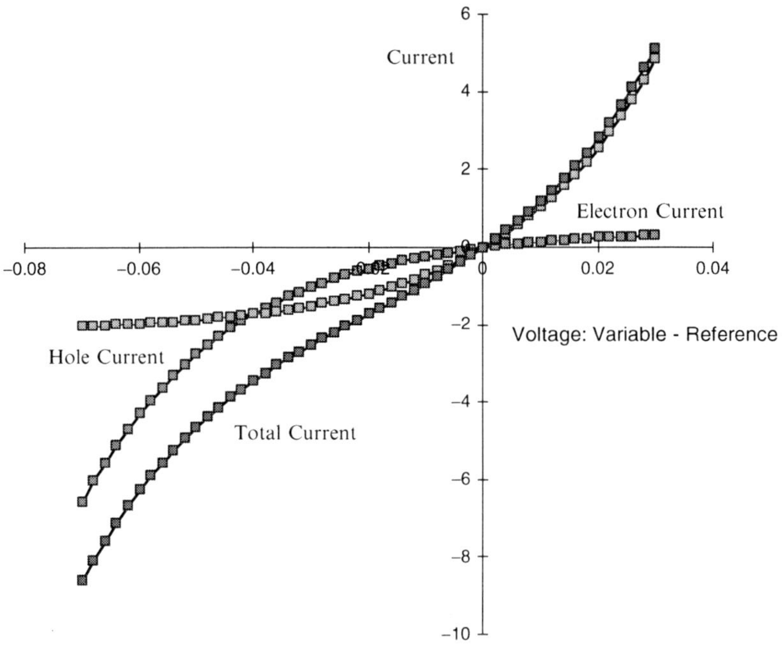

Fig. 17A.5 Example of the dependence of the electron, hole, and total currents upon the relative potentials of the reference and variable electrodes when the reference electrode potential is 20 mV positive of where the electron and hole concentrations are equal

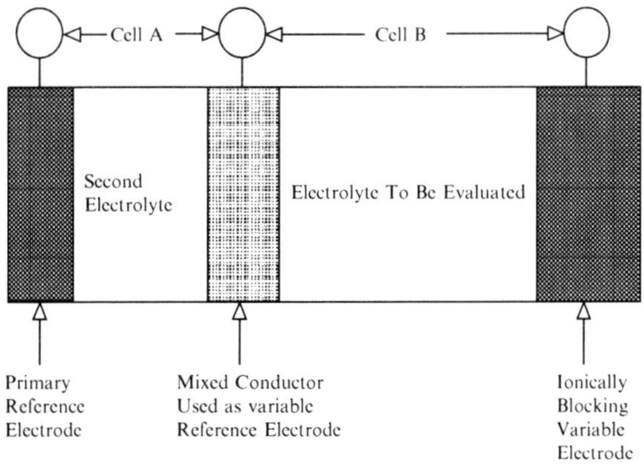

Fig. 17A.6 Double-cell arrangement that can be used to vary the potential of the reference electrode in a Hebb–Wagner-type experiment

The electron and hole conductivities can be simply determined by plotting the experimental results as $I_{total}/[\exp(u) - 1]$ vs. $\exp(u)$. The value of σ_e is calculated from the slope, and σ_h from the intercept.

Two examples of the use of this method are shown in Figs. 17A.7 and 17A.8, from [7].

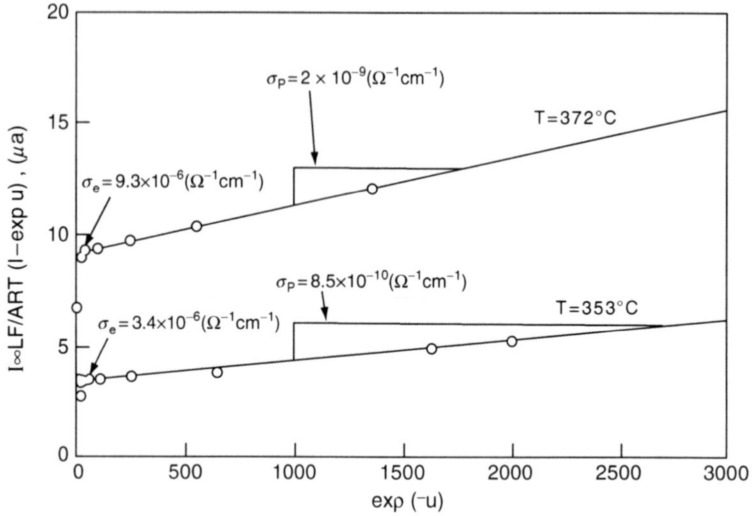

Fig. 17A.7 Replotted data for AgBr at 353 and 372°C

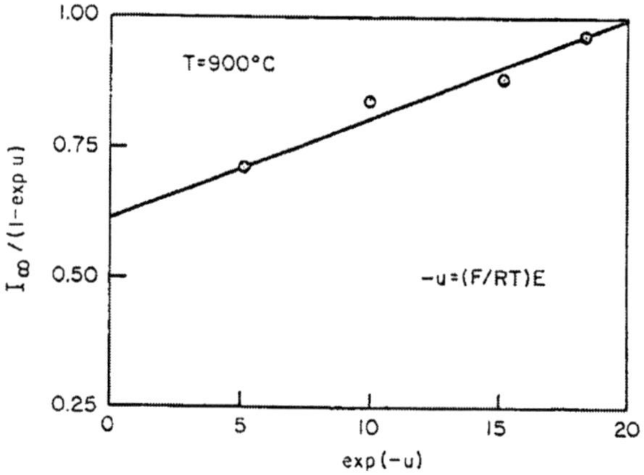

Fig. 17A.8 Measurements of the minority electronic conduction in yttrium-doped thorium oxide

In order to produce meaningful results by use of this DC asymmetric polarization procedure, careful consideration must be given to several additional factors.

1. The voltage that is placed across the electrochemical cell must not exceed the decomposition voltage of the electrolyte that is being evaluated. If this is not the case, electrolytic decomposition of the electrolyte will produce an additional ionic current.
2. The potentials of both the reference electrode and the ionically blocking variable electrode must remain within the chemical potential range in which ionic defect species are dominant. What happens when this is not the case has been discussed in [8].

3. As the partial conductivities of the different species vary with the potential, and thus the composition of the material being measured, the results that will be obtained depend upon the compositional range over which the measurement is made.

There are two parts to this matter. One is the composition, and therefore the potential, of the reference electrode. The other is the range of composition investigated by variation of the potential difference between the variable potential electrode and the reference electrode, as well as the polarity of the cell.

4. It is necessary to wait until steady state is obtained. Since placing a voltage across the cell results in the imposition of a gradient in the chemical potentials of the components within it, there must be a corresponding rearrangement of the defect species concentrations. A substantial time may be required before this new equilibrium state is attained in some cases.

17A.6 DC Open Circuit Potential Method

A voltage is expected if a difference in the chemical potential of the electroactive species is imposed across an electrolyte under open circuit conditions such that neither chemical species nor electronic current can be transported externally from one electrode to the other. Under those conditions, the measured voltage E_{out} is equal to the thermodynamic voltage E_{th} given by

$$E_{out} = E_{th} = -\left({\Delta G_r}^{\circ} / zq \right), \qquad (17A.34)$$

where ΔG_r° is the change in Gibbs free energy of the virtual reaction, z is the charge number of the mobile species, and q is the magnitude of the elemental charge. If, on the other hand, there is electronic leakage through the electrolyte, the measured voltage is given by

$$E_{out} = t_i E_{th} = -\left(t_i \Delta G_r^{\circ} / zq \right), \qquad (17A.35)$$

where t_i is the ionic transference number [9, 10]. This method requires that well-defined and known thermodynamic conditions are maintained at each of the two electrodes, which can sometimes be difficult to achieve.

It has been shown [11] that this result is also readily obtained by the analysis of the electrical response of the simple equivalent electrical circuit presented in Chap. 1 that represents the properties of an electrochemical cell with different potentials at the two electrodes, as is the case with batteries and fuel cells. This is shown in Fig. 17A.9, where E_{th} is the thermodynamic voltage, I_{out} the output current, Z_i the internal impedance to the transport of ionic species, and Z_e the impedance related to internal electronic leakage through the electrolyte. Under open circuit conditions I_{out} is 0. Impedances are used here instead of resistances, as they are often frequency, or time, dependent.

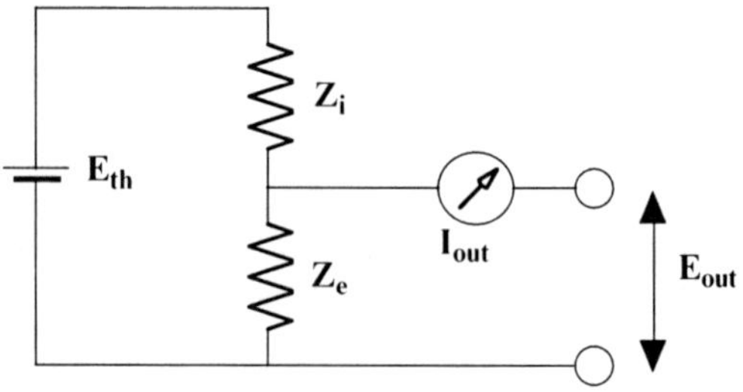

Fig. 17A.9 General equivalent circuit of battery or fuel cell

From this simple circuit it can be seen that

$$E_{out} = [Z_e/(Z_i + Z_e)] = t_i E_{th} \tag{17A.36}$$

or

$$E_{out} = E_{th}(1 - t_e). \tag{17A.37}$$

17B Experimental Determination of the Critical Properties of Potential Electrode Materials

17B.1 Introduction

In order to determine the response of electrodes to different requirements one needs to have basic information about the relevant parameters of the electrochemically active materials involved. The most important properties of electrode materials are as follows:

1. The potential, and its dependence upon the state of charge
2. The capacity, i.e., the maximum amount of charge that can be stored or supplied
3. The kinetic behavior under various conditions

The actual behavior of electrodes may deviate from the properties of their electrochemically active components, of course. This may be due to nonuniform reaction under dynamic conditions or microstructural inaccessibility of the active material due to the presence of blocking constituents, for example. It is also possible that charge transfer reactions at the electrolyte/electrode interface may play a role at high charge–discharge rates, especially if the electroactive species is complexed in the electrolyte. Thus, information about the basic properties of the electroactive components themselves represents the limiting case. Actual performance may be less attractive.

Several electrochemical methods have been developed by which information about insertion-reaction electroactive components can be obtained. These include the following:

1. The galvanostatic intermittent titration method (GITT) [12]
2. The potentiostatic intermittent titration method (PITT) [13]
3. The Faradaic intermittent titration method (FITT) [13]
4. The alternating current (Wechselstrom) intermittent titration method (WITT) [14]

In each case, stepwise measurement of the electrochemical titration curve is accompanied by an evaluation of the kinetic behavior after each step. Thus, one can simultaneously obtain both thermodynamic and kinetic information as a function of electrode composition (the extent of reaction).

The kinetic measurement parts of these four methods have been discussed in detail elsewhere [13, 15], and can be represented schematically as shown in Fig. 17B.1 [15], in which the controlled (independent) variables and the type of response of the relevant dependent variables are shown.

The design of these different methods and the analysis of the data obtained are both based upon the analytical solution of the diffusion equations under various relevant initial and boundary conditions.

The most important parameter relating to the transient behavior of a solid material that operates by an insertion reaction mechanism is the chemical diffusion coefficient. This quantity can be much greater than the self, or component, diffusion coefficient that is often evaluated by the use of radiotracer or nuclear magnetic measurement techniques because of the enhancement factor W. This factor results from the influence of a concentration gradient upon the interaction between the different particle fluxes involved in the mass transport process in the solid. This will not be discussed here, as it can readily be found elsewhere [12, 16]. The important point

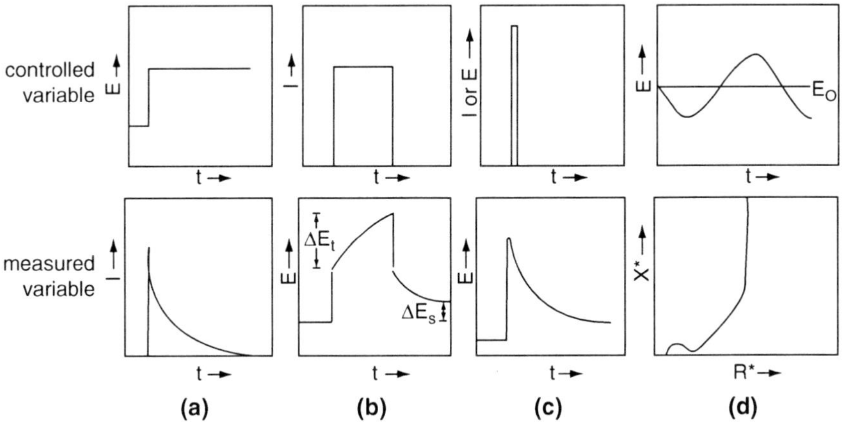

Fig. 17B.1 Schematic of four experimental methods that can be used to determine the chemical diffusion coefficient in a mixed-conducting electrode: (**a**) PITT, potential jump, (**b**) GITT, current pulse, (**c**) FITT, relaxation after Faradaic pulse, and (**d**) WITT, steady state ac measurement

is, however, that W can be very large in some mixed conductors, so that one has to be careful about using data obtained by the use of experimental methods that do not employ concentration gradients. In one example, it was shown to reach values of about 70,000 [12].

17B.2 The GITT Method

In the GITT method, the time dependence of the potential after a current step can be expressed as

$$\frac{dE}{dt^{1/2}} = \left[\frac{2IV_m}{zFSD^{1/2}p^{1/2}}\right]\left[\frac{dE}{dy}\right] \text{ if } t \ll (x^2/D). \tag{17B.1}$$

Here, V_m is the molar volume, z the charge carried by the electroactive species, F the Faraday constant, y the stoichiometric parameter, and S the surface area.

Alternatively, D may be evaluated from an experiment in which a current step of finite length is applied, using the simplified expression

$$D = \left[\frac{4x^2}{pt}\right]\left[\frac{\Delta E_s}{\Delta E_t}\right]^2 \text{ if } t \ll (x^2/D), \tag{17B.2}$$

where t is the time duration of the current step, ΔE_s is the change in the equilibrium open circuit voltage resulting from the current pulse, and ΔE_t is the total transient voltage change, after eliminating the IR drop.

For a fixed value of dE/dy, the dependence of E on t will be linear at long times and the diffusion coefficient can be evaluated from

$$E - E(t=0) = \frac{IV_m}{zFS}\left[\frac{x}{3D}\right]\left[\frac{dE}{dy}\right] \text{ if } t \gg (x^2/D). \tag{17B.3}$$

Experimental data obtained by the use of this method are shown in Fig. 17B.2 [15].

17B.3 The PITT Method

In the case of the PITT method, the current varies with time after the imposition of a potential step. At short times, the time dependence of the current can be approximated by

$$I = \frac{QD^{1/2}}{2xp^{1/2}}[t^{-1/2}] \text{ if } t \ll x^2/D, \tag{17B.4}$$

where Q is the total charge passed following the potential step, and x is the distance parameter into the solid.

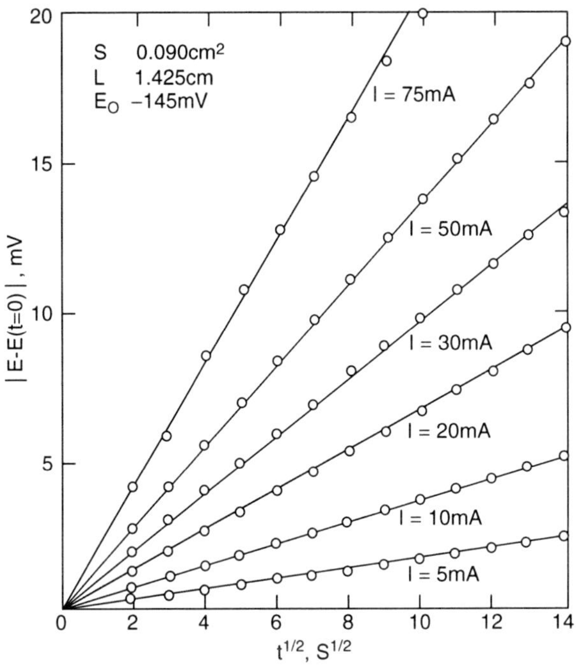

Fig. 17B.2 Experimental measurement of the time dependence of the potential of an insertion reaction electrode upon the application of a current step. The GITT method

At long times it becomes

$$I = \frac{2QD}{x^2}\exp\left[\frac{p^2Dt}{4x^2}\right] \text{ if } t \gg (x^2/D). \tag{17B.5}$$

Experimental data obtained on the same material by the use of this technique are shown in Figs. 17B.3 and 17B.4 [15].

17B.4 The FITT Method

In the FITT method, a given amount of electroactive solute material is deposited onto, or into, the surface of the electrode material. As it diffuses into (or out of) the underlying bulk material under open circuit conditions, the surface composition will gradually approach the value characteristic of a uniform solute distribution. The time dependence of the potential is thus

$$\frac{dE}{d(1/t^{1/2})} = \frac{QV_m}{zFSD^{1/2}p^{1/2}}\left[\frac{dE}{dy}\right] \text{ if } t \ll (x^2/D). \tag{17B.6}$$

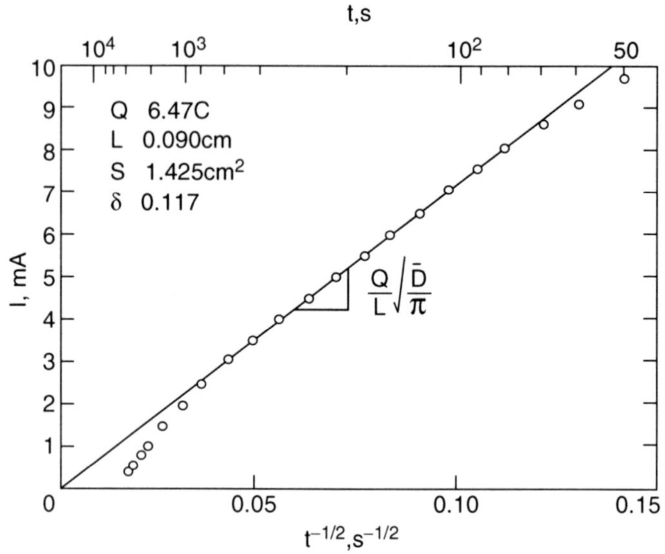

Fig. 17B.3 Experimental measurement of the time dependence of the current into an insertion reaction electrode following a potential step, short-time behavior. The PITT method

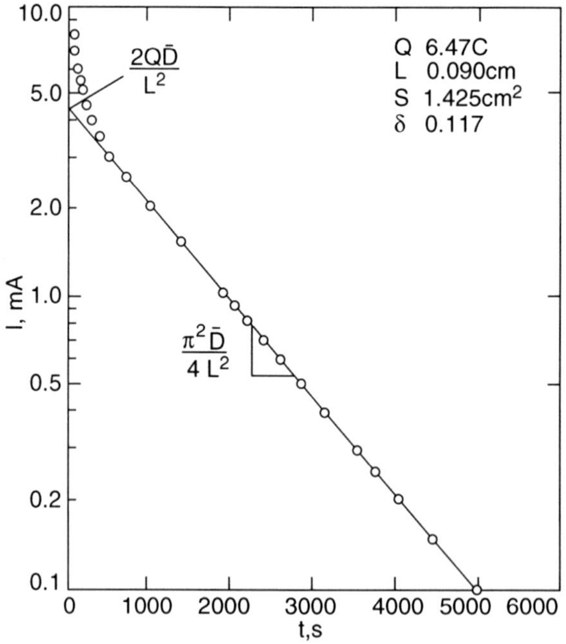

Fig. 17B.4 Time dependence of the current after a potential step, long-time behavior. The PITT method

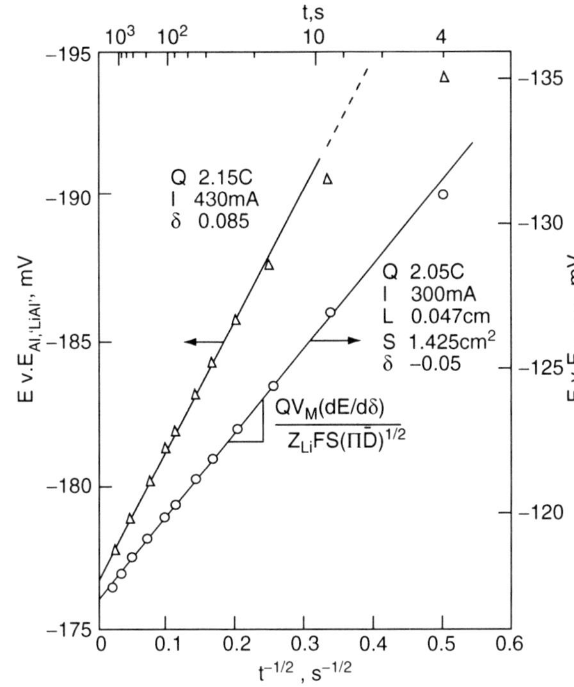

Fig. 17B.5 Time dependence of the potential of an insertion reaction electrode during relaxation under open circuit conditions after the deposition of a fixed amount of solute. The FITT method

Experimental data from this type of experiment, again on the same material, are shown in Fig. 17B.5 [15].

17B.5 The WITT Method

The WITT method involves the use of small signal alternating current methods to evaluate diffusion in a solid. The origin of the label WITT comes from the German word for alternating current, *Wechselstrom*. Without going into the details of the method, which can be found elsewhere [14, 15], the type of result that is obtained is illustrated in Fig. 17B.6 [14].

Under ideal conditions and one-dimensional diffusion into a semi-infinite solid, the response is dominated by the diffusional admittance at low frequencies. In the case of thin films, the range of frequency in which diffusion dominates the observed behavior can be limited. At even lower frequencies, below a critical value given by

$$\omega = 2D/x^2, \tag{17B.7}$$

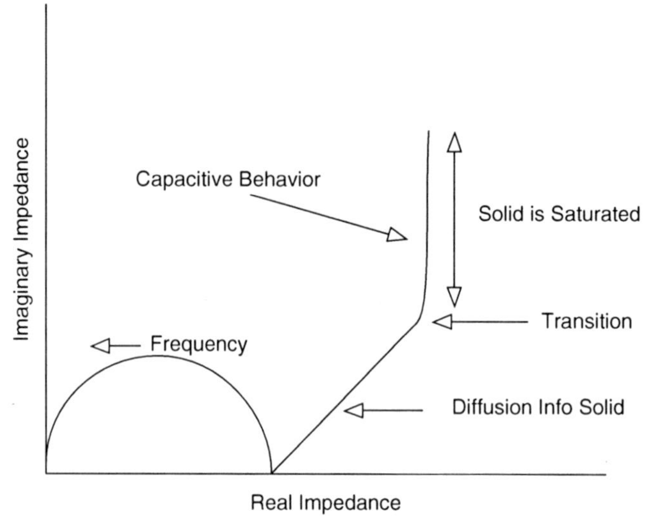

Fig. 17B.6 Schematic of the frequency dependence of the complex impedance of an insertion reaction electrode. The WITT method

where D is the chemical diffusion coefficient, and x the dimensional parameter, there is sufficient time for the solid composition to stay saturated at the value related to the time-varying imposed potential. When that is the case, the sample electrode has the characteristics of a capacitor, for its composition varies with the applied potential.

The analytical solution to this problem has been obtained for two important cases, one-dimensional diffusion in a slab whose thickness is large compared with the depth of penetration of the diffusion profile, i.e., $x^2 \gg D/\omega$, and one-dimensional diffusion into a thin layer of material under conditions that compositional saturation is achieved, i.e., $x^2 \ll D/\omega$.

In the first (thick layer) case, the phase difference between the current and the potential is independent of the frequency, and is equal to 45°. The impedance related to diffusion of the electroactive species is then

$$|Z| = \left| \frac{V_m}{zFSD^{1/2}} \frac{dE}{dy} (\omega^{-1/2}) \right|. \tag{17B.8}$$

In the thin layer case the current is 90° out of phase with the potential, and is independent of the diffusion coefficient. The real part of the complex impedance (R) is

$$R = \frac{|Z|\omega x^2}{3DS} \tag{17B.9}$$

and the imaginary part (X) is

$$X = |Z| = \left| \frac{V_m}{zF\omega xS} \frac{dE}{dy} \right|. \tag{17B.10}$$

The limiting low-frequency capacitance C_l and resistance R_l are given by

$$C_l = \omega X = \omega |Z| \tag{17B.11}$$

and

$$R_l = \left| \frac{V_m}{zFS} \frac{1}{3D} \frac{dE}{dy} \right|. \tag{17B.12}$$

The experimental variation of the real and imaginary components of the impedance in the phase LiAl in the intermediate frequency range within which the impedance is dominated by the solid-state diffusion process is shown in Fig. 17B.7 [15].

The good correspondence obtained by the use of these four different methods is shown in Fig. 17B.8 [15], in which the composition dependence of the measured values of the chemical diffusion coefficient in the phase LiAl is plotted. Also shown is the measurement reported by L'vov et al. [17].

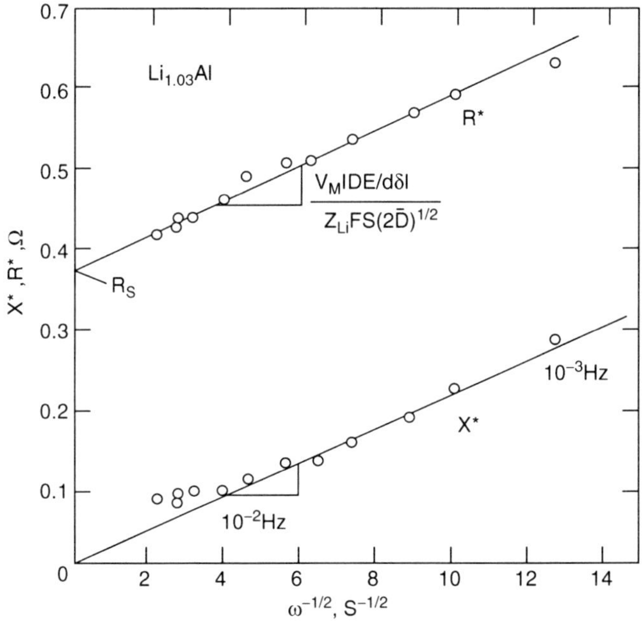

Fig. 17B.7 Experimental data showing the frequency dependence of the real and imaginary components of the impedance of an insertion reaction electrode in the intermediate frequency range. The WITT method

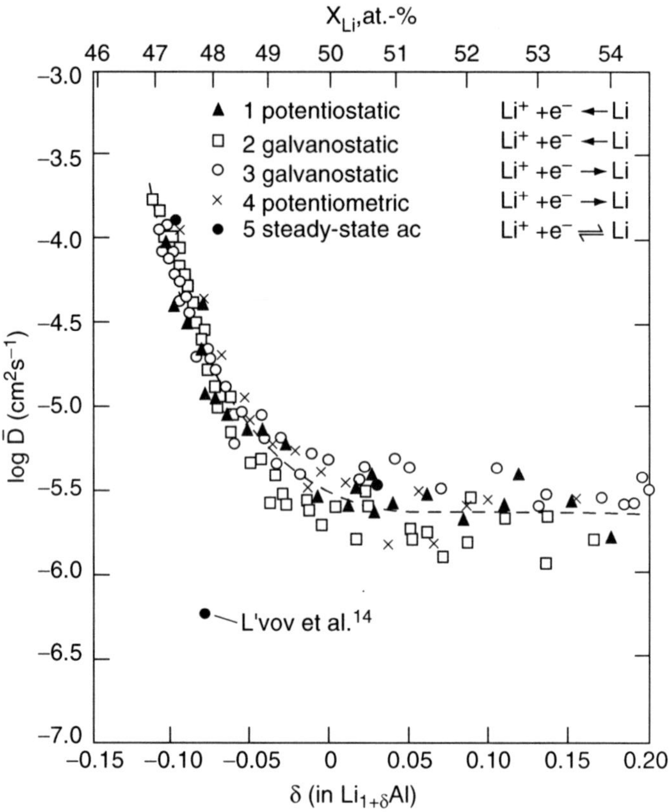

Fig. 17B.8 Composition dependence of the chemical diffusion coefficient in the phase LiAl, evaluated by the use of four different methods

17C Use of AC Methods to Determine the Electronic and Ionic Components of the Conductivity in Solid Electrolytes and Mixed Conductors

17C.1 Introduction

Variable frequency AC methods are now often used to evaluate the charge transport properties of potential solid electrolytes. They involve the use of inert electronically conducting electrodes on two sides of the solid to be investigated. The experimental arrangement generally looks like that shown schematically in Fig. 17C.1.

It is generally assumed that the electrical current within the solid phase between the electrodes is carried by ionic species, but that they are blocked at the electrolyte/electrode interface, whereas the transport of electrons is not. Both theoretical and experimental considerations have shown that such an interface between

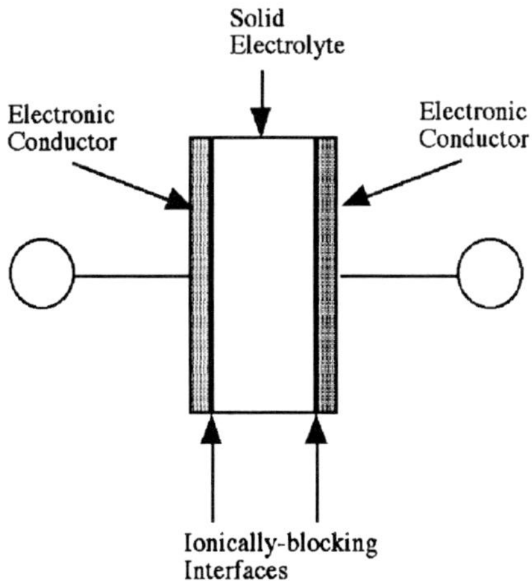

Fig. 17C.1 Simple experimental arrangement generally used for variable frequency AC measurements to evaluate ionic transport in a solid electrolyte

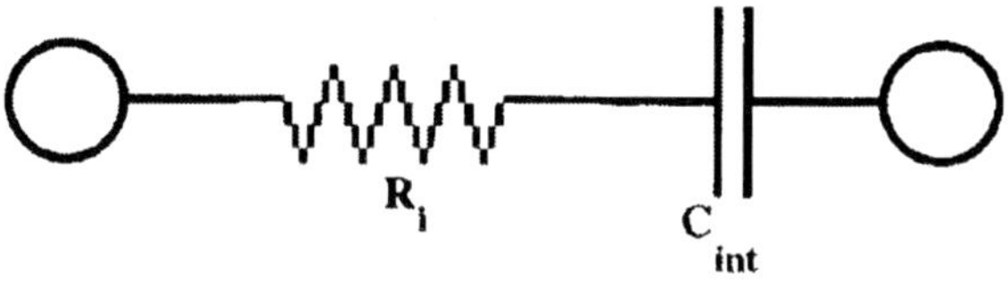

Fig. 17C.2 Simple equivalent circuit for the low-frequency behavior of an electrochemical cell with an ionic conductor between two ionically blocking, but electronically conducting, electrodes

a solid electrolyte, considered to be a purely ionic conductor, and a purely electronic conductor can be simply modeled as a parallel plate capacitor. The excess ionic charge on one side of the ionically blocking interface is balanced by excess electronic charge in the adjacent metal electrode.

Since charge passes from one electrolyte/electrode interface to the other by passing through the electrolyte, which has a finite resistance, R_i, this physical configuration can be represented by a simple electrical equivalent circuit shown in Fig. 17C.2 at low frequencies. The capacitive properties of the two electrolyte/electrode interfaces are combined into a single capacitance, C_{int}.

At higher frequencies, however, the experimental configuration shown in Fig. 17C.2 begins to exhibit the effects of an additional geometrical capacitance due to the presence of a material (the electrolyte) with a finite dielectric constant between the two parallel metallic electrodes.

This parallel plate capacitance due to this geometric arrangement, C_{geom}, acts across the whole configuration, and typically has very small values. Thus, the equivalent circuit becomes that shown in Fig. 17C.3, which is often called the *Debye circuit.*

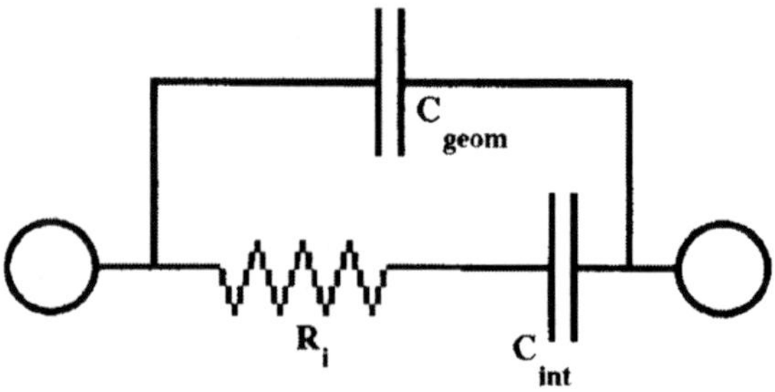

Fig. 17C.3 *Debye equivalent circuit* corresponding to the physical configuration of a solid electrolyte between two electronically conducting, but ionically blocking, electrodes

This equivalent circuit has the desirable characteristic that it also corresponds closely to the physical arrangement. Typical values of C_{geom} and C_{int} are about 10^{-12} and 10^{-6} $\mathrm{F\,cm^{-2}}$, respectively.

The experimental results are typically expressed in terms of a plot of the imaginary and real parts of the impedance, or the imaginary and real parts of the admittance, against each other on the complex plane. Since this topic may be unfamiliar to many readers, and one often sees misinterpretation of experimental data in the literature, a brief introductory explanation is included here.

17C.2 Representation of the Properties of Simple Circuit Elements on the Complex Impedance Plane

The impedance of a resistance Z_R is simply its resistance R, which is in phase with an applied AC voltage signal. One the other hand, the impedance of a capacitance Z_C is out of phase with the applied AC signal, and is called the reactance X, where

$$X = -j/\omega C \tag{17C.1}$$

and

$$\omega = 2\pi f. \tag{17C.2}$$

The symbols ω and f are the angular frequency and the simple AC frequency, respectively. The representation of these simple elements on the complex impedance plane, sometimes called an *Argand diagram*, is illustrated in Figs. 17C.4 and 17C.5, in which the negative value of the imaginary part of the impedance, the reactance, is plotted vs. the real part, the resistance over a range of frequency.

If there is a resistance in series with a capacitance the impedance of this combination Z_{RC} is given by

$$Z_{RC} = R + jX = R - j/\omega C \tag{17C.3}$$

and thus varies with the measurement frequency. The complex impedance plot will appear as shown in Fig. 17C.6.

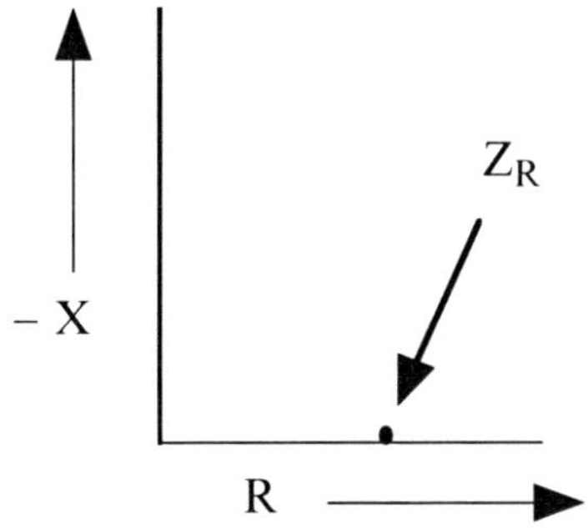

Fig. 17C.4 Impedance of a simple resistance on the complex impedance plane. The value of Z_R is independent of frequency

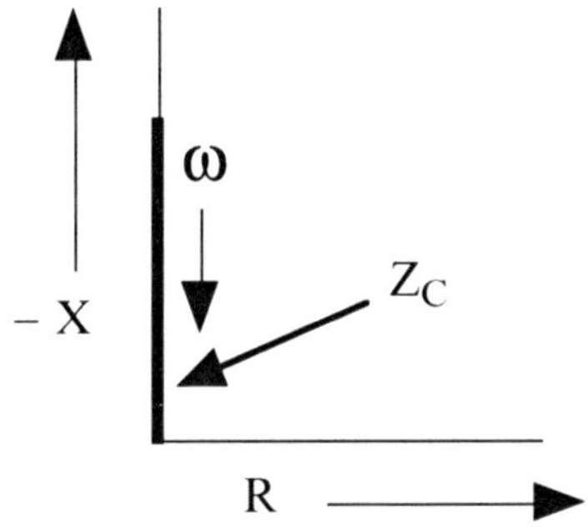

Fig. 17C.5 Impedance of a simple capacitance on the complex impedance plane. The value of the reactance Z_C of the capacitor varies with the frequency. Its absolute value approaches 0 as the frequency becomes larger. In the absence of a series resistance, its values lie along the abscissa

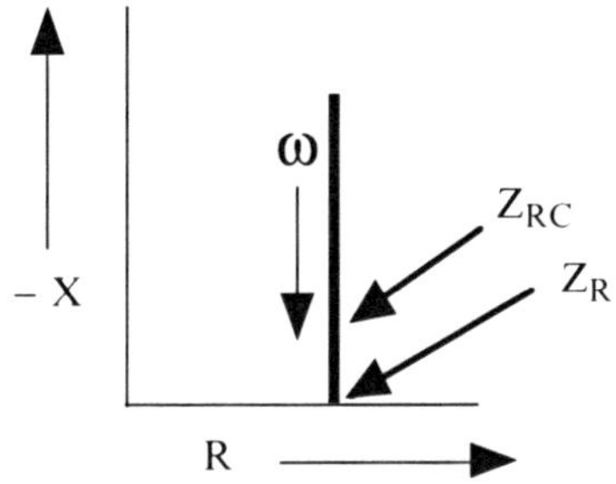

Fig. 17C.6 Appearance of the impedance of a series combination of a resistance and a capacitance upon the complex impedance plane

On the other hand, the impedance of a parallel configuration of a resistance and a capacitance, $Z_{R\text{-}C}$, has the shape of a semicircle on the complex impedance plane, as shown in Fig. 17C.7.

One can readily understand this behavior qualitatively. At very low frequencies the impedance of the capacitor is very large. Thus, essentially all the current flows

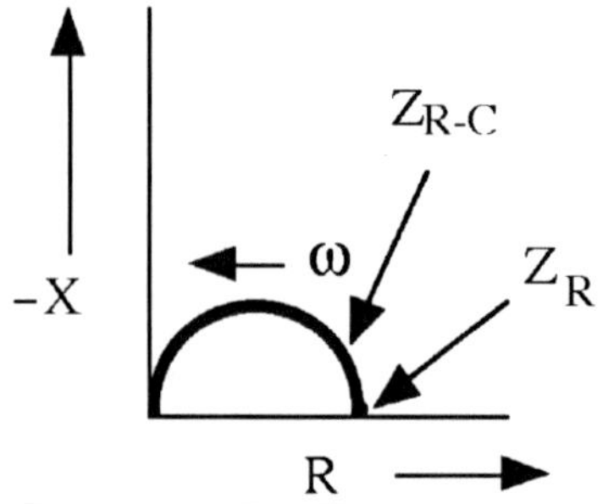

Fig. 17C.7 Appearance of a parallel combination of a resistance and a capacitance upon the complex plane. The direction in which the impedance varies with increasing frequency is indicated by the arrow

through the resistor and its properties dominate the behavior. At very high frequencies the impedance of the capacitor becomes very small, so that it effectively shorts out the resistor and the total impedance tends toward 0. The two impedances are equal, and equal current flows through the two legs, when the resistive component is one half of Z_R. This is the center of the circle, so that it is also the value of the reactance of the capacitor. Thus,

$$Z_{R-C} = Z_R/2 = -j/\omega C \tag{17C.4}$$

and therefore

$$\omega RC = 1 \tag{17C.5}$$

at the angular frequency corresponding to the top of the semicircle. It is possible to determine the value of the capacitance from this relation and the experimental data.

The interpretation of physical processes by the use of alternating frequency methods and the interpretation of the results in terms of equivalent circuits is not always straightforward. In general, more than one equivalent circuit can be devised that will describe the experimental results. Figure 17C.8 shows two circuits that have the same form of response at all frequencies. One is essentially a parallel arrangement of series units, and the other is a series arrangement of parallel units. The values of the component magnitudes are not the same in the two cases, however. As an example, it is easy to see that R_1 is equal to $r_1 + r_2$.

It is most useful to use a circuit that most directly reflects the physical arrangement and processes in the experiment. For instance, parallel networks would obviously not be appropriate for two physical processes that are known to occur in sequence or in series. Nor would a series configuration be appropriate for physical processes that are known to occur in parallel.

In the simple case of measurements on a solid electrolyte using an experimental configuration such as that shown in Fig. 17C.1, assuming that there is no significant electronic conductivity in the solid electrolyte phase and that the ionic species are completely blocked at the electrolyte/electrode interfaces, the data can be readily

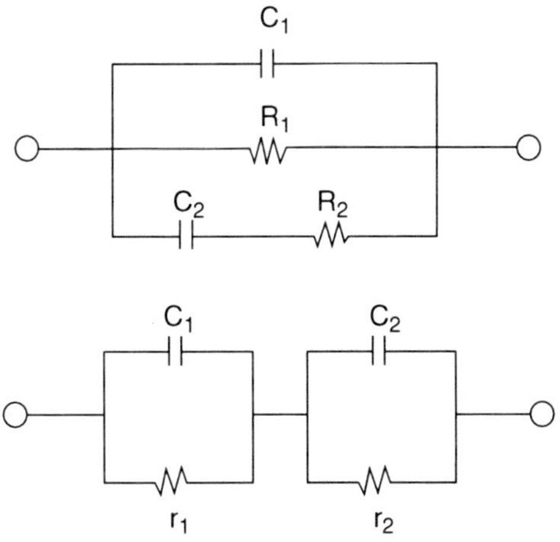

Fig. 17C.8 An example of two equivalent circuits that, by modification of the component values, will have the same frequency-dependent impedance

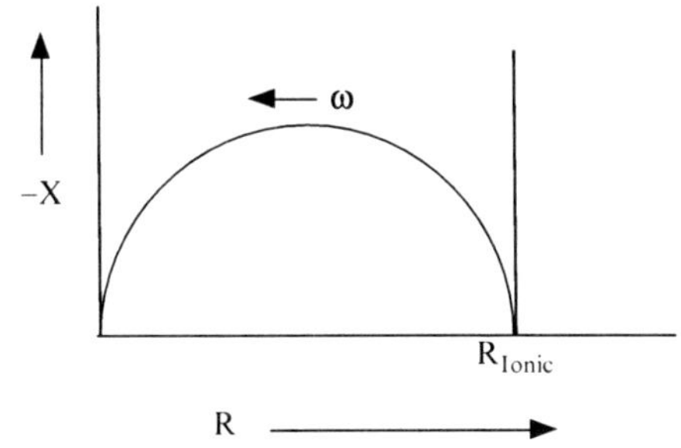

Fig. 17C.9 Schematic complex plane plot for the experimental arrangement shown in Fig. 17C.1 that corresponds to Debye equivalent circuit shown in Fig. 17C.3

interpreted by use of the standard Debye circuit discussed earlier and shown in Fig. 17C.3. The corresponding complex plane plot is shown in Fig. 17C.9. If the temperature is varied, the time constants of the various processes change and the experimentally observable portion of the total frequency response (the frequency window) shifts.

This is a combination of the behavior illustrated in Figs. 17C.6 and 17C.7. At very low frequencies no current flows through the geometric capacitance C_{geom}, because of its very small value, typically of the order of picofarads, which gives it a very large impedance. Thus it is not visible in the measurements. Therefore, the data show a *low-frequency tail* characteristic of the series R_i-C_{int} configuration. At higher frequencies more and more current moves through C_{geom}, so the data appear as though the equivalent circuit becomes a parallel configuration of C_{geom} and R_i.

The figure gets distorted, and the impedance does not go all the way down to the real axis at intermediate frequencies if the values of C_{geom} and C_{int} are not

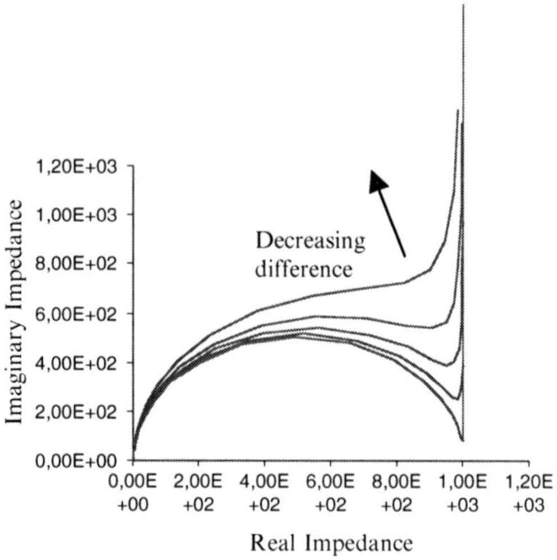

Fig. 17C.10 Typical changes in the complex plane figure if the values of the two capacitances are not sufficiently different from each other

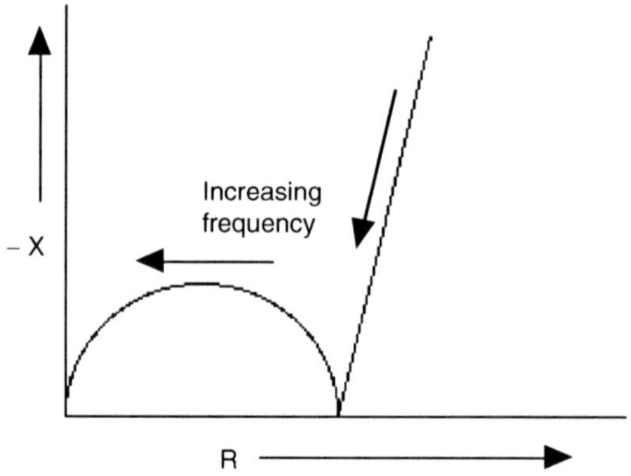

Fig. 17C.11 Complex impedance plane plot showing tipped straight line at low frequencies

sufficiently different from each other. This is shown in Fig. 17C.10. It generally results from the use of electrodes that only poorly contact the solid electrolyte, leading to a reduced value of C_{int}. The values of C_{int} are typically 15–40 μF cm^{-2} [18] for an interface with complete contact between the electrolyte and the electronically conducting electrode. Poor contact can lead to much smaller values.

A common experimental observation is that the low-frequency tail in complex impedance plots is not truly vertical, but is a straight line inclined at a finite angle from the vertical, at lower frequencies and/or higher temperatures, as illustrated schematically in Fig. 17C.11. An early example of experimental data showing this behavior is presented in Fig. 17C.12 [19]. In the case of the corresponding figure on

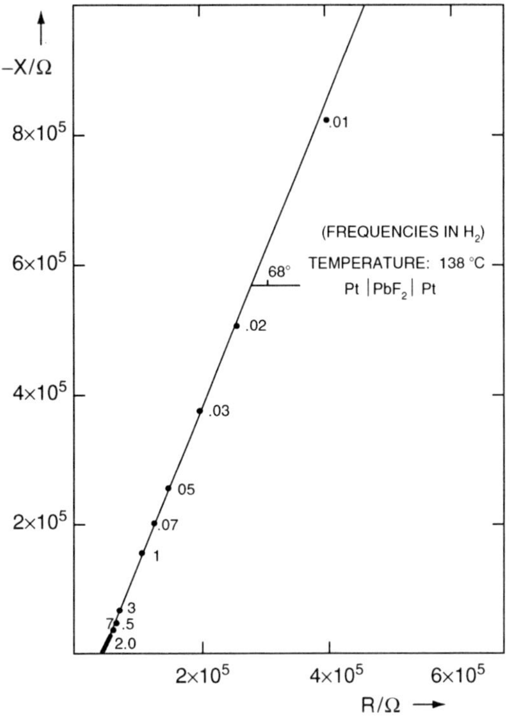

Fig. 17C.12 Low-frequency portion of the experimental data measured on β-pbf$_2$ at 138°C

the complex admittance plane the center of the semicircular portion of the figure is lowered below the real axis.

It was shown that this deviation from ideal *Debye circuit* behavior relates to processes that occur at the solid electrolyte/electrode interface [20], and that the electrical response in that range of frequency can be represented by an impedance of the form

$$Z = A\omega^{-\alpha} - jB\omega^{-\alpha} \tag{17C.6}$$

that acts in series with the bulk ionic impedance.

This is different from the behavior of a completely blocking electrode interface that is assumed to act like a frequency-independent electrical double-layer capacitance, so that the low-frequency tail should be vertical, as shown in Fig. 17C.9.

If the electroactive species is not blocked at the interface, but diffuses into or out of one or both of the electrodes as a neutral reaction product, according to Fick's second law the low-frequency data will fall onto a straight line with a slope of 45° and so the value of α in that case will be $^1/_2$ if the interface is planar. This involves the introduction of a so-called "Warburg admittance", or "impedance", into the equivalent circuit. This has been treated fully elsewhere [14], and since it is not directly related to the subject of this chapter, it will not be further discussed here.

17C.3 The Influence of Electronic Leakage Through an Ionic Conductor

If the material between the electrodes is a mixed conductor, with both ionic and electronic contributions to charge transfer, the equivalent circuit must be modified to include an electronic current path in parallel with the ionic current path. This is shown in Fig. 17C.13.

If the electronic resistance is much lower than the ionic resistance the corresponding complex plane plot is shown in Fig. 17C.14. Because there is no blockage of the charge transport due to electronic species, there is no interface capacitance, and thus no capacitive tail at very low frequencies. Also, the electronic current will be much larger than the ionic current, even at frequencies that are sufficiently high that the interfacial capacitance is shorted out.

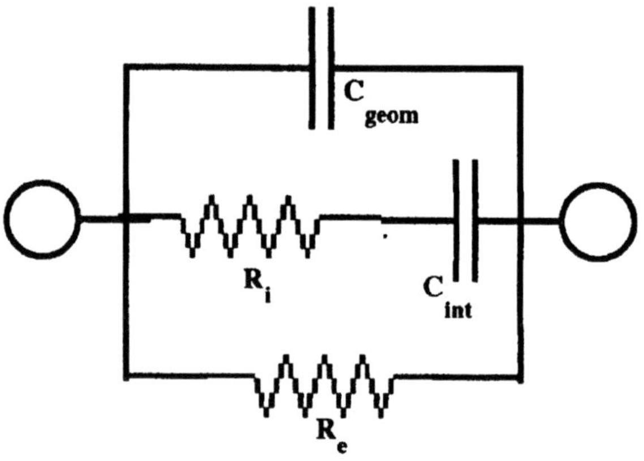

Fig. 17C.13 Equivalent circuit for a Debye system with electronic current, as well as ionic current, through the electrolyte

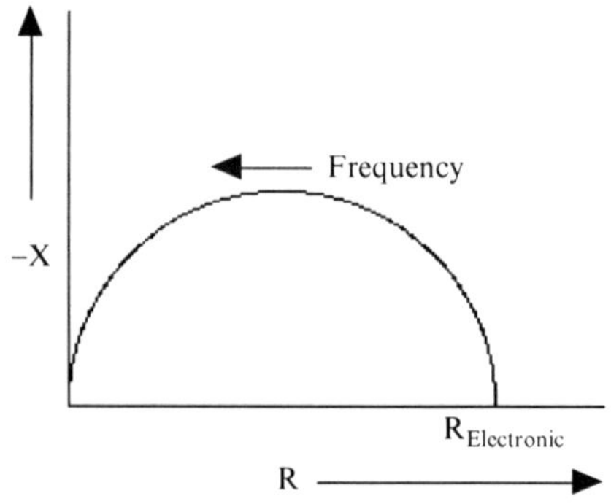

Fig. 17C.14 Complex impedance plane figure for the case of an ionic conductor with dominant electronic leakage

The semicircle in this figure is only due to the interaction of the electronic resistance and the geometrical capacitance. It has nothing to do with the ionic resistance or the interfacial capacitance. Because the geometrical capacitance is so small, if the electronic resistance is also small, the frequency range of the data in the semicircle will be extremely high. This can be seen by evaluation of the frequency at the top of the semicircle, where

$$\omega R_e C_{geom} = 1. \tag{17C.7}$$

As a result, experiments on materials that are predominantly electronic conductors often do not show the semicircular part of the overall data, but only a single point at the low-frequency end of the figure.

17C.4 Case in Which Both Ionic and Electronic Transport Are Significant

If there is a substantial amount of ionic transport, as well as an electronic shunt at low frequencies, the complex plane plot will look like that shown in Fig. 17C.15. There are two semicircles, representing the parallel combination of the electronic resistance and the geometrical capacitance at low frequencies, and the parallel combination of the ionic resistance and the geometrical capacitance at the higher frequencies. There is no low-frequency tail, as the interfacial capacitance is shunted by the electronic current at low frequencies, as discussed earlier.

There are two intercepts on the resistance axis, designated as R_1 and R_2. The one at low frequencies, R_2, is the same as the DC value, and thus is merely the electronic resistance, R_e. At higher frequencies the impedance due to the relatively large interfacial capacitance becomes insignificant and is not observed in the measurements. The equivalent circuit of Fig. 17C.13 then becomes a parallel arrangement with

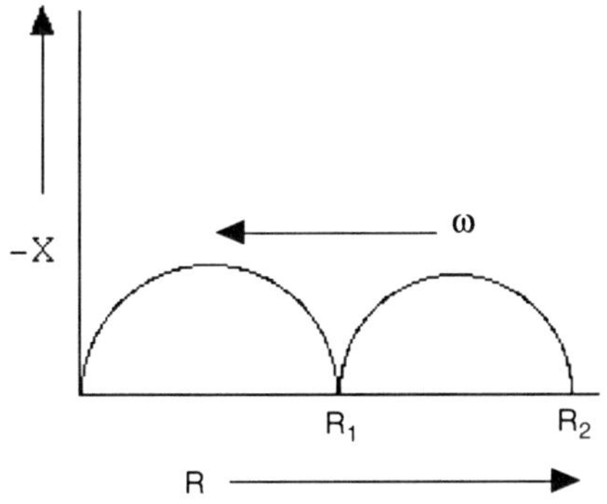

Fig. 17C.15 Complex impedance plane plot for the case that both ionic and electronic transports are significant

three legs, the geometric capacitance and the two resistances, which are in a parallel configuration. This leads to the second semicircle in the impedance plane. The resulting intercept on the resistance axis at the higher frequency, R_1, is a combination of the two parallel resistances, R_i and R_e. Thus

$$1/R_1 = 1/R_i + 1/R_e, \tag{17C.8}$$

which can be rewritten to give

$$R_1 = (R_e R_i)/(R_e + R_i). \tag{17C.9}$$

The ionic and electronic fractions of the total conductance, and thus of the conductivity, can be found.

The ionic conductance G_i is

$$G_i = 1/R_i = 1/R_1 - 1/R_e = (R_2 - R_1)/(R_1 R_2), \tag{17C.10}$$

and the electronic conductance G_e is

$$G_e = 1/R_e = 1/R_2. \tag{17C.11}$$

The total conductance G_{total} is

$$G_{total} = G_i + G_e = (R_2 - R_1)/(R_1 R_2) + 1/R_2 = 1/R_1. \tag{17C.12}$$

The transference numbers of the ionic and electronic species are equal to their respective partial conductances:

$$t_i = G_i/G_{total} = (R_2 - R_1)/R_2 \tag{17C.13}$$

and

$$t_e = G_e/G_{total} = R_1/R_2. \tag{17C.14}$$

Thus, it can be seen that the transference numbers of the ionic and electronic species can be obtained by simple evaluation of the relative values of the two intercepts on the real axis in a complex impedance plane plot.

If the ionic transference number is relatively large, and the electronic transference number is relatively small, as is desired for useful electrolytes, the right hand, lower frequency, semicircle is considerably larger than the higher frequency semicircle. In the extreme case of very little electronic conduction, R_1 approaches 0. On the other hand, if the electronic conduction is dominant, the low-frequency semicircle becomes quite small, and R_1 approaches R_2. There is no low-frequency capacitive tail on the impedance plot in either case.

17C.5 Influence of an Additional Impedance Due to Transverse Internal Interfaces

Grain boundaries and interphase boundaries in polycrystalline materials can have a significant influence upon their observable transport properties. In the case that transport along such internal interfaces is faster than that through the bulk crystals in the microstructure, the presence of such features oriented in the direction of the charge flux can result in an enhancement of the related species transport. Transverse interfaces of this type will have little effect. On the other hand, if species transport in such boundaries is slower than in the crystalline portions of the microstructure, transport along them will have little effect. However, high-impedance grain boundaries or interphase boundaries that are oriented transverse to the charge flux direction can act to impede transport processes, and can introduce additional resistive and capacitive effects into the measurements.

If ionic transport through the solid is impeded by the presence of internal interfaces, e.g., grain boundaries, across which species must travel, the additional ionic impedance must act in series with the bulk ionic resistance. This partial blocking of the ionic flux will also typically have an associated parallel local capacitance similar to the capacitive effect at an ionically blocking exterior electrode. This additional feature causes the Debye equivalent circuit shown in Fig. 17C.3 to be modified to become that shown in Fig. 17C.16.

This circuit thus has two parallel R/C configurations, so that there should be two semicircles on the complex impedance plane figure. It will look like the one shown in Fig. 17C.15, except that there is a low-frequency tail. This is seen in Fig. 17C.17.

However, in this case the meanings of R_1 and R_2 are different. At low frequencies the impedance of Cg is very large, and the behavior is determined by the series arrangement of R_i, R_{gb}, and C_{int}. This produces the tail, and the value of R_2 is thus $(R_i + R_{gb})$.

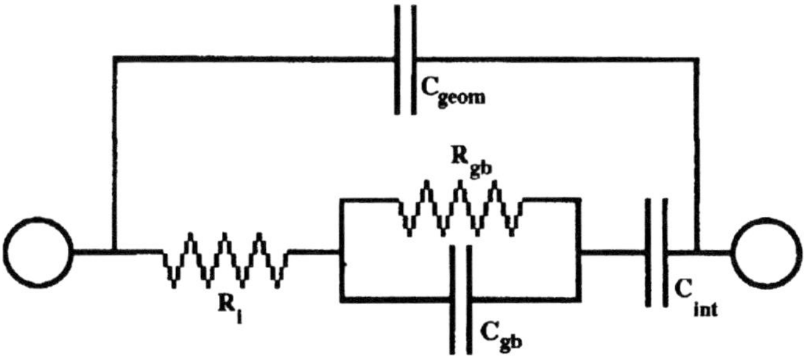

Fig. 17C.16 Equivalent circuit containing a parallel R/C combination, R_{gb} and C_{gb}, in series with the ionic resistance due to the presence of transverse internal interfaces, e.g., grain boundaries, that impede ionic transport

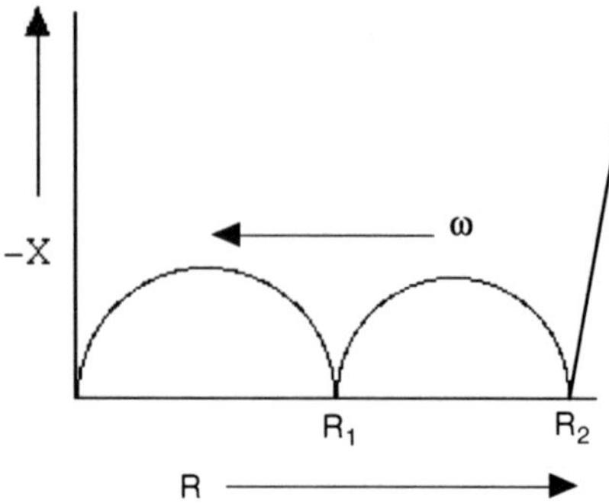

Fig. 17C.17 Schematic complex impedance plane figure corresponding to the equivalent circuit shown in Fig. 17C.16

At high frequencies the impedance of the grain boundary capacitance approaches 0, so that it shorts out the grain boundary resistance. Thus, the circuit becomes the same as the Debye circuit, and the behavior is dominated by the parallel arrangement of R_i and C_{geom}.

Thus, when there is an internal interface impedance in addition to the bulk ionic impedance and no appreciable electronic conduction the value of R_1 is that of R_i and the value of R_2 becomes that of $R_i + R_{gb}$.

17C.6 Behavior When There Is Internal Transverse Interface Impedance as well as Partial Electronic Conduction

If a material has these same general properties, ionic transport through the bulk crystallites and some transverse interface impedance, and also some electronic conductivity, one must make a further modification of the equivalent circuit. This simply involves putting an electronic resistance in parallel with the ionic conduction path. The result is shown in Fig. 17C.18.

In this case, as well as any other that has significant electronic charge transport, there will be no low-frequency tail in the complex impedance plane figure. Instead, there will be a semicircle at very low frequencies due to the parallel action of R_e and C_g. But there will also be two additional semicircles, as there are two additional parallel R/C combinations that can contribute to the experimental observations, R_{gb} and C_{gb}, and R_i and C_{geom}.

The result is that there will, in principle, be three semicircles in the complex impedance plane figure, as illustrated schematically in Fig. 17C.19. The two at the higher frequencies reflect the influence of the ionic transport part of the system.

The values of the intercepts on the real impedance axis can now be identified. The one at the lowest frequency, R_3, is equal to the electronic shunt resistance R_e. The one at the next higher frequency range is due to the parallel combination of R_e

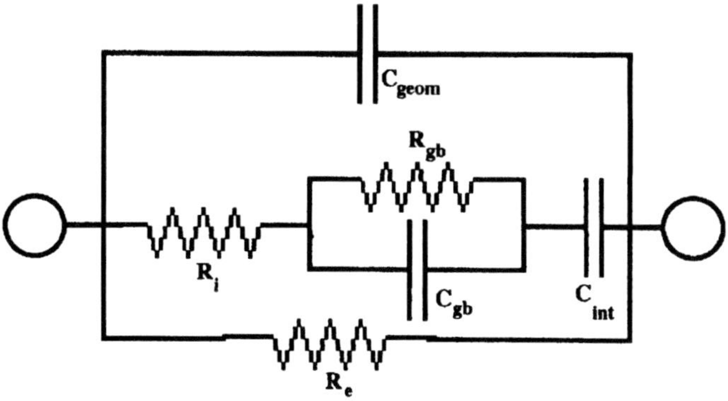

Fig. 17C.18 Schematic equivalent circuit representing the same physical processes as that shown in Fig. 17C.16, but also including the effect of electronic leakage

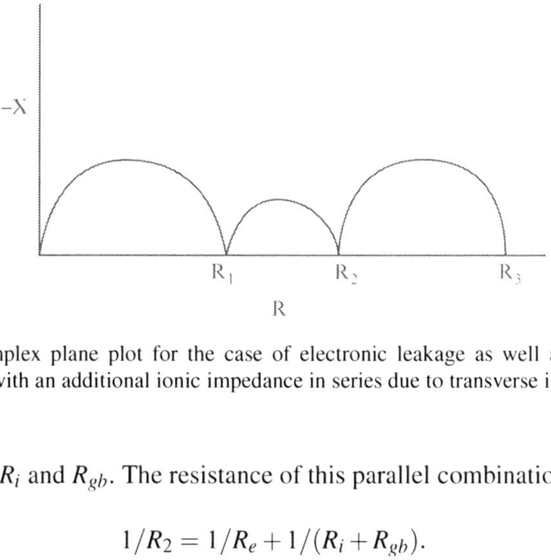

Fig. 17C.19 Complex plane plot for the case of electronic leakage as well as ionic transport through the bulk with an additional ionic impedance in series due to transverse internal interfaces

and the sum of R_i and R_{gb}. The resistance of this parallel combination can be found from

$$1/R_2 = 1/R_e + 1/(R_i + R_{gb}). \tag{17C.15}$$

Thus R_2 is given by

$$R_2 = \frac{R_e(R_i + R_{gb})}{R_e + R_i + R_{gb}}, \tag{17C.16}$$

and the one at the highest frequencies is due to the parallel combination of R_i and R_e, as R_{gb} is shunted out by the grain boundary capacitance C_{gb}. As before, R_1 can be found to be

$$R_1 = (R_e R_i)/(R_e + R_i). \tag{17C.17}$$

The transference numbers can now be found by first calculating the conductances, as was done earlier for the simpler case.

The ionic conductance is

$$G_i = 1/R_{ionic} = 1/(R_i + R_{gb}). \tag{17C.18}$$

The electronic conductance is

$$G_e = 1/R_e = 1/R_2 - 1/(R_i + R_{gb}). \tag{17C.19}$$

And the total conductance is

$$G_{total} = G_i + G_e = 1/(R_i + R_{gb}) + 1/R_e = 1/R_2. \tag{17C.20}$$

The ionic transference number can be found from

$$t_i = G_i/G_{total} = R_2/(R_i + R_{gb}) = R_2(1/R_2 - 1/R_e) = (R_3 - R_2)/R_3. \tag{17C.21}$$

And the electronic transference number can be found from

$$t_e = G_e/G_{total} = (R_i + R_{gb})/(R_e + R_i + R_{gb}) = R_2/R_3. \tag{17C.22}$$

Thus, it can be seen that the presence of an electronic leak through the electrolyte produces important changes in experimentally observed complex impedance plots obtained from small-signal AC measurements using ionically blocking electrodes. This electronic shunt leads not only to a difference in the shape of the complex plane figures, with no low-frequency tail, but also necessitates different interpretations of the data. The relationships between the values of the intercepts of the data with the real impedance axis and the specific resistances within the sample for several cases are given in Tables 17C.1 and 17C.2.

The values of the ionic and electronic transference numbers in the presence of transverse internal interfaces in addition to electronic leakage are analogous to those for the simpler case without the additional internal ionic impedance. In both cases

Table 17C.1 Interpretation of the intercepts on the real axis

Type of conduction	R_1	R_2	R_3
Ionic conduction only	R_i		
Electronic conduction only	R_e		
Ionic with internal interface impedance	R_i	$R_i + R_{gb}$	
Ionic with electronic shunt	$R_eR_i/(R_e + R_i)$	R_e	
Ionic + interface impedance, elect. shunt	$R_eR_i/(R_e + R_i)$	$(R_e)(R_i + R_{gb})/$ $(R_e + R_i + R_{gb})$	R_e

Table 17C.2 Evaluation of the different resistances

Type of Conduction	R_e	R_i	R_{gb}
Ionic conduction only		R_1	
Electronic conduction only	R_1		
Ionic with internal interface impedance		R_1	$R_2 - R_1$
Ionic with electronic shunt	R_2	$R_2R_1/(R_2 - R_1)$	
Ionic + interface impedance, elect. shunt	R_3	$R_3R_1/(R_3 - R_1)$	

Table 17C.3 Transference numbers

Type of conduction	Ionic transference number	Electronic transference number
Ionic conduction only	1.0	0.0
Electronic conduction only	0.0	1.0
Ionic with electronic shunt	$(R_2 - R_1)/R_2$	R_1/R_2
Ionic + interface impedance, electronic shunt	$(R_3 - R_2)/R_3$	R_2/R_3

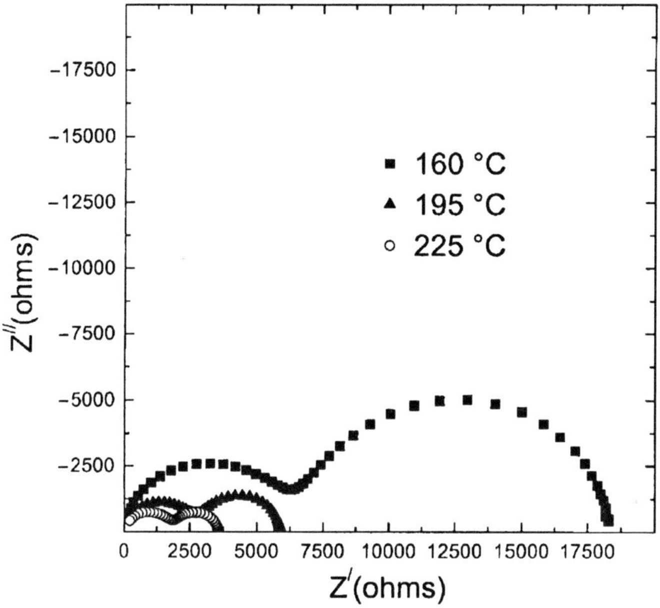

Fig. 17C.20 Plot of the frequency dependence of the complex impedance of $La_{0.4}Pr_{0.4}Sr_{0.2}In_{0.8}Mg_{0.2}O_{2.8}$ at several temperatures

the respective transference numbers can be obtained merely by observation of the intercepts of the data on the real impedance axis. These results are summarized in Table 17C.3.

17C.7 An Example

Experimental results on an indium-containing analog of the well-known perovskite solid electrolyte LSGM, $La_{0.4}Pr_{0.4}Sr_{0.2}In_{0.8}Mg_{0.2}O_{2.8}$, at three different temperatures are shown in Fig. 17C.20, and the resultant data are shown in Table 17C.4 [21].

Table 17C.4 Results of experiments on $La_{0.4}Pr_{0.4}Sr_{0.2}In_{0.8}Mg_{0.2}O_{2.8}$

Temp.(°C)	R_1	R_2	R_i	R_e	t_i	t_e
160	6,000	18,000	9,000	18,000	0.67	0.33
195	2,500	5,900	4,338	5,900	0.58	0.42
225	1,875	3,750	3,750	3,750	0.50	0.50

17C.8 Summary

The impedance spectroscopy method can be used to evaluate a number of parameters related to battery materials, both ionic conductors and mixed conductors, even when there is not a lot of difference between the ionic and electronic impedances. This general method does not require high ionic conductivity and does not involve passing a significant amount of material through the sample. Nor does it require the imposition of different thermodynamic conditions upon the two sides of the sample and waiting for slow equilibration of species within it. The whole sample can be surrounded by the same atmosphere, so data can be obtained as a function of the thermodynamic environment.

One can easily recognize the presence of a significant amount of electronic leakage through the sample by the absence of a capacitive tail in complex impedance plots at low frequencies. When that is the case, evaluation of the relative values of the intercepts upon the real impedance axis provides information about the ionic and electronic resistances, as well as their respective transference numbers.

When substantial electronic conduction is present, the meanings of these intercept values are distinctly different from that when the charge transport is only ionic. This has been shown for the situation in which there is no appreciable impedance due to the presence of transverse internal interfaces, and also for the case in which internal, partially blocking, transverse interfaces are present.

In contrast to resistance measurements, and conductivity values that result from them, transference number values are relative, and thus independent of sample dimensions. Therefore, it is possible to obtain useful transference number information using this AC method even on samples containing significant amounts of porosity.

On the other hand, it is important to recognize the assumption that the electrodes are completely blocking to ionic species when desiring to evaluate the electronic conduction. If this is not the case, the apparent value of the transference number for the electronic species will be too high.

References

1. W. Hittorf, Ann. Phys. 106, 543 (1859)
2. C. Tubandt, Z. Electrochem. 26, 338 (1920)
3. M. Hebb, J. Chem. Phys. 20, 185 (1952)

4. C. Wagner, Z. Elektrochem. 60, 4 (1956)
5. C. Wagner, Proc. Int. Comm. Electrochem. Thermo. Kinet. (CITCE) 7, 361 (1957)
6. R.A. Huggins, Solid State Ionics 143, 3 (2001)
7. J.W. Patterson, E.A. Bogren and R.A. Rapp, J. Electrochem. Soc. 114, 752 (1967)
8. S. Crouch-Baker, Solid State Ionics 45, 101 (1991)
9. V.B. Tare and H. Schmalzried, Z. Physik. Chem. NF 43, 30 (1964)
10. C. Wagner, Adv. Electrochem. Electrochem. Eng. 4, 1 (1966)
11. R.A. Huggins, Ionics 5, 269 (1999)
12. W. Weppner and R.A. Huggins, J. Electrochem. Soc. 124, 1569 (1977)
13. W. Weppner and R.A. Huggins, Electrochemical Methods for Determining Kinetic Properties of Solids in *Annual Review of Materials Science*, R.A. Huggins, Ed., Annual Reviews, Inc., Palo Alto, CA (1978), p. 269
14. C. Ho, I.D. Raistrick and R.A. Huggins, J. Electrochem. Soc. 127, 343 (1980)
15. C.J. Wen, C. Ho, B.A. Boukamp, I.D. Raistrick, W. Weppner and R.A. Huggins, Int. Metals Rev. 5, 253 (1981)
16. C. Wagner, J. Chem. Phys. 21, 1819 (1953)
17. A.L. L'vov, et al. Elektrokhimiya 11, 1322 (1975)
18. B.E. Conway, J. Electrochem. Soc. 138, 1539 (1991)
19. I.D. Raistrick, C. Ho, Y.W. Hu and R.A. Huggins, J. Electroanal. Chem. 77, 319 (1977)
20. I.D. Raistrick, C. Ho and R.A. Huggins, J. Electrochem. Soc. 123, 1469 (1976)
21. V. Thangadurai, R.A. Huggins and W. Weppner, Determination of the Electronic and Ionic Partial Conductivities in Several Mixed Conductors Using a Simple AC Method, in *Electrochem Society Proceedings*, Vol. 2001-28, A. Nazri, Ed., Electrochem Society, Pennington, NJ (2001), p. 109

Chapter 18
Use of Polymeric Materials As Battery Components

18.1 Introduction

The announcement of the unusually rapid transport of ions in materials with the beta alumina crystal structure [1,2], and the invention of the Na/Na_xS battery based upon the use of that material as a solid electrolyte [3] initiated a lot of research aimed at understanding of fast ionic transport in solids. There were also efforts in many laboratories to find other materials that also had this kind of behavior.

It was found that this unusual property was directly related to features of the crystal structure. This focused much attention upon the crystallographic aspects of such materials.

Although it had been suggested earlier [4], and it had been shown that poly(ethylene oxide) salt complexes exhibit ionic conductivity [5], it was a surprise when it was demonstrated that organic polymers can also be used as solid electrolytes in batteries [6–8]. This was at about the same time as it was shown that it is possible to insert electroactive species into electronically conducting organic polymers [9].

The idea that polymeric materials could be used as battery components appeared to be quite attractive, for they generally tend to be rather inexpensive, and a number of polymer products can be readily fabricated as thin layers or films, and in large quantities. The idea of using this kind of technology was very appealing, and there were visions of such things as shape-conforming power sources for many types of products.

Thus, the field of *solid state ionics* suddenly had two new directions, the use of polymeric materials as electrolytes and as electrodes. Their use as electrolytes is discussed in the following section.

R.A. Huggins, *Advanced Batteries*: Materials Science Aspects,

18A Polymer Electrolytes

18A.1 Introduction

The demonstration by Armand and his coworkers that polymeric materials could act as solid electrolytes elicited a lot of excitement and activity in the research community. It is now recognized that there are several general types of ionically conducting polymeric materials. One is based on the type of materials whose properties were initially demonstrated, in which salts with mobile ions are introduced as components of the structure of high molecular weight polar polymers, such as polyethylene oxide (PEO).

A different approach involves materials that are often called *ionic rubbers*, in which much larger amounts of salt are present. They are hybrids, and a number of their properties can be significantly different from those of the other types of polymeric materials.

Another, multistep hybrid approach is to make an electrolyte as a two-component system, with an electrically inert polymer matrix containing a conducting liquid plasticizer.

A further variant is to make a gel electrolyte by dissolving a salt in a polar liquid, and then adding a sufficient quantity of an inert polymeric material to provide mechanical stability.

More detailed reviews and information about the use of polymeric materials as electrolytes can be found in a number of other places in addition to the references specifically mentioned in the text of this short chapter [8, 10–15].

18A.2 High Molecular Weight Polymers Containing Salts

A typical material of the first type is a high molecular weight polymer into which a lithium salt LiX has been dissolved, where X is a large soft anion. These can be formed into a solvent-free membrane, and are often called *dry polymers*.

Examples are polyethylene oxide (PEO) and polypropylene oxide (PPO). Later examples that have received attention are poly(methoxyethoxyethoxyphosphazene) (MEEP), and polyacrylonitrile (PAN) containing a solution of propylene carbonate containing a salt. Another is polyvinylidene fluoride (PVFC) containing a carbonate solution of a salt.

The result is that these materials generally have the mechanical properties of a polymer, but the electrochemical properties of a liquid electrolyte.

A particularly interesting feature of the use of these solvent-free polymeric materials as electrolytes is that they can be used with elemental lithium negative electrodes without the problem of formation of filaments or dendrites that is endemic with organic solvent liquid electrolytes that form SEI layers [16]. This led to the development of the all-solid Li/polymer/V_6O_{13} batteries that were commercialized by Hydro-Quebec.

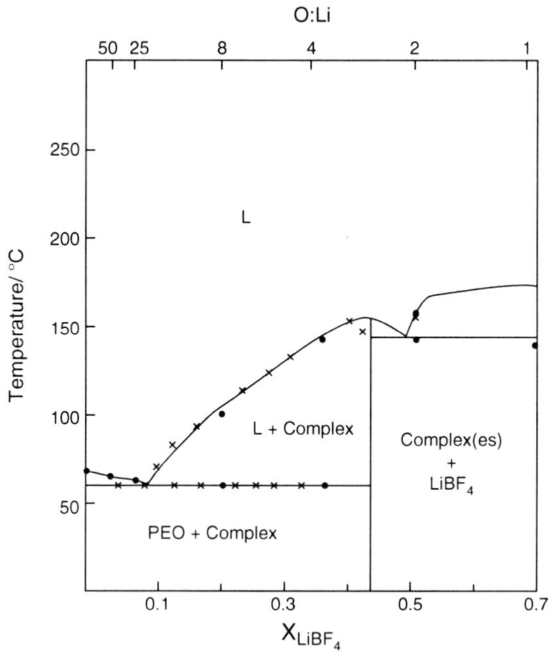

Fig. 18A.1 Phase diagram for the system PEO–$LiBF_4$

Their ionic conductivities are generally quite low at ambient temperatures. Thus, they are generally used at slightly elevated temperatures. This can be understood by consideration of the relevant phase diagrams, an approach introduced by Sørensen and Jacobsen [17] to help explain the dependence of the ionic conductivity upon both the composition and temperature.

An example of a phase diagram is shown in Fig. 18A.1 [19].

It can be seen that there is a eutectic reaction at about 65°C. This is typical of PEO-based polymers. Below that temperature a wide range of compositions tend to crystallize into a two-phase structure with essentially pure PEO and a second phase with a composition of about 0.43 $LiBF_4$. Ionic mobility in both of these phases is very limited.

However, there is a liquid (amorphous) region above the eutectic temperature, whose composition range increases with temperature. It was found that ionic transport occurs predominantly through the amorphous part of the structure [20].Thus, good ionic conductivity is found only within the composition and temperature range of this phase. However, the mobility of the structural components in the amorphous phase also leads to a low creep strength, so steps must be taken to provide mechanical support.

Temperatures in the range of 80–120°C are typically used in order to get satisfactory values of conductivity in materials of this general type. Care must be taken, however, that the temperature does not go above the melting point of lithium (180°C) so as to avoid thermal runaway, with disastrous consequences.

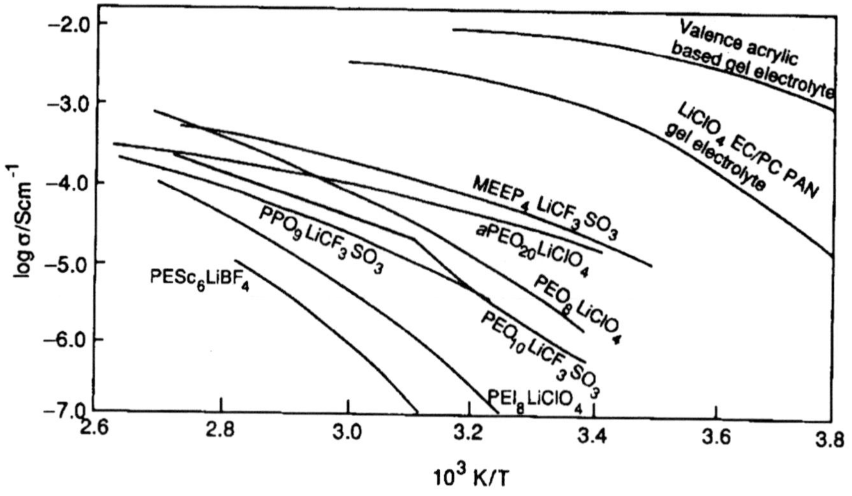

Fig. 18A.2 Temperature dependence of the ionic conductivity of a number of polymeric electrolytes

Figure 18A.2 shows data on the temperature dependence of the ionic conductivities of a number of polymeric electrolytes [12].

It can be seen that the relationship between the logarithm of the conductivity and the reciprocal of the temperature does not give straight lines in the case of these materials, whereas that is what is usually found for crystalline ionic conductors.

Such curved lines are generally found for ionic transport in materials that undergo crystalline-to-amorphous reactions. Instead of the usual Arrhenius relation, the temperature dependence of the conductivity in such materials can be expressed as

$$\sigma = \sigma_0 \exp\left[\frac{-B}{(T - T_g)}\right] \tag{18A.1}$$

Where σ is the conductivity, T the absolute temperature, T_g the *crystalline-to-amorphous*, or *glass transition temperature*, and σ_0 and B are constants. This relation is generally called the Vogel-Tammann-Fulcher, or TVF, equation [22–24].

18A.3 Particle-Enhanced Conductivity

It has been shown that the dispersion of fine particles of a solid oxide, such as TiO_2, Al_2O_3, or SiO_2, can enhance the conductivity of the high molecular weight polymers [25–27]. The presence of the inorganic particles evidently acts to prevent long-range order in the polymer structure, extending the higher-conductivity amorphous range to lower temperatures.

18A.4 Ionic Rubbers

Another interesting new direction was pioneered by C.A. Angell and his coworkers [28, 29].They noted that the liquidus (line in the phase diagram above which the structure is liquid) rises rapidly above the eutectic composition in the regime that had been normally studied by researchers interested in ionically conducting polymers, such as PEO. Instead of looking at this salt-in-polymer region of the phase diagram, they set out to investigate the very high-salt composition region.

They found quite different properties in this polymer-in-salt regime. A number of materials have been studied, often with a mixture of ionic salts to which modest amounts of high molecular weight PEO has been added. These materials have a combination of the properties similar to those of crystalline salts, with only a minor decrease in ionic conductivity, yet the mechanical behavior is more like that of rubber.

18A.5 Hybrid Electrolytes Containing an Ionically Conducting Plasticizer

An interesting innovation was to make materials with microstructures of an inert polymer containing a liquid plasticizer, and then exchanging the plasticizer for an ionically conducting liquid [30,31]. A practical example was to use poly(vinylidene fluoride), PVDF, containing hexafluoropropylene, HFP, as the plasticizer [31]. The presence of the plasticizer causes the polymer matrix to swell considerably, up to 60% by volume. Subsequently, the HFP plasticizer can be removed and replaced by a lithium-containing liquid electrolyte [32, 33]. The presence of inorganic fillers, such as SiO_2, can help maintain the geometry during this process. These hybrid materials can attain values of ionic conductivity up to $3\,mS\,cm^{-1}$, considerably greater than values typical of the high molecular weight polymers and close to those of organic liquid electrolytes, as seen in Fig. 18A.3. The influence of the presence of an inorganic filler on the conductivity is shown in Fig. 18A.4.

Replacing the liquid plasticizer by the moisture-sensitive lithium-containing liquid electrolyte late in the production process, after the polymer structure is formed, is advantageous, for all prior steps can be performed in air, rather than in an atmosphere-controlled environment.

18A.6 Gel Electrolytes

There are two different ways in which gel electrolytes can be formed. One is to add some material to an organic liquid solvent electrolyte to increase its viscosity. This could involve a soluble polymer, such as PEO, poly(methyl methacrylate) (PMMA),

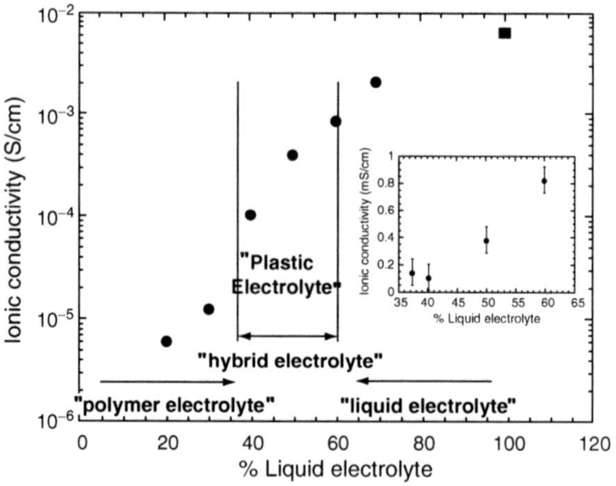

Fig. 18A.3 Ionic conductivity of materials with various amounts of liquid electrolyte [31]

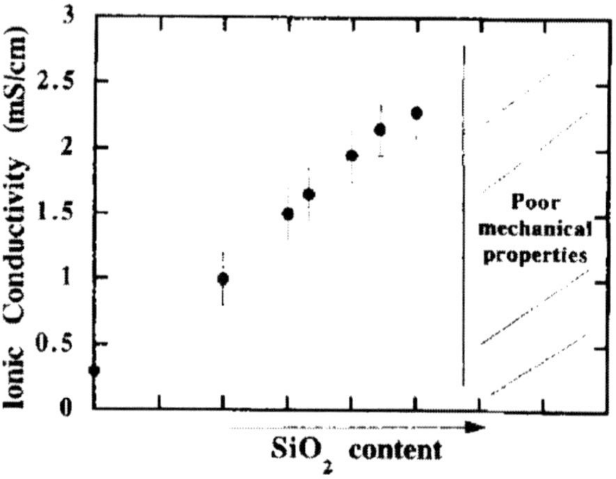

Fig. 18A.4 Influence of the presence of an inorganic filler on the ionic conductivity of hybrid electrolyte layers based on (PVDF–12% HFP) in 1 M $LiPF_6$ in EC/PC electrolyte [31]

poly(acrylonitrile) (PAN), or poly(vinylidene fluoride) (PVDF). This is similar to the polymer-in-salt strategy mentioned earlier in the section on ionic rubbers.

The mechanical properties are determined by the extent of entanglement developed by the polymer fragments. It is also possible to add a further plasticizing solvent in order to increase particle mobility and reduce the glass-forming temperature. It is obvious that a wide range of structures and properties are possible by following this approach [34, 35].

Another alternative would be to put a liquid electrolyte into a polymer such as polyethylene with a microporous structure [36, 37].

References

1. Y.F.Y. Yao and J.T. Kummer, J. Inorg. Nucl. Chem. 29, 2453 (1967)
2. R.H. Radzilowski, Y.F. Yao and J.T. Kummer, J. Appl. Phys. 40, 4716 (1969)
3. N. Weber and J.T. Kummer, Proc. Ann. Power Sources Conf. 21, 37 (1967)
4. M.B. Armand, in *Fast Ion Transport in Solids*, ed. by W. van Gool, North-Holland, Amsterdam (1973), p. 665
5. D.E. Fenton, J.M. Parker and P.V. Wright, Polymer 14, 589 (1973)
6. M.B. Armand, J.M. Chabagno and M. Duclot, Presented at Second International Meeting on Solid Electrolytes, St. Andrews, Scotland, Sept. 1978
7. M.B. Armand, J.M. Chabagno and M. Duclot, in *Fast Ionic Conduction in Solids*, ed. by P. Vashishta, J.N. Mundy and G.K. Shenoy, Elsevier, New York (1979), p. 131
8. M. Armand, Polymer Electrolytes, *in Annual Review of Materials Science*, Vol. 16, ed. by R.A. Huggins, Annual Reviews, Palo Alto, CA (1986), p. 245
9. C.K. Chiang, C.R. Fincher, Jr., Y.W. Park, A.J. Heeger, H. Shirakawa, E.J. Louis, S.C. Gau and A.G. MacDiarmid, Phys. Rev. Lett. 39, 1098(1977)
10. F.M. Gray, *Solid Polymer Electrolytes*, VCH, New York (1991)
11. M. Alamgir and K.M. Abraham, Room Temperature Polymer Electrolytes, in *Lithium Batteries, New Materials, Developments and Perspectives*, ed. by G. Pistoia, Elsevier, Amsterdam (1994), p. 93
12. F. Gray and M. Armand, Polymer Electrolytes, in *Handbook of Battery Materials*, ed. by J.O. Besenhard, Wiley-VCH, Weinheim (1999), p. 499
13. B. Scrosati, Lithium Polymer Electrolytes, in *Advances in Lithium-Ion Batteries*, ed. by W.A. van Schalkwijk and B. Scrosati, Kluwer Academic, New York (2002), p. 251
14. J.B. Kerr, Polymeric Electrolytes: An Overview, in *Lithium Batteries, Science and Technology*, ed. by G-A. Nazri and G. Pistoia, Kluwer Academic, Boston, MA (2004), p. 574
15. C.A. Angell, Mobile Ions in Amorphous Solids, *Annu. Rev. Phys. Chem.* 43, 693 (1992)
16. M. Gauthier, A. Belanger, B. Kapfer and G. Vassort, in *Polymer Electrolyte Reviews*, Vol. 2, ed. by J.R. MacCallum and C.A. Vincent, Elsevier, New York (1989), p. 285
17. P.R. Sørensen and T. Jacobsen, Polym. Bull. 9, 47 (1982)
18. M.Z.A. Munshi and B.B. Owens, Appl. Phys. Commun. 6, 279 (1987)
19. F.M. Gray, *Solid Polymer Electrolytes*, VCH, New York (1991), p. 76
20. S.M. Zahurak, M.L. Kaplan, E.A. Rietman, D.W. Murphy and R.J. Cava, Macromolecules 21, 654 (1988)
21. W. Gorecki, M. Minier, M.B. Armand, J.M. Chabagno and P. Rigaud, Solid State Ionics 11, 91 (1983)
22. H. Vogel, Phys. Z. 22, 645 (1921)
23. G. Tammann and W. Hesse, Z. Anorg. Allg. Chem. 156, 245 (1926)
24. G.S. Fulcher, J. Am. Ceram. Soc. 8, 339 (1925)
25. F. Croce, G.B. Appetecchi, L. Persi and B. Scrosati, Nature 394, 456 (1998)
26. F. Croce and B. Scrosati, Ann. N.Y. Acad. Sci. 984, 194 (2003)
27. F. Croce, L. Settimi and B. Scrosati, Electrochem. Commun. 8, 364 (2006)
28. C.A. Angell, C. Liu and E. Sanchez, Nature 362, 137 (1993)
29. C.A. Angell, C. Liu and E. Sanchez, in *Solid State Ionics III*, Vol. 293, ed. by G.-A. Nazri, J.-M. Tarascon and M. Armand, Materials Research Society Symposium Proceedings, Pittsburgh, PA (1993)
30. G. Feuillade and Ph. Perche, J. Appl. Electrochem. 5, 63 (1975)
31. J.-M. Tarascon, A.S. Gozdz, C.N. Schmutz, F.K. Shokoohi and P.C. Warren, Solid State Ionics 86–88, 49 (1996)
32. T. Gozdz, C. Schmutz, J.-M. Tarascon and P. Warren, U.S. Patent 5,418,091 (1995)
33. T. Gozdz, C. Schmutz, J.-M. Tarascon and P. Warren, U.S. Patent 5,456,000 (1995)
34. J.M.G. Cowie, in *Polymer Electrolyte Reviews*, Vol. 1, ed. by J.R. MacCallum and C.A. Vincent, Elsevier, New York(1987), p. 69

35. F.M. Gray, in *Polymer Electrolyte Reviews*, Vol. 1, ed. by J.R. MacCallum and C.A. Vincent, Elsevier, New York (1987), p. 139
36. T. Itoh, K. Saeki, K. Kohno and K. Koseki, J. Electrochem. Soc. 136, 3551 (1989)
37. K. Koseki, K. Saeki, T. Itoh, C.Q.Juan and O. Yamamoto, in *Second International Symposium on Polymer Electrolytes*, ed. by B. Scrosati, Elsevier, Amsterdam (1990), p. 197

Chapter 19
Transient Behavior of Electrochemical Systems

19.1 Introduction

Much of the discussion in this text has had to do with materials and phenomena under conditions that are at or near equilibrium. That is the best place to start in order to understand the driving forces and mechanisms involved, but actual behavior involves deviations from equilibrium.

The passage of current, either during discharge, or during recharge, results in the displacement of the system from thermodynamic equilibrium. In many cases the displacements are rather minor, and the results are easy to understand.

However, there is currently a great deal of interest in the storage of electric energy under very transient conditions, at very high powers and relatively short times. Such applications, therefore, emphasize power, rather than the amount of energy that can be stored. A very large cycle life is also very important. One of the most popular and obvious applications of such systems would be as energy buffers in hybrid automobiles, but there are many more.

This calls for different materials and different configurations from those that are common in most current battery systems. Thus, the labels *supercapacitors* and *ultracapacitors* have become common. Conventional capacitors in which the electrical charge is stored in the electrical double layer at the electrolyte/electrode interface cannot store sufficient energy to be useful for many of the potential applications, despite the development of electrode structures with extremely large surface areas. What is needed is a compromise in which very rapid kinetics can be attained, along with sufficient energy storage capacity. This is a very popular topic, and these matters have been discussed in many places. Among them are [1, 2].

An introduction to these matters, and useful experimental measurement methods and their interpretation are included in the following sections.

R.A. Huggins, *Advanced Batteries*: Materials Science Aspects,

19A Transient Behavior Under Pulse Demand Conditions

19A.1 Introduction

There is increasing interest in the use of electrochemical systems under conditions in which the electrical power demand is highly time-dependent, often in the form of short-term pulses. This has led to efforts to develop electrical devices that are optimized in terms of their short-term power output, rather than total energy content.

Because of the general observation that capacitors can be used to provide short electrical pulses, attention has been focused upon the concept of the use of electrochemical systems that exhibit capacitor-like characteristics. These can typically store much more charge than the conventional type of physical dielectric capacitor.

The range of both current and potential future requirements and applications is very broad. Typical examples that are now highly visible include digital communication devices that require pulses in the millisecond range, implanted medical devices that require pulses with characteristic times of the order of seconds, and hybrid vehicle traction applications, where the high-power demand can extend from seconds up to minutes, and the ability to absorb large currents upon braking is also important.

All these require output characteristics that are different from those that are normally considered for electrochemical systems, which are typically more oriented toward maximizing energy storage and output at lower power levels.

A typical pulse output requirement for a digital communication device involves a number of millisecond-length pulses with currents up to 2.5 A, on top of a 15-ms square wave at about 2 A. This is generally followed by a *recovery period* of 75 ms at a much lower current, only 0.25 A.

The requirements for batteries that are to be used for subcutaneous heart defibrillators are different. In one such case, a group of four 10-s pulses with current densities of $20\,\mathrm{mA\,cm^{-2}}$, scaled to the macroscopic surface area of the battery electrodes, is repeated every 30 min.

The characteristics needed to meet the requirements for electric vehicle propulsion depend greatly upon the type of duty cycle that is assumed to represent the typical usage demand. One of these, known as the ECE-15 cycle, was developed a number of years ago for full-electric vehicles. It was composed of two parts, an urban part that simulated the needs during local travel, and a suburban part that required higher power levels, such as what is needed for travel at higher velocities and greater distances. These were both expressed in terms of power–time profiles, and are shown in Figs. 19A.1 and 19A.2.

The introduction of full-electric automobiles was generally not successful, due to their high cost and limited performance. This was due primarily to the weight and cost of the batteries required in order to provide sufficient driving range.

More recently, hybrid vehicles have been introduced, in which the propulsion is provided by a combination of a modest internal combustion engine and a relatively small battery. Several of these have proved to be very successful. In this case, the

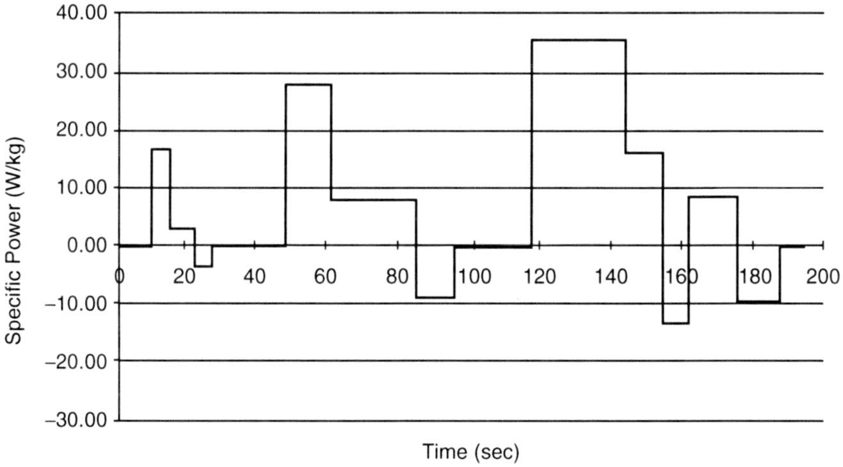

Fig. 19A.1 Power demand profile for the ECE-15 reference vehicle in urban travel simulation

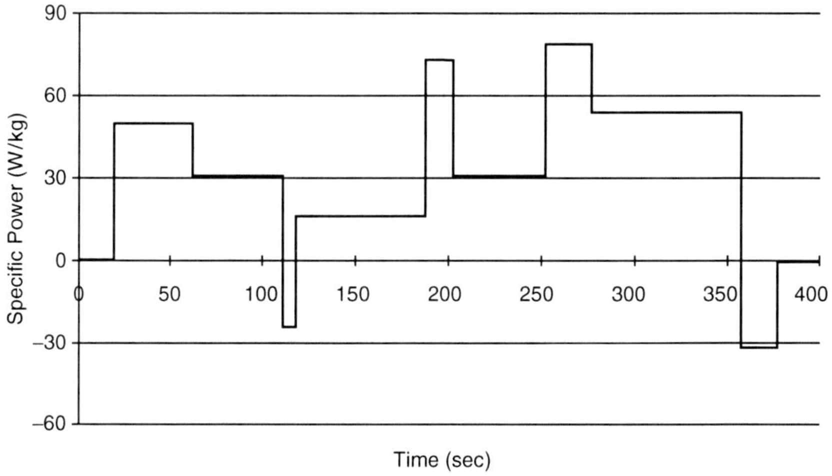

Fig. 19A.2 Power demand profile for the ECE-15 reference vehicle in suburban travel simulation

control system allows the engine to operate at relatively high efficiency, and the battery also gets some charging during braking. The large number of cycles and the high power that must be absorbed during braking place requirements on the battery portion of the system that are more akin to those expected for capacitors than from traditional batteries.

There is currently increasing interest in what are called plug-in hybrid vehicles. In this case, it is expected that the battery part of the system can be electrically recharged, perhaps overnight at home, to provide sufficient energy to propel the vehicle a modest distance – say 20–50 miles – without the use of the internal combustion engine at all. This sounds attractive, as the cost of electricity per mile of travel is less than the cost of the equivalent amount of liquid fuel. However, the type of battery that is needed for this application must be optimized for its specific energy, rather than for its behavior under transient conditions. This means added weight and cost, and it will be interesting to see how this is handled by the manufacturers of such vehicles.

Thus, there is a great variation in the requirements for transient power sources, and it is unrealistic to assume that any one type of device, or any one design, will be able to optimally fulfill such diverse needs.

It is now quite common to think in terms of hybrid systems that include components that can meet two different types of needs: a primary energy source, and a supplemental source that can meet the transient needs for higher power levels than can be handled by the primary source. This combination can be represented schematically in terms of the commonly used Ragone type of diagram, in which the specific power is plotted vs. the specific energy, both on logarithmic scales, as shown in Fig. 19A.3.

A possible strategy to consider in order to accomplish this is to use a high-energy system that operates at a relatively high voltage when the power demand is low. The output voltage of such energy sources typically falls off as the output current is increased. If a second high-power, but low-energy source that operates at a lower voltage is placed in parallel, it will take over under the conditions that drive the output voltage of the primary high-energy system down into its range of operating voltage, and meet the high-power demand for a short period. When this demand is no longer present, the voltage will rise again, and the high-power component of the system will be recharged by the high-energy component. This combination is shown schematically in Fig. 19A.4.

The way that the properties of batteries are typically described, such as by a graphical display of the discharge (voltage vs. state of charge) curves at different constant current densities, or in terms of the change of extractable capacity as a function of the number of discharge cycles, cannot be considered to provide

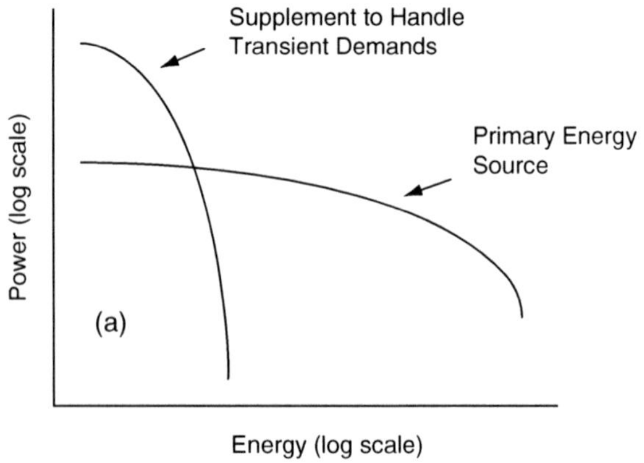

Fig. 19A.3 Typical hybrid system characteristics

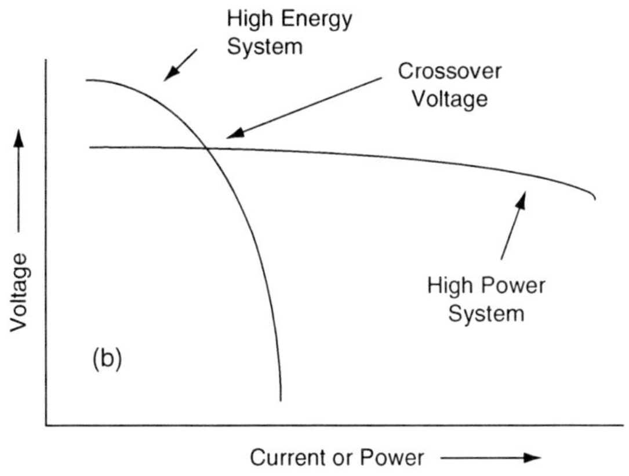

Fig. 19A.4 Schematic of possible hybrid system strategy

a satisfactory description of behavior in these very different types of applications. Likewise, the value of the capacitance at a single frequency is also certainly not a satisfactory description of the behavior of a capacitor over such a wide range of potential uses.

In order to approach the development of useful devices for this type of application one should consider the several types of charge storage mechanisms that can be employed, their thermodynamic and kinetic characteristics, and the basic properties of candidate materials, as well as the relationships that determine system performance.

19A.2 Electrochemical Charge Storage Mechanisms

There are two general categories of energy storage mechanisms in electrochemical systems. One of these involves the storage of charge in the electrical double layer at or near the electrolyte/electrode interface, and the other is related to the reversible absorption of atomic species into the crystal structure of bulk solid electrodes. They have quite different fundamental characteristics.

19A.2.1 Electrostatic Energy Storage in the Electrical Double Layer in the Vicinity of the Electrolyte/Electrode Interface

The interface between an electrode and an adjacent electrolyte can function as a simple capacitor with charge separation between two parallel plates. The amount of charge stored, Q, is related to the potential difference ΔV and the capacitance C by $Q = C \times \Delta V$, and the amount of free energy stored $\Delta G = 1/2C(\Delta V)^2 = 1/2Q \times \Delta V$.

In this approach the electrode is considered to be a chemically inert electronic conductor. The amount of charge that can be stored is generally of the order of

Table 19A.1 Characteristics of some double-layer electrode materials

Electrode material	Specific capacitance (F g^{-1})
Graphite paper	0.13
Carbon cloth	35
Aerogel carbon	30–40
Cellulose-based foamed carbon	70–180

$15-40\,\mu F\,cm^{-2}$ of interface [3,4]. Thus, one tries to optimize the amount of interface in the device microstructure. Techniques have been devised to produce various types of carbon, as well as some other electronically conducting, but chemically inert, materials in very highly dispersed form, leading to very large interfacial areas. Typical values of specific capacitance, Farads per gram, of a number of double-layer electrode materials are included in Table 19A.1.

The potential difference is limited by the decomposition voltage of the electrolyte, which is 1.23 V for aqueous electrolytes but can be up to 4–5 V for some organic solvent electrolytes. The result is that the specific energies of aqueous electrolyte systems are generally $1-1.5\,Wh\,kg^{-1}$, whereas those that use organic solvent electrolytes can be $7-10\,Wh\,kg^{-1}$.

A device in which this is the dominant charge storage mechanism will behave like a pure capacitor in series with its internal resistance. The time constant is equal to the product of the capacitance and the series resistance. Thus, it is important to keep the resistance as low as possible. Organic electrolytes characteristically have much lower ionic conductivity, and thus provide greater resistance than aqueous electrolytes. The conductivity of acids, such as H_2SO_4, is somewhat greater than that of aqueous bases. Furthermore, the larger the capacitance, the greater the time constant, slower the device, and lower the power level.

Another important feature of devices that operate by the double-layer mechanism is that the amount of charge stored is a linear function of the voltage. The voltage therefore falls linearly with the amount of charge extracted. Thus, voltage-dependent applications can only utilize a fraction of the total energy stored in such systems. The power supplied to resistive applications is proportional to the square of the instantaneous voltage, so this can be an important limitation.

Although there is some confusion in the literature about terminology, devices of this type that have been developed with large values of such capacitance have generally been called either EDLC (electrical double-layer capacitive) devices [5], or *ultracapacitors* [3,4]. The label *supercapacitors* is now becoming more and more common, and will be discussed later.

Ultracapacitor devices utilizing the storage of charge in the electrochemical double layer have been developed and produced in large numbers in Japan for a considerable period of time [5]. These are primarily used for semiconductor memory backup purposes, as well as for several types of small actuators. Information about the five largest producers is given in Table 19A.2.

Table 19A.2 Early producers of electrochemical double-layer *ultracapacitors*

Company	Started development	Products on market
Matsushita	1971	1978
NEC	1973	1978
Murata	1982	1986
Elna/Asahi Glass	1985	1987
Isuzu/Fuji	1988	1990

19A.2.2 Underpotential Faradaic Two-Dimensional Adsorption on the Surface of a Solid Electrode

Because of the characteristics of the electrolyte/electrode surface structure and its related thermodynamics, it is often found that modest amounts of *Faradiac electrodeposition* can occur at potentials somewhat removed from those needed for the bulk deposition of a new phase. This results in the occupation of specific crystallographic surface sites. This mechanism typically results in only partial surface coverage, and thus the production of an *adsorption pseudocapacitance* of some 200–400 mF cm^{-2} of interface [4]. This is significantly larger than the amount of charge stored per unit area in the electrochemical double layer. However, materials with which this mechanism can be effectively used are rare.

19A.2.3 Faradaic Deposition That Results in the Three-Dimensional Absorption of the Electroactive Species into the Bulk Solid Electrode Material by an Insertion Reaction

As discussed in earlier chapters, many materials are now known that can act as solid solution electrodes. In such cases, the *electroactive species* diffuses into and out of the interior of the crystal structure of the solid electrode. Since the amount of energy stored is proportional to the amount of the electroactive species that can be absorbed by the electrode, this bulk storage mechanism can lead to much higher values of energy storage per unit volume of electrode structure than any surface-related process. Because it makes no sense to express this bulk phenomenon in terms of the capacitance per unit interfacial area, values are generally given as *Farads per gram*.

This mechanism has been called *redox pseudocapacitance* by the Conway group [3, 4], who also first started the use of the term *supercapacitors* to describe devices utilizing this type of charge storage. In this way, bulk storage *supercapacitors* can be distinguished from double-layer storage *ultracapacitors*.

The bulk storage *supercapacitor* mechanism is utilized in the devices that are most interesting for energy-sensitive pulse applications. Since the kinetic behavior of such devices is related to the electrolyte/electrode area, it is important that they have very fine microstructures.

During investigations of the *dimensionally stable positive electrodes* that are used as positive electrodes in the *chlor-alkali process* it was noticed that RuO_2 seemed to behave as though its interface with the electrolyte has an unusually large capacitance. This swiftly led to several investigations of the capacitive behavior of such materials [6–9].

The possibility of the development of RuO_2-type materials as commercial capacitors began in Canada about 1975, the key players being D.R. Craig [10, 11] and B.E. Conway. This soon evolved into a proprietary development program at the Continental Group, Inc. laboratory in California, which subsequently was taken over by Pinnacle Research Institute. The products that were developed and manufactured were all oriented toward the military market. This orientation is changing now, and activities are being undertaken by several firms to produce products for the civilian market. Activities in this area have also been initiated more recently in Europe. These include work at Daimler-Benz (Dornier) and Siemens in Germany, and at Thompson-CSF in France.

It was originally thought that charge storage is due to *redox reactions* at or very near the interface between the electrolyte and RuO_2. Careful measurements showed that the measured capacitance is large, and proportional to the surface area [12].

It is now known, however, that the charge is stored in the bulk of the RuO_2, and not just on the surface [13]. At relatively short times the diffusion depth will be limited, and the amount of charge stored in the bulk will not reach its ultimate value. Measurements of the variation of the potential with the hydrogen content have also shown that the electrochemical titration curve is quite steep. This leads to apparent capacitor-like behavior.

Experiments [14] showed that the chemical diffusion coefficient of hydrogen in bulk crystalline RuO_2 is about $5 \times 10^{-14}\,cm^2\,s^{-1}$, which is consistent with what was found by others [7].

Thus, the penetration into the bulk crystalline solid is rather shallow at the relatively high frequencies typically used in capacitor experiments. It has been shown in later work [15] that the apparent solubility is considerably higher in amorphous RuO_2 than in crystalline material. Experiments on hydrated RuO_2 [13, 16, 17] demonstrated that it has a substantially larger charge storage capacity than anhydrous RuO_2.

It has also been found [17] that the amount of charge stored in hydrated RuO_2 is independent of the surface area, but proportional to the total mass. This is shown in Fig. 19A.5. Over 1 hydrogen atom can be reversibly inserted into the structure per Ru atom [13]. The coulometric titration curve is shown in Fig. 19A.6.

Crystalline RuO_2 is a very good electronic conductor. Its electronic resistivity is about 10^{-5} ohm cm. This is about a factor of 100 lower than that of bulk carbons. The hydrated material, on the other hand, has a considerably higher resistivity, and it has been found to be advantageous to add some carbon to the microstructure in order to reduce the series resistance in that case.

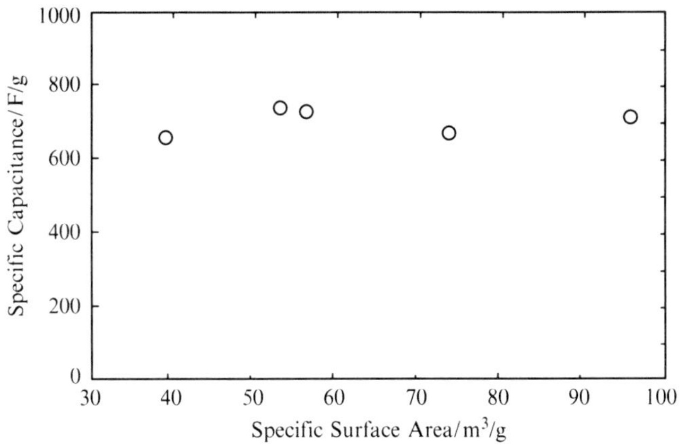

Fig. 19A.5 Apparent capacitance of RuO_2 hydrate as a function of surface area

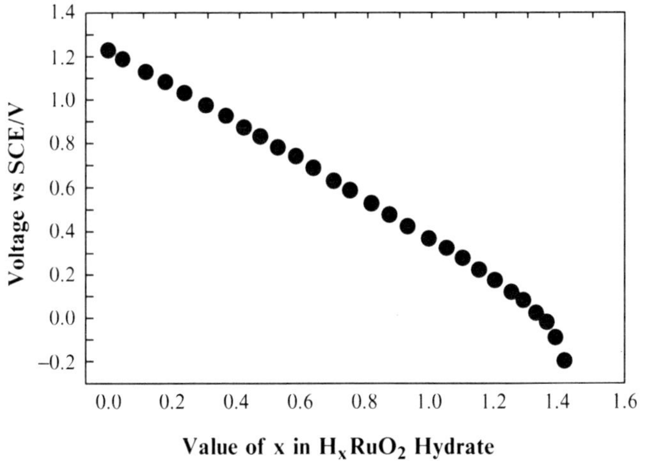

Fig. 19A.6 Dependence of the potential of RuO_2 hydrate upon the amount of inserted hydrogen

Table 19A.3 Characteristics of some insertion reaction electrodes

Electrode material	Specific capacitance (F g^{-1})
Polymers (e.g., polyaniline)	400–500
RuO_2	380
RuO_2 hydrate	760

Several other materials that are *electrochromic* show similar pseudocapacitive behavior, such as NiOOH and IrO_2. This clearly indicates the insertion of species into the bulk crystal structure.

Typical values of specific capacitance of a number of insertion reaction electrode materials are included in Table 19A.3.

The insertion of guest species into the host crystal structure in such insertion reactions generally results in some change in the volume. This can lead to morphological changes and a reduction in capacity upon cycling. The volume change is generally roughly proportional to the concentration of the guest species. As a result, it is often found that the magnitude of this degradation depends upon the depth of the charge–discharge cycles.

19A.2.4 Faradaically Driven Reconstitution Reactions

The electrodes in many battery systems undergo *reconstitution reactions*, in which different phases form and others are consumed. In accordance with the *Gibbs phase rule*, this often results in an open-circuit electrode potential that is independent of the state of charge. As discussed in an earlier chapter, the amount of charge storage is determined by the characteristics of the related phase diagram, and can be quite large. Some reactions of this type can also have relatively rapid kinetics. However, there is a potential difficulty in the use of this type of reaction in applications that require many repeatable cycles, for they generally involve microstructural changes that are not entirely reversible. Thus, the possibility of a cycle-life limitation must be kept in mind.

A special strategy whereby this microstructural irreversibility may be avoided or reduced in certain cases has been proposed [18]. This involves the use of an all-solid electrode in which a mixed-conducting solid matrix phase with a very high chemical diffusion coefficient surrounds small particles of the reactant phases.

19A.3 Comparative Magnitudes of Energy Storage

The maximum amount of energy that can be stored in any device is the integral of its voltage-charge product, and cannot exceed the product of its maximum voltage and the maximum amount of charge it can store. On this basis, one can make a simple comparison between these different types of energy storage mechanisms.

The results are shown schematically in Fig. 19A.7, in which the relationship between the potential and the amount of charge delivered is plotted for three different types of systems: a double-layer electrode, an insertion reaction electrode, and a reconstitution reaction electrode. Electrodes that involve two-dimensional *Faradaic underpotential* deposition are not included, as they do not constitute a practical alternative.

In the case of a true capacitor, the amount of charge stored is a linear function of the applied voltage. Thus, as shown on the left side of Fig. 19A.7, the voltage falls off linearly with the amount of charge delivered.

A single-phase solid solution insertion-reaction type of electrode characteristically has a potential-charge relation of the type shown in the middle. The

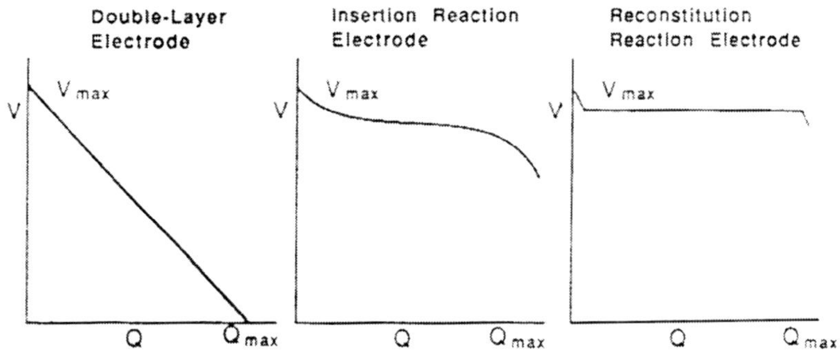

Fig. 19A.7 Comparison of the variation of the potential with the amount of charge extracted for different types of energy storage mechanisms

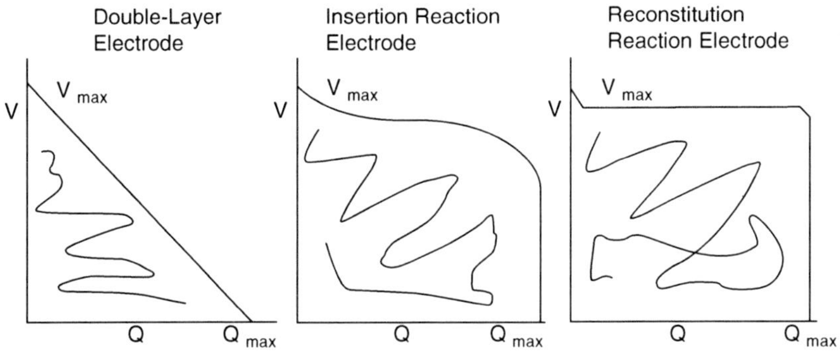

Fig. 19A.8 The amount of energy available for materials with different types of storage mechanisms, indicated by the area under their curves

thermodynamic basis for this shape, in which the potential is composition dependent, and thus state-of-charge dependent, was discussed earlier in Chap. 6.

The characteristic behavior of a reconstitution-reaction electrode system is shown on the right side. In this case, it is assumed that the temperature and pressure are fixed, and that the number of components is equal to the number of phases, so that from a thermodynamic point of view there are no degrees of freedom. This means that all of the intensive variables, including the electrode potential, are independent of the overall composition, and thus independent of the amount of charge delivered. Thus, the discharge characteristic consists of a voltage plateau.

As mentioned earlier, the maximum amount of energy that is available in each case is the area under the V/Q curve. This is indicated in Fig. 19A.8 for the three cases of interest. It is seen that the maximum amount of energy that can be stored in an electrode that behaves as a capacitor is $1/2(V_{max})(Q_{max})$. The actual amount of available energy will, of course, depend upon the power level, due to unavoidable losses, such as that due to the inevitable internal resistance of the system.

19A.4 Importance of the Quality of the Stored Energy

As mentioned in Chap. 1, the quality of heat is a commonly used concept in engineering thermodynamics. High-temperature heat is generally much more useful than low-temperature heat. Thus, in considering a practical thermal system one has to consider both the amount of heat and its quality (the temperature at which it is available).

One can consider an analogous situation in the application of energy storage devices and systems. In such cases, in addition to the total amount of energy that can be stored, one should also consider the voltage at which it is available. Thus, it is useful to consider the quality of the stored energy.

If this factor is taken into account, an additional difference between systems that utilize electrodes that operate by these three different types of mechanisms can be seen. This is indicated in Table 19A.4, in which the amount of higher value energy available in the different cases is compared. In that case, only a simple distinction is used. Energy at a potential above $V_{max}/2$ is considered to be high-value energy.

Thus, there are a number of parameters that determine important properties of a transient storage system. These are listed in Table 19A.5.

Table 19A.4 Maximum amount of high-value energy available

Type of electrode	High-value energy (%) where ($V > V_{max}/2$)
Double-layer electrode	37.5
Insertion reaction electrode	About 80
Reconstitution reaction electrode	About 90

Table 19A.5 Parameters that determine the values of maximum potential, maximum charge, and maximum energy stored

Type of electrode	V_{max} determined by
Double-layer electrode	The electrolyte stability window
Insertion reaction electrode	Thermodynamics of guest–host phase
Reconstitution reaction electrode	Thermodynamics of polyphase reactions
Type of electrode	Q_{max} determined by
Double-layer electrode	Electrode microstructure, electrolyte
Insertion reaction electrode	Mass of electrode, thermodynamics
Reconstitution reaction electrode	Mass of electrode, thermodynamics

19B Modeling Transient Behavior of Electrochemical Systems Using Laplace Transforms

19B.1 Introduction

It has been seen in earlier chapters that there are several electrochemical methods that have been developed in order to experimentally determine the critical parameters of materials that might be employed in energy storage systems used to supply pulse power.

The quantitative understanding of the application of these components in actual devices requires knowledge of the relationship between component properties and system behavior, for systems typically involve more than one component. As an example, in addition to the electrode impedances, there are almost always resistive, and/or capacitive, impedances present, both relating to internal phenomena and to external factors.

As discussed earlier, it is often very useful to utilize *equivalent electrical circuits* whose electrical behavior is analogous to the behavior of physical systems as *thinking tools* to obtain insight into the important parameters and their interrelationships. This allows the use of the methods that have been developed in electrical engineering for circuit analysis to evaluate the overall behavior of interdependent physical phenomena.

A useful way to do this is based upon the simple concept of a relation between a *driving force* and the *response* of a device or system to it. This relation can be written very generally as:

$$Driving\ function = (transfer\ function) \times (response\ function).$$

In an electrochemical system, the driving function represents the current or voltage demands imposed by the application, and the response function is the output of the electrochemical system in response to these demands. The key element of this approach is the determination of the (time-dependent) transfer function of the device or system, for that determines the relationship between application demand and system output.

19B.2 Use of Laplace Transform Techniques

The general method that has been developed for electrical device and circuit analysis involves the use of *Laplace transform* techniques. There are several basic steps in this analysis. They involve the following:

1. The determination of the transfer function of the individual equivalent circuit components
2. The calculation of the transfer function of the total system

3. The introduction of the driving function determined by the application, and finally
4. The calculation of the system (energy source) output

Some readers of this chapter may not be familiar with Laplace transform methods. But they can be readily understood by the use of an analogy. Consider the use of logarithms to multiply two numbers, e.g., A and B. The general procedure is to find the logarithms (transforms) of both A and B, to add them together, and then to use antilogarithms to reconvert the sum of the logarithms (transforms) into a normal number.

This method has been applied to some simple electrochemical situations, including the influence of the presence of series resistance upon the rate of charge accumulation in an insertion reaction electrode [19] and the electrical response of electrochemical capacitors [12].

As described in Sect. 17A.6, the calculation of the transient electrical response of an insertion, or solid-solution, electrode involves the solution of the diffusion equation for boundary conditions that are appropriate to the particular form of applied signal. In addition, the relation between the concentration of the electroactive mobile species and their activity is necessary. This approach has been utilized to determine the kinetic properties of individual materials by employing current and/or voltage steps or pulses.

But in real electrochemical systems or devices one has to consider the presence of other components and phenomena, i.e., other circuit elements. As a simple example, there is always an electrolyte, and thus a series resistance present, and the behavior of the electrolyte/electrode interface may have to be also considered. Thus, the simple solution of Fick's diffusion laws for the electrochemical behavior of the electrode alone may not be satisfactory.

Examples of the Laplace transforms of several common functions are given in Table 19B.1.

Table 19B.1 Examples of Laplace transforms

Function	Laplace transform
General impedance function	$Z(p) = E(p)/I(p)$
Fick's second law	$pC - c(t=0) = D\frac{d^2C}{dx^2}$
Current step d(t)	$I(p) = 1$
Potential vs. time	$E(p) = V(dE/dy)$
Impedance of insertion reaction electrode	$Z(p) = Q/Da$

where $Q = \frac{V(dE/dy)}{nFs}$, $a = (p/D)^{1/2}$, dE/dy = slope of coulometric titration curve, y = composition parameter, n = stoichiometric coefficient, F = Faraday constant, s = surface area, p = complex frequency variable, x = positional coordinate, V = molar volume, $q(t)$ = charge accumulated in electrode, $i(t)$ = instantaneous current, and $F(t)$ = instantaneous electrode potential

19B.3 Simple Examples

To illustrate this method, the response of an insertion electrode under both a step in potential and a step in current, as well as a system consisting of a simple series arrangement of a resistance and an insertion reaction electrode that has a diffusional impedance will be described.

(A) Upon the imposition of a step in potential F_0, the time dependence of the current i(t) is given by

$$i(t) = \frac{F_0}{Q}\left(\frac{D}{pt}\right)^{1/2}. \tag{19B.1}$$

(B) The time dependence of the charge accumulated (or produced) q(t) is

$$q(t) = \frac{2F_0}{Q}\left(\frac{t}{p}\right)^{1/2}. \tag{19B.2}$$

(C) For the case of a step in current i_0, the time dependence of the electrode potential F(t) is given by

$$F(t) = 2Q\left(\frac{t}{pD}\right)^{1/2}. \tag{19B.3}$$

(D) The time dependence of the current after the imposition of a step potential of F_0 for the more complicated case of a resistance in series with an insertion reaction electrode is found to be

$$i(t) = (F_0/R)\exp\left[\left(\frac{Q}{R}\right)^2 t\right] erfc\left[\frac{Qt^{1/2}}{R}\right]. \tag{19B.4}$$

(E) The charge accumulated (or produced) in the case of this series combination is found to be

$$q(t) = \frac{F_0 R}{Q^{1/2}}\left[\exp\left[\left(\frac{Q}{R}\right)^2 t\right] erfc\left[\frac{Qt^{1/2}}{R}\right] - 1\right] + \frac{2F_0}{Q}\left(\frac{t}{p}\right)^{1/2}. \tag{19B.5}$$

The influence of the value of the series resistance can readily be seen in Fig. 19B.1 [19].

The parameters used in the calculation illustrated in Fig. 19B.1 are as follows:
$Q^{-1} = 6.33\,\text{ohm}\,\text{s}^{-1/2}$

Derived from $D = 10^{-8}\,\text{cm}^2\,\text{s}^{-1}$, $V_m = 30\,\text{cm}^3\,\text{mol}^{-1}$, $dE/dy = -2\,\text{V}$, $s = 1\,\text{cm}^2$, and $n = 1$.

The applied voltage step was 0.5 V.

Examination of the solutions for the behavior of single components under the first two sets of conditions obtained by this method shows that they are equivalent

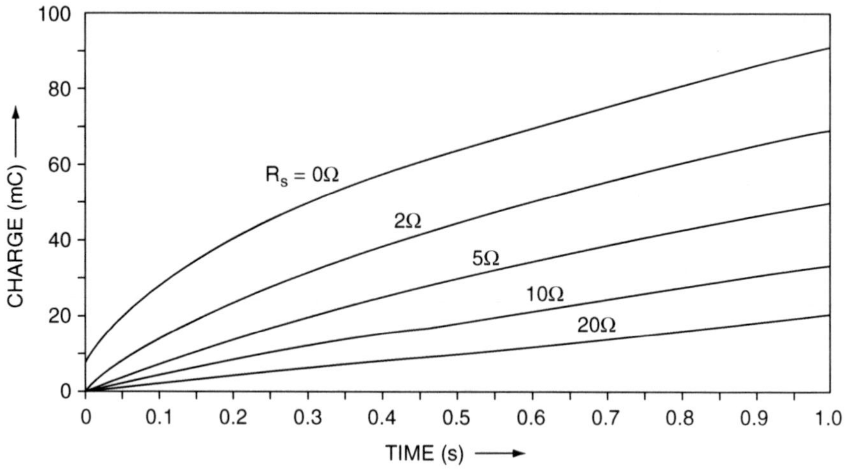

Fig. 19B.1 The charge accumulated in an insertion reaction electrode as a function of time for various values of series resistance

to those presented in Sect. 17A.6 for the analytical solution of the diffusion equation under equivalent experimental conditions.

The impedance of an insertion reaction electrode alone under an ac driving force has also been described in [20].

This Laplace transform procedure is thus an alternative to the more conventional analytical approach. But the real value of the Laplace transform approach becomes clearer, however, under more complex conditions, such as when more than one component is present, and in which the normal procedures become quite cumbersome.

References

1. B.E. Conway, *Electrochemical Supercapacitors: Scientific Fundamentals and Technological Applications,* Plenum, New York (1999)
2. R.A. Huggins, Solid State Ionics 134, 179 (2000)
3. B.E. Conway, J. Electrochem. Soc. 138, 1539 (1991)
4. B.E. Conway, in *Proceedings of Symposium on New Sealed Rechargeable Batteries and Supercapacitors,* B.M. Barnett, E. Dowgiallo, G. Halpert, Y. Matsuda and Z. Takeharas, Eds., The Electrochemical Society, Pennington, NJ (1993), p. 15
5. A. Nishino, in *Proceedings of Symposium on New Sealed Rechargeable Batteries and Supercapacitors,* B.M. Barnett, E. Dowgiallo, G. Halpert, Y. Matsuda and Z. Takehara, Eds., The Electrochemical Society, Pennington, NJ (1993), p. 1
6. L.D. Burke, O.J. Murphy, J.F. O'Neill and S. Venkatesan, J. Chem. Soc. Faraday Trans. 1, 73 (1977)
7. T. Arikado, C. Iwakura and H. Tamura, Electrochem. Acta 22, 513 (1977)
8. D. Michell, D.A.J. Rand and R. Woods, J. Electroanal. Chem. 89, 11 (1978)
9. S. Trasatti and G. Lodi, in *Electrodes of Conductive Metallic Oxides,* Part A, ed. by S. Trasatti, Elsevier, Amsterdam (1980), p. 301

10. D.R. Craig, Canadian Patent 1,196,683 (1985)
11. D.R. Craig, European Patent Application 82,109,061.0 (submitted September 30, 1982) Publication No. 0 078 404 (1983)
12. I.D. Raistrick and R.J. Sherman, 'Electrical Response of Electrochemical Capacitors Based on High Surface Area Ruthenium Oxide Electrodes,' in *Proceedings of Symposium on Electrode Materials and Processes for Energy Conversion and Storage,* S. Srinivasan, S. Wagner and H. Wroblowa, Eds. Electrochemical Society, Pennington, NJ (1987), p. 582
13. T.R. Jow and J.P. Zheng, J. Electrochem. Soc. 145, 49 (1998)
14. J.E. Weston and B.C.H. Steele, J. Appl. Electrochem. 10, 49 (1980)
15. J.P. Zheng, T.R. Jow, Q.X. Jia and X.D. Wu, J. Electrochem. Soc. 143, 1068 (1996)
16. J.P. Zheng and T.R. Jow, J. Electrochem. Soc. 142, L6 (1995)
17. J.P. Zheng, P.J. Cygan and T.R. Jow, J. Electrochem. Soc. 142, 2699 (1995)
18. B.A. Boukamp, G.C. Lesh and R.A. Huggins, J. Electrochem. Soc. 128, 725 (1981)
19. I.D. Raistrick and R.A. Huggins, Solid State Ionics 7, 213 (1982)
20. C. Ho, I.D. Raistrick and R.A. Huggins, J. Electrochem. Soc. 127, 343 (1980)

Chapter 20
Closing Comments

20.1 Introduction

Anyone who has looked at even a modest part of this text will have noticed that it is quite different from almost any other book related to batteries. The objective is not to describe battery technology, and there is relatively little in the way of descriptive treatment of current commercial batteries and their properties.

The major emphasis here is upon the fundamental phenomena that determine the properties of the components, i.e., electrodes and electrolytes, of electrochemical cells that are applicable to many of the advanced systems, not just to a particular one.

20.2 Terminology

Some aspects of the terminology that is used here may seem a bit unconventional, or even awkward, to those who are familiar with the literature relating to battery technology.

This is partly because this is a book on the materials science aspects of electrochemical materials and phenomena, not a book on electrochemistry. But it is also intentional, the result of a deliberate effort toward consistency.

One example is the use of the terms *positive* and *negative electrodes*, rather than *cathode* and *anode*. This is done deliberately, for according to the traditional definition, the cathode is the electrode at which reduction occurs. Thus, the positive electrode is the cathode when the cell is discharging, and it is the anode when it is being charged. During charging the negative electrode is the cathode.

If the terms *positive* and *negative electrode* are used, there is never any confusion. The facts can simply be checked by the use of a voltmeter. When an electrochemical cell is being charged, the difference between the potentials of the electrodes, the cell voltage, becomes greater, and when the cell is discharged the difference becomes smaller, but the negative electrode is still negative, and positive still positive, with respect to the other.

R.A. Huggins, *Advanced Batteries*: Materials Science Aspects,

Another example is a deliberate consistency in the way that cells are described, reactions written, and figures drawn. In all cases, potentials are always higher on the right hand side.

It will also be noted that some of the standard terminology that is often used in discussing electrochemical systems does not appear at all. For example, the outdated terms *depolarizer* and *depolarization* are not used.

Overvoltage is generally defined as the difference between the potential under imposed current and the equilibrium potential. This is zero under open circuit conditions. Unfortunately, one often finds this term in the electrochemical literature, even when its sign is negative, and it should be called *underpotential*. These terms are just not used here.

The related quantity, *polarization*, defined as the magnitude of the change in the potential from the equilibrium value that is caused by the passage of current, is also not used here.

20.3 Major Attention Is Given to the Driving Forces and Mechanisms That Determine the Potentials, Kinetic Properties, and Capacities of Electrodes

A major point, which is not generally found elsewhere, is that the driving forces that determine the voltages of electrochemical cells are related to the thermodynamic properties of neutral chemical species, and have little to do with electrochemical mechanisms.

This can be easily seen by consideration of hydrogen/oxygen fuel cells that have electrically neutral gases, hydrogen and oxygen, on their two sides. The open circuit voltage is determined by the standard Gibbs free energy change related to the formation of water from neutral hydrogen and neutral oxygen. Some such fuel cells operate by the transport of positively charged species, H^+, or the related hydrated species, through the electrolyte, and others by the transport of negatively charged species, OH^-. The electrochemical reactions at the electrodes are quite different in these two cases, but regardless of which species carries the charge, and which electrochemical reactions occur at the electrodes, the open circuit voltage is always the same, and is related to the standard Gibbs free energy change involved in the cell reaction, the Gibbs free energy of formation of the neutral reaction product, H_2O, from neutral hydrogen and oxygen.

Thus, one can calculate the open circuit voltage of such a fuel cell just from the thermodynamic properties of neutral species, regardless of what the electrochemical reactions at the electrodes might be.

This theme is found throughout the text. The driving forces for electrochemical reactions are always chemical reactions between electrically neutral species, and it can be seen that this approach works very well and produces not only information about electrode potentials and cell voltages, but it also can provide information about the relevant reactions.

Electrochemical features of the reactions have to do with mechanisms, and are often important factors in determining kinetic behavior. But relatively little attention is given here to the kinetic aspects of current battery technology: the influence of the discharge rate and state of charge upon the voltage, changes upon cycling, etc. Electrode and battery kinetic aspects depend to a large extent upon technological features such as particle size, binders, electronically conductive additives, and minor changes in the chemical constitution. Such matters are better addressed elsewhere.

20.4 Thinking Tools

There are a number of thinking tools that are used in the text to help understand, and in some cases provide quantitative information, about relevant phenomena in battery systems, in some cases without the necessity to know all the mechanistic details.

These include the following:

Binary phase diagrams
Defect equilibrium diagrams
Equivalent circuits
Ellingham diagrams
Pourbaix diagrams
Ternary phase stability diagrams – isothermal Gibbs triangles

Some of these are used in various parts of materials science, but are not common currency in the electrochemical literature.

20.5 Major Players in This Area

Batteries are no longer the province of electrochemists. Many of the major players in recent developments have been people whose backgrounds are in materials science, physics, solid-state chemistry, and even polymer chemistry, rather than traditional electrochemistry.

A significant change is the recognition that much of what is important in advanced batteries has to do with what goes on inside the electrode materials, not on their surfaces. As a result, some of the traditional tools, such as the table of the standard electrode potentials, have little to do with modern batteries.

As an example, insertion reactions, which play such a critical role in lithium-ion batteries, did not even appear in discussions of batteries until just a few years ago, and are still not included in much of the conventional electrochemical literature.

20.6 The Future

The storage of energy is an important component of the total energy and environmental picture that is getting a great deal of attention at present. Battery technology is playing, and will continue to play, a very important role.

Significant progress in this area will most likely come from the discovery of new materials and mechanisms, rather than incremental improvements in existing systems.

The author would be very pleased if the understanding of the relevant factors related to the critical materials that is presented here is instrumental in helping to make that progress possible.

Index

$(CH_3)_4NOH.5H_2O$ 211
$(La,Li)TiO_3$ 370
A-A-A-A-A stacking 129
A-B-A-B-A stacking 128
AB_2 alloys 203, 208
AB_2 alloys - structures 208, 209
AB_2 materials - oxidation 210
AB_2O_4 169
AB_5 alloys 203
ABO_3 369
absolutely stable phases 41, 93
ABX_3 structure 237
AC methods 414
acid salts in graphite 250
activation energy versus ionic size 359
activity coefficient 270
activity of electrons 270
addition - methods 190
addition - of hydrogen 190
addition - of oxygen 190
adsorption pseudo-capacitance 447
advantages of electrochemical methods 330
Ag,AgCl 285
Ag/AgCl - reference electrode 192
Ag_2S 308, 347, 354
AgBr 404
AgCl 348
AgI 348, 352, 354
Al_2O_3 151, 436
$AlCl_4$ groups 367
aliovalent species 272
alkali metals and water 189
alternating current intermittent titration method 407
altervalent species 272
aluminum electrodeposition 319
ambient temperature electrolytes for lithium 321
ambient temperature Li alloys 137
amorphization 41, 180
amorphization - by insertion reactions 98
amorphization - of alloys 147
amorphous materials 147
amorphous RuO_2 448
amorphous silicon 113
amorphous silicon precursors 147
amorphous structure stability 96
amphoteric behavior of hydrogen 330
amphoteric materials 162
Angell 437
anionic electrodes 285
anti-fluorite structure 362, 363
Anti-Frenkel disorder 346
Anti-Schottky disorder 346
anti-structure disorder 346
aqueous electrolytes for hydrogen 324
Argand diagram 416
Armand 293, 434
assembly in discharged state 162

B_2O_3 151
BaF_2 363
band model of electrons in solids 271
Barnard 226
BCC anion lattice 353
BCC iodide ions 352
Bell Laboratory patents 128
Berlin blue 237
Berlin green 242
beta alumina 71, 340, 365
binary electrolyte phases 376
binary phase diagrams 41
binary potential scale relations 280

binding energy 269
birnessite 214
bis(oxalato)borate 323
bismuth oxide phases 381
blocking electrodes 305
blocking interface 415
blocks 105
Bode 217
Born and Mayer 357
both ionic and electronic transport 423
bridging layers 365
bromine 296
brucite 217
bulk absorption 447
Butler 283

C-Rate 22
C-Rate influence upon discharge curves 22
$C_2H_2Li_2$ 135
$Ca_3Al_2Si_3O_{12}$ 370
cadmium electrode 199
cadmium electrode - kinetics 201
cadmium electrode - microstructure 201
cadmium environmentally hazardous 200
cadmium/nickel cells 199
CaF_2 362, 363
calculation of voltages 74
calomel electrode 285
capacitor development programs 448
capacity density 78
$CaSi_2$ 147
catechole carbonate 388
cation vacancy model 214
cationic electrodes 284
cavities 105
CC 387
$Cd(OH)_2$ 200
$Cd(OH)_3^-$ 201
Cd/Ni cells 226
CdI_2 structure 103, 365
CdO 200
CdO formation 201
CeF_3 366
ceria-based solid electrolyte 381
CH_3CN 296
changes in host structure 111
characteristic properties 347–350
charge capacity 12
charge curves 14
charge number 298
charge storage mechanisms 445
Cheetham 364
chemical binding energy 268
chemical diffusion 257
chemical diffusion coefficient 363, 407
chemical driving force 4
chemical model 341
chemical potential 266, 297
chemical potential - decomposition into ions and electrons 289
chemical potential - definition 180
chemical potential difference 309
chemical potential gradient 309
chemical potential of neutral species 298
chemical potential of oxygen and electrical-potential 181
chemical potential probe 307
chemical reaction between neutral species 4
chemical reactions 1
chemical short-circuiting 327
chemical stability window 380
chemically mimic electrochemical behavior 294
chimie douce 96
chlor-alkali process 448
chlorides 362
chlorine gas sensor 319
circuit analysis 453
CO poisoning of Pt catalysts 327
co-insertion of solvent species 110
Coleman 214
color changes as function of potential 249
columnar growth 146
complex impedance plane 416
component diffusion coefficient 407
composite electrochemical cells 309
composite structures 380
composition distributions within components of cells 297
composition parameter F 302
composition path during discharge of HNi_2O_3 231
composition trajectory 68
concentration dependence of chemical potential 117
conduction band 115
conductivities of polymers 436
conductivity drop at melting point 352
conductivity of aqueous electrolytes 324
configurational entropy of guest ions 116
constitutional supercooling 125
contact potential 274
conversion from crystalline to amorphous structure 111
convertible oxides 157
Conway 448
CoO 86
correspondence between different methods 413

Coulombic interaction 357
Coulombic term 364
Coulometric titration 49
Coulometric titration in Li-Co-O system 86
Coulometric titration technique 36
covalently-bonded slabs 103
Craig 448
critical role of water 325
CrO_3 in graphite 293
cross-over of methanol 327
crumbling 143
crystallization of amorphous structure 113
crystallographic changes 36
Cu_2S 354
Cu_3N 381
CuCl 72
$CuCl_2$ 72
cycling behavior 15

Dahn 191
DC assymetric polarization method 395, 396
DC methods 393
DC open circuit potential method 405
DDQ 296
Debye circuit 415
DEC 384
decomposition of organic solvent electrolytes 383
decrepitation 143, 205
DED 396
defect concentrations 301
defect equilibrium diagram 297, 303, 396
defect model 341
Delmas 218, 258
dendrites 125
depolarizer 460
deposition in unwanted locations 124
deviations from equilibrium for kinetic reasons 98
diethyl carbonate 322, 384
different ionic conduction mechanisms 349
diffusion coefficient maximum 348
diffusion coefficients 61
diffusion through outer phase 258
dimensionally stable positive electrodes 448
dimethyl carbonate 322, 384
dipolar polarization 357
discharge curves 14
discharge curves - shape 30
dislocations 341
displacement reaction 8, 151
disproportionation 204
dissociation in water 283
DMC 384
domino-cascade model 258
dopants 116, 272
double electrochemical cell 401
double layer electrode materials 445
driving forces 460
driving function 453
dry polymers 434

E vs pH diagram showing difference between ZDF and non-ZDF electrodes 290
EC 383
ECE-15 cycle 442
EDLC 446
effective charge 342
Eigen cation 324
electonic work function 270
electrical double layer 441, 445
electrical double layer capacitive devices 446
electrical mobility 299
electrical potential in metallic solid solutions 116, 118
electricaly neutral species 282
electrochemical behavior - lithium in amorphous graphite 134
electrochemical behavior - lithium in graphite 132
electrochemical interface 220
electrochemical mechanisms 3
electrochemical potential 263, 266, 298
electrochemical potential of electrons 114
electrochromic materials 449
electrodes - of the first kind 284
electrodes - of the second kind 285
electrodes with mixed-conducting matrices 292
electrodes, other topics 235
electroless Cu and Ni 210
electroless plating 206
electrolyte additives 387, 388
electrolyte as filter 3
electrolyte stability window 375–391
electrolytic manganese dioxide 214
electron cloud displacement 167
electron energy band model 115
electron transference number 427
electron work function 269
electronegativity 316
electroneutrality condition 272, 302
electronic and ionic components of conductivity 393
electronic conductance 428
electronic leakage 18, 422
electronic probe 307
electronic transport blocked 308

electronically conducting polymers 433
electrostatic driving force 4
electrostatic energy 268
electrostatic macropotential 265
elevated temperature electrolytes for alkali metals 317
Ellingham diagrams 56
Ellingham difference diagrams 57
Ellingham integral diagrams 56
EMC 384
EMD 214
energies of charged species 267
energy balance 4
energy band model 271
energy gap 115
energy quality 10
energy relations for electron 268
energy storage mechanism comparison 450
energy transduction 1
enhancement factor 407
epitaxy 61, 102
equilibrium oxygen pressure 86
equivalent circuit of battery 406
equivalent circuit of electrochemical cell 17
equivalent circuits 415, 453
ES 388
escaping tendency 180
ethyl methyl carbonate 384
ethylene carbonate 322, 384
ethylene sulfite 388
eutectic melting point 55
evaluation of properties 391
Everitt's salt 242
external chemical potential 298
external sensors 307
extrinsic semiconductor 272
extrusion 189

Faradaic deposition 447
Faradaic electrodeposition 447
Faradaic intermittent titration method 407
Faraday 340
Faraday's law 395
fast ionic conductors 341
FCC anion lattice 354
FCC oxide ions 167
FCC packing 164
$Fe(OH)_2$ 103
Fe/Ni cells 226
Fe_2P 178
Fe_3O_4 88
$Fe_4[Fe(CN)_6]_3$ 239
$FeFe(CN)_6$ 242
$FeFe(CN)_6Ax$ 242
Fermi level 115, 272
Fermi level equilibration 274
FeS 160
FeS_2 160
filamentary growth 125
fine particle electrodes 139
FITT method 407, 409–411
floating layered structures 104
flow batteries 252
flow battery configuration 252
flow battery electrical output 255
flow battery redox systems 254
fluorides 362
fluorine anion conductors 354
fluorite structure 362–364
flux relations inside phases 299
force balance 299
force balance upon ions 4
Ford Scientific Laboratory 340
formal valence of species 116
formation of metals and alloys from oxides 151
formation of second electrolyte in situ 388
formation reactions 2, 6
fracture toughness 144
framework structures 164, 175
free electron theory 115
Frenkel defects 346
Frenkel disorder 346
frequency-dependent double layer capacitance 421
frozen-in concentrations 276
Fuji Photo Film Co. 151
functions of electrolyte 340
future 462

galleries 105
gallery space in graphite 130
Galvanic potential difference 274
galvanostatic intermittent titration method 407
garnet structure 367, 370
gas electrodes 278
gas evolution during SEI formation 387
gel electrolyte 434, 437, 438
general mobility 299
geometrical capacitance 415
Gibbs free energy of formation data 281
Gibbs phase rule 31, 42, 45, 119, 287, 378
Gibbs triangle 181
GIRIO 203
GITT 370
GITT method 408
glass transition temperature 436
Goodenough 162

grain boundaries 425
graphene layers 128
graphite amphoteric behavior 250
graphite exfoliation 384
graphite felt 252
graphite is amphoteric 128
graphite structure 127
graphite structure variations 130
graphite sub-grains 261
graphite, influence of flake thickness 385
graphitizing carbons 130
Grot 325
Grotthus mechanism 324
growth of protrusions 125
guest species configurations 105

H-Cd-O phase stability diagram 200
H-Cd-O system 200
H-Mg-Al system 80, 83
H-Mg-Cu system 80, 83
H-Mg-Ni system 80
H-Mn-O system 215
H-Ni-O Gibbs triangle 225, 229, 231
H-Zn-O phase stability diagram 198
H-Zn-O system 198
H_2/Ni cells 226
H_2/O_2 fuel cell 279
$H_2NiO_2/HNiO_2$ electrode 189
$H_5O_2^+$ 324
$H_9O_4^+$ 324
hard carbons 130
HCl 331
HCP oxide ion materials 175
Hebb-Wagner experiment 395
Hess's law 182
heterophase structure 61
heterosite 178
hexacyanometallates 237
hexacyanometallates, cations on A sities 242
hexafluorophosphate 296
hexafluoropropylene 437
hexagonal tungsten bronze structure 236
HFP 436
Hg,Hg_2Cl_2 285
Hg,Hg_2SO_4 285
Hg/HgO 285
Hg/HgO reference electrode 226
high impedance boundaries 425
high molecular weight polymers 434
historical classification of electrodes 284
Hittorf transference number 312, 394
$HMnO_2$ 216
HNi_2O_3 230
hollandite structure 110, 132
hopping model of defect transport 350, 352
H_2NiO_2 382
H_xNiO_2 electrode 199
hybrid electrolytes 437
hybrid ion cells 180
hybrid strategy 445
hybrid system characteristics 444
hybrid vehicles vii, 442
hydrated acidic polymers 325
hydrated cation radii 242
hydrated RuO_2 448
hydride ion salts 333
hydride theoretical capacities 211
hydride-conducting electrolyte 332
Hydride/Ni cells 226
hydrides - pressure vs volume of unit cell 206
Hydro-Quebec 434
hydrogen and water in positive electrode materials 188
hydrogen pressure vs temperature 206
hydrogen storage viii
hydrogen/oxygen fuel cells 460
hydrophilic regions 326
hysteresis 259
hysteresis - due to mechanical work 206
hysteresis - in hydrogen-containing carbons 135
hysteresis - in lithium-graphite 133

immobile anionic species in graphite 250
immunity 331
impedance spectroscopy 305
impedances to species transport 18
implications for safety of lithium cells 320
importance of electrical contact 145
importance of energy quaiity 451
importance of pH 283
important parameters 453
incorporation relation 302
induction effect 176
influence of effective mass on potential 118
influence of environment on potential 166
influence of ion size 358
influence of temperature on ordering 111
initial extraction 114
inner electrical potential 298
inorganic filler 437
insertion 61
insertion into graphite 250
insertion into graphite from polymeric electrolyte 128
insertion into parallel linear tunnels 110
insertion of either cations or anions 235
insertion of guest species into polymers 250

insertion reaction 8, 160
insertion reaction electrodes 101
insertion reactions to modify materials 293
inside and outside quantities 299
intensive thermodynamic parameters 67
intercalation 61, 102
intercalation of solvated species in graphite 385
interface motion 5
interior of solids 298
internal ZDF chemical reaction 292
interstitial mechanism 61
interstititals 341
intrinsic semiconductor 272
inverse spinels 169
iodine in acetonitrile 296
ion exchange 189
ionic conductance 427
ionic conductivity unusually high 350
ionic liquids 323
ionic model 342
ionic rubbers 434, 437
ionic transference number 428
ionomers 325
ions inserted into graphite 128
irreversible capacities 155
isolated tetrahedra 367
isothermal Gibbs triangle 66, 287
isothermal phase stability diagram 181
Ives and Janz 283

Jahn-Teller distortion 171
Joule heating due to self discharge 21
jump frequency 350
junctions between different metals 273
junctions between metals and semiconductors 275

$K_2FeFe(CN)_6$ 242
$KAlCl_4$ 367
KCl 348
Keggin and Miles 238
$KFe_2(CN)_6$ 237
$KFeFe(CN)_6$ 239
Kröger-Vink notation 301, 344
$K_xV_2O_5$ 108

L/S/L 159
L/S/L configuration 252, 340
L/S/L systems 58
La-Pr-Sr-In-Mg-O 429
LaF_3 365
LaPlace transform examples 454
LaPlace transform techniques 453
LaPlace transforms 453
law of mass action 301
layer stacking 103
layer structure materials 364–367
layered structures 167
lever rule 44
Li-Ag 147
Li-Al 137
Li-Al-Cl system 319
Li-Bi 137
Li-Bi system 51
Li-Cd 137
LI-Cd system 90
Li-Cd-Sn system 90
Li-Co-O Gibbs triangle 182
Li-Co-O system 86, 182
Li-Cu-Cl system 72
Li-Fe-O system 182
Li-Fe-P-O quaternary system 178
Li-Ga 137
Li-H-O phase stability diagram 190
Li-H-O system 190, 280
Li-I cell 27
Li-In 137
Li-Mn-Fe phosphate 165
Li-Mn-O system 89, 182
Li-N-O system 389
Li-Pb 137
Li-S-O system 187
Li-Sb 137
Li-Sb system 7, 46
Li-Si 137
Li-Si alloy 292
Li-Si system 333
Li-Si, Li-Sn composites 140
Li-Sn 137
Li-Sn alloy 292
Li-Sn system 90, 152
Li-Sn-Cd system 140
Li-Sn-O system 153
Li-Ti-O system 34
Li/copper chloride cells 77
Li/CuCl cell 74
Li/$CuCl_2$ cell 75
Li/$LiNO_3$/C cell 390
Li/polymer/V_6O_{13} batteries 434
Li/SO_2 186
Li/SO_2 system 186
Li/$SOCl_2$ 186
Li/$SOCl_2$ system 188
$Li_{0.29}La_{0.57}TiO_3$ 369
$Li_{0.7}Sn$ 152
$Li_{1-x}CoO_2$ 162
$Li_{1-x}NiO_2$ 162

$Li_{1.33}Ti_{1.67}O_4$ 174
$Li_{1.71}Si$ 140
$Li_{2.33}Sn$ 152
$Li_{2.6}Sn$ 140
$Li_2M_2O_3$ 377
Li_2O 152, 154, 187, 280, 384, 389
Li_2O formation 86
Li_2S 187
$Li_2S_2O_4$ 186
Li_2Sb 46, 49, 51
$Li_2V_2O_5$ 113
Li_3Bi 51
Li_3N 366, 376, 381, 389
Li_3N structure materials 147
Li_3PO_4 367
Li_3Sb 46, 49
$Li_3V_2(PO_4)_3$ 165, 176
$Li_3V_2O_5$ 113
$Li_{4.4}Sn$ 140, 152
Li_4GeO_4 367
Li_4SiO_4 367
$Li_4Ti_5O_{12}$ 120, 174
Li_5AlO_4 363
Li_5GaO_4 363
$Li_5La_3M_2O_{12}$ 370
$Li_6BaLa_2M_2O_{12}$ 370
Li_6ZnO_4 363
LiAl 277, 413
$LiAlCl_4$ 318, 367
$LiAsF_6$ 322
$LiBF_4$ 322
LiBOB 323
LiBr 296
LiCl 72
LiCl-KCl molten salt 53, 85, 318, 333,
LiCl-KCl-$AlCl_3$ salt 319
LiCl-NaCl-KCl-$AlCl_3$ salt 319
$LiClO_4$ 322
$LiCoO_2$ 86, 296
LiF 384
LiF-LiBr-KBr salt 318, 333
LiF-LiCl-LBr-LiI salt 318
LiF-LiCl-LiBr 333
LiF-LiCl-LiBr-LiI 333
LiF-LiCl-LiI 318, 333
$LiFe_2(SO_4)_3$ 165
$LiFe_5O_8$ 88
$LiFeO_2$ 88
$LiFeO_2$ system 87
$LiFePO_4$ 164, 177
$LiFePO_4OH$ 179
LiH 331
LiI 295, 376
$LiM_2(PO_4)3$ 381
limiting oxygen pressure in glove box 320
$LiMn_{1-y}Co_yO$ 168
line phases 46
$LiNO_3$ 389
$LiNO_3$-KNO_3 molten salt 54, 388
$LiPF_6$ 322
liquid binary alloys 57, 138
liquid electrode cells 252
liquid electrolytes 315
liquid junction potential 286
liquid positive electrode reactants 186
liquidus 42
Lisicon 367
lithium - tin alloys 152
lithium - tin oxide system 153
Lithium - transition metal oxides 85
lithium activity and potential 86
lithium alkyl carbonate 385
lithium alloys 136
lithium carbon alloys 127
lithium cells in aqueous electrolytes 282
lithium extraction from aqueous solutions 193
lithium in graphite - discharge curve 133
lithium in hydrogen-containing carbons 134
lithium ion conductors 368–371
lithium metal electrodes 120
lithium phases stable in water 191
lithium positive electrodes 159, 163
lithium reference electrode 277
lithium salts 322
lithium-aluminum alloys 136
lithium-silicon alloys 136
lithium/carbon cell 390
LiV_3O_8 160
$LiVO_2$ 296
$LiVPO_4F$ 179
Li_xCoO_2 164, 167
$Li_xCu_xMn_{2-x}O_4$ 171
$Li_xFe_2(SO_4)3$ 167
Li_xFePO_4 120
Li_xFePO_4 - electronic conductivity 178
$Li_xMn_{0.5}Ni_{0.5}O_2$ 169
$Li_xMn_{1/3}Ni_{1/3}Co_{1/3}O_2$ 169
$Li_xMn_2O_4$ 164, 170
$Li_xMn_2O_4$ in aqueous electrolyte 192
Li_xMnO_2 168
$Li_xNa_{0.4}WO_3$ 119
$Li_xNi_{0.5}Mn_{1.5}O_4$ 172
Li_xNiO_2 168
Li_xTiS_2 95, 160
Li_xVS_2 296
load levelling ix
localized doping 309
low frequency tail 419

lower temperature alkali halide molten salts 318
LSGM 429
Ludi 240

M_2X_3 groups 175
major players 461
$MAl_{11}O_{17}$ 348
manganese oxides 57
materials containing fluoride ions 179
materials with spinel structure 169
maximum theoretical specific energy 13, 50
mechanical strain energy 259
MEEP 434
memory effect curing 227, 232
memory effect in "nickel" electrodes 226
metal hydride alloys 203
metal hydride battery production 203
metal hydride electrodes 202
metal hydride systems 280
metal hydrides 78
metal hydrides containing magnesium 79
metastable 61
metastable microstructures 41
metastable model 93
metastable phases 41
metastable state 93
methods for study of hydrogen storage materials 329
$Mg(OH)_2$ 103
$Mg_{2.35}Ni$ 80
Mg_2Ni 80
MgF_2 360
MgH_2 79
$MgNi_2$ 80
microencapsulation 210
minimum energy path 357, 363
minority electronic current 400
Mischmetal 204
mixed-conductor as reference 401
mixed-conductor matrix electrode 138, 292
mixed-ion materials 180
$Mn(OH)_2$ 216
MnO_2 213
MO_6 octahedra 175
modeling transient behavior 453–456
modest temperature molten salts 318
molten sub-lattice model 353
monolithic structure 309
monopole-dipole interaction energy 360
more than one equivalent circuit 418
more than one type of site 161
Mössbauer experiments 240
moving interface reconstitution reaction 36
MTSE 13
MTSE calculations 77
multi-phase reactions 34
Murphy 296
MX 341
M_xWO_3 236

n-butyl lithium 294
Na-S phase diagram 59
Na-S system 58
Na/Na_xS 160
Na/Na_xS cell 340
$Na/NiCl_2$ 159
$Na/NiCl_2$ system 71
Na_2O 280
$Na_3V_2(PO_4)F_3$ 180
$Na_3Zr_2Si_2PO_{12}$ 175
$NaAl_{11}O_{17}$ 159
$NaAlCl_4$ 71, 367, 382
$NaAlEt_4$ 80, 335
NaCl 71
$NaFeO_2$ structure 168
Nafion 280, 324
Nafion - alternatives 328
Nafion - platinum catalysts 327
NaH 80
nanowires 146
Nasicon 165, 367
Nasicon structure 175
Na_xS 58
Nb - O system 55
negative electrodes in aqueous systems 197
Nernst equation 27, 264, 279, 330, 398
Nernst Glower 340
Nernst relation 79, 207
Nernst-Einstein relation 348, 351
network structures 175
NHE 283
Ni, Ni_xO system 279
$Ni(OH)_2$ 103, 216
Ni/Cd cells 199
Ni_3O_4 230
NiCad cells 199
nickel electrode 216, 381
nickel electrode - mechanism 218
nickel electrode - phases 218
nickel electrode - potential vs hydrogen content 224
nickel electrode - second plateau 227
nickel electrode - thermodynamic data 224
$NiCl_2$ 71
NiO 230
NiO_2 slabs 217
NiOOH 216

$NiPS_3$ 108
$NiSi_2$ 147
Ni_xO 279
NO_2 389
NO_2PF_6 296
non-aqueous electrolytes for hydrogen 328
non-blocking electrodes 101
normal hydrogen electrode 283
normal spinels 169
nuclear magnetic methods 407
nucleation overshoot 140

off-center positions 204
olivine 164
olivine structure materials 177
optical excitation 116
ordered cation distributions 164
organic solvent liquid electroytes 321
organic solvents - melting points, dielectric constants 322
organic-anion molten salt 335
Ostwald ripening 139
other binders 211
other hydride alloys 210
outer potential 267
output voltage related to transference numbers 20
overcharge 223
overcharging 228
overlap repulsion 357
overvoltage 460
oxidation of water 192
oxygen evolution 221
oxygen evolution problem 180

P-C-T methods 329
p-n junctions 272
PAN 434, 438
partial electronic conduction 426
particle flux density 299
particle-enhanced conductivity 436
particles breathe 145
Patterson 401
Pauling 331
PbF_2 363
PC 384
PEO 433, 437
PEO-$LiBF_4$ 435
perfluorovinyl ether groups 325
perovskite structure 237, 368
peroxide ions 185
pH definition 284
pH of aqueous electrolytes - measurement 291
phase diagrams 6, 41
phase stability diagram 66, 287, 389
Philips Laboratory 203, 301
phosphoric acid 328
pillared layer structures 104
pillars 105
PITT method 408
Pitzer and Brewer book 180
plasticizer 434, 437
plug-in hybrid vehicles 443
PMMA 437
point defects 341
point phases 66
polarization 460
polarization catastrophy 360
poly(acrylonitrile) 433, 438
poly(benzimidazole) 328
poly(ethylene) 438
poly(methoxyethoxy ethoxyphosphazene) 434
poly(methyl methacrylate) 437
poly(vinylidene fluoride) 434, 437, 438
polyacetylene 251
polyanion structures 175
polyethylene oxide 434
polymer electrolytes 434–438
polymer-in-salt regime 437
polymeric materials 433
polypropylene oxide 434
polythiophene 251
positive electrodes in aqueous systems 213
positive ions in water 283
potassium evaporation 86
potential and lithium activity 86
potential and oxygen pressure relation 319
potential distributions within components of cells 297
potential overshoot 140
potential plateaus 46
potential profiles 356
potential ranges of metal chloride systems 331
potential scales 265
potential versus oxygen pressure 183
potential-pH plots 276, 286
potentials 263
potentials - chemical reactions 293
potentials - chemical redox equilibria 294
potentials - in and near solids 264
potentials - Li-H-O system 281
potentials - Li-metal-oxygen systems 158
potentials - lithium reaction materials 296
potentials - redox reactions 166
potentiostatic intermittent titration method 407
Pourbaix 286
Pourbaix diagrams 276, 286
Pourbaix diagrams - general form 286

PPO 433
practical parameters 9
prediction of potentials and capacities 85
predominantly electronic conductors 423
pressure - composition relation in hydrides 205
properties of electrode materials 406
propylene carbonate 322, 384
proton absorption from water vapor 192
proton conductivity in Nafion 327
protons in MnO_2 214
Prussian blue 237
Prussian blue - “insoluble” 239
Prussian blue - “soluble” 239
Prussian blue - batteries 246
Prussian blue - catalytic behavior 247
Prussian blue - cyclic voltammetry 242
Prussian blue - electrochemical behavior 241
Prussian blue - electrochromic behavior 248
Prussian blue - electronic properties 246
Prussian blue - potentials 242
Prussian blue - structure 239
Prussian blues - other cationic species 244
Prussian white 242
Prussian yellow 242
pseudo-binary system 228
pseudo-capacitance 447
pseudo-ternary stability diagram 315
pulse demand conditions 442
pulse output requirements 442
PVDF 437
PVFC 434
pyrolucite 214
pyrolysis of hard carbons 131

quaternary ammonium groups 323

radiotracer methods 407
Ragone-type diagram 444
ramsdellite 214
$RbAg_4I_5$ 347
$RbAg_4I_5$ family 367
reacted particles 260
reaction coordinate diagram 94
reaction front 260
real potential 269
recent changes ix
rechargeability of elemental electrodes 120
reconstitution reaction 6, 61, 257, 450
reconstitution reaction potentials 119
Redox Frenkel disorder 346
redox in vanadium/vanadium system 255
redox ion solutions 252
redox ion valences 165
redox potentials in spinels 172
redox pseudo-capacitance 447
redox reactions 116, 252
reference electrode - importance of ZDF or not 290
reference electrodes 276
reference electrodes - aqueous systems 283, 287
reference electrodes - non-aqueous systems 287
reference electrodes - oxide systems 278
reference electrodes - polyphase solid 279
relative charge 342
relevant approximations 304
relics in the microstructure 139
ReO_3 structure 237
replacement of Mn 171
residual capacity 227
residual value of F 31
response function 453
reversible hydrogen electrode 285
RHE 285
Robin and Day 237
rocksalt phase 35
Ruetschi 214
Ruetschi defect 214
RuO_2 448
Ruthenium purple 248
rutile structure 358
RVO_4 materials 147

S/L/S 159
S/L/S configuration 340
S/L/S systems 58
salt bridge 286
salt-in-polymer regime 437
SCE 285
Schottky defects 346
Schottky disorder 346
SEI 317, 321
SEI formation 383
SEI formation on graphite 384
SEI thickness 385
selective equilibrium 95, 276
self diffusion coefficient 407
self discharge 16, 221
selvedge 266
sequential insertion reactions 107
series and parallel arrangements 418
shape change 124
SHE 264, 283
sheets 104
SiB_3 147
silicon nanowires 147
silver iodides 340

SiO 147
SiO_2 436
skeleton structures 164, 175
slabs 105
SLI applications viii
small particle electrodes 259
small perturbations 61
small temperature dependence 350
small value of pre-exponential factor 350
$Sn_2P_2O_7$ 151
SnO 151, 154
SnO_2 154
SO_2 187
SO_2F functional groups 325
SO_3H^+ sulfonic acid groups 325
sodium beta alumina 58, 159, 347
sodium-napthalene 294
sodium-sulfur battery 138
soft carbons 130
soft chemistry 96, 218, 293
solid electrolyte and molten salt 388
solid electrolyte in negative electrode 211
solid electrolyte interphase 317, 321, 383
solid electrolyte, aqueous electrolyte in series 381
solid electrolyte, molten salt in series 381
solid electrolytes 338
solid solution reactions 102
solidus 42
solubility of alkali metals in their halides 124
solubility of oxides in ionic liquids 323
solvation energies 264
solvation of protons 326
solvent-free polymers 434
SONY 162
SONY battery 128
Sorenson and Jacobsen 435
sparingly soluble salt 285
specific capacity 78
specific energy 50
spinel 164
spinel phase 35
spinel structure 170, 365
spinel structure materials 120
stability limits - relative to lithium 378
stability limits - relative to oxygen 379
stability ranges of phases 51
stability window 332, 375
stable tie lines 378
stacks 104
stage structure in graphite 261
stage structure transition 261
stage structures 110
stages 128
staging 108
staging in lithium-graphites 131
standard calomel electrode 285
standard electrode potential 285
standard Gibbs free energy - of formation 4
standard Gibbs free energy - of reaction 4
standard hydrogen electrode 264, 283
stoichiometric composition 347
Stotz and Wagner 190
structural dependence of ionic conduction 347
structure terminology 104
structure-dependent model 356
sub-grain boundary transport 261
sub-grains in graphite structure 131
sub-lattices 95
sub-triangles 67
substitutional solid solid solution 60
sulfonate groups 325
supercapacitors 441, 446
superionic conductors 341
supervalent ion doping 178
surface potential 266
surface shape 146
symbols for isolated defects 345

tavorite 179
Teflon backbone 325
temperature dependence - cell voltage 29
temperature dependence - ionic conductivity 348
temperature dependence - Li-LiAl reference 277
temperature dependence - of potential 53
terminal particle size 144
terminology xii, 459, 460
ternary diagram rules 73
ternary electrodes 65
ternary electrolyte phases 377–380
ternary oxides 236
ternary phase diagrams 65
ternary phase stability diagram 377
ternary phase stability diagram - Li-Cd-Sn system 90
ternary phase stability diagram - Li-Co-O system 87
ternary phase stability diagram - Li-Fe-O system 88
ternary phase stability diagram - Li-Mn-O system 89
ternary solvated species 385
ternary systems - two binary metal alloys 90
ternary titanates 367
tetrahedral anions 367
Thackeray 170

thermal batteries 136
thermal excitation 116
thermal runaway 126, 181
thermodynamic enhancement factor 363
thermodynamic potential 265
thermodynamic properties - binary lithium alloys 136
thermodynamic properties - individual species 25
thick layer approximation 412
thin layer approximation 412
thinking tools xi, 41, 461
Ti-H system 334
Ti_2Ni 210
$TiCl_2$ 331
time constant 446
TiNi 210
TiNi ferroelastic 210
TiO_2 360, 436
TiS_2 103
$TMAH_5$ 211
topotactic formation of electrolyte 382
topotactic reactions 61, 102
topotaxy 61
total interaction energy 357
transfer function 453
transference numbers 297, 394, 424
transference numbers - of species 19, 311
transference numbers can vary 394
transient behavior 441
transition metal oxide bronzes 236
translation of two-phase interface 257
transverse internal interfaces 425
transverse transport 259
trapped lithium 134
triphylite 178
Tubandt 395
Tubandt method 395
tungsten bronze 119
tunnels 105
turbostratic disorder 130
TVF equation 436
two different redox potentials 294
two electrolytes in series 382
two-phase lithium alloy reference electrodes 277
two-phase tie line 66
types of disorder 346
types of notation 342

ultracapacitors 441, 446
underpotential 460
underpotential charge storage 447
unidirectional channels 361

V-H system 334
V_2O_5 112, 160
V_6O_{13} 112, 160
vacancies 60, 341
vacuum level 263, 266
valence band 115, 263
valences of redox ions 165
van der Waals forces 103, 365
Van't Hoff plots 206
variation of composition with potential 300
VC 388
VCl_2 331
vinylene carbonate 388
$VO_2(B)$ 192, 282
Vogel-Tammann-Fulcher equation 436
Volta potential difference 274
voltages in semiconductor techology 12
volume change by lithium 145

Wagner 340, 395, 399, 401
Warburg admittance 421
Warburg impedance 421
water in Nafion 326
wechselstrom 411
Whittingham 160
Willis 364
windows 105
WITT method 407, 411–413
Wood's metal alloy 136

XO_4 tetrahedra 175

yttrium-doped thorium oxide 404

ZDF conditions 287
ZDF electrodes 276
Zebra battery 70, 382
zero-degree-of-freedom electrodes 276
zero-strain insertion reaction 174
zinc electrode problems 199
zinc in aqueous systems 197
zincate ions 199
$Zn(OH)_4{}^{2-}$ 199
Zn/MnO_2– alkaline cell 197
Zn/MnO_2– technology 197
Zn/Ni cells 226
ZnO 198
ZrO_2 339, 361
Zundel cation 324
α $Ni(OH)_2$ 217
β $Ni(OH)_2$ 217
β NiOOH 217
γ NiOOH 217
λ MnO_2 spinel 193

Lightning Source UK Ltd.
Milton Keynes UK
19 February 2010

150364UK00001B/21/P

9 780387 764238